Fundamentals of

MICROFLUIDICS AND LAB ON A CHIP FOR BIOLOGICAL ANALYSIS AND DISCOVERY

Fundamentals of
MICROFLUIDICS AND LAB ON A CHIP FOR BIOLOGICAL ANALYSIS AND DISCOVERY

Paul C. H. Li

CRC Press
Taylor & Francis Group
Boca Raton   London   New York

CRC Press is an imprint of the
Taylor & Francis Group, an **informa** business

CRC Press
Taylor & Francis Group
6000 Broken Sound Parkway NW, Suite 300
Boca Raton, FL 33487-2742

© 2010 by Taylor and Francis Group, LLC
CRC Press is an imprint of Taylor & Francis Group, an Informa business

No claim to original U.S. Government works

Library of Congress Cataloging-in-Publication Data

Li, Paul C. H., 1962-
    Fundamentals of microfluidics and lab on a chip for biological analysis and discovery / Paul C.H. Li.
      p. cm.
    Includes bibliographical references and index.
    ISBN 978-1-4398-1855-8
    1. Biochips. 2. Microfluidics. I. Title.

R857.B5L515 2010
610.28--dc22
                                       2009035035

Visit the Taylor & Francis Web site at
http://www.taylorandfrancis.com

and the CRC Press Web site at
http://www.crcpress.com

To the late Maria So, who taught me Chinese wisdom and medicine, and supported my scientific herbal research

Contents

Preface

The microfluidic lab-on-a-chip has provided a platform to conduct chemical and biochemical analysis in a miniaturized format. Miniaturized analysis has various advantages, such as fast analysis time, small reagent consumption, and less waste generation. Moreover, it has the capability of integration, coupling to sample preparation, and further analysis.

The book is divided into several chapters, which include (1) fundamentals of microfluidics, such as micromachining methods, microfluidic flow, sample introduction, sample preconcentration, and detection technology, and (2) applications of lab-on-a-chip to chemical and biological analysis and discovery, such as chemical separations, cellular analysis, nucleic acid analysis, and protein analysis.

This book is derived from the book *Microfluidic Lab-on-a-Chip for Chemical and Biological Analysis and Discovery* (Boca Raton, FL: CRC Press, Taylor & Francis Group, 2006). Beginners in the field should find this book useful to navigate the vast literature of the technology, while experienced researchers will find the compiled information useful for easy comparison and references. Although some comparisons among different approaches are made here, the readers should judge on the suitability of a certain technology for their needs. Moreover, most citations refer to complete studies.

Many of the principles of "operations" on the chip are adapted from conventional wisdom. This adaptation not only results in the miniaturization advantage, but also leads to enhanced effects because of favorable scaling down (e.g., plasma emission detection and large DNA molecules separation) and even to scientific discovery (e.g., single-cell study). However, the original citations of these theories or principles, which have already been given in the original research articles, will not be repeated here, to avoid increasing the number of microchip references. Therefore, readers are encouraged to consult the original work from the references cited.

Acknowledgments

The author is indebted to the following individuals without whom this book could not have been realized: Xiujun Li, Wei Xiao, Alice Koo, Margaret Wong, Maisy Chow, and Wai Lan Chan (Faria). Partial funding support from University Publication Fund of Simon Fraser University is gratefully acknowledged.

About the Author

Dr. Paul Li obtained his M.Sc. and Ph.D on chemical sensors with Professor Michael Thompson at the University of Toronto in 1995. He then worked with Professor Jed Harrison at the University of Alberta during 1995–1996 in the field of microfluidic lab-on-a-chip. During 1996–1998, Dr. Li started his first independent research career at the City University of Hong Kong.

After joining Simon Fraser University in 1999, Dr. Li has continued to conduct research in the areas of microfluidics for single-cell analysis. Dr. Li is also interested in integrating microfluidics with the DNA microarray for pathogen detection and DNA-based diagnostics. Dr. Li is the inventor of four granted patents and five pending patents. He is also interested in analyzing chemical compounds from herbs, especially those with antitumor properties. He has authored a monograph entitled "Microfluidics lab-on-a-chip for chemical and biological analysis and discovery" in 2006, and has coauthored a book chapter on microfluidic lab-on-a-chip in *Ewing's Analytical Instrumentation Handbook, 3rd edition*, 2005.

1 Introduction

A miniaturized gas chromatographic (GC) column with a thermal conductivity detector (TCD) on silicon (Si) was first constructed in 1979[1] and a high-performance liquid chromatographic (HPLC) column with a conductometric detector was constructed on Si-Pyrex in 1990.[2] The first demonstration of a liquid-based miniaturized chemical analysis system was based on capillary electrophoresis (CE).[3] In this work, CE separation with laser-induced fluorescence (LIF) detection was achieved on a glass chip with a 10 μm deep and 30 μm wide channel in 6 min. The success is attributed to the use of electroosmotic flow (EOF) to pump reagents inside small capillaries that could develop high back pressure, preventing the use of HPLC. Since then, different CE modes have been demonstrated and different analyses (i.e., cellular, oligonucleotide, and protein analyses) have been achieved by numerous research groups, as mentioned in subsequent sections. This book is focused on chemical/biological analysis conducted on the microfluidic lab-on-a-chip (coined in 1992)[582] leading to discovery. Applications other than analysis, such as on-chip chemical synthesis, are covered elsewhere.[4]

Recently, numerous review articles summarizing research performed in the area of the microfluidic chip have been published.[5,6] Reviews on specific topics have also appeared (see Table 1.1). Whereas these reviews summarize exciting research and propose new directions, this book is focused on an overview of the available technology, its limitations, and its breakthroughs over the years.

TABLE 1.1
Reviews on Microfluidic Chip Technology

Review Topics	References
History	5, 7–11
Miniaturization	5, 12–17, 19, 21, 31
μTAS	5, 8, 11, 16, 18–25
Separation	6, 9, 21, 26–34, 44, 46, 47, 53, 189
Clinical analysis	6, 35–40
Cellular analysis	6, 40–43, 73
Protein analysis	4, 6, 42, 44–50
Nucleic acids analysis	6, 36, 42, 44, 45–47, 49, 51–59
Detector	6, 8, 44, 46, 53, 60–70
Sample introduction	6, 8, 10, 32, 65, 71
Microfabrication	5, 7, 8, 26, 42, 44, 72–78, 189
Microfluidic flow	5, 10, 18, 39, 52, 53, 79
Micropumps	5, 8, 11, 18, 21, 23, 80–82
Microvalves	5, 11, 18, 21, 23, 80–82
Micromixers	6, 11, 80

2 Micromachining Methods

Micromachining methods, which include film deposition, photolithography, etching, access-hole drilling, and bonding of microchip, were first achieved on silicon (Si).[3] Thereafter, glass (Pyrex glass, soda lime glass, fused silica) was used as the micromachining substrate. Recently, polymeric materials have become widely used as substrates. The micromachining methods of these substrates differ and are given in separate subsequent sections.

2.1 MICROMACHINING OF SILICON

The Si substrate was micromachined based on photolithography; see Figure 2.1 for the process steps of a standard one-mask Si-micromachining process.[3] After photolithography, the exposed portions of photoresist were dissolved by a developer solution, and the remaining unexposed areas were hardened by heating (baking). The exposed areas of the Si substrate were etched subsequently.

Meanwhile, a Pyrex glass plate was patterned with metal electrodes. Then the Pyrex plate was bonded to the Si wafer using the anodic bonding process. The etching, access-hole drilling, and bonding processes are described in more detail in subsequent sections.

2.1.1 ETCHING OF SI MICROCHANNELS

Isotropic wet etch using HF-HNO$_3$, whose etch rate is independent of the etching direction, has been employed to produce approximately rectangular grooves, oriented in any direction on the Si <100> wafer, using thermal SiO$_2$ (~1 μm thick) as the etch mask.[1]

Anisotropic etch, whose etch rate depends on the etching direction, can be achieved using KOH on <100> Si or <110> Si. This method allows V-grooves or vertical walls to be formed on the substrate. Apparently, the anisotropic etch rate of Si at the <111> crystal plane was very small in comparison with the <100> plane.[827] Therefore, the etch depth will be dependent on the opening width, rather than the etch time. However, this method will produce the desired groove profiles only if the groove lies along specific crystallographic axes on the wafer.[1] For instance, in Figure 2.2, anisotropic KOH etch on <100> Si will produce a rectangular cross section, if the channel (A-A', B-B', and C-C') is 45° relative to the wafer flat. But if the channel (D-D') is perpendicular to the wafer flat, a V-groove cross section will result due to the slanted <111> planes that define the stop etch conditions.[84] However, certain shapes (square corner and circle) cannot easily be realized,[1] except by corner compensation.[281]

Isopropyl alcohol (IPA) is sometimes added to KOH in order to reduce the etchant surface tension. This in turn assists in releasing H$_2$ bubbles and in dissolving organic contaminants. So the addition of IPA avoids the micromask effect and produces smoothly etched channels.[303]

Anisotropic etch on Si can also be achieved by reactive ion etching (RIE), which is a dry (plasma) etch process. However, in deep RIE (DRIE), cyclic etch characteristics could be produced on the etched wall.[85,86]

Different etch methods used to produce microfluidic chips on Si are tabulated in Table 2.1.

Since Si is a semiconductor, it is not electrically insulating. However, electrical insulation (<500 V) of Si can be provided by a film of SiO$_2$/Si$_3$N$_4$ (~200 nm)[304] or low-temperature oxide (LTO).[87] It was found that insulation by the SiO$_2$ film was better achieved by plasma deposition rather than by thermal growth; a 13 μm thick plasma-deposited SiO$_2$ film would withstand an operation voltage of 10 kV.[88,706] For better insulating properties, glass (e.g., Pyrex, fused silica), which sustains at least an electric field of 10^5 V/cm without dielectric breakdown, is used.

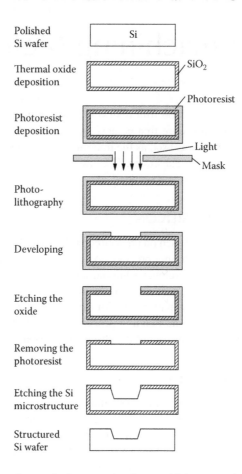

Polished
Si wafer

Thermal oxide
deposition

Photoresist
deposition

Photo-
lithography

Developing

Etching the
oxide

Removing the
photoresist

Etching the Si
microstructure

Structured
Si wafer

FIGURE 2.1 Process steps of a standard one-mask micromachining procedure to etch a channel structure into silicon.

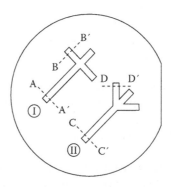

FIGURE 2.2 Microstructures defined on a <100> silicon wafer for chemical separation. Design I, cross-type separator; design II, 45° type separator.

TABLE 2.1
Etching of Si Substrates

Etch Conditions	Etch Results	References
Isotopic Wet Etch		
HF-HNO$_3$	Unspecified etch rate	1
Anisotropic Wet Etch		
TMAH and water (1:4)	28 μm in 30 min (90°C)	856
	175 μm (~80°C)	299
EDP	100 μm/h	90, 390
	43 μm in 100 min, 98°C	1011
	2.5 μm/min, 115°C	300
	Unspecified etch rate	301
KOH with IPA	~0.4 μm/min, 10% IPA, 76°C	302
	3% IPA, 70°C	303
	5% IPA, 60°C	223
KOH	380 μm in 4–5 h, 85°C	304
	~36 μm/h, 70°C	215
	17 μm, 75°C	209
	Unspecified etch rate	1, 213, 249, 305, 825
Anisotropic Dry Etch		
RIE	Unspecified etch rate	85, 307, 458, 459, 825
DRIE (SF$_6$ plasma)	5 μm/min	860
	2–3 μm/min	308, 309
	Unspecified etch rate	847

After etching, access holes can be created on the Si substrate by wet etch-through[1,89,90,281,306] or by drilling.[442,495] A two-mask process was also used to create channel access holes on the Si wafers.[90]

2.1.2 BONDING OF SI CHIPS

Anodic bonding of Si to Pyrex was commonly used to create a sealed Si chip.[1,281,442,820,836] Various bonding conditions used are summarized in Table 2.2. A positive voltage is applied on the Si wafer with respect to that of the Pyrex wafer (negative).[91,92,281]

If anodic bonding between glass and Si fails (due to a thick oxide layer [1 μm] on Si or platinum electrode on Si), a low-melting spin-on-glass (SOG) can be applied as an adhesive. SOG, which is a methylsilsesquioxane polymer, flows very well at the temperature between 150 and 210°C, and therefore it fills the grooves (i.e., around the Pt electrodes) for effective bonding.[93]

Bonding of Si to glass was also achieved using UV-curable optical adhesives. Benzocyclobutene has been used to bond Si and Pyrex.[94–99,942]

On the other hand, bonding of Si to Si was achieved by a low-temperature curing polyimide film,[100] or by using an intermediate deposited layer of borophosphosilicate glass and subsequent anodic bonding (300 V, 350°C).[101]

2.2 MICROMACHINING OF GLASS

Similar to the procedure for Si micromachining, micromachining for glass also includes thin-film deposition, photolithography, etching, and bonding (see Figure 2.3).[102]

TABLE 2.2
Various Conditions of Anodic Bonding of Si and Pyrex Plates

Temperature	Voltage	References
200–300°C	200–400 V	91
350°C	1,100 V for 2–5 min	304
	700 V (under N$_2$ at atmosphere pressure)	310
375°C	1,000–1,500 V for 30 min	309
400°C	400 V	390
	600 V	1
	800 V	820
	1,600 V for 40 min	308
430°C	1,600 V	306, 311
450°C	800 V	92
	850 V	312
	1,000 V	281, 299
485°C	800 V	90

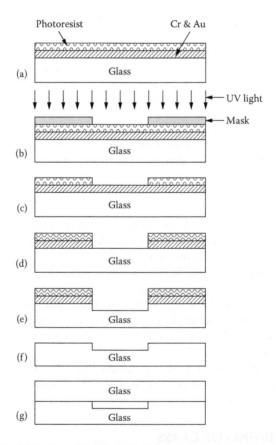

FIGURE 2.3 Sequence for fabrication of the glass microfluidic chip. (a) Cr and Au masked glass plate coated with photoresist; (b) sample exposed to UV light through a photomask; (c) photoresist developed; (d) exposed metal mask etched; (e) exposed glass etched; (f) resist and metal stripped; (g) glass cover plate bonded to form sealed capillary.[102] Reprinted with permission from American Chemical Society.

2.2.1 ETCHING OF GLASS MICROCHANNELS

Etching of glass is commonly achieved by isotropic wet etch. Various etching conditions based on HF are tabulated in Table 2.3. A typical glass channel created by wet etch is shown in Figure 2.4.

The etch rate of Pyrex depends on the HF concentration. The lateral etch rate was found to be greater than the vertical etch rate, and the rates were reduced from 12 μm/min to 0.5 μm/min when the HF concentration decreased from 49% to 25%.

Before photolithography, a photoresist film is usually spin coated on glass. To improve the adhesion of photoresist on glass, it is first spin coated with a film of hexamethyldisilazane.

In the photolithographic process, a photomask is needed. The photomask is usually generated by photoablation (by a laser) or e-beam ablation. It is found that e-beam ablation has produced smoother edges (ten times better than the UV laser-ablated mask) in the photomask, thus leading to smoother channel walls after etching, and higher efficiency in CE separations.[106] Uniformity in channel depth and width can be experimentally verified by examining the linearity of a plot of separation efficiency (N) vs. channel length (L).[107]

Scratch effects in glass etch can be reduced by pre-etching the glass in HF before deposition of the metal etch mask.[108]

For mechanically polished glass plates, thermal annealing should be performed before etching to improve etch quality.[96,102,105,109,430,529]

To improve channel smoothness, wet HF etch was performed in an ultrasonic bath[105,529,626] or wafted by hand.[136] The glass being etched was also agitated every 10 min for 2 to 3 h, followed by lateral stirring with a plastic rod.[110,136] Borofloat glass does not require agitation during HF etch and leaves no reaction products, as is seen with soda lime glass.[108]

In one report, surfactant was added to yield better dimensional consistency and smoothness of channel edges.[626] In another case, glycerol was added to the HF etching bath.[729]

To remove precipitated particles in the buffered oxide etch (BOE) process, the substrates were removed every 2 to 5 min and dipped in a 1 M HCl solution for 10 to 20 s. Adding HCl directly into the HF etch bath will have a similar effect.[105,109–111]

Because wet chemical etch on glass results in an isotropic etch, the resulting channel cross section is trapezoidal or semicircular.[112,136,137] In order to form a circular channel, two glass plates were first etched with a mirror image pattern of semicircular cross section channels. Then the two plates were aligned and thermally bonded to form a circular channel.[113,114,456,802,813]

Normally, a one-mask process was employed for glass etching, but for specific applications, a two-mask process was used to create the shallow channel (1 to 6 μm deep) and the deep channel (20 to 22 μm deep).[115,117] In order to fabricate shallow (18 μm) and deep (240 μm) glass channels, we etched them separately on the bottom and top plates using two masks. Alignment was achieved during bonding.[116,117] A two-level glass etch was also performed, first over photoresist for shallow etch (10 μm), and then over black lithographical masking tape for deep etch.[117]

It was reported that double etch (shallow + deep etch) resulted in a cross section leading to less dispersion in chemical separations (by simulation) than the usual single deep-etched cross section.[118,119]

Channel etching using HF was usually performed before bonding, but etching could also be achieved after glass bonding (for enlarging some channels).[120,121,671,1065]

A deep wet HF etch of borosilicate glass to >500 μm depth was achieved. This was carried out using an anodically bonded Si wafer (patterned) as a mask. A conventional Cr/Au etch mask can only produce shallow etch (<50 μm) before the problem of pinholes appeared. After etching, the Si mask was sacrificed by KOH etching, thus releasing the etched glass plate. The glass plate could then be sealed with a Si cover plate by anodic bonding for a second time.[122]

In the selection of etch mask for deep glass etching, thick SU-8 is a choice, but SU-8 cannot be used in an HF bath (48%) because SU-8 does not adhere well to SiO_2.[123] However, with a polycrystalline amorphous Si seed layer SU-8 adheres very well. For instance, with a 1.5 μm thick polished polysilicon, a 50 μm thick SU-8 can be deposited as the etch mask, leading to a

TABLE 2.3
HF-Based Wet Etching of Glass

Etch Conditions	Etch Results	References
BOE[a] (10:1)	~10 nm/min	302
	Unspecified etch rate	102
BOE (7:1)	0.03 μm/min	313
	0.9 μm/min	314
BOE (6:1)	36 μm in 40 min	105
	10 μm in 20 min	103
	0.66 μm/min	747
	Unspecified etch rate	110, 447, 529
BOE (unspecified ratio)	8 μm in 15 min	315
	5.2 μm in 20 min (stirred)	316, 317, 564
32% HF/NH₄F (1:1)	2 μm/min	125
HF/NH₄F (1.7%:2.3%)	~0.7 μm/min	318
	Unspecified etch rate	273
5% NH₄Cl/1% HF	Unspecified etch rate	136
HF (1%)/NH₃	100 μm in 1 h (70°C)	284
Conc. HF	30 μm in 4 min	980
	20 μm for 3 min	610
	7 μm/min	114
50% HF	~10 μm/min	1017
	13 μm/min	319, 430
	~8 μm/min	615
49% HF	9 μm/min	313
	7 μm/min	95, 96, 98, 244, 320, 409, 536, 548, 557
25% HF	30 μm for 15 min	739
	Unspecified etch rate	321
20% HF	~1.1 μm/min (21°C)	92
10% HF	With flow (1.55 μm/min soda lime glass, 0.786 μm/min borosilicate); no flow (0.85 μm/min soda lime)	125
50% HF/H₂O (1:5)	12.9 μm/h	122
HF/HNO₃ (7:3)	6.5 μm/min	322, 323, 394
HF/HNO₃	Unspecified etch rate	324, 330, 844
HF/HNO₃/H₂O (20:14:66)[b]	25 μm in 6 min	818
	2 μm/min	117
	0.8 μm/min	325
	~0.3 μm/min	139
	Unspecified etch rate	102, 292, 326, 327, 1078
5% HF/70% H₃PO₄	Unspecified etch rate	145, 328
HF/HCl/H₂O (1:1:2)	16 μm/2.5 min	329
	Unspecified etch rate	244, 282
BOE (6:1)/HCl/H₂O (1:2:4)	Unspecified etch rate	670

[a] BOE represents buffered oxide etch with various ratios of 40% NH_4F/49% HF.
[b] $HF/HNO_3/H_2O$(20:14:66) is identical to 49% HF/69%HNO_3/H_2O (2:1:2).

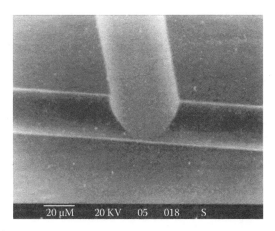

FIGURE 2.4 Electron micrograph of a T-intersection on a Pyrex glass microchip.[102] Reprinted with permission from the American Chemical Society.

maximum etch depth of 320 μm. Usually photoresist (2 μm thick) is only useful for shallow etch, less than 50 μm.[123]

Polished or unpolished polysilicon (by low-pressure chemical vapor deposition [CVD] at 620°C) is another etch mask option. Utilizing 2.5 μm thick unpolished polysilicon, a maximum etch depth of 160 μm was reached using an $HF/H_2O/HNO_3$ (6:40:100) solution. Further etching causes large pits (1.5 to 2.2 mm diameter) to form on the glass. With polished polysilicon (1.5 μm thick), etch depth up to 250 μm can be achieved. When amorphous Si is used as the etch mask, a maximum etch depth of 170 μm can be reached.[123]

Deep glass etching is useful in creating special glass structures such as a glass diaphragm. Forturan is another glass material used to make chips. When treated in a high-temperature oven (500°C for 1 h and then 600°C for 1 h), the exposed regions on the Forturan wafer will be transformed into a ceramic that has a much higher etch rate (e.g., in 10% HF) than the surrounding glass.[124]

Instead of photolithography, glass chips have been patterned using the poly(dimethylsiloxane) (PDMS) channel mask. Glass etching can be achieved directly in the regions defined by PDMS channels. On the other hand, glass etching is achieved using an etch mask (Ni) that is formed by electrodeless deposition in the regions as defined by the PDMS channels. Wet HF etch has been carried out on soda lime glass (microscope slides) or borosilicate glass (coverslips).[125] Figure 2.5 shows the comparison of wet etch (10% HF) in two types of glass using PDMS channels for patterning. It was found that the etch depth of the coverslip was less, indicating borosilicate glass was more resistant to HF etch than soda lime glass. In addition, there was a reduced channel broadening (or undercut) in the borosilicate coverslip, probably because of a less lateral etch rate (vs. vertical etch rate) in borosilicate glass. A lower concentration of HF was also found to produce smoother etched channels, as shown in Figure 2.6, in which a PDMS channel was used as an etch mask. On the other hand, glass etching is achieved using an etch mask (Ni) that is formed by electrodeless deposition in the regions as defined by the PDMS channels. It was found that the etch rate on the glass slide is slower in the case of Ni metal mask vs. the PDMS channel mask; see Table 2.3 for the results of 10% HF etching. This is because the use of a flow of HF in the PDMS channel facilitates etch product removal. In addition, the Ni mask produced a much better etch profile than the PDMS channel mask (see Figure 2.7a). This may be due to etchant leakage and the parabolic nature of HF flow, causing more replenishment of HF in the center (hence faster etch rate in the center).

Other than wet etch, dry etch of glass by RIE has been reported.[813] Dry plasma etch has also been carried out on soda lime glass (microscopic slides) or borosilicate glass (coverslips).[125] When dry etch was compared with wet etch, it was found that dry etch produced a vertical profile (see

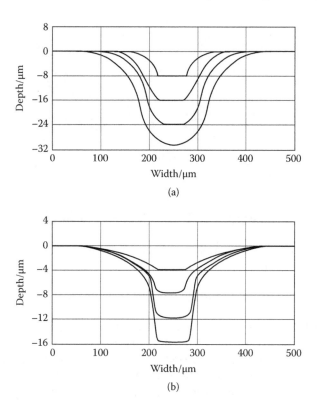

(a)

(b)

FIGURE 2.5 Etch profiles of channels as a function of etching time for a microscope slide (a) and a coverslip (b). The PDMS channels used as etch masks were 50 μm in width. The profiles were taken every 5 min (i.e., at 5, 10, 15, and 20 min).[125] Reprinted with permission from Elsevier Science.

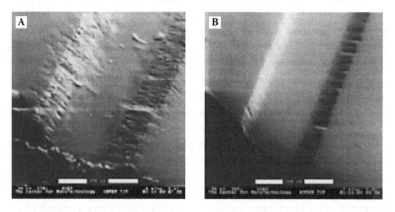

FIGURE 2.6 SEM images of etched channels illustrating the dependence of the smoothness of the etched surface on the etching solution. (A) SEM image of the channel etched with a high concentration of buffered HF (32% HF buffered with a 1:1 ratio of HF to ammonium fluoride) (scale bar: 150 μm). (B) SEM image of the channel etched with a low concentration of nonbuffered HF solution (10% HF) (scale bar: 100 μm).[125] Reprinted with permission from Elsevier Science.

Figure 2.7b). Note that Figure 2.7b was obtained after a long time, 28 h![125] This is due to the difficulty in removing the involatile reaction products.[111]

There are other dry micromachining processes for glass. Powder blasting[126–128,820,860,1127] has been reported. Powder blasting using alumina (9 μm) powder was used to etch a 25 μm deep channel in a glass chip.[129] Powder blasting can also be used to fabricate freestanding monolithic glass structures

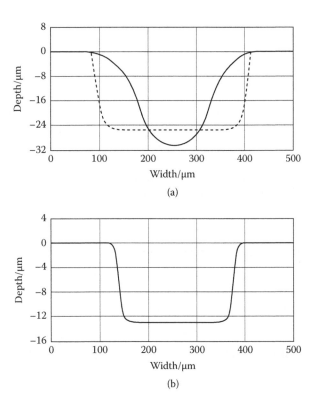

FIGURE 2.7 Etch profiles of microchannels obtained by wet etching (a) and dry reactive ion etching (b). (a) The more rounded profile was obtained with direct wet etching using a PDMS channel mold (50 μm in width), whereas the trapezoidal profile (dotted curve) was made with the deposited nickel layer as the etch mask (150 μm in width). (b) Dry etch was obtained with a Ni etch mask.[125] Reprinted with permission from Elsevier Science.

(see Figure 2.8). This is achieved by using an oblique powder beam and the mask underetching effect. The eroding powder consists of 30 μm sized alumina particles.[130]

Finally, microchannels in glass have been mechanically machined using a wafer saw.[131,792]

2.2.2 Drilling of Glass for Access-Hole Formation

In order to provide liquid access to the microfluidic chips, holes are usually created on the cover plate. These access holes can be created on glass by drilling using diamond drill bits. However, other methods have also been used, and these are summarized in Table 2.4.

For protection against chipping during drilling, the cover plate was sandwiched between two protecting glass slides and were glued with Crystal Bond.[844] After hole drilling, the hole could be etched by HF to provide a smooth conical channel exit.[104,670]

Drilling has also been performed on the bonded plate, rather than on the cover plate before bonding. To avoid plugging the channel in the bonded plate with glass particles during drilling, the channel was filled with Crystal Bond.[132]

In some cases, drilled holes are created on the same plate where the channels are etched. To achieve this, the holes were drilled either on the channel plate or on a plain plate with subsequent channel etching.[13,134] In the latter case, the photoresist can only be spray coated[134] or dip coated,[103–105] rather than spin coated on the drilled glass plate.[134]

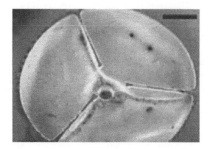

FIGURE 2.8   SEM picture of the 115 µm wide and 2 mm long suspended glass cantilever beams fabricated by deep glass etching.[130] Reprinted with permission from the Institute of Physics Publishing.

TABLE 2.4
Access-Hole Formation on Glass Chips

Methods	References
Diamond-bit drilling	316, 330, 331, 548, 747, 974
HF etching (4 min, 1 mm diameter, 145 µm deep)	329
CO_2 laser drilling	332, 931
Ultrasonic abrasion	102, 333, 430, 717, 718
Electrochemical discharge machining	95, 96, 135, 146, 334, 615, 725, 738
Powder blasting	129, 314, 335–338, 820, 1127

Drilling by a diamond-tipped drill bit (15 s per hole) is much faster than ultrasonic drilling (15 min per hole). On the other hand, ultrasonic drilling allows all holes to be drilled simultaneously with a multitipped ultrasonic drill bit.[108]

The appearances of the holes formed by ultrasonic drilling, laser machining, and sand blasting are compared in Figure 2.9.[104]

Another hole-forming method is electrochemical discharge machining (ECDM). In this method, an electric voltage is applied to a metal point touching the glass surface in an electrolyte solution. The point is slowly pushed through the glass as drilling continues.[738] As shown in Figure 2.10, the cathode and anode are immersed in a highly conductive aqueous electrolyte (e.g., NaOH). When a voltage (20 to 60 V) is applied between them, electrolysis occurs and H_2 is produced at the cathode. When the cathode is brought into contact with the glass substrate at the drilling location, an electric discharge can produce localized heating or ignition of H_2 gas or local sputtering. The removed glass debris might then be complexed and solubilized with NaOH. However, the principle of ECDM is not well understood. It was found that an increase in the NaOH concentration increased the drilling rate, possibly caused by an increase in the solution conductivity. It has been found that when a 0.5 mm diameter Pt wire is used as the cathode and a 25.4 mm^2 Ni gauge as the anode, a solution of 10% NaOH and an electric voltage of 47.8 V produced the best glass etch rate of 0.133 mm/min. To assist in the electrolyte access into the hole and removal of glass debris, the cathode could be moving vertically in and out of the drilled hole.[135]

In some devices no drilling was performed, but a circular coverslip was used with the end of the channel protruding out of the coverslip to reach the solution reservoir.[136,137]

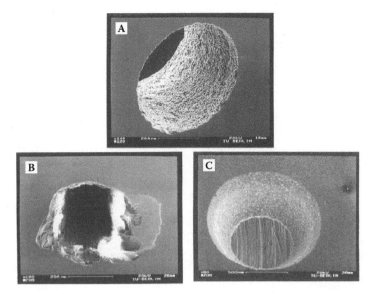

FIGURE 2.9 SEM micrographs of micromachined holes in borosilicate glass using three different methods. (a) Ultrasonically drilled hole; hole diameter, 500 μm. (b) Laser machined hole; hole diameter, 300 μm. (c) Sand-blasted hole; hole diameter, 1,000 μm.[104] Reprinted with permission from the Institute of Physics Publishing.

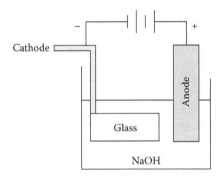

FIGURE 2.10 Physical setup for electrochemical discharge machining of a glass plate.[135] Reprinted with permission from the Institute of Physics Publishing.

2.2.3 GLASS BONDING

After channel etching and access-hole forming, the channel plate and cover plate are bonded. Thermal bonding is the major method used, though alternative schemes exist. All glass bonding methods are summarized in Table 2.5.

Normally, the glass plates are cleaned and dried before thermal bonding. In one report, a special microchip washer was designed that allows simultaneous cleaning of the channel and cover plates.[138] In another report, the two glass plates were brought into contact while wet, then the plates were bonded thermally with pressure (50 g/cm²).[105,747]

In thermal bonding, the glass plates were sandwiched between two polished alumina flats so that weights could be applied on top, and the glass surfaces would remain smooth after bonding.[105] To ensure even heat distribution, the two bonding plates were sandwiched between two Macor ceramic plates.[139]

TABLE 2.5
Thermal Bonding of Glass Chips

Bonding Conditions	References
Up to 440°C	340
500°C	136, 339, 439, 564
560°C (3 h)	329
570°C (3.5 h)	244
580°C (20 min)	105, 109, 529
595°C (6 h)	671, 1086
600°C	725
Up to 600°C using a program	316
610°C	314, 321
620°C (3 h), 10°C/min	818
620°C (4 h)	340
623°C (3.5 h) (Borofloat)	341,342
650–660°C (4–6 h) (Pyrex)	324, 343
650°C (4 h) (Borofloat)	320
680°C (Borofloat glass)	980
690°C (Borofloat glass)	331
3°C/min to 620°C (30 min), cool to room temperature (3°C/min)	747
500°C (3 h), 580°C (3 h), cooled for at least 12 h	729
665°C (4 h), cool slowly (5 h)	626
280°C/h to 400°C (4 h), 280°C/h to 588°C (6 h), cooled to room temperature (soda lime)	344
60°C to 500°C (1 h), to 570°C (5 h), cooled to 60°C and left for 12 h	739
Equilibrated at 500°C, 550°C, 600°C, then annealed at 625°C for 2 h (Pyrex)	345
440°C (0.5 h), 473°C (0.5 h), 605°C (6 h), 473°C (0.5 h), cooled overnight	879
500°C (1 h), 550°C (0.5 h), 620°C (2 h), 550°C (1 h), cooled overnight	582
40°C/min to 550°C (0.5 h), 20°C/min to 610°C (0.5 h), 20°C/min to 635°C (0.5 h), 10°C/min to 650°C (6 h), cooled to room temperature	102
10°C/min to 620°C (3.5 h), ramp down at 10°C/min (Borofloat)	108
10°C/min to 640°C (8 h), 10°C/min to room temperature (under vacuum)	139
20°C/min to 400°C, 10°C/min to 650°C (4 h), 10°C/min to 450°C, cooled to room temperature (Borofloat glass)	576
550°C (1 h), 580°C (5 h), 555°C (1 h), cool for at least 8 h	697
To 400°C (280°C/h) (4 h), 280°C/h to 588°C (6 h), cooled to room temperature	447

To achieve thermal bonding of a glass chip to thin glass (0.17 mm thick), which is not very planar, a matrix of underpinning posts was employed.[111]

Surface charge (due to silanol) on glass was found to be modified during the thermal bonding process.[532] After thermal bonding of glass plates, some defects showed up that might be due to surface damage, organic contaminants, dust, and incomplete metal mask stripping.[140]

Bonding of glass plates consisting of metal electrodes can be achieved in two ways: (1) between an etched glass plate (containing patterned electrodes) and a drilled cover plate or (2) between an etched glass plate and a drilled cover plate (containing patterned electrodes).[745]

Usually metal electrodes cannot withstand high temperatures, except in the case of Pt/Ta in which Pyrex plates can be fusion bonded up to 650°C.[141] Therefore, a low-temperature bonding method should be adopted. This was accomplished using potassium silicate with the following thermal treatment: 90°C (1 h), 0.3°C/min to 200°C (12 h).[590,925] Low-temperature bonding by polysiloxane was also reported,[142] as was the use of water glass (or sodium silicate). Bonding was carried out at 1.5°C/min to 80°C and 1 MPa pressure and held for 6 min, followed by cooling to room temperature. An ultrasonic microscope was employed to image the bonding performance. High sodium content (due to sodium silicate) at the bonding region was reported. This effect of high sodium on the microfluidic operation has been evaluated.[143]

Anodic bonding can be modified to bond glass plates (i.e., Pyrex to Pyrex). To achieve this, a silicon nitride or Si_3N_4 film (200 nm) was used as an adhesive layer, and the anodic bonding conditions of 1,500 V, 450°C, 10 min were used.[144] In another example, a 160 μm Si_3N_4 film was used and the bonding conditions were 100 V, 400°C for 1 h.[143]

There are nonthermal methods for bonding. For instance, bonding of glass was carried out by 1% HF.[109,529] Moreover, optical UV adhesive was used to bond glass to glass,[125,129,323] glass to silicon,[146,322,392,394,615] or glass to quartz.[392] To avoid filling the fluidic channels with the UV-curable glue, a stamping process was developed that allows the selective application of a thin glue layer.[129]

It was also reported that the UV glue can be degraded with a thermal procedure (100°C [6 min], 550°C [240 min], 100°C [30 min]), all at 1°C/min rate.[1078] So in one report, such an adhesive-bonded device was separated after heating up to 500°C and then rebounded.[147]

Bonding of glass plates can also be achieved at room temperature (20°C). This is based on hydrogen bonding at the glass interface, which has been achieved only after rigorous cleaning. This method allows bonding of different types of glass substrates having different thermal expansion coefficients.[844] In another report, pressure sealing of glass plates by clamping was used as a non-thermal reversible bonding method to avoid thermal degradation of a chemically modified (octadecylsilane [ODS]) layer.[430]

2.3 MICROMACHINING OF FUSED QUARTZ (OR FUSED SILICA)

2.3.1 FUSED QUARTZ CHANNEL ETCHING AND HOLE DRILLING

Various wet and dry etch methods have been employed to micromachine fused quartz chips in a manner similar to that in glass etching. However, the RIE method can be more effectively employed to etch quartz than can glass. Both wet and dry etch methods for fused quartz are summarized in Table 2.6.

Normally, one-level etching was performed, but in some applications (e.g., prechannel filter), a two-level etching method was also performed.[148] BOE etch was chosen over HF etch for quartz because BOE produced a smoother etch (0.23 μm relative standard deviation [RSD]) than did HF (0.55 μm RSD).[149] The cross section of an etched channel in a quartz chip is shown in Figure 2.11.[1006]

With RIE dry etch, the etched surface is very uneven, showing a typical rough cauliflower-like surface, but a subsequent wet etch by 1% HF/2% HCl solution for 10 min will smoothen the rough surface.[150]

In order to provide access holes on fused quartz cover chips, laser drilling[151,152,828,1006] or ultrasonic drilling[349,1041] has been used. In order to remove particulates caused by drilling, the drilled quartz plate was rinsed in 0.5% HF for 5 to 10 min in an ultrasonic bath.[153,658] However, in some cases no drilling was performed for liquid access holes, with the end of the channel protruding out of the coverslip.[148]

2.3.2 BONDING OF FUSED QUARTZ CHIPS

The etched quartz chips are sealed using various bonding methods, as summarized in Table 2.7. In thermal bonding for fused quartz, a much higher temperature than in thermal bonding glass (e.g., 1150°C) is required, but such a high temperature will distort nano-sized channels. So HF bonding (at room temperature) must be used in these applications.[676]

In one report, no bonding plate was needed to create the sealed quartz channels. These were achieved by first depositing a layer of poly-Si on the quartz substrate, which was then wet etched through a tiny hole to create a quartz channel and a thin SiO_2 roof.[154]

TABLE 2.6
Fused Quartz Etching

Etch Conditions	Etch Depth	References
Wet HF Etch		
BOE (10:1)	5 µm	148
BOE (6:1)	7 µm/h (room temperature)	263
	Unspecified etch rate	149, 814
BOE (1:1)	28 µm (50°C)	152, 1006
NH$_4$F/HF (36.3%:4.7%)	40 µm in 8 h	275
Dilute, stirred BOE	8 µm (50°C)	346
49% HF	20 µm in 16.7 min	347
HF	19 µm	840
Dry Etch		
Fast atom beam	10 µm	348, 846, 1021
DRIE	10 µm (CHF$_3$)	349, 676, 676
	<10 µm (C$_3$F$_8$/70% CF$_4$)	350
	<1.2 µm (C$_3$F$_8$/CF$_4$)	351
	10 µm (SF$_6$)	846
	14–39 nm/min (SF$_6$)	150
Laser ablation	Excimer laser	425, 426, 676
	CO$_2$ laser	846, 1021
Sandblasting (SiC particles)	Unspecified etch rate	676

Note: BOE represents buffered oxide etch with various ratios of 40% NH$_4$F/49% HF.

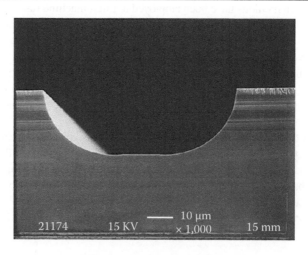

FIGURE 2.11   Electron micrograph of the cleaved edge of a quartz chip showing the channel cross section.[1006] Reprinted with permission from the American Chemical Society.

2.4   MICROMACHINING OF POLYMERIC CHIPS

Polymeric chips have been fabricated with various methods, as developed from the plastics industry. These processes include casting, injection molding, laser ablation, imprinting, and compression molding. The physiochemical properties of various polymeric materials and their molding methods have been summarized in a recent review article.[7]

TABLE 2.7
Bonding of Fused Quartz Chips

Bonding Conditions	References
Thermal Bonding	
1,110°C	346
1,110°C, 3 h	351
1,110°C, 5 h	352
10°C/min to 800°C, 3°C/min to 1,000°C, 3°C/min cool to 800°C, natural cooling	347
200°C (2 h) then 1,000°C (overnight)	152, 658, 1006
1,150°C	425, 426, 846, 1021
Other Bonding Methods	
NaOH, 500°C, 10^{-4} bar vacuum, 60 min	150
1% HF with 12 kg weight pressure for 12 h at room temperature	676
1% (w/v) HF with 0.16 MPa pressure, at room temperature for 24 h	264, 353
1% HF with 1.3 MPa pressure	354, 355
1% HF at 60°C for 4 h	263
1% HF, at room temperature, clamped for 24 h	275
CYTOP adhesive or poly(1,1,2,4,4,5,5,6,7,7-decafluoro-3-oxa-1,6-heptadiene)	356
Sealed to a PDMS layer	840
Potassium silicate (95°C for 10 h, 200°C for 10 h)	148

2.4.1 Casting

The procedure for micromolding in capillaries (MIMIC) for fabricating the microstructure of polymeric materials has been reported.[155–157] Vacuum-assisted MIMIC was later reported. These methods are essentially casting a soft curable polymer against a mold.[158]

The casting method has mostly been employed to fabricate poly(dimethylsiloxane) (PDMS) chips, but has also been used with partially polymerized polystyrene (PS) liquid preparations.[85]

A procedure for casting of PDMS replicas from an etched Si master with positive relief structures (in Si) has been developed for chemical analysis (see Figure 2.12).[159]

The PDMS replica is a negative image (see Figure 2.13) of the positive relief structure. A mixture of PDMS prepolymer and its curing agent was poured over the master mounted within a mold. The mixture should be degassed, e.g., under low vacuum (~20 to 30 Torr).[160] After curing, the PDMS replica was peeled off from the Si master.[159] To facilitate peeling off, the Si master was silanized or treated with various releasing agents, as tabulated in Table 2.8. The master can be used for ~100 casting procedures without observed degradation.[161]

To create liquid access, holes were punched through the replica. Then it was placed on another thin slab of PDMS for sealing. Because an anisotropic KOH-based etching method was employed to etch the Si master without corner compensation, the channel intersections in the replica were limited in shape by the <111> plane of the Si master.[159]

Rapid prototyping (<24 h) was suggested to produce PDMS chips. A transparency photomask was used to create a master consisting of positive relief structures (made of photoresist SU-8) on a Si wafer. PDMS was then cast against the master to yield numerous polymeric replicas (see Figure 2.14). This method allows the channel width of >20 μm to be made. Channel depths of 1 to 200 μm can be cast by using different thicknesses of photoresist. This method allowed >30 replicas to be made per master.[1033] However, in another report the SU-8 master could only be used for one or two PDMS castings, unless a special coating was used on the SU-8-on-Si master. This coating is an

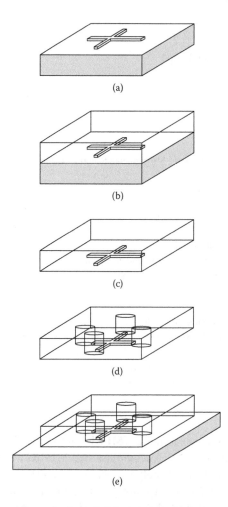

FIGURE 2.12 Fabrication procedure of a PDMS chip: (a) silicon master wafer with positive surface relief, (b) premixed solution of Sylgard 184 and its curing agent poured over the master, (c) cured PDMS slab peeled from the master wafer, (d) PDMS slab punched with reservoir holes, and (e) ready-to-use device sealed with another slab of PDMS.[159] Reprinted with permission from the American Chemical Society.

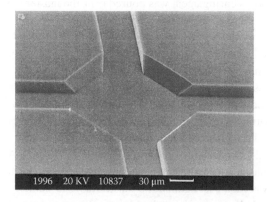

FIGURE 2.13 Electron micrograph showing the sample injection region of the PDMS device. Channel cross section, 50 μm (width) × 20 μm (height). The geometrical features of the channel junction result from the anisotropy of the wet etching process of the Si master.[159] Reprinted with permission from the American Chemical Society.

TABLE 2.8
Releasing Reagents for Casting

Molding Master	Releasing Reagent	References
Si	3% (v/v) dimethyloctadecylchlorosilane	159, 181, 260
	Trichloro(3,3,3-trifluoropropyl)silane	249
	Dichlorodimethylsilane	134
	Coated with 200 nm Teflon	161
	Au	362
Photoresist on Si	Hexamethyldisilazane on SU-8	357
	Tridecafluoro-1,1,2,2-tetrahydrooctyl-1-trichlorosilane on SU-8	182, 262, 358, 359
	Plasma-polymerized fluorinated layer on SU-8	247
	5% SDS on SPR-220-7	360
	1H,1H,2H,2H-perfluorodecyltrichlorosilane on photoresist (AZ 9260)	164
Glass	Treated by CHF_3 plasma in a RIE machine (SU-8)	361
	Chlorotrimethylsilane	174
	Au	362

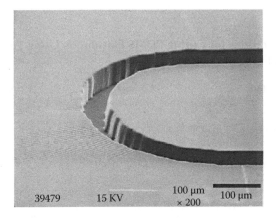

FIGURE 2.14 Electron micrograph of a turn in the channel fabricated in a PDMS chip, created by casting the polymer against a positive relief, which is made of photoresist patterned on a glass substrate. The roughness in the sidewall arises from the limited resolution of the transparency used as a photomask in photolithography.[1033] Reprinted with permission from the American Chemical Society.

amorphous diamond-like carbon/poly(dimethylsiloxane) hybrid, which significantly improves the lifetime of the master for PDMS replica production (no degradation after ten castings).[162]

To form a thick SU-8 layer, it was poured onto a Si wafer until a 25.4 mm diameter puddle was formed. The wafer was spun at 500 rpm for 10 s, ramped up to 3,000 rpm (or 250 µm/s), and held for 10 s. A 140 µm high SU-8 structure was formed.[163] The reflow of the photoresist (AZ 9260) on Si occurred during the hard bake final step (115°C for 30 min), rounding of the positive relief photoresist structure occurred, and the semicylindrical PDMS channel molded from this structure was achieved.[160] Similarly, SU-8 softens as temperature increases, unlike other photoresists, which harden at high temperature.[164]

PDMS chips have also been cast from a glass (not Si) master containing photoresist as the positive relief structure.[556,960] PDMS was also cast against an acrylic master that was hot embossed by an aluminum "mother" milled with the channel features (>10 µm).[166] In one case, PDMS was cast directly against an Al master.[166]

PDMS can be cast from two commercially available kits: (1) General Electric RTV 615[167,890,985] or (2) Dow Corning Sylgard 184 (more commonly used). Both kits contain the PDMS elastomer (component A) and the curing agent (component B). Upon curing, the PDMS elastomer or prepolymer is cross-linked by the curing agent, which is a proprietary platinum-based catalyst. Components A and B from either source are mixed in a definite ratio (10:1). The mixture is either spin coated to obtain a thin film (10 to 12 μm) or poured into a mold to obtain a slab of greater thickness (2 mm). The mixture is then incubated for either curing or bonding. RTV 615 needs a higher temperature than Sylgard 184 to cure, whereas Sylgard 184 is stiffer and more chemically inert to acids and bases. Various curing conditions in the use of these two commercial kits are summarized in Table 2.9.[890]

To ensure a better reversible seal between PDMS with glass, a mixing ratio of 20:1 is used, because this produces a less rigid polymer replica.[168,169] On the other hand, to fabricate a more rigid PDMS chip a ratio of 8:1 (curing at 70°C for 48 h) was used. It was found that even with this more rigid PDMS chip, the sealing to another PDMS remained good.[162] Such a ratio of 8:1 was also used to create a PDMS/poly(methylmethacrylate) (PMMA) chip.[178]

In order to homogenize the elastomer surface properties, thermal aging of PDMS was sometimes carried out (at 115°C for 12 h).[170]

TABLE 2.9
Various Conditions for Curing PDMS Elastomers Using Two Commercial Kits

Curing Temperature	Curing Time	References
Sylgard 184		
Room temperature	For 24 h	160
60°C	For 60 min	955
	For 60 min, then at 200°C for 1 h	85 for use as a mold
	For 3 h	363
	For 6 h	160
	For 12–16 h	85 for use as a mold
65°C	For 1 h	359, 364, 960, 1033
	For >3 h	365, 366
	For 3 h, then at 25°C for 21 h	164
70°C	For 12 h	857
	At least 72 h	800
75°C	For 1 h	367
	For 3 h	302
80°C	For 25 min	890
	1 h	134, 250
100°C	For 3 h	175
110°C	For 8 h	174
150°_C	For 1 h	160
200°C	For 2 h	817, 821
RTV 615		
60°C	For 5 h	270
80°C	For 1 h	890
	For 1.5 h	167, 357, 368
90°C	For 2 h	985
Unspecified temperature	Unspecified time	780

A two-mask process has been used to create a glass master containing two levels of positive photoresist relief structures. The master was used to cast a PDMS chip consisting of the channel/chamber (25 to 30 μm deep) and weir (7 to 12 μm clearance).[960] Another two-mask process was used to create a Si molding master consisting of 3 μm high Si relief structures and 25 μm high photoresist relief structures.[364]

The PDMS channel slab should be bonded to a cover substrate in order to create a sealed chip. This can be achieved by irreversible bonding after oxidation of the PDMS layer to create the silanol groups.[170,960,1033,1043] The PDMS surfaces can be oxidized in an O_2 plasma,[1033] Tesla coil,[171,172] RIE system,[173] ozone generator,[171] or ozone plasma created by a mercury vapor lamp.[174] PDMS can be sealed irreversibly to various substrates, such as PDMS, glass, Si, SiO_2, quartz, Si_3N_4, polyethylene, glassy carbon, and oxidized polystyrene. But this method of irreversible bonding did not work well with PMMA, polyimide, and polycarbonate.[1033] Moreover, PDMS (after O_2 plasma treatment) has been bonded to Au, glass, and Si-SiO_2 substrates.[1025]

A study of the effect of O_2 plasma parameters on PDMS bonding shows that low RF power and short treatment time are better than higher RF power and long treatment time, respectively, for bonding instantly after plasma treatment. Moreover, the bonding has to be carried out within 15 min after the plasma treatment to ensure complete bonding.[175]

Irreversible bonding of two PDMS layers was also achieved by first allowing the bottom cured layer to have an excess component of the elastomer base (i.e., 30A:1B) and the upper cured layer to have an excess curing agent (i.e., 3A:1B). Then the two layers were put into contact. On further curing, the excess components in each layer allowed new PDMS polymer to form at the interface to facilitate bonding.[167]

Another method to achieve bonding of a polymeric membrane with a glass plate for cell experiments was to first decorate the membrane with avidin and modify the glass plate surface with biotin. The avidin-coated membrane was then tightly bonded with the biotin-coated glass plate.[176]

On the other hand, reversible bonding of PDMS was used. In this case, no oxidation of the PDMS layers was needed.[748,858] Although the surface free energy of PDMS is not particularly high (~22 mN/m), the smoothness of the cast surface of PDMS in combination with its elastomeric nature ensures good adhesion of the PDMS layer on both planar and curved substrates. These properties of the PDMS elastomer simplify the sealing method (i.e., no thermal bonding needed). Pressure (up to 1 bar) can be applied to the channels without the use of clamping to prevent liquid leakage. Moreover, since the sealing is not permanent, after use, the two PDMS slabs can be peeled off for more thorough cleaning, if needed.[159]

As compared to irreversible bonding, channels formed by reversible bonding have more stable electroosmotic (EOF), and are easier to clean (after chip removal). However, reversible bonding is prone to leakage and cannot withstand pressure greater than 5 psi.[748] Reversible sealing was more effective after rinsing the PDMS and glass plates with methanol and then drying in an oven at 65°C for 10 min.[365] It was found that no cleaning effect was obtained with deionized water or isopropyl alcohol, but a dilute HCl solution was very effective for cleaning contaminated PDMS surfaces, even on aged chips.[175]

Reversible bonding was also made between a PDMS layer and a gold-coated glass slide that had been deposited with thiolated DNA capture probes.[955]

Reversible bonding of PDMS to PMMA was also achieved.[177,178,364] A PDMS replica containing microchannels (<100 μm deep) was sealed against a PMMA plate (or a PDMS replica of it) that had deep (300 to 900 μm) solution reservoirs machined in it.[1042] PDMS was also sealed against a patterned hydrophobic fluorocarbon film.[179]

A three-dimensional (3D) microfluidic channel system was also fabricated in PDMS. The system was created using a "membrane sandwich" method. In a three-layer PDMS device, the middle layer had channel structures molded on both faces (and with connections between them). This middle layer was sandwiched between two thicker PDMS slabs to provide channel sealing and structural support.[180] For instance, a three-layer structure was created to produce a basket-weave pattern consisting of crossing but nonintersecting channels (see Figure 2.15). In addition, a five-layer PDMS channel system was fabricated with a straight channel surrounded by a coiled channel (see Figure 2.16).[180]

A 3D microfluidic structure was also constructed with two horizontal PDMS layers and the third top layer.[181] In one report, a simple stacking of two PDMS channel layers was also performed to realize a Z-shaped 3D interface.[302]

Stacking of five thin PDMS layers (100 μm thick) also produced a special 3D microfluidic structure (see Figure 2.17).[175] During casting, a transparency film was used to flatten the thin PDMS prepolymer layer against the molding master by surface tension effect. This transparency film also helped to peel off the cured PDMS layer. Methanol was used to prevent instant PDMS bonding and promote smooth movement between each layer during stacking and alignment of the layers. The chip made by this multilevel soft lithographic method has been employed to perform 2D protein separations.[182]

Besides PDMS, 3D structures for mass spectrometry (MS) analysis have been cast in epoxy (Epofix) using two negative masters (in PDMS), which in turn were created from Lucite positive relief structures. Epofix was selected as the chip substrate, among acrylic-polyester resin (Casolite) and epoxy resin (Araldite), because of its having the best mechanical properties and the least chemical interference needed for fabricating the MS chip.[780] In another report, PDMS was chosen over epoxy to fabricate MS chips because of its less chemical noises (interferents) in MS, and over

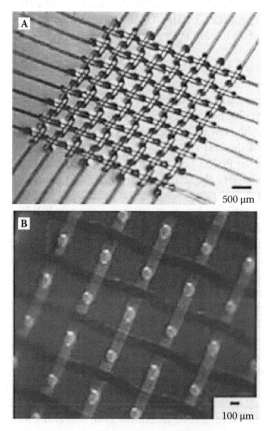

FIGURE 2.15 A microchannel system in a basket-weave pattern fabricated by three PDMS layers. (a) Optical micrograph of the middle PDMS layer, which contains the entire 8 × 8 channel system. The channels are 100 μm wide and 70 μm high. (b) Optical micrograph of a portion of the enclosed, fluid-filled channels. Channels in the up–down direction are filled with a solution of fluorescein and channels in the left–right direction are filled with a solution of Meldola's Blue dye.[180] Reprinted with permission from the American Chemical Society.

100 μm

FIGURE 2.16 A coiled channel surrounding a straight channel fabricated by five PDMS layers. The two channels were filled with fluorescein.[180] Reprinted with permission from the American Chemical Society.

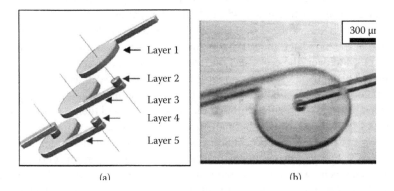

Layer 1
Layer 2
Layer 3
Layer 4
Layer 5

300 μι

(a)   (b)

FIGURE 2.17 A five-layer stacked PDMS microchannel system. (a) Schematic of a channel path structure. (b) High-magnification optical photograph of a vertical channel path.[175] Reprinted with permission from the Institute of Physics Publishing.

polyurethane because of good adhesion properties. Even so, in the use of PDMS its curing (at 70°C) should be carried out for at least 72 h to further reduce the chemical noise.[800]

Casting of a PDMS chip was also achieved using a master fabricated from a printed circuit board. A board with a calculated Cu foil thickness of 34.3 μm gave 29 μm deep channels that were measured after etching. With the etching solution of ammonium peroxydisulfate at 38°C, it took ~20 min to etch. Since the Cu layer was slightly rough, the cast PDMS layer did not adhere well to glass. However, it did adhere well to another PDMS layer.[183]

A solid-object printer was used to produce a plastic molding mask for casting PDMS chips.[184] A melted thermoplastic build material (m.p. 80 to 90°C) was used as the "ink" for printing. This method provides an alternative to photolithography in fabricating the molding master, but only creates features of >250 μm. Since the mold is easily damaged by deformation, only about 10 replicas can be made from one master. The roughness of the master is ~8 μm, which can be reduced by thermally annealing it. In this case, a negative master is first made and annealed, and a positive master is created from the smoothed negative master.[184]

Reduction photolithography using an array of microlenses (40 μm diameter) was demonstrated. This method achieved a lateral size reduction of 10^3, and generated a feature size down to 2 μm over a large area (2 × 2 cm^2).[363] Reduction can also be achieved using 35 mm film photography (8×) or microfiche (25×).[185]

Photolithography using gray-scale photomasks enables the generation of micropatterns that have multilevel and curved features on the photoresist layer upon a single exposure.[363]

Instead of using a planar molding master, a fused silica capillary tube (50 μm i.d. and 192 μm o.d.) was used as a template for casting PDMS channels, and as the fluid inlet/outlet tubes. After PDMS curing, the middle prescored section (4 cm) of the capillary was removed to reveal the PDMS channel (192 μm wide and deep).[817] Similarly, a capillary was used to mold a PDMS channel, and to produce an electrospray emitter. In this case, after PDMS curing, the last 0.5 cm section of the capillary was removed to create a channel.[821]

2.4.2 Injection Molding

Injection molding is a common procedure in the plastics industry. Therefore, this method has been adapted to fabricate acrylic chips. First, the channel pattern was fabricated by wet etching a Si master. Second, a Ni electroform "mother" was made out of the Si master. Third, many Ni electroform "daughters" were made. Finally, the electroform daughter was used to create acrylic chips via injection molding. One Ni electroform daughter could produce five hundred chips.[186] In order to mold the channel patterns on the acrylic plates, a negative image should be formed on the electroform daughter. This was fabricated via the electroform mother using the negative image formed on the Si master because it is easier to create a 45 μm high relief pattern in Si than to etch a 45 μm deep channel in Si. Although the width (45 μm) and depth (45 μm) of the Ni electroform mother followed the relief of that of the Si master, the acrylic molded channels produced only 35 μm deep channels (see Figure 2.18). Moreover, the cross section indicated that there were asymmetry and rounding of some features. This may be caused by unoptimized conditions, such as temperature used during mold injection, cooling duration before chip ejection from the mold, and postinjection annealing. Access holes (3 mm) were first drilled through the acrylic chip, and then it was sealed with a cover film (2 mil thick Mylar sheet) using a thermally activated adhesive.[186]

Besides acrylics or PMMA, injection molding was also carried out on regular and optical-quality polycarbonate (PC) chips.[187] Furthermore, three PS plastic plates were injection molded to create the microfluidic channels, which were employed for a biomolecular interaction study. The upper plastic plate with ridge patterns was ultrasonically welded to the middle plate. The middle plate was sealed to the bottom plate with a molded silicone rubber layer.[188]

2.4.3 Ablation

Ablation using radiations of various wavelengths (IR, visible, UV, or x-ray) has been employed to fabricate plastic chips. For instance, photoablation using pulsed UV lasers (193 nm) has been used to fabricate plastic chips out of polyethylene terephthalate (PET; 100 μm thick)[189,190,258,758] and polycarbonate (PC; 125 μm thick).[189,258] Channels as narrow as 30 μm and as deep as 100 μm can be made.[258,758] The cross section of a photoablated PET channel plate laminated with another PET using a thin PE adhesive layer was shown in Figure 2.19.[191]

An excimer laser was also used to machine a PC chip (6 mm thick) to create 160 μm wide channels (60 μm deep),[811] or on a polyimide sheet.[192,811] Another UV excimer laser (248 nm) was used to ablate microstructures within PC channels (fabricated by imprinting).[193]

A UV excimer laser (193 nm) was used to ablate channels in polystyrene (PS; 1.2 mm), PC (1 mm), cellulose acetate (100 μm), and PET substrates (100 μm).[194,195] Low-temperature (125°C) lamination was used for fast bonding (<3 s) the ablated substrates to PET films (35 μm thick) with PE adhesive (5 μm thick).[189,194]

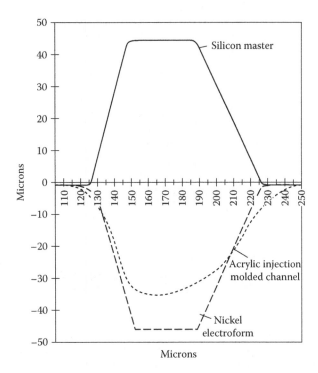

FIGURE 2.18 Profilometer scans across the microchannel structure of the silicon master (negative image), the nickel electroform mother (positive image), and injection-molded acrylic chip (positive image). The nickel electroform daughter (negative image) is not shown.[186] Reprinted with permission from the American Chemical Society.

A XeCl excimer laser (308 nm) was used to photoablate biodegradable polymers (Poly(d-lactic acid) [PDLA] and polyvinyl alcohol [PVA]) into channels (10 to 50 μm deep).[196]

A molecular fluorine excimer laser (157 nm) was also used to ablate PMMA chips to a depth of 500 μm.[197] It was also reported that acrylic (or PMMA) plates were processed using a laser-cutting machine.[198]

Photoablation (diode-pumped Nd:YVO₄ laser, λ = 532 nm) was used to create a master on the PMMA layer coated on a Si wafer (see Figure 2.20a). The PMMA layer was doped with rhodamine B to facilitate the absorption of the laser radiation. The width of the ablated features depends on the diameter and the position of the laser focal point. The best aspect ratios were obtained with the laser beam focused 3 to 4 μm into the PMMA film. The ablated PMMA-Si master was used to cast a PDMS layer (see Figure 2.20b), which appeared to have smoother surfaces than the PMMA master.[367]

X-ray micromachining (7 to 9 Å synchrotron radiation) was used to ablate PMMA channel structures (20 μm wide and 50 μm deep).[199] Bonding was achieved by first heating both PMMA plates on a hot plate (150°C) for 5 to 10 min, then they were aligned, pressed together, and the two-plate assembly was allowed to cool. This surface heating prevents outgassing and bubble inclusions, which occur in bulk heating.[199] In contrast, conventional (rather than x-ray) machining on PMMA can only create channels of large depths (from 125 μm to 3 mm) and width (>125 μm).[1042]

In another report, instead of direct x-ray ablation, an x-ray mask, which was 10 μm thick Au on a Kapton film, was employed to ablate a PMMA wafer. Then the ablated PMMA wafer was developed. For sealing, a PMMA cover was first spin coated with a 2 μm layer of poly(butyl methacrylate-co-methylmethacrylate) and then thermally bonded to the PMMA plate (120°C for 1 h).[200]

An infrared CO_2 laser (1,060 nm) was used to cut through a polycarbonate (black, carbon-coated) wafer of 250 μm thickness to create microfluidic channels. The laser-machined black PC

FIGURE 2.19 SEM representation of the cross section of a PET microchannel made by photoablation. The trapezoidal form is typical for this type of microfabrication procedure. The lamination on the top is evidenced by the interface between the polymer layers. The channel dimensions are 30 μm wide in the bottom, 40 μm wide in the top, and 40 μm deep.[191] Reprinted with permission from Elsevier Science.

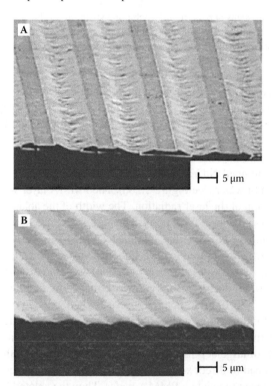

FIGURE 2.20 Electron micrograph of (a) the rough channel features on a PMMA master as generated by laser ablation; (b) the smooth channel features on the PDMS replica cast from the PMMA master.[367] Reprinted with permission from the American Chemical Society.

wafer was then thermally bonded between two transparent PC wafers at 139°C under two tons of pressure for 45 min.[937] A CO_2 laser was also used to ablate Mylar sheets (with adhesive). Then the machined Mylar sheets were laminated together.[1051] Moreover, a PMMA substrate was machined by a CO_2 laser.[201,202] However, the microchannel (~200 μm deep) has a Gaussian-like cross section and a certain degree of surface roughness.[201] The CO_2 laser has also been used to machine PET substrates.[203]

2.4.4 WIRE IMPRINTING

Plastic channels could also be fabricated by the hot wire imprinting method. A channel was imprinted on a piece of Vivak (copolyester) by a nichrome wire (50 μm diameter). It was clamped between the plastic and a glass slide (see Figure 2.21), and the assembly was heated at 80°C for 15 min. Then a perpendicular channel was similarly imprinted on a second piece of Vivak. The two pieces were then sealed at 75°C for 25 min (note that the two channels are not coplanar!).[1011] The channels were filled with water for storage.[204,685]

A chromel wire (13 or 25 μm diameter) was used to hot imprint microchannels on PMMA or Plexiglas substrates (1.6 mm thick). The plastic substrate was heated for 10 min at 105°C (softening temperature of PMMA). Although another imprinting procedure can be carried out on the same substrate using a wire perpendicular to the first imprinted channel, the plastic substrate appeared to be more rigid following the first heating cycle. Therefore, the second channel was imprinted on a second PMMA plate. Then the two plates were bonded together at 108°C for 10 min using a press with the application of a uniform pressure. Sometimes, the PMMA plate was not well sealed due to bubble entrapment. These failed devices could generally be salvaged by reheating the device at the same temperature, but for a longer time.[1011]

Microchannels were also wire imprinted on Plexiglas using a 90 μm diameter tungsten wire at 175°C. The channels were triangular-like, typically 200 μm wide and 75 μm deep.[205]

Wire imprinting was also achieved on PS, PMMA, and a copolyester material (Vivak) using chromel wire.[206] Low-temperature bonding was applied for channel sealing. The EOF was found to be the highest in the copolyester chip and the lowest in the PS chip. The chromel wire was stretched taut and placed on a plastic plate (for 7 min) heated to a temperature higher than the polymer softening point.[206]

Instead of imprinting two cross-channels on two separate PMMA chips, the two wires have been stretched and crossed over on the same PMMA substrate for imprinting.[207]

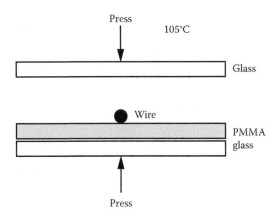

FIGURE 2.21 Fabrication protocol for wire-imprinted devices.[1011] Reprinted with permission from the American Chemical Society.

2.4.5 COMPRESSION MOLDING

To imprint more complex structures, a Si master with positive relief structures was first created. Then the master was used to imprint the channel pattern on a PMMA plate (135°C for 5 min) (see Figure 2.22).[1011] This method is called compression molding or embossing. In using this method, a vacuum chamber is needed to remove air bubbles trapped at small features and to remove water vapor from the polymer.[7]

For proper embossing, the PMMA was baked at 80°C for 8 h to reduce the absorbed water to 0.1% (from ~0.4%). During embossing, the molding die was heated to 150°C and pressed into the PMMA wafer with a force of 4,000 lb for 4 min at 160°C. During demolding, the PMMA wafer was maintained at 85°C.[208]

Besides PMMA, compression molding was also used to fabricate microstructures on PC chips (1 mm thick). High temperature (188°C) and pressure (11 metric ton pressure applied by a hydraulic press) were used. Before bonding, the hydrophobic channel surface was treated with UV irradiation (220 nm) to increase surface charge, which would assist aqueous solution transport. The molded chip was thermally bonded to another PC wafer. During use, the bonded chip did not yield to a liquid pressure up to 150 psi (134°C, 4 metric ton, 10 min).[938]

Moreover, a Zeonor plastic plate, normally used to manufacture CDs and DVDs, was hot embossed (130°C, 250 psi) using a Si master. The embossed chip (with microchannels 60 μm wide and 20 μm deep) was thermally bonded to another Zeonor plate (85°C, 200 psi, 10 to 15 min).[808] In another report, a 2 mm thick cyclo-olefin (Zeonor 1020 R) substrate was embossed using a Si master to create 20 μm wide and 10 μm deep channels.[788]

Hot embossing of Zeonor using different molding masters was evaluated in terms of pattern transfer quality, surface roughness, and mold reusability. Different materials (DRIE Si <100>, wet etch Si <110> and <100>, and SU-8) were used as the embossing masters. The thermal expansion coefficients (master and polymer) and glass transition temperatures (polymer) were carefully evaluated in the selection of embossing and de-embossing temperatures.[209] To help smooth out the scalloping features on the sidewalls of Si embossing master (caused by DRIE), a layer of Teflon or SiO_2 was deposited.[209]

Hot embossing was also used to make polymer chips on PMMA,[149,211,215] polystyol,[103,210] or PC.[211,212]

Besides using a Si master, hot embossing was also carried out using a Ni master (300 μm). This was electroformed from a Si master using a Ti/Ni seed metal layer on Si. After electroplating Ni on the Si, the Si was etched away, exposing the opposite relief of the Ni master. The Ni master was used to emboss on PC and PMMA substrates.[213]

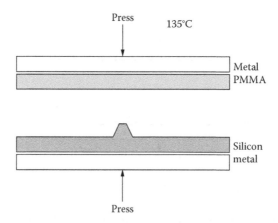

FIGURE 2.22 Fabrication protocol for silicon-template-embossed devices.[1011] Reprinted with permission from the American Chemical Society.

In another report, the Ni master was electroformed on a developed PMMA layer (supported on a stainless steel plate). The PMMA layer (3 mm) was first exposed to x-ray, then developed in developer solution (60% BEE, 15% morpholine-2-(2-butoxyethoxyl)ethanol, 5% ethanolamine, 20% water), and rinsed in a rinse solution (80% BEE and 20% water). Thereafter, the PMMA layer was dissolved away by methylene chloride.[208]

Bonding is usually conducted thermally. For bonding PMMA plates, the plates were immersed in a boiling water bath for 1 h to eliminate residual air. It was found this method was more effective than the convection oven bonding method.[215] Alternative bonding methods were also employed. For instance, poly(ethylene terephthalate glycol) (PETG) was also used to bond embossed PC substrates, because PETG has a lower glass transition temperature (81°C) than does PC (150°C).[193] Moreover, solvent bonding using isopentylacetate was employed for sealing two embossed PMMA plates.[214]

Room temperature embossing on PMMA (Lucite) and copolyester (Vivak) plates was achieved using a Si master. A hydraulic press was employed to apply the pressure (450 to 2,700 psi). Such a room temperature operation will prevent breakage of the Si master due to the differences in the thermal expansion coefficients of Si and PMMA. This room temperature procedure improves the lifetime of the master, so that one hundred, instead of ten, devices per master can be embossed.[177] For comparison, when embossing was performed at 140°C in a convection oven using G-clamps to apply the pressure, the Si template could be used to emboss up to sixty-five PMMA chips.[215]

Copolyester sheets were also embossed using a Si master (room temperature, 1,600 psi, 5 min) to produce microchannels (30 μm deep and 100 μm wide).[821]

A PS substrate was embossed by a Si template at room temperature using pressure (890 psi or 6.1×10^6 Pa) applied by a hydraulic press for 3 min. The embossed PS chip was then sealed with a PDMS lid.[216]

When an elevated temperature (110°C) is used to bond PMMA chips, hot embossing must be used to create channels on the PMMA substrate, 5.1×10^6 Pa or 740 psi for 1 h. If room temperature embossing is used, the PMMA channel will be distorted during the thermal bonding process.[216,259]

Low-pressure embossing of PS sheets or pellets was achieved (see Figure 2.23). By using a Si template, a PDMS master was first fabricated by casting. The curing procedures of the PDMS masters are 60°C on-mold for 12 to 16 h. Then hot embossing (210°C) was carried out for 20 to 60 min on a PS thin sheet (0.1 mm thick). Binder clips were used to produce the low pressure needed for embossing (~360 N/cm²). In order to get rid of the air and to ensure complete filling of fine features during embossing, *in vacuo* hot embossing was employed. In this second method, the PDMS master was fabricated by curing at 60°C in-mold for 1 h, and then 200°C off-mold for 1 h. Weights were used to create the low pressure (0.18 to 0.52 N/cm²). The substrate was heated to 225°C for 3 h to complete the process. A polished Al block was used so that the embossed PC structure was smooth and so transparent. A special PS substrate, which was biologically inert, was used.[85]

Channels can be fabricated on a Zeonor substrate by embossing the substrate against a fused silica capillary (360 μm o.d.) that is used as a template.[217]

2.4.6 PHOTOPOLYMERIZATION

Liquid-phase photopolymerization was used to fabricate plastic chips.[218,219] To create a microchannel, a UV photomask was used so that the masked channel areas were prevented from polymerization, while the exposed areas were photopolymerized. Subsequent suction and flushing removed the unexposed monomer mixtures.[218]

Thiolene-based optical adhesive was patterned by a photomask to form the microfluidic channel. This material, in contrast to PDMS, is solvent resistant.[220]

In addition to fabrication of microchannels, photopolymerization was also employed to fabricate microstructures within microchannels. Fabrication of hydrogel structures in microchannels

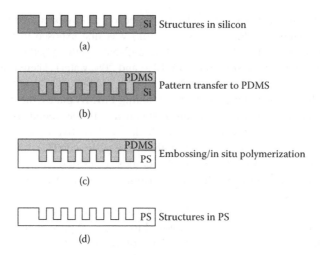

FIGURE 2.23 Scheme for fabrication of plastic microdevices from silicon master using an intermediate soft mold. (a) Silicon structures are fabricated using conventional photolithography and reactive ion etching. (b) PDMS is cured *in situ* over the silicon master. (c) Polystyrene is hot embossed onto the PDMS mold or polymerized *in situ* from partially polymerized styrene. (d) Polystyrene replica is separated from the mold.[85] Reprinted with permission from Springer Science and Business Media.

was first achieved by functionalizing the glass-PDMS channel with 3-(trichlorosilyl)propyl methacrylate (TPM). Then the channel was filled with a solution consisting of the monomer, 50% (v/v) poly(ethylene glycol) diacrylate (PEG-DA), and the photoinitiator, 1% 2-hydroxy-2-methylpropiophenone (HMPP).[960,1056] PEG-DA was first purified using an alumina column to remove the stabilizer and impurities. Free-radical polymerization was initiated by UV (365 nm) through a photomask to dissociate HMPP into methyl radicals that attacked the acrylate functionalities in both PEG-DA in the solution and TPM at the channel wall, allowing formation and attachment of the hydrogel structure, respectively.[960]

In another report, transparent channels (of width 500 to 200 µm and depth 50 to 180 µm) were filled with a photopolymerizable liquid mixture consisting of acrylic acid and 2-hydroxyl methacrylate (1:4 molar ratio), ethylene glycol dimethacrylate (1 wt%), and a photoinitiator (3 wt% Irgacure 651 or 2,2-dimethoxy-2-phenylacetophenone). Polymerization was completed in less than 20 s to produce the hydrogel structures.[221]

In another report of hydrogel formation, 2,2-bis(hydroxymethylpropionic acid) was used as the photoinitiator.[222]

A liquid preparation with solid polystyrene (0.6 g) dissolved in liquid styrene monomer (1.5 ml) was cast against a mold. Polymerization was accomplished with UV irradiation (21°C, 18 h). Solid PS was included to reduce the degree of shrinkage that occurred when monomeric styrene was photopolymerized.[85] In a similar manner, PMMA dissolved in MMA was cast against a Si master. Upon UV polymerization (with BME as the photoinitiator), a PMMA chip is formed. Nearly one hundred PMMA chips can be replicated using a single Si master.[223]

UV-curable polyurethane (PU) precursors were cast into a mold (made of PDMS) and cured by UV for 10 min. Several cured PU layers were stacked together to form a 3D structure (see Figure 2.24). Microcomputer tomographic (CT) imaging was used to visualize the stacked microstructures.[160]

A photocurable resin (hydroxyethylmethacrylate [HEMA]) was injected (by N_2 gas) into two aligned PMMA channel plates for bonding and surface modification. This method can bond PMMA plates well and produce a hydrophilic surface coating.[224]

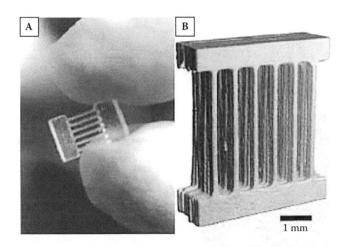

FIGURE 2.24 Example of a stack of micromolded PU structures viewed by (a) optical microscopy or (b) microcomputer tomography. Scale bar = 1 mm.[160] Reprinted with permission from Springer Science and Business Media.

2.4.7 OTHER PLASTIC MICROMACHINING PROCESSES

Fluoropolymers (polytetrafluoroethylene [PTFE], fluoroethylenepropylene [FEP]) were used to fabricate microstructures, which were etched by Ar ion beam etching (IBE).[225]

Channels were milled on a PMMA disk by a computer numerically controlled (CNC) machine using end mills (of size 127 to 762 µm).[226,638] Channels were milled in PC wafers using a milling machine or direct laser writer.[227]

An all-polyimide microchannel chip was created by patterning two photosensitive polyimide layers, and then by laminating the two layers based on a partially imidized interfacing layer.[228,229]

Polyimide microcavities (53 µm diameter and 8 µm deep) have been fabricated using two 4 µm layers sandwiched by Au/Cr electrodes.[230]

Parylene C plastic microstructures were formed by an additive process. A sacrificial layer of photoresist was used to define the channel regions. The structures were supported on a PC substrate[139,231] or Si substrate.[231,691]

Microchannels have also been directly patterned on SU-8 photoresist.[232–234,892] Multilevel structures were fabricated using the SU-8 photoresist.[232] In another report, the SU-8 photoresist was spun (1,250 rpm for 30 s) on an indium-tin oxide (ITO)-coated glass plate, which was first treated by an O_2 plasma to increase the adhesion of SU-8 on ITO. The photoresist channel was of ribbon-like structure with triangular ends (40 µm height, 10 mm width, and 90 mm length). The channel was bonded by hot pressing to another ITO glass on top of the photoresist structures.[892] SU-8 photoresists have also been used to create multilayered structures, i.e., as the channel wall materials, which were sandwiched between Si and Pyrex or between quartz and quartz.[235]

The photoresist SU-8 has been used directly to fabricate microstructures. In order to create 400 µm high micropillars, a 400 µm thick SU-8 layer was used. To reduce the risk of bubbles and to ensure uniformity in the SU-8 layer, two layers of 200 µm SU-8 were consecutively deposited, each with a postexposure soft-bake step. The exposed SU-8 layer was developed using propylene glycol methyl ether acetate.[236]

Ultrathick microfluidic structures (up to 1.5 mm high) were fabricated using SU-8 photoresist. Instead of using a spin coater, a constant-volume injection method was used to apply thick photoresist for patterning.[237]

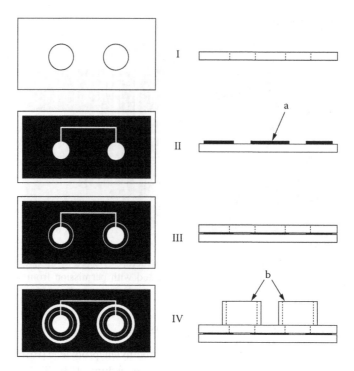

FIGURE 2.25 Schematic of the main parts generated by the single toner layer (STL) microfabrication process: I, perforated PET polyester cover; II, printed polyester base (a, toner layer); III, cover and base laminated together; IV, the final device (b, liquid reservoirs).[238] Reprinted with permission from the American Chemical Society.

Toner particles printed on a transparency (PET) film by a laser printer were used to fabricate a microfluidic device. It was then laminated with a blank transparency film to form a closed device.[238,239] The complete process is shown in Figure 2.25. The PET film is 100 μm thick and has optical transmittance above 80% in the range 400 to 800 μm. The toner layer is 6 μm deep. Approximately twice this depth can be obtained by laminating two printed films. Lamination was carried out by a commercial laminator at a heating temperature of 130°C and translation spread of 45 cm min^{-1}. The toner is composed of a polystyrene/PMMA copolymer (45 to 55 wt%) and iron(III) oxide (45 to 55 wt%). This material will soften and remelt in the range of 110 to 150°C without significant thermal decomposition of the toner or softening of the PET polyester film. The microchip was found to be compatible with various aqueous solutions, including 20% MeOH or CH_3CN. CE separation of 100 μM K^+, Na^+, and Li^+ was carried out using conductivity detection by copper tape electrodes.[238]

An acrylic chip was fabricated by stereolithography without an assembly process such as bonding. This is a 3D fabrication method that solidifies a photopolymerizable resin layer by layer via the scanning of a UV laser beam. A special double-controlled surface method was adopted in order to produce a smooth and transparent surface for high-quality optical detection.[240]

A PDMS network was constructed in which the microfluidic function was defined after assembly with square capillary (open and plugged) (see Figure 2.26). In Figure 2.26a, a top plugged capillary and two side open capillaries allow left and right flow to drain to the bottom channel. In Figure 2.26b, two plugged capillary tubes on the left and right produce a vertical through channel, as shown by the fluorescent solution (Coumarin 343). The capillary can be functionalized to form a Ca ion sensing and pH-sensing capillary, which serve as the detection zone. Ca and pH testing in the range of 10^{-5} to 1 M and pH 4 to 8 were achieved.[241]

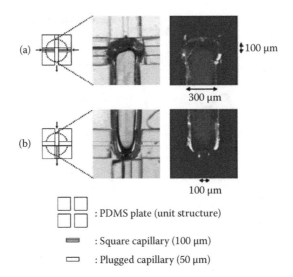

2.4.8 PROBLEMS ENCOUNTERED IN POLYMERIC CHIPS

Polymeric materials may have some unfavorable properties for microfluidic operations, and these properties are described in subsequent sections.

2.4.8.1 Optical Properties

Different polymeric materials show different UV transmittance, as shown in Figure 2.27. PDMS, which is optically transparent down to ~230 nm, has a lower UV cutoff than glass.[159,242] Since the UV transparency of olefin polymers is down to 300 nm, these materials may be better substrates than PMMA or PC when UV radiation is used for detection.[243] PDMS has a relatively low refractive index ($n = 1.430$), and this reduces the reflectance and hence the amount of reflected excitation light and lowers the background.[159]

There is autofluorescence emitted from the plastic substrates and bonding materials. The fluorescent background of different substrate materials was shown in Figure 2.28. PMMA and PDMA have low fluorescent backgrounds similar to that of glass.[224] The use of a longer excitation wavelength (530 nm) is one way to alleviate such a problem in fluorescent detection.[186] The PDMS was reported to be over thirty times more fluorescent than an equal thickness of borosilicate glass.[244] The background autofluorescence of the polyolefin chip was very low, and was determined to be only ~2.3 times that of borofloat glass at 520 μm.[245]

2.4.8.2 Electrical Properties

Polymeric materials usually have low dielectric breakdown voltages. Fortunately, the electrical insulation property of PDMS is sufficient ($R > 10^{15}$ Ω/cm).[159] Moreover, the use of a lower electric field (~1,100 V/cm) helps alleviate this problem of dielectric breakdown.[556]

2.4.8.3 Thermal Properties

The thermal conductivity (in W m⁻¹ K⁻¹) of PDMS (0.15) appears to be sufficient, although it is lower than those of PC (0.16), PET (0.2), glass (0.7 to 1.0), fused silica (1.38), and silicon (124).[159,246]

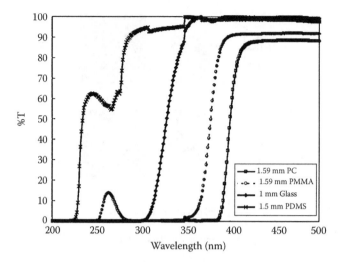

FIGURE 2.27   UV transmittance spectra of PC, PMMA, glass, and PDMS substrates.[224] Reprinted with permission from the American Chemical Society.

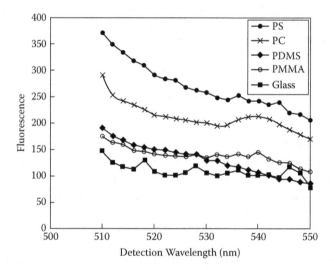

FIGURE 2.28   Fluorescence background (excitation at 490 nm) of various substrates.[224] Reprinted with permission from the American Chemical Society.

Since the channels in the plastic chip are usually narrow (i.e., with high surface-to-volume ratio), the heat dissipation properties of the plastic (e.g., acrylic) channel compared favorably with those of a fused silica capillary (75 μm i.d.).[186]

2.4.8.4   Mechanical Properties

Deep (>1.5 μm) PDMS channels were prone to collapse, either spontaneously (because of gravity) or during suction.[1025] In one report, support pillars were put within channels to avoid bowing of the deep PDMS channels (100 μm wide and 3 μm deep) during bonding to glass.[985] Furthermore, lateral collapse of PDMS channels was observed when they were not spaced sufficiently apart, as shown in Figure 2.29.[247]

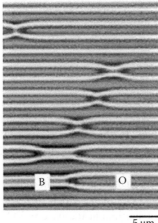

5 μm

FIGURE 2.29 An optical micrograph showing the lateral collapse and pairing of PDMS walls in a microfluidic channel network (1.2 μm high, 0.8 μm wide). Channels are blocked (B) or left open (O) after the stresses of forming, handling, and using the channels.[247] Reprinted with permission from the American Chemical Society.

2.4.8.5 Surface Properties

The surface of native polymeric (e.g., PDMS) chips is hydrophobic. This surface property has caused problems in filling aqueous solutions in the PDMS channels, and in supporting solution transport based on electroosmotic flow (EOF).

To fill hydrophobic channels, especially in PDMS, there is a need to prime the channels. For instance, the addition of ethanol (up to 5%)[160,247,1025] or 0.1% v/v Tween-20[181] assisted in channel filling. When PDMS was sealed against a hydrophilic surface, coating the channel by bovine serum albumin (BSA) (1% in phosphate buffered saline [PBS]) for 1 h appeared sufficient to provide good filling.[247] The liquid filling in thin PDMS channels is also facilitated by a large filling pad plus a second large flow-promoting pad.[1025] A PDMS-glass chip could be filled by immersing the chip in the buffer solution in an ultrasonic bath for 5 to 10 min. The solution was filled by ultrasonic vibrations. It was reported that no bubbles were present in the channels.[248]

The PDMS channels can also be primed with CO_2 gas, which is readily dissolved in aqueous solution, and so no bubbles will be formed after solution filling.[160] It was reported that gas bubbles in PDMS channels, if formed, can be removed by blocking the output port, and pressurizing the channel via the input port. This method is successful because of high gas permeability of PDMS.[249]

The PDMS channel can be rendered hydrophilic by oxidation (O_2 plasma treatment).[250,1025,1033] However, the hydrophilic surface in air has a short useful lifetime (~15 min) because the surface easily becomes hydrophobic again. Storing the treated PDMS under water maintained the channel hydrophilicity for >1 week.[1025] This loss of hydrophilic PDMS surface was possibly caused by the migration of un-cross-linked PDMS oligomers to the surface.[251]

PDMS channels can also be rendered hydrophilic by silanization with 3-aminopropyltriethoxysilane (APTES) (1% v/v) for 45 min.[250] It was found that derivatizing the oxidized PDMS surface by APTES produced a hydrophilic surface for more than 10 days.[171]

PDMS channels can also be rendered hydrophilic by acid treatments:

0.01% HCl, pH 2.7, 43°C for 40 min[985]
5% HCl, 5 min[170]
9% HCl for 2 h[1027]

1 M HCl[250]
0.001 M HCl at 80°C for several minutes[252]
30% HCl at 25°C for 4 h[253]

The contact angles of native PDMS (110°) have been decreased to 5° (O$_2$ plasma treatment), 45°(HCl treatment), and 75° (APTES treatment).[250]

Graft polymerization on the PDMS channel surface (not channel lumen) in enclosed channels was carried out by first adsorbing the photoinitiator (benzophenone) on the PDMS surface, followed by UV-mediated photopolymerization. In this manner, the EOF stability of the PDMS channels lasts for 45 h.[254] Besides PDMS chips, PC chips have also been chemically modified by sulfonation (using SO$_3$ gas) to produce hydrophilic surfaces.[255]

The EOF in PDMS is tenfold less than that found in glass or fused silica,[159] but the oxidized PDMS chip yields negatively charged channels, which then support EOF.[256,1033] It was found that the oxidized PDMS chip produced a significant increase in EOF compared with the native PDMS chip (at low ionic strength solution near neutral pH). However, in the presence of high ionic strength buffer (e.g., biological buffer) the increase of EOF by oxygen plasma treatment was minimal, suggesting this treatment is not useful to increase the rate of EOF. If not kept in an aqueous solution, the PDMS surface would lose its hydrophilicity with a half-life of 9 h. The EOF can be regenerated by exposing the PDMS channels to a strong base (e.g., 1 M NaOH), though a long and inconvenient 3 h incubation time is needed. Such a loss of hydrophilicity was illustrated by the disappearance of the OH band in the absorbance total internal reflection–infrared (ATR-IR) spectra, as shown in Figure 2.30.[256]

However, in one report, it was mentioned that even native PDMS supported EOF, which was stable based on flow and contact angle measurements. It was postulated that the surface charge of PDMS was probably dictated by the amount of silica fillers present in the PDMS prepolymer formulations.[302]

Hydrophilicity of polymeric channels can also be increased by photoablation. For instance, polymeric channels (37 μm deep) were photoablated through a copper foil mask. Relative to the original polymer, the photoablated surface is rougher and has increased hydrophilicity. The EOF increases in the following order: PC < PS < cellulose acetate < PET.[194] The excimer laser ablation has also

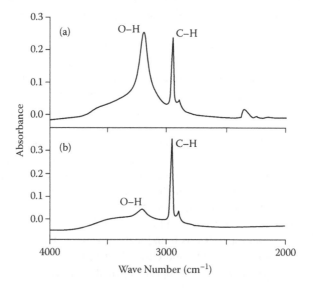

FIGURE 2.30 Measurement of the infrared absorbance of PDMS by ATR. (a) A PDMS substrate immediately after treatment with an oxygen plasma. (b) A PDMS substrate treated with an oxygen plasma and then incubated in air for 3 h.[256] Reprinted with permission from Elsevier Science.

been exploited to increase the surface charge of the PMMA channel at a 90° turn, reducing band broadening (plate height) by 40%.[257]

In composite plastic microchannels, there is an additional problem of extra dispersion (Taylor dispersion) in EOF that is caused by the difference in zeta potentials of the different materials forming the channels.[258] Caged fluorescent dye (fluorescein bis[5-carboxymethyoxy-2-nitrobenzyl]ether dipotassium salt) was used to visualize the greater dispersion obtained in acrylic or composite channels due to nonuniformity in the surface charge density.[259]

But if a hydrophobic channel is desired, the oxidized PDMS chip can be placed at 90°C in order to recover the hydrophobicity of the channels.[260]

2.4.8.6 Solvent-Resistant Properties

Although PDMS is swollen by many organic solvents, it is unaffected by water, polar solvents (e.g., ethylene glycol), and perfluorinated compounds.[367] A detailed study of solvent compatibility of PDMS has been reported. It was found that nonswelling solvents include water, nitromethane, DMSO, ethylene glycol, acetonitrile, perfluorotributylamine, perfluorodecalin, and propylene carbonate.[251] The compatibility of PDMS to other organic solvents can be improved by coating the PDMS surface by sodium silicate.[261]

Chemical vapor deposition (CVD) of poly(para-xylylene carboxylic acid pentafluorophenolester-co-para-xylylene) (PPX-PPF) on PDMS can prevent it from solvent swelling.[262]

Unlike PDMS, Zeonor chips have strong chemical resistance to alcohols, ketones, and acids.[808] In addition, Zeonor has lower water absorption (< 0.01%) than PC (0.25%) and PMMA (0.3%).[808]

2.5 METAL PATTERNING

Metal layers have been deposited on microchip substrates. The metal layer was used either as an etch mask in micromachining or for detection (e.g., as metal electrodes for electrochemical detection). Various metals have been used as the overlayer and adhesion layer, as summarized in Table 2.10.

As an etch mask, Cr/Au was usually employed in which Cr was used as an adhesion promotion underlayer. In some cases, only Cr was used as the etch mask, mainly on quartz.[150,263,264] In addition, a thick Au film (200 nm) was used as an etch mask to reduce the number of pinholes.[108]

In the case of using Ti/Au as an etch mask for HF etching, the exposed Ti edges in Figure 2.31 were passivated by electroplating in a gold sulfite bath (at 55°C) to produce a 250 nm Au coating. This will prevent attack of Ti by HF, causing undesirable undercutting.[123]

To ensure sufficient adhesion of photoresist on the Au/Cr etch mask on a glass wafer, an adhesion layer of hexamethyldisilazane was coated.[265]

As metal electrodes, various combinations, such as Cr/Au, Ti/Au, Au only, Pt/Cr, and Pt/Ti, have been used. A thin 10 nm Au layer was sometimes used because it is transparent to the visible wavelength (absorbance 0.2–0.3).[133,363]

The lift-off process is usually employed to fabricate metal electrodes. This method, as opposed to the wet etch process, allows the dual-composition electrode to be patterned in a single step.[747] In order to achieve well-defined metal electrodes in a channel recess using the lift-off technique, the metal (Pt/Ta) will not be deposited onto the sidewalls of the photoresist structure (see Figure 2.32). This discontinuity of the deposited metal layer around the sidewalls allows metal on the resist to be removed cleanly from the surface without tearing away from the metal on the surface. Thus, negative resists were used because they can be easily processed to produce negatively inclined sidewalls. To achieve this, the photoresist is subjected to underexposure, followed by overdevelopment.[141]

To fabricate an array of one thousand nonintersecting Au electrodes on glass, multilayer lithography was used, and Si_3N_4 was used as the insulating layer.[266]

TABLE 2.10
Metal Patterning on Microchips

Au on PMMA	Au (10 nm)/Cr (5 nm)	199
	Au (200 nm)/Cr (20 nm)	178
Au on Si	Au (30 nm)/Cr (30 nm)	249
	Au (1 μm)/Cr (50 nm)	727
	Au (50 nm)/Ti (10 nm)	459
	Au (45 nm)/Cr (1.5 nm)	955
Au on glass	Au (200 nm)/Cr (20 nm)	281
	Au (250 nm)/Cr (50 nm)	717, 718
	Au (400 nm)/Cr (60 nm)	96
	Au (400 nm)/Cr (60 nm)	98
	Au (38 nm)/Ti (5 nm)	367
	Au (50 nm)/Ti (5 nm)	369
	Au (500 nm)/Ti (25 nm)	370
	Au (100 nm)/Ti (5 nm)	365
	Au (450 nm)/Ti (100 nm)	311
Au on polycarbonate	Au (350 nm)/Cr (20 nm)	371
Au on polyurethane	Au (12 nm)/Ti (1.5 nm)	372
Au on SU-8 vertical wall		373
C on plastic	Conductive ink (20–100 μm)	374
Pd on glass	Pd (200 nm)/Ti (20 nm)	375
Pt on PMMA	Pt (200–300 nm)/Cr (40–60 nm)	376
	Pt (200 nm)/Cr (50 nm)	377
Pt on glass	Pt (90 nm)/Ti (10 nm)	270
	Pt (~150 nm)/Ti (5 nm)	131
	Pt (200 nm)/Ti (30 nm)	323, 615
	Pt (200 nm)/Ti (20 nm)	670
	Pt (200 nm)/Ti (10 nm)	282
	Pt (260 nm)/Ti (20 nm)	745
	Pt (300 nm)/Ti (10 nm)	747
	Pt (150 nm)/Ta (20 nm)	141, 280
Pt on Si	Pt (300 nm)	458
	Pt (150 nm)/Ti (10 nm)	100
	Pt (500 nm)/Ti (30 nm)	378
	Pt (100 nm)/Ti (30 nm)	96, 98
	Pt (200 nm)/Ti (10 nm)	146
Ti on PMMA	Ti (1 μm)	379

In PDMS chips, metal electrodes cannot easily be made on PDMS because of its pliable nature. So the Au/Cr microband electrodes have first been formed on a glass plate, which was then aligned and sealed with a PDMS channel plate.[748]

Usually, photolithography was employed to create a metal etch mask for wet etch. However, PDMS channels have also been used to confine the etchant (aqua regia) to pattern gold-coated glass plates.[267] In this method, special treatment was needed to ensure good sealing of the PDMS layer to the gold surface by rendering the sealing surface to be hydrophobic, and yet retaining the hydrophilicity of the channels for easy liquid filling (see Figure 2.33).

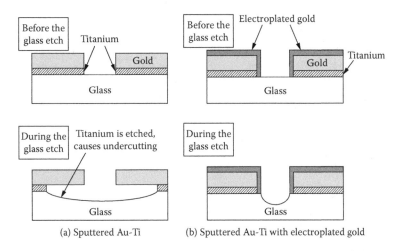

FIGURE 2.31 Glass substrates before and after etching by 48% HF when the etch masks are (a) sputtered Ti (50 nm)/Au (200 nm) and (b) sputtered gold titanium passivated with electroplated gold.[123] Reprinted with permission from the Institute of Physics Publishing.

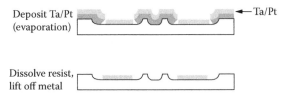

FIGURE 2.32 The electrode formation on glass by metal lift-off technique without deposition on the channel sidewalls.[141] Reprinted with permission from Wiley-VCH Verlag.

Other nonconventional masking methods include etching as defined by (1) marker ink masking and (2) laminar flow of etchants in channels. These methods were used to fabricate metal electrodes for EC detection.[268]

In addition to etching, electroplating has also been used to create metal electrodes. Gold electrodes have been prepared from Pt wire by electroplating the wire tip using a plating solution (20 mM, AuCN, 23 mM KCN).[120] In the construction of a Pt thin film heater (200 nm Pt/20 nm Ti), gold was electrodeposited (to 5 μm) onto a certain region of the Pt/Ti layer to form low-resistance heater leads.[269] Pt black was deposited on Pt to reduce its impedance by using a solution of 2.5% chloroplatinicacid/0.05% lead acetate.[270] Chloroplatinic acid (H_2PtCl_6) (and lead acetate as an additive) was used for depositing Pt nanoparticles on the Au electrode.[242]

Electrodeless electrodeposition has been performed to produce silver and Ni electrodes[125,271] and Au electrodes.[272]

In addition to metal, other materials, such as ITO[134,273–275,727,869] or graphite,[291] were used as electrodes on microfluidic chips.

To facilitate leak-free fusion bonding of chips consisting of metal electrodes, the metal layers were deposited in pre-etched recesses on chips.[141,276–280,747,1127]

Thermal degradation of the Au/Cr metal layers has been reported. In one report, room temperature bonding was used to avoid the grain-boundary diffusion (from the Cr adhesion layer to the Au overlayer).[748] This phenomenon has been studied using Auger electron spectroscopy (AES). It was found that when the glass plate was heated at 200 to 800°C, diffusion of Cr into the upper Au layer easily occurred, and Cr got oxidized at the surface.[281] However, Pt and Ti have been proven to be among the few metals capable of withstanding thermal bonding without significant resistivity change due to oxidation.[282]

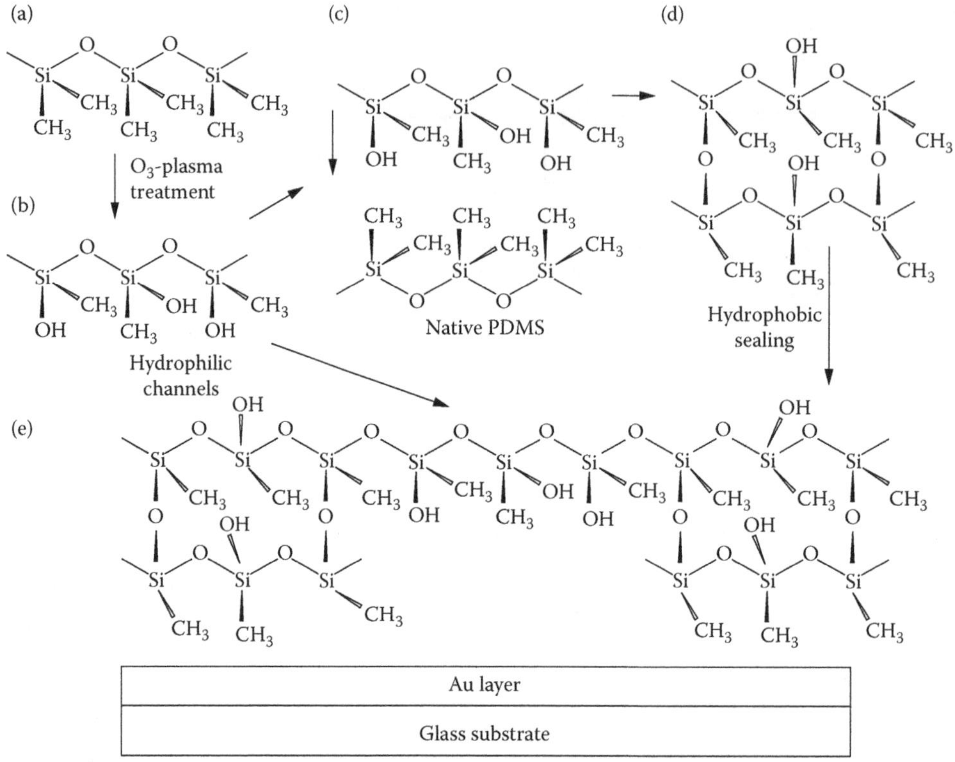

FIGURE 2.33 Creation of a hydrophilic PDMS channel with hydrophobic sealing. (a) The native PDMS surface with methyl groups. (b) After treatment with an O_2-plasma, a hydrophilic surface with exposed OH-groups is obtained. (c) The hydrophilic PDMS surface is in mechanical contact with a native PDMS surface. (d) Rearrangement of the PDMS surface leads to its hydrophobic recovery. (e) A hydrophilic channel with hydrophobic sealing is obtained.[267] Reprinted with permission from Elsevier Science.

2.6 WORLD-TO-CHIP INTERFACE

After glass bonding, solution reservoirs were created over access holes to hold reagents. The reservoirs were formed by various methods. Most commonly, short plastic or glass tubings were glued to the access holes using epoxy resin. Septa have also been used to define reservoirs around drilled holes. The septa were compressed on the chip using Plexiglas support plates.[804,806,809,812] In some reports, a thick coverglass plate (17 mm!) with drilled holes (3 mm i.d.) as reservoirs was thermally bonded to the channel plate to create 17 mm deep solution reservoirs.[33,283,284]

On the other hand, thick PDMS slabs (with punched holes) have been directly placed on the chip and aligned with the access holes to create solution reservoirs.[285,329,557,610,745,1017] Usually, no sealant is needed, but in one report, a silicone sealant was used to attach the PDMS slab.[813]

In order to avoid any change in the buffer concentration caused by solution evaporation, the solution reservoir was sealed with a thin rubber septum, in which there was a small hole in the septum for Pt wire electrode insertion.[107,148,286] With this strategy, good reproducibility (RSD of peak area and migration time are less than 1.3% and 1.2%, respectively) can be maintained.[107]

Moreover, a liquid overflowing structure was constructed, consisting of a semicircular trough holding sample overflow and maintaining a constant sample level in the reservoir.[287] Another strategy for alleviating the solution evaporation problem is to create a replenishing well that is connected to the reagent well.[288] Change of buffer concentration due to electrolysis was alleviated by using a porous bridge that separated the electrode from the buffer solution.[289] A porous membrane was placed in the solution reservoir to isolate the electrode from the fluidic channel and to prevent air bubbles from entering the channel.[290]

When high voltages are used, the solution reservoirs should be sufficiently apart to avoid electric arcing.[107] PDMS layers have also been employed to prevent arcing between adjacent electrodes that are inserted in the access holes.[1007] A graphite ink electrode was integrated on a COC chip at the reservoir locations for application of electric voltage for CE.[291]

Pt electrodes for CE high voltage were embedded into PDMS prepolymer cast against a mold. Upon curing, the Pt electrodes were well positioned at the solution reservoirs of the PDMS chip.[292] Metal electrodes were pierced through the PDMS well when the chip was mounted.[293]

A miniaturized high-voltage power supply (powered by batteries) was constructed for portable microchip work.[293]

A reversible tubing connection on the Pyrex cover plate can be made on access holes fabricated by electrochemical discharge drilling.[135] In this case, hourglass holes can be formed by drilling from both sides of the glass plate, as shown in Figure 2.34. This hourglass hole allows the attachment of a rubber tubing. To make a threaded hole, a cylindrical hole was first created (see Figure 2.35a). Then a coiled nickel wire was used as a cathode during drilling to transfer the threaded pattern (see Figure 2.35b).[135] Subsequently, a thermoplastic was melted into the hole to form a mating plastic screw for creating the reversible interconnect. Teflon capillary tubing was attached to a glass plate by first inserting the tube into the access hole on the plate, and then thermally deforming the tube's end and gluing it with high-temperature epoxy.[294]

A multiport flow control system was constructed on a glass chip. The port interface, which has both an electrical contact and a pressure connection, is shown in Figure 2.36.[295,296]

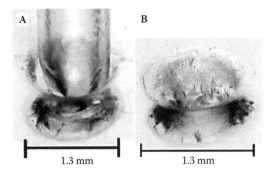

FIGURE 2.34 (a) An hourglass-shaped interconnect on a glass plate with tubing attached, and (b) a detailed view of the hourglass-shaped interconnect.[135] Reprinted with permission from the Institute of Physics Publishing.

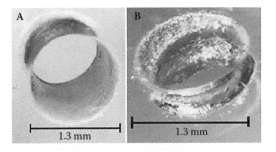

FIGURE 2.35 (a) A cylindrical glass hole and (b) the final threaded interconnect.[135] Reprinted with permission from the Institute of Physics Publishing.

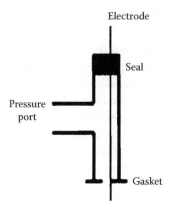

FIGURE 2.36   A schematic diagram of port interface showing both the pressure port and electrical connection.[295] Reprinted with permission from Springer-Verlag.

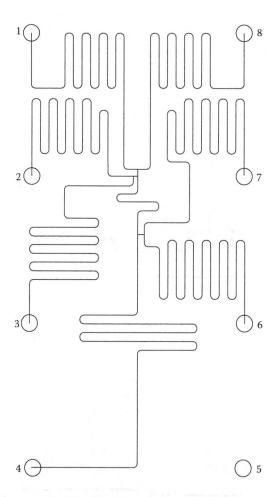

FIGURE 2.37   Chip design optimized for experiments using hydrodynamic flow control. The chip has the pair-well design suitable for dilution and enzymatic studies.[295] Reprinted with permission from Springer-Verlag.

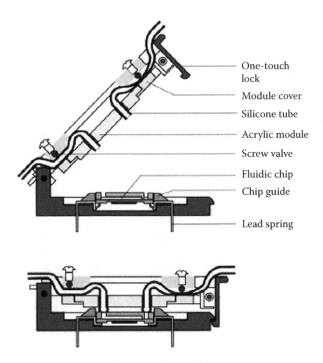

One-touch
lock

Module cover

Silicone tube

Acrylic module

Screw valve

Fluidic chip

Chip guide

Lead spring

FIGURE 2.38 Schematic drawings of the cross section of the socket for quick mount and dismount of a fluidic chip (not to scale). The upper diagram shows the structure of the socket in the open state. The silicone tubes are extended slightly out of the acrylic plate. In the screw valve, a ball is inserted to avoid the damage directly from screw to silicone tube. The fluidic chip is aligned easily by following the chip guide. The bottom diagram shows the socket in the closed state for testing. The extended silicone tubes are deformed and serve as O-rings to seal the flow channel ports on the fluidic chip. The lead spring for electrical connection is contacted firmly to the lead pads formed on the backside of the chip. Both fluidic and electrical connections are adhesive- and solder-free.[298] Reprinted with permission from Elsevier Science.

In order to reduce variations in flow due to secondary forces such as capillary forces within reservoirs, the flow resistance at each port is increased by using a long serpentine path as a flow restrictor (see Figure 2.37). This chip design also contains the so-called pair-well design so that two channels from the pair wells (e.g., wells 1 and 8) are connected to a common node. Since the pair channels have the same flow resistance, the change of flow in one channel could be easily compensated by that from another channel, and so keeping the main channel flow constant. For instance, a 40% flow in the main channel can be achieved by either −4% from well 1 and +44% from well 8, or 4% from well 1 and 36% from well 8.[295]

A fluidic interface was constructed to allow fast chip positioning, channel washing, and electrolyte filling.[297] A socket was designed for the chip-to-world interface, in which the open and closed states are shown in Figure 2.38.[298]

2.7 PROBLEM SETS

1. Name three kinds of glass substrates commonly used to fabricate microchips. Describe the process of photolithography used in glass micromachining. (5 marks)
2. What is the difference between a photomask and an etch mask in glass micromachining? (2 marks)
3. Compare and contrast positive and negative photoresists. (4 marks)

4. Describe the use of NH_4F in the wet HF etch process for glass micromachining. (2 marks)

5. One of the reasons for high aspect ratio (deep glass) etching is to increase the optical detection sensitivity. Why? (2 marks)

6. Describe briefly the RIE process. (2 marks)

7. Describe four reasons why it is desirable to carry out room temperature bonding of glass chips.[844] (4 marks)

8. Compare and contrast hot wire imprinting and hot embossing for fabrication of polymeric chips.[1011] (4 marks)

9. In the injection molding apparatus for the fabrication of polymeric chips, why is a vacuum pump needed?[7] (2 marks)

10. In injection molding using a Ni mold insert (positive relief), it is obtained by creating a Ni electroform daughter from its Ni electroform mother (negative relief), which is in turn created from a Si master (positive relief) (see Figure 2.18). Isn't it simpler to create a negative relief Si master and then produce a positive relief electroform mold insert?[7] (2 marks)

11. Describe six problems associated with the use of polymeric chips. (6 marks)

12. What kinds of laser are used in laser ablation of PMMA chips? (2 marks)

13. Why is there a maximum thickness for an electrode coated on a substrate to ensure its good sealing with PDMS? State the maximum value.[273] (2 marks)

14. What are the purposes of soft bake and hard bake after photoresist is coated on a wafer?[321] (4 marks)

15. What are the different methods used to deposit a metal layer on a substrate?[271] (2 marks)

16. What are the metal adhesion layers usually used to pattern metal on wafers? (2 marks)

17. How is electrodeless deposition conducted for the following metals?
    a. Silver[271]
    b. Nickel[125,271]
    c. Copper[318]
    d. Gold[272,318,893] (8 marks)

18. What is the developer solution used for the photopatternable epoxy-based photoresist (SU-8)?[236,271,748,762] (2 marks)

19. What are the chemical compositions of the following metal etchants?
    a. Chromium[92,152,156,319,333,430,439,604,798,1006, 1168, 1169]
    b. Gold[156,210,319,430,604,748,798,1169]
    c. Silver[156,1169] (3 marks)

20. Why would the use of sonication during etching help produce smoothly etched channel walls?[105] (2 marks)

21. What are the mold-releasing agents used in PDMS casting? (2 marks)

22. PDMS absorbs water. How is the amount of water absorbed in PDMS calculated and measured?[249] (2 marks)

23. How is PDMS irreversibly bonded to another substrate? (2 marks)

24. Native PDMS channels are hydrophobic. What are the different methods used for filling these channels? (2 marks)

25. For making a transparency photomask using a printer of 3,600 dpi, show that the minimum line width is 20 μm.[800] (2 marks)

26. Two solutions of 200 ml of 49% HF and 140 ml 70% HNO_3 were mixed and diluted to 1 L with water. Calculate the final percent of HF and HNO_3.[324] (2 marks)

27. Why should not the glass chips that contain Au/Cr metal electrodes be thermally bonded?[748,762] (2 marks)

28. In Figure 2.31, during glass micromachining, gold and titanium metals are sputtered on a glass substrate to serve as an etch mask for wet HF etch. (a) State the purpose of the

titanium layer, (b) describe the problems encountered during HF etch, and (c) provide a solution to the problems found in (b). (4 marks)

29. What are the differences between isotropic and anisotropic wet etch of silicon? (2 marks)

30. Sketch the design layout mask to create the given channel structure as shown in the Figure 4.4 when a positive photoresist is used. Sketch the design mask when a negative photoresist is used.

31. In fabricating a glass microchip, why are drilling and bonding required? (2 marks)

32. Describe three methods of hole drilling on glass that are used to produce liquid access. (3 marks)

33. Write down the chemical equation for HF etch on a fused quartz substrate. How can the addition of HCl help to create a smooth channel surface in a glass substrate during HF etching? (2 marks)

34. HF is used to etch a glass substrate to create a channel using a 10 μm line opening on an etch mask. If the etch rate is 4 μm/min, what will be the channel depth and the channel width after etching for 6 min?[818] (2 marks)

35. In fabricating a microchip, why cannot quartz be thermally bonded to glass? Name two methods to solve this problem.[844] (4 marks)

36. Describe three advantages of using PDMS to fabricate microchips. (6 marks)

37. List six polymeric materials commonly used to fabricate plastic chips. (3 marks)

38. Describe five micromachining methods for making polymeric chips. (5 marks)

3 Microfluidic Flow

The liquid pumping in the microfluidic chip is mostly achieved by using electroosmotic flow (EOF).[324] Other liquid pumping methods have also been employed for microfluidic flow. Flow has been employed for fraction collection and generation of concentration gradient. Laminar flow in the microfluidic channel allows liquid–liquid extraction and microfabrication to occur within the channels. Moreover, valving and mixing are needed in order to achieve a better flow control. All these microfluidic flow operations are further described in subsequent sections.

3.1 LIQUID PUMPING METHODS

3.1.1 ELECTROOSMOTIC FLOW (EOF)

Electroosmotic flow has been the main liquid pumping method because it does not require moving parts. The EOF in a network of channels has been modeled by the electric current flowing in a network of resistors using Kirchoff's rules. On the basis of this study, a better control of liquid flow can be achieved by designing channels of different solution resistances and by applying different voltages.[324] Surface modification of the microchannel is sometimes necessary to manipulate the direction and magnitude of EOF (see Section 3.2.1).

For a better flow control using only EOF, secondary hydrodynamic flow (HDF) should be avoided. This could be achieved by ensuring that all solution reservoirs are filled to the same liquid level to avoid HDF.[159] In addition, HDF could be prevented by closing the inlet reservoir to the atmosphere using a valve. In this way, better EOF control and more reproducible capillary electrophoresis (CE) separation (relative standard deviation [RSD] of migration time decreased by ten to thirty times) can be achieved.[631]

Liquid pumping by EOF with less pH dependence was fabricated on a glass chip (see Figure 3.1). The liquid was pumped from two 500 μm wide side channels, which contain anionic exchange beads and cationic exchange beads (5 μm diameter), to the field-free main channel (100 μm wide). At low pH, the cationic exchange beads are protonated, and thus this channel is pumping, but the anionic exchange beads are neutral at low pH, and so this channel is nonpumping. The reverse occurs at high pH. In this manner, the flow dependences at low and high pH compensate each other, which produces an enhanced flow rate at low pH. However, there is some backflow from the pumping channel into the nonpumping channel. To remedy this, the 1.5 mm section of a 50 μm wide channel was filled with smaller-sized beads (0.5 μm diameter). This prevents backflow into the nonpumping channel, thus enhancing the flow rate at higher pH and resulting in less pH dependence in the EOF speed.[325]

3.1.2 PRESSURE-DRIVEN FLOW

Hydrodynamic flow (HDF) in microchannels can be driven by pressure, and this is normally achieved by using a pump as in HPLC or by using suction.

Pressure-driven liquid flow can also be achieved by a piezoelectric actuator and a pivoted lever for linear displacement amplification (ninefold) on a poly(methylmethacrylate) (PMMA) chip. A flow rate of 1 nl/min has been attained.[380]

Hydrostatic flow (based on liquid level difference) has been used to introduce beads into microchannels. This method has been found to be superior to the use of HDF.[959]

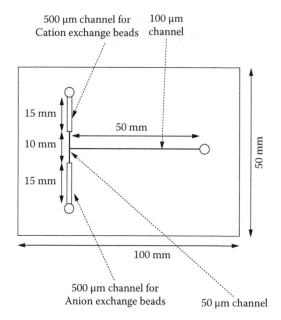

FIGURE 3.1 Schematic diagram of the microchip design of an EOF pump, which consists of a 100 μm wide main channel and two 500 μm side channels. The side channels are filled with cation exchange and anionic exchange beads (5 μm). At the appropriate pH (e.g., low pH), only one side channel is pumping (e.g., cationic) and the other one is nonpumping (e.g., anionic). The 50 μm constriction of the pumping channels reduces backflow through the nonpumping section at pH extremes by increasing the resistance to hydrodynamic flow into the nonpumping section.[325] Reprinted with permission from the American Chemical Society.

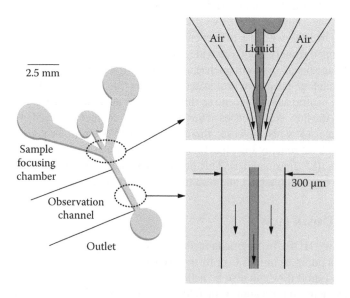

FIGURE 3.2 Basic mechanism of aerodynamic focusing by air sheath flow. Undisturbed interaction of injected liquid flow with the channel walls and air sheath flow produces a stable two-phase flow configuration throughout the microchannel. Channel height is 100 μm.[382] Reprinted with permission from Springer Science and Business Media.

A reduced pressure was employed to create liquid flow so as to facilitate filling of poly-(dimethylsiloxane) (PDMS) channels with aqueous solutions. Better results, without trapped air bubbles, are achieved, as compared with the methods by capillary force or pressure gradient. This technique also allowed the filling of a three-dimensional (3D) microchip when a single solution entry was used, and a reduced pressure was simultaneously applied to all eleven reservoirs. This was achieved by placing the whole chip under a reduced pressure.[381]

A focused pressure-driven liquid flow in a PDMS chip has been achieved.[382] By applying a vacuum (45 mmHg) at the outlet reservoir, a stably focused air–water HD flow is formed, as shown in Figure 3.2. The focused flow of liquid is very stable in a hydrophobic channel, as shown by the cross section images of the fluorescein solution in the channel (Figure 3.3b). However, in a hydrophilic channel, the focused flow of liquid is not stable, as shown in Figure 3.3a. This observation can be explained by the high aqueous solution affinity of the hydrophilic channel wall, which pulls the water column toward the top and bottom channel walls and splits the column at the "neck" region.

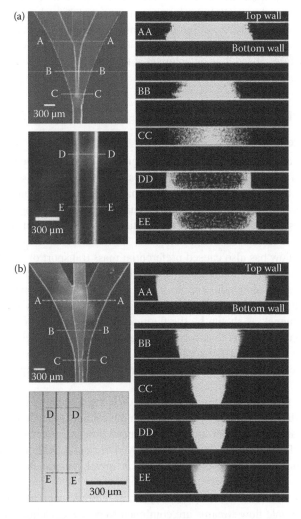

FIGURE 3.3   Cross-sectional images of liquid column in a (a) hydrophilic channel and a (b) hydrophobic PDMS channel. Water containing fluorescein was used as a sample liquid in confocal microscopic imaging.[382] Reprinted with permission from Springer Science and Business Media.

3.1.3   CENTRIFUGAL PUMPING

Besides EOF and HDF, another pumping mechanism is employed to generate the microfluidic flow. For instance, centrifugal pumping was used to drive fluids through microchannels in a rotating plastic disk. The fluids are loaded at the center of the disk. Various flow rates (5 nl/s to > 0.1 ml/s) were achieved in channels of different dimensions and at different rotation speeds (60 to 3,000 rpm). This pumping method provides a wider range of flow rates than EOF (10 nl/s to 0.1 µl/s) or HDF (10 nl/s to 10 µl/s). In addition, the centrifugal flow is insensitive to various physiochemical properties (e.g., pH, ionic strength) of the liquids and works well even in different conditions of the channel (e.g., wall adsorption, trapped air bubbles). Therefore, this method can be used to pump aqueous, biological, and organic liquids. However, one limitation of the centrifugal flow is that the flow direction cannot be reversed.[1042] A recent review on centrifuge-based fluidic platforms reported the use of centrifugal pumping to transport fluids inside plastic microchips.[383]

Besides pumping, centripetal acceleration is created. A maximum fluid rotational velocity of up to 12 m/s and a corresponding radial acceleration in excess of 10^6 g have been produced within a diamond-shaped microchamber (55×55 µm). This notch chamber was constructed along the sidewall of an otherwise straight channel (30 µm wide, 30 µm deep), which was fabricated on a PDMS chip. This microstructure caused flow detachment at the opening of the notch, leading to recirculating flow of the microvortex inside the notch.[384]

3.1.4   ALTERNATIVE PUMPING PRINCIPLES

3.1.4.1   EOF-Induced Flow

The EOF can induce a liquid flow in a field-free region, and this is termed the EOF-induced flow. This flow can be generated at a T-intersection[816] or near a thin gap.[385,386] At a T-intersection, EOF was first initiated to a side arm, rather than to the main channel. However, when the side arm was coated to reduce EOF, a flow was induced in the field-free main channel, resulting in an EOF-induced flow (see Figure 3.4). This strategy has been used to maintain a stable electrospray for mass spectrometry (MS) analysis.[816]

The EOF-induced flow has also caused preferential mass transport of anions (relative to cations and neutrals) into the field-free main channel (see the electropherogram in Figure 3.5).[387] This is because the EOF (after its reduction) can no longer overcome the negative electrophoretic mobility of the anions in the side channel. Therefore, the anions can only flow in the field-free channel, but not in the side channel, as shown by the fluorescent images in Figure 3.6. This effect is even greater for doubly charged anions than for a singly charged anions.[387] Selective ion extraction was enhanced by using an additional HDF applied to the field-free channel. An increase in the pressure applied at the field-free channel increased the amount of extraction of negatively charged ions into this channel.[388]

The EOF-induced flow can be achieved without differential coating on the glass channel. The flow can be formed when the solution flowing in the main channel is at a high concentration and that in the side stream is at a low concentration. The induced flow is greater for a higher concentration difference.[389] The EOF-induced flow has been employed to eject or withdraw reagents via apertures (rather than side channels) for the development of an artificial synapse chip (to eject neurotransmitter molecules).[344]

The EOF-induced flow has been amplified by using multiple capillary channels (of width 1 to 6 µm), so that the multiple flow streams are combined to produce adequate hydraulic pressure for liquid pumping (see Figure 3.7).[115] The multiple channels (one hundred) ensure the generation of sufficient flow rate (10 to 400 nl/min), while the small dimensions (of depth 1 to 6 µm) result in the necessary hydraulic pressure to prevent pressurized backflow leakage (up to 80 psi).[115] Based on a similar approach, a narrow-gap EOF pump was constructed to produce ~400 Pa pressure with 850

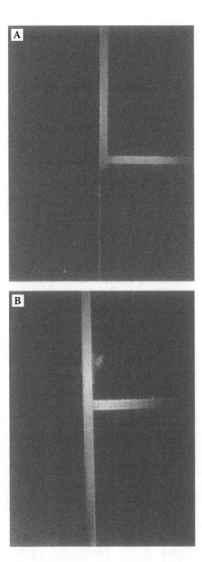

FIGURE 3.4 CCD images of 100 μM rhodamine B in water obtained at the intersection of a side-arm channel with a main channel on a microchip, applying a positive potential at the top channel relative to the side arm and using (a) all native glass surfaces and (b) a native glass main channel and a linear polyacrylamide surface coated side-arm channel.[816] Reprinted with permission from the American Chemical Society.

nm deep channels cascaded in three stages to produce a 200 μm/s flow velocity.[390] Another pump was constructed with 130 nm thin channels cascaded in ten stages to produce 25 kPa pressure.[264]

3.1.4.2 Electrochemical Bubble Generation

Liquid pumping or dispensing was achieved on a Si-Pyrex chip based on electrochemical generation of gas bubbles (H_2 and O_2) at Pt electrodes.[309,391] To ensure accuracy in the dispensing volume, an alternating current (AC) impedance measurement (at 200 kHz) of gas volume using interdigital electrodes was adopted. This method is superior to the simpler method of coulometric charge measurement that did not account for the catalytic back reaction (by Pt) of H_2 and O_2 to water and for the dissolution of gases.[309] Figure 3.8 shows two such devices (A and B) fabricated on the same chip. Each of them comprises two connected bubble reservoirs filled with a redox inactive electrolyte solution (e.g., 0.5 M KNO_3). This reservoir is in hydraulic contact (at the filling port) with the

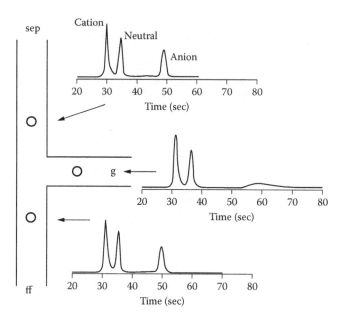

FIGURE 3.5 Electropherograms showing the separation of a cation (TRITC-Lys-Lys), a neutral (TRITC-Arg), and an anion (TRITC-Gly) at detection points 0.05 cm down each arm of the T-intersection.[387] Reprinted with permission from the American Chemical Society.

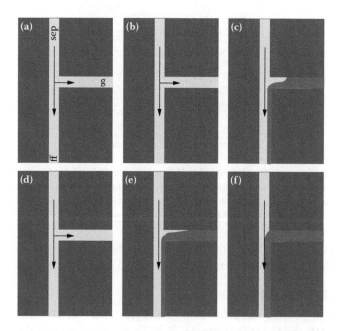

FIGURE 3.6 A neutral (TRITC-Arg; (a) and (d)), a singly charged anion (TRITC-Gly; (b) and (e)), or a doubly charged (TRITC-Asp; (c) and (f)) anion is continuously electrophoresed. Preferential transport of anions into the field-free (ff) channel at a tee intersection. The electroosmotic velocity in the side or ground channel (g) has been reduced relative to that in the separation channel (sep) by selectively coating the ground channel with a viscous polymer.[387] Reprinted with permission from the American Chemical Society.

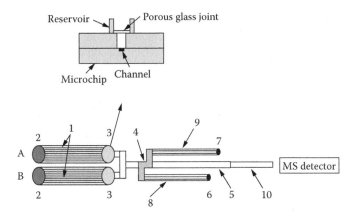

FIGURE 3.7 Diagram of the microfabricated electroosmotic pumping system. (1) Open-channel electroosmotic pump, (2) micropump inlet reservoir, (3) micropump outlet reservoir, (4) double-T sample injection element, (5) channel for sample infusion or separation, (6) sample inlet reservoir, (7) sample waste reservoir, (8) channels for sample inlet, (9) channels for sample outlet, (10) ESI emitter to an MS detector. The inset shows an expanded view of the micropump outlet reservoir (3) containing the porous glass disk.[115] Reprinted with permission from the American Chemical Society.

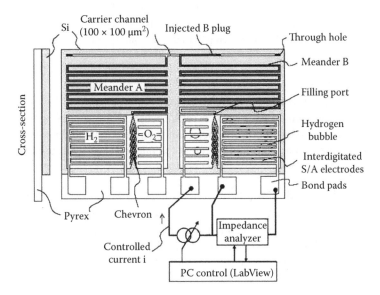

FIGURE 3.8 Proposed dual-liquid dosing system layout, geometry, and electrical connections based on electrochemical bubble generation.[309] Reprinted with permission from the Institute of Physics Publishing.

meander structure (with a volume of 1.5 µl) containing the liquid to be pumped or dispensed. The other end of the meander structure is connected to the carrier channel to which the liquid is to be delivered. The interdigital sensor/actuator (S/A) electrodes act both as an actuator (producing gas bubbles) and as a sensor (for impedance measurement). To limit diffusion of dissolved gases, and yet to provide the ohmic contact between the O_2 and H_2 bubble reservoirs, a chevron-shaped channel structure is located between the two reservoirs to act as a diffusion barrier. Since H_2 is produced at a volume twice as large as the O_2 production, the volume of the H_2 bubble reservoir (1.0 µl) is twice as large as that of the O_2 reservoir (0.5 µl). The 100 µm channels were etched on the Si wafer, coated with 250 nm thick thermal SiO_2 for insulation. The Pyrex cover is sputtered with the electrode pattern (25 nm Ti/500 nm Pt).[309]

3.1.4.3 Thermally Induced Pumping

The thermal effect is exploited to produce liquid pumping. For instance, a thermocapillary pump was constructed. The pumping method is based on surface tension change due to local heating created by heaters fabricated on a Si-Pyrex chip. This method was used to pump reagents to perform successive polymerase chain reaction (PCR), gel electrophoresis, and detection.[392,393]

Thermopneumatic pressure was generated on a Si-glass chip to pump discrete droplets (see Figure 3.9). Trapped air (100 nl) was heated by a resistive metal heater (by tens of degrees Celsius) to generate an air pressure of about 7.5×10^{-3} Pa. A flow rate of 20 nl/s could be obtained on channels using a heating rate of ~6°C/s. Hydrophobic patches were used to define the location of the discrete droplet and to prevent the liquid from entering the zones of the trapped air and the vent.[394]

A thermal-bubble-actuated micropump was constructed in which pumping was caused by periodically nucleating and collapsing air bubbles.[395] As shown in Figure 3.10, the pumping chamber was connected to two nozzle-diffuser flow controllers. When the resistive heater below the chamber heated up, the air bubble expanded and more liquid flowed out to the outlet (left) than to the inlet (right). On the other hand, when the bubble collapsed, more liquid flowed in from the inlet (right) than from the outlet (left). With this arrangement, a maximum flow rate of 5 μl/min is generated when the heating pulse is 250 Hz at a 10% duty cycle.

Thermoresponsive hydrogels based on *N*-isopropylacylamide were incorporated in a glass chip as actuators for liquid pumping. A PDMS membrane (15 to 40 μm) should be used to cover the hydrogel actuators, which are situated at right angles to the channel direction, because the actuator cannot be used to displace liquid with which it is in direct contact. The response time in gel swelling is slower than that in gel shrinking.[396]

3.1.4.4 Surface Energy

Capillary forces were used to pump reagents through Si microchannels.[397,459] Additional gradients in surface pressure, which could be created by electrochemically generating and consuming surface-active species at the two ends of a channel, have been used for liquid pumping.[369]

Surface energy present in a small liquid drop at the inlet reservoir was used to pump the liquid through a PDMS microchannel.[398]

3.1.4.5 Pneumatic Control

Pneumatic (N_2) pumping is also used to introduce liquid samples (e.g., probe DNA) in microchannels via a ten-port gas manifold. This method appears to be more effective than the use of EOF.[959]

Bidirectional liquid flow can be controlled on a PMMA chip by using pneumatic control air coupled with a suction structure (a constriction with converging and diverging cross sections) and an exclusion structure.[399]

3.1.4.6 Magnetohydrodynamic (MHD) Pumping

An AC MHD micropump was constructed on a glass plate patterned with gold structures (30 μm) to define the channels. The glass plate was then laminated with another glass plate. A maximum flow of 40 μm/s can be achieved using a 1 M KCl solution. The AC frequency was optimized to be 2.44 kHz because if the frequency was too high, eddy current in the gold film generated too much heat and raised the solution temperature to 150°C! If the frequency was too low, undesirable electrolysis occurred.[400]

Different electrode pairs were fabricated at different arms of a channel network in a Si-Pyrex chip for AC MHD pumping (see Figure 3.11). These electrode pairs allow fluid switching to occur. In AC MHD pumping, Lorentz force is used to pump electrolyte solution (e.g., 1 M NaCl). When an AC current (189 mA) flows through an electrode pair in the presence of a magnetic field (0.024 T), a flow of 0.3 mm/s is obtained.[401]

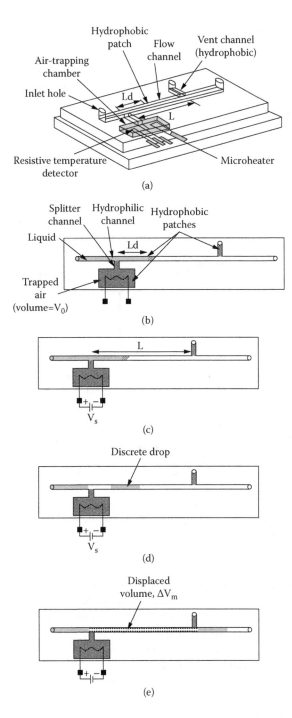

FIGURE 3.9 Thermopneumatic pressure generated for discrete liquid drop pumping. (a) Schematic of the discrete drop pumping device. (b–e) Operation of the discrete drop pump.[394] Reprinted with permission from the American Chemical Society.

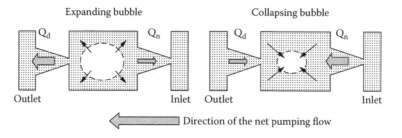

FIGURE 3.10 Principle of the thermal bubble-actuated pump. Qd, flow at the diffuser; Qn, flow at the nozzle.[395] Reprinted with permission from the Institute of Physics Publishing.

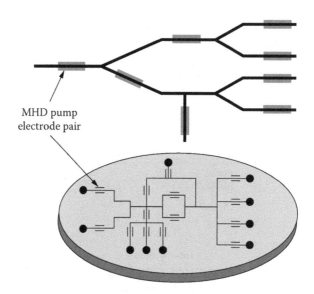

FIGURE 3.11 Complex microfluidic circuitry of a magnetohydrodynamic (MHD) micropump. The inset shows a close-up of the circuitry of a slightly different channel design.[401] Reprinted with permission from Springer Science and Business Media.

3.1.4.7 Evaporation

Liquid evaporation was employed for liquid pumping in Si-glass microchannels. With hydrophobic patterning at the outlet reservoir, the evaporation rate at the liquid meniscus was controlled to produce a flow rate of 5 nL/min. The hydrophobic region was patterned by an Al mask using a silane solution (fluorododecyltrichlorosilane [FDTS]).[146]

Evaporation-based pumping was conducted on a Plexiglas device using a sorption agent (molecular sieve). A flow rate of 35 nL/min was obtained.[402]

Liquid flow in a glass chip was generated by evaporation at the channel end. A transport speed of 2.25 mm/s in 110×28 μm channels was attained. With a fan, the liquid flow speed was increased because of forced evaporation.[403]

3.1.4.8 Miscellaneous Pumping Methods

Ultrasonic vibration (49 kHz) of a lead zirconate titanate (PZT) disk adhered to a Si chip through a thin Si diaphragm was used for degassing in a portable dialysis system. When the dissolved O_2 in the dialysate is reduced to 6 ppm (from 8 ppm), no O_2 gas bubbles will be generated inside the dialyzer during normal operation.[404]

A ferrofluidic magnetic micropump was constructed on a Si chip. Operation relies on the use of magnetically actuated plugs of ferroliquid (a suspension of nano-sized ferromagnetic particles), which is immiscible with the pumped fluid.[301]

Additional pumping principles based on magnetohydrodynamics[405–407] and electrodynamics[408] have also been exploited for microfluidic pumping.

3.1.5 MICROFLUIDIC FLOW MODELING STUDY

Mathematical modeling and computer simulation have been applied for various flow studies in rectangular microchannels (see Table 3.1). An equation to describe the flow in a rectangular channel has been given.[124] Simulation of fluid flow can be conducted by solving the coupled Poisson and Navier-Stokes equation for fluid velocity.[532] However, this complicated computation has been simplified by solving the Laplace equation for the electric field because it is proportional to fluid velocity.[321]

Modeling study has been applied to investigate the dispersion occurring at the channel turns. Resolution enhancement for microchip separation requires longer serpentine channels to be fabricated. This creates dispersion problems due to the turn geometry, as seen in the simulation and experimental results in channels containing constant-radius turns shown in Figure 3.12. Various approaches have been employed to reduce the dispersion. For instance, compensating turns, which have constrictions to even out the differences in the path lengths and electric field strengths of the inner and outer tracks at the turns, have been proposed.[535,971] Simulation and experiments (using photobleached or caged fluorescence dye visualization) were employed to illustrate the improvement obtained by using compensating turns (see Figure 3.13).[535] The advantage of the use of tapered turns over noncompensating (90 or 180°) turns was applied to CGE separation of DNA samples.[409]

Caged fluorescent dyes have been used to measure EOF in microchannels.[410] A nanosecond N_2 laser pulse first activated a caged fluorophore, and its subsequent single-molecule detection was used to measure velocity of liquid flows, which ranged from slow laminar flow to fast near-turbulent flow.[411] Unfortunately, it was found that the dyes increased EOF in PMMA, PC, and PDMS microchannels, though the dyes did not affect the flow in a fused silica capillary.[410]

Other methods for flow measurement include the use of fluorescence correlation spectroscopy to trace the hydrodynamic flow of a fluorophore in a Si microchannel.[412] Velocity of flowing particles

TABLE 3.1
Various Mathematical Modeling and Stimulation Methods to Study Microfluidic Flow

Topics of Study	References
Electrokinetic Flow	
EK flow in microchannels	519–527
EK flow in structured channels	526, 528
EK sample injection	529–531
Electrokinetic focusing	532
Field-amplified sample stacking	579
Dispersion	258, 533, 534
Dispersion at channel turns	535–543
Hydrodynamic Flow	
Hydrodynamic flow in microchannels	442, 445, 544, 545
HDF in structured channels	236, 471
Sample injection	546
Dispersion	118

FIGURE 3.12 Simulation (left column) and photobleached-fluorescence visualization (right column) of an analyte band traveling around a constant-radius corner. In both cases the channel is 250 μm wide. In the experiments, the channels were ~40 μm deep.[535] Reprinted with permission from the American Chemical Society.

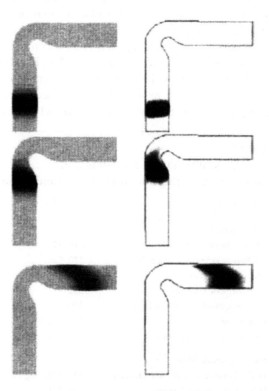

FIGURE 3.13 Simulation (left column) and caged fluorescence visualization (right column) of an analyte band traveling through a compensating turn design. In both cases, the channel is 250 μm wide. In the experiments, the channels were ~40 μm deep. The channel boundaries have been drawn onto the images.[535] Reprinted with permission from the American Chemical Society.

was measured using Shah convolution Fourier transform (SCOFT) detection as described in Chapter 7, Section 7.1.3.[413]

Flow measurement has been achieved without the use of beads or dyes. A short heating pulse generated by a CO_2 infrared laser (10.6 µm) was delivered through the infrared (IR)-transparent Si wafer into the channel. The radiative image of the hot liquid plug was recorded by an IR camera.[414]

3.2 MICROFLUIDIC FLOW CONTROL

3.2.1 SURFACE MODIFICATIONS FOR FLOW CONTROL

In some applications, an alteration in the microchannel surface charge is needed in order to modify EOF, i.e., to increase, suppress, or stabilize EOF. This charge alteration can be achieved by surface modification.

EOF can be suppressed by coating the channel surface with methylhydroxyethylcellulose (MHEC)[631] or photopolymerized polyacrylamide.[415] Reversal of EOF was also achieved; e.g., the PMMA surface was modified with N-lithioethylenediamine or N-lithiodiaminopropane,[1063] or the PDMS surface was coated by hydrophobic interaction with TBA^+.[302]

To coat selectively only at some sections of a channel network, outflow of solvent at other sections should be used to prevent them from being wetted by the chemical modification reagents.[148,766]

Polyelectrolyte multilayers (PEMs) have been used to alter surface charges, thus controlling the direction of EOF in chips made of PS,[155,216,416] PMMA,[216] and poly(ethylene terephthalate glycol) (PETG).[416] PEM deposition was carried out by exposing the microchannel alternatively to solutions of positively charged PEM poly(allylamine hydrochloride) (PAAH) and negatively charged PEM poly(styrene sulfonate) (PSS).[417] Moreover, by depositing the PEM at different parts of the PMMA channels, special flow control patterns can be achieved (Figure 3.14).[216]

PDMS channels were also coated by PEM using alternating cationic layers (polybrene [PB]) and anionic layers (dextran sulfate [DS]). Stable EOF can be achieved for up to 100 runs.[365] Since these polyelectrolytes have high and low pKa values, respectively, in the usual solution pH of 3 to 10, PB is essentially positive and DS is essentially negative. For comparison, in glass channels, the silanol groups have intermediate pKa values, and so the surface charge strongly depends on pH.[365]

Surface modification can be achieved by O_2 plasma oxidation. For instance, because of the increase in surface negative charges after oxidation, the oxidized PDMS surface supports EOF. However, because of the instability of the charge created on the polymer surface, EOF was unstable. Better stability can be achieved by immediately filling the PDMS channel with liquids, rather than letting it be exposed to air. The useful lifetime of these devices for quantitative CE analysis, which requires EOF stability, is probably 3 h.[1033]

UV irradiation can also increase surface charge (and hence hydrophilicity) in PC chips.[418] The surface charge on the glass channel can also be modified by UV irradiation, but the surface is first coated with a TiO_2 film.[419]

Ultraviolet polymerization has been used to graft acrylic acid, acrylamide, dimethylacrylamide, 2-hydroxylethyl acrylate, or poly(ethyleneglycol) monomethoxyl acrylate on the PDMS channel in order to create a hydrophilic surface. The magnitude of the EOF of the grafted PDMS chip is intermediate between that of native PDMS and oxidized PDMS. Unlike oxidized PDMS, the EOF of grafted PDMS remains stable upon exposure to air. The grafted PDMS channels are more readily filled with liquids than are native PDMS. The grafted surfaces also show reduced adsorption of solutes (e.g., peptides), compared to oxidized PDMS.[420]

Either laser-ablated PETG or hot-imprinted PETG followed by NaOH hydrolysis has produced surface-bound carboxylated groups, leading to enhanced EOF to similar extents. These groups can be further chemically modified to give amine moieties.[421]

For Vivak polyester channels, alkaline hydrolysis of surface groups (e.g., ester) to ionizable groups (e.g., carboxylate) has produced a more reproducible EOF. In combination with a dynamic

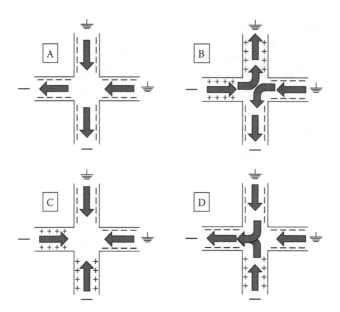

FIGURE 3.14 Four flow patterns achieved in the intersection of the PS chip derivatized to have different surface charges on the various arms. The applied voltages are the same in all cases.[216] Reprinted with permission from the American Chemical Society.

coating (e.g., cetyltrimethylammonium bromide [CTAB]), EOF may be either eliminated or reversed in direction, depending on the CTAB concentration.[204]

Valving can be achieved by field-effect control using an electric field perpendicular to the channel in order to alter the zeta potential for EOF.[371,422–424] For instance, the transverse electric field was applied through the microchannel wall of a PC/PDMS chip by Cr/Au electrodes embedded in Parylene C coating. As shown in Figure 3.15, differential pumping can occur when different electric fields are applied to different sections of the channels. For instance, in Figure 3.15a, EOF is higher (because of greater negative zeta potential) in the left channel than the right one. Owing to mass conservation at the intersection point, there is a liquid flow down the central field-free (ff) channel. A flow rate of 2 nL/min was achieved in the field-free central channel. On the other hand, in Figure 3.15b, a greater flow in the right channel causes the liquid to flow up the central channel.[371] External electrodes were located 50 μm away from a glass microchannel for surface charge control, leading to EOF control.[424] Other than the use of surface chemical modifications, field-effect flow control was achieved in a PDMS-Si chip using a 2.0 μm electrically insulated SiO_2 layer connected to ground.[422]

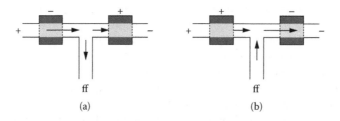

FIGURE 3.15 Differential EOF pumping rates were achieved from different gate voltages applied to the field-effect flow control electrodes. This produced (a) positive pressure for flow down and (b) negative pressure for flow up the field-free (ff) microchannel.[371] Reprinted with permission from the American Chemical Society.

3.2.2 Laminar Flow for Liquid Extraction and Microfabrication

Laminar flow is prevalent in microfluidic channels, and this phenomenon is exploited for various applications. For instance, a two-phase flow between an organic and an aqueous stream has been achieved in microchannels for the ionophore-based ion-pair extraction. This was carried out for the detection of potassium and sodium[425] and iron.[426] In such a system, degradation in the signal response, which can be caused by leaching of ion-sensing components as found in conventional ion-selective optodes, is not an issue. This is because a fresh organic phase can be used in every measurement. In addition, a low-viscosity organic solvent can be used to increase solute diffusion, which is usually slow in polymeric membrane-based ion sensors involving aqueous solutions.[425]

To further stabilize multiphase flow, the liquid–liquid interface can be created at the boundary formed at a constricted opening.[427] Moreover, guide structures (5 μm high) were fabricated in a microchannel (20 μm deep) to stabilize a three-phase flow, as shown in Figure 3.16. These structures were etched on quartz using a photomask consisting of three closely spaced parallel lines (10 μm wide, 35 μm apart). The multiphase flow (aqueous/aqueous, aqueous/organic, aqueous/organic/aqueous) became more stable in such microchannels, allowing extraction reactions to proceed over a long distance along the channel.[428]

The conventional procedure (with forty steps!) for the determination of Co became simpler using the microfluidic chip with guide structures to stabilize the multiphase flow (see Figure 3.17). Co^{2+} in the sample (aqueous phase 1) was first mixed with a chelating agent (2-nitroso-1-naphthol [NN]) (aqueous phase 2) to form a complex, which was then extracted into m-xylene (organic phase 1) along a microchannel. The Co chelates were then separated from the chelates of other metal interferences (Cu^{2+}, Ni^{2+}, Fe^{2+}), which were decomposed by HCl and then extracted into aqueous phase 3. The decomposed NN was extracted into aqueous phase 4 (NaOH). Finally, the undecomposed Co

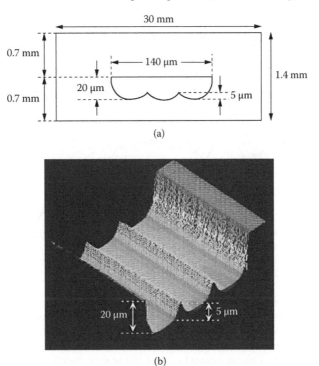

(a)

(b)

FIGURE 3.16 (a) Schematic cross-sectional view of the guide structures for stabilizing a three-phase liquid flow. (b) 3D image of the guide structures.[428] Reprinted with permission from the American Chemical Society.

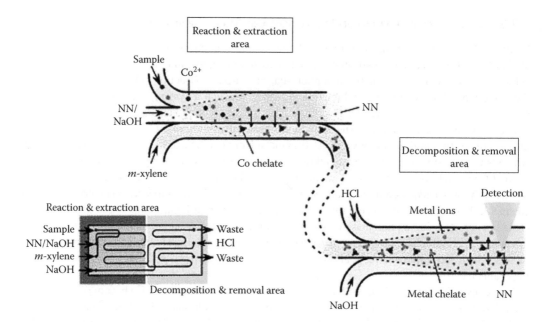

FIGURE 3.17   Schematic illustration of Co(II) determination in a quartz chip fabricated with guide structures in the channel for multiphase flow. For details in operation, see text.[428] Reprinted with permission from the American Chemical Society.

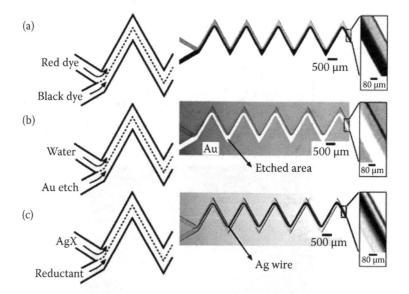

FIGURE 3.18   (a) Optical micrograph of a two-phase laminar flow: two aqueous phases, each colored with a different dye (black ink and Congo Red) brought together with a Y-junction in an elastomeric microfluidic system (PDMS). Only diffusional mixing over distances of several centimeters is observed. (b) Optical micrograph of a pattern etched in Au in a zigzag channel (PDMS) that is sealed to a Au-covered glass slide. (c) A silver wire deposited in a zigzag channel at the laminar flow interface between solutions containing the components of an electrodeless silver plating solution.[432] Reprinted with permission from the American Association for the Advancement of Science.

chelates (still in organic phase 1) were detected.[428] In a similar manner, Al was analyzed by extraction at the laminar flow interface (water/1,4-dioxane) as the Al–2,2'-dihydroxyazobenzene (DHAB) complex and detected fluorescently.[103] Similar extraction in a three-phase flow was achieved for determination of zinc and yttrium ions (0.1 mM) in a Pyrex chip.[429] A two-phase flow of liquid–liquid (water–nitrobenzene) crossing flow has been demonstrated in glass chips to mimic countercurrent flow for liquid–liquid extraction. The flow was stabilized by a silanized (octadecylsilane [ODS]) channel.[430] A two-phase flow was utilized to perform liquid ion exchange, which is applied for conductivity suppression employed in ion chromatography. Tetraoctylammonium hydroxide (TOAOH) or Amberlite LA-2 (a secondary amine) was dissolved in an organic solvent (butanol) to exchange away H^+ for conductivity detection of heavy metal ions (Ni, Zn, Co, Fe, Cu, Ag).[431]

Laminar flow that is prevalent in the microchannel has also been utilized for microfabrication (see Figure 3.18). For instance, using a laminar flow of Au etchant flowing parallel to a water flow, a gold-coated (250 Å thick) glass-PDMS channel was patterned with a zigzag Au wire (Figure 3.18b). In addition, a silver wire was formed chemically at the interface of two parallel laminar flows of silver salt and reductant (Figure 3.18c).[432] Moreover, glass etching was also achieved at the interface of two parallel laminar flows of HCl and KF, i.e., creating HF for etching at the interface. A polymeric structure was precipitated at the interface of two flows of oppositely charged polymers. The classic example was the fabrication of a three-electrode system inside a channel (see Figure 3.19). First, a Au etchant was flowed to separate a gold patch, creating two isolated Au electrodes. Subsequently, a silver wire was formed in the middle, and 1% HCl then converted silver on its surface to AgCl, creating a silver/AgCl reference electrode.[432]

In another report, the self-assembled monolayer (SAM) together with multistream laminar flows were used to pattern flow streams in glass chips (see Figure 3.20). For instance, an aqueous stream was confined to the central hydrophilic region by two virtual walls, which were two organic streams flowing in hydrophobic regions patterned by two organosilanes, OTS and heptadecafluoro-1,1,2,2-tetrahydrodecyltrichlorosilane (HFTS). An increase in the liquid pressure allowed the central liquid stream to burst into the adjacent hydrophobic regions with successively lower surface free energy. This behavior is like a pressure-sensitive valving system.[433]

The virtual wall created can also be employed to carry out gas–liquid reactions. For instance, acetic acid vapor was allowed to react with a pH indicator solution at the virtual wall, allowing kinetic studies to be carried out (see Figure 3.21).[433] Moreover, using photocleavable SAM (with the 2-nitrobenzyl photosensitive group), the SAM can be patterned through a UV photomask, giving rise to complex flow patterns.[433,434]

Nylon-6,6 membrane was formed at the solution interface of adipoyl chloride (0.01 M in 1,2-dichlorocthane solution) and hexamethylenediamine (0.1 M in NaOH solution) within a Pyrex glass microchip (treated with APTES) (see Figure 3.22). The membrane was used in a permeation study to examine diffusion of dissolved NH_3 gas through the membrane to a phenolphthalein-containing solution.[435] The membrane can also be modified with horseradish peroxidase on only one side for carrying out an enzymatic reaction. H_2O_2 permeates through the membrane and enzymatically reacts with N-ethyl-N-(2-hydroxy-3-sulfopropyl)-m-toluidine and 4-AAP (4-aminoantipyrine) to form a dye.[435]

3.2.3 GENERATION OF CONCENTRATION AND TEMPERATURE GRADIENTS

Concentration gradients can be generated by mixing a solution with a solvent. Mixing dilution between two liquid streams is feasible by controlling their relative EOF speeds.[324,436]

Gradients of different concentrations have been generated by parallel mixing and serial mixing.[437–440,442–445,840] In parallel mixing, seven different concentrations, including zero concentration, were generated by mixing three different sample concentrations with three buffer streams in parallel.[439] In serial mixing, five concentrations were generated by consecutively mixing one sample concentration with one buffer stream.[439] More complex channel networks for parallel mixing even

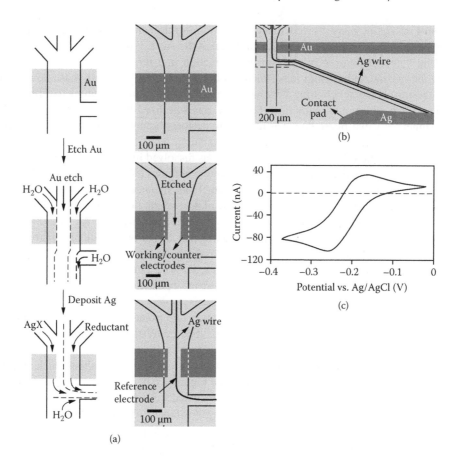

FIGURE 3.19 (a) Optical micrographs of the stepwise fabrication of a three-electrode system inside a 200 μm wide PDMS channel. Two gold electrodes (counter and working) are formed by selectively etching the gold stripe with a three-phase laminar flow system. A silver reference electrode is fabricated at the interface of a two-phase laminar flow. (b) Overview picture of the three-electrode system including the silver contact pad. The dashed box corresponds to the last picture shown in (a). The silver wire widens to a large silver contact pad. (c) Cyclic voltammogram of 2 mM $Ru(NH_3)_6Cl_3$ in a 0.1 M NaCl solution (scan rate: 100 mV/s).[432] Reprinted with permission from the American Association for the Advancement of Science.

produced various types of concentration gradients (e.g., linear, parabolic, or periodic) on a PDMS chip (see Figure 3.23).[440,441] The gradient has been maintained over a long period of time.[440]

Concentration gradients were generated along the interface of two parallel HDF streams.[442–444] A concentration gradient of HF generated in this manner has been used to etch Si differentially.[441] Similarly, a concentration gradient was generated along a long parallel dam at the border of two parallel channels, where the sample and buffer flowed separately. Diffusion across the dam created a concentration gradient in the same direction of the flow.[445,840]

A concentration gradient was created by mixing two streams of liquid at a Y-junction and allowing diffusional mixing to occur downstream for a fixed distance (i.e., 2 cm) (see Figure 3.24).[447] Thereafter, these laminar flow streams of various concentrations were separated by dividing the main channel into a series of ten[447] or twenty-three small tributaries.[110] Depletion effects in capillary filling of PDMS channels were employed to generate patterned protein gradients, which have been used to conduct immunoassays.[446] A concentration gradient of solvent B (65% acetonitrile/10 mM acetic acid) and solvent A (10 mM acetic acid/10% methanol) was generated on-chip for gradient frontal separation of peptides using a C18 cartridge.[803]

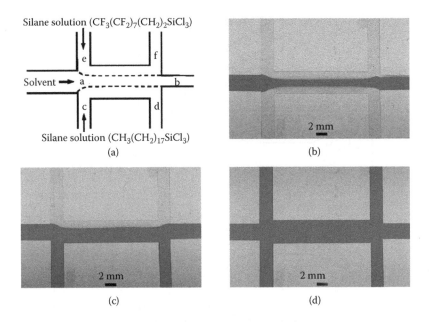

FIGURE 3.20 Pressure-sensitive valves. The laminar flow scheme for patterning surface free energies inside channels with two different trichlorosilanes, OTS and HFTS, is shown in (a). Images of flow patterns of a rhodamine B solution obtained in increasing pressures from (b), (c), to (d).[433] Reprinted with permission from the American Association for the Advancement of Science.

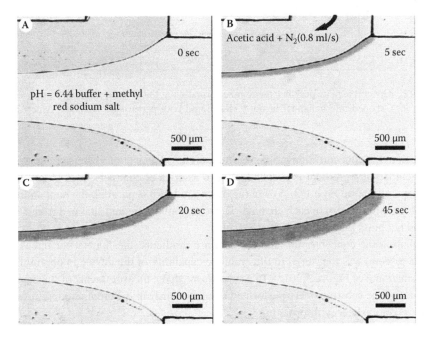

FIGURE 3.21 Optical micrographs of a gas–liquid reaction. The liquid was a pH = 6.44 phosphate buffer solution containing the acid-base indicator methyl red (0.007 w/w%). Acetic acid vapor was carried into the channel by bubbling N_2 gas (0.8 ml/s) through glacial acetic acid. Shown are micrographs of the channel (a) before acetic acid vapor was introduced into the channel (0 s), and at (b) 5 s, (c) 20 s, and (d) 45 s after initiating flow of acetic acid vapor.[433] Reprinted with permission from the American Association for the Advancement of Science.

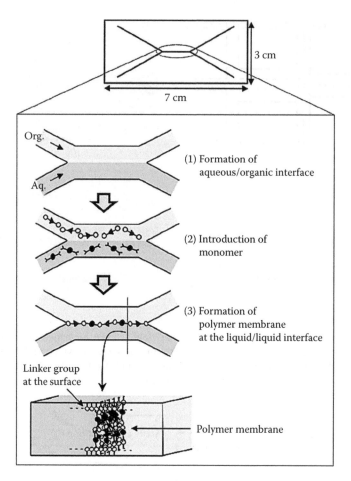

FIGURE 3.22 General idea of polymer membrane formation at the interface of a two-phase organic/aqueous flow in an X-shaped microchannel layout.[435] Reprinted with permission from the American Chemical Society.

In addition to the concentration gradient, a temperature gradient has been created on the chip.[447] When the temperature gradient is perpendicular to a series of channels, each channel is held at a discrete temperature (see Figure 3.25). On the other hand, if it is parallel, the temperature varies as the liquid flows downstream in each channel. The temperature gradient spanned from 36 to 77°C by the cold and hot brass tubes separated by 3.4 mm.[447]

By combining the concentration and temperature gradients, the fluorescent intensity of a dye (carboxyfluorescein) was measured to illustrate the capability of the device to obtain three-dimensional information (see Figure 3.26).[447] In a subsequent study, measurements of the dephosphorylation of a fluorogenic substrate were performed simultaneously at different temperatures in order to extract the activation energy information.[448]

3.2.4 FLOW SWITCHING

For better flow control and switching, microfluidic valves are designed. For instance, a latex membrane (150 μm thick) has been used to construct a valve. A sample is loaded by liquid pressure (10 to 12 psi) through the microfluidic valve opened by applying vacuum (30 mm Hg) on the opposite side of the valve diaphragm. To facilitate sample filling, while the valve is open, air is simultaneously

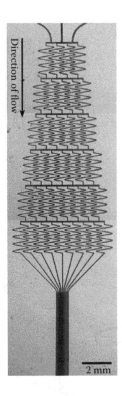

Direction of flow

2 mm

FIGURE 3.23 Photograph showing a microfluidic device used for generating concentration gradients of green and red dyes in solution. The three incoming channels (top part of the photograph) were connected to syringes via tubings (not visible). After combining the nine streams into a single, wide channel (bottom of the photograph), a gradient was formed across the channel, perpendicular to the direction of flow.[440] Reprinted with permission from the American Chemical Society.

evacuated through a hydrophobic membrane vent (1.0 μm pore size). Both the valve and vent have dead volumes of about 50 nL.[926]

Multiple valves have been constructed on a PDMS valve control layer (see Figure 3.27).[167] Multilayer soft lithography has been used to generate the valve control layer (4 mm thick for strength) plus a fluid layer (40 μm thick) on PDMS. The small Young modulus (~750 kPa) of PDMS allows a large deflection (1 mm) to be produced using a small actuation force (~100 kPa on a 100 × 100 μm valve area). The response time is on the order of 1 ms. Round channels were more easily sealed (~40 kPa) than trapezoidal channels were (>200 kPa). Round channels were cast from rounded photoresist positive relief (after baking it at 200°C for 30 min[167] or 150°C for 30 min).[449] By pneumatic actuation, the valving rate was attained at ~75 Hz, and this produced a pumping flow rate of 2.35 nL/s.[167]

It was found that Sylgard 184 provided a higher tensile strength and elongation than RTV 615, and so the former was selected for fabrication of the microvalve control layer.[449] Moreover, the control channel must be filled with water before use, in order to prevent bubble formation in the fluidic channel due to the gas permeability of PDMS.[449]

There are other reports on the use of PDMS valves.[124,357,368,450] A thin PDMS valve membrane (3 μm) allowed for a high valve density (300 valves/cm²). Actuation of the valves was caused by expansion of the air in the air-filled cavities below the valves. PDMS prepolymer diluted with toluene reduced the solution viscosity, allowing for spin coating of very thin layers.[124]

A similar PDMS valve control layer was used to achieve rotary liquid pumping for PCR.[357] In another report, a similar valving method was used to deliver cells and to introduce reagents for cell reactions. Solutions were pumped at 5 to 60 Hz to achieve a linear flow rate of ~300 to 1,000 μm/s.[368]

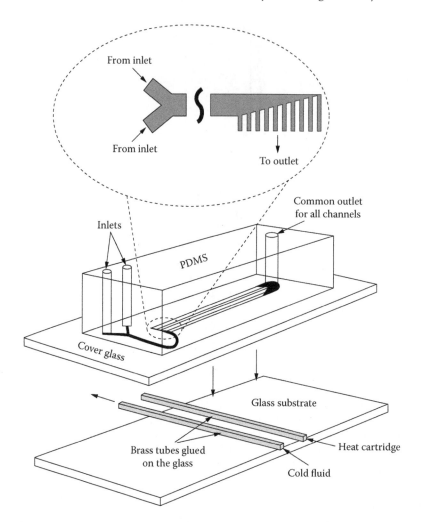

FIGURE 3.24 Schematic diagram of a microfluidic device that creates a concentration gradient and a temperature gradient. The inset shows the channel layout design.[447] Reprinted with permission from the American Chemical Society.

Valving on a PDMS sheet has also been achieved after an elliptical hole was punched in the PDMS, and an elliptical needle was inserted in the hole. Then valve actuation was achieved by having the elliptical needle rotated for 90°, thus compressing the fluidic channel below for valve closure.[1043] A PDMS membrane was fitted on a glass wafer to create a valve. Three valves placed in series form a diaphragm pump.[244]

Pluronics F127 has been exploited to construct microfluidic valves. This material, which is an uncharged triblock copolymer $((EO)_{106} (PO)_{70} (EO)_{106})$, is a commercially available surfactant. But within a certain concentration (18 to 30%), the material, which is a liquid of low viscosity (<2 cP) at low temperature (0 to 5°C), will form a self-supporting gel (cubic liquid crystalline) at a higher temperature (e.g., room temperature).[937] This material has been used as a one-shot, phase-change valve for PCR. The Pluronics valve can hold up to 20 psi pressure, which is above the holding pressure (6.8 psi) required for PCR (up to 94°C).[937]

Hydrogel valves were also created within microfluidic channels to provide fluid control (see Figure 3.28). The swelling and contraction provide valve-close and valve-open actuation, respectively, due to chemical stimuli (e.g., pH change). To increase the mechanical stability, the hydrogel

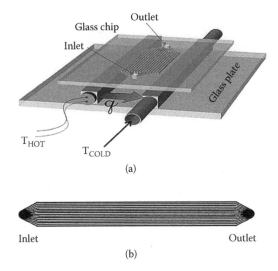

FIGURE 3.25   (a) Schematic diagram of a temperature gradient microfluidic device. The temperature gradient platform is formed by fixing two square brass tubes onto a glass substrate with epoxy. The left tube houses a cartridge heater that serves as the heat source. The right tube allows liquid to flow through it and serves as the cooling element. A detachable chip containing microchannels is placed directly onto the platform, with the microchannels perpendicular to the temperature gradient. The arrow (q) represents the direction of heat flow. (b) Geometry of the glass channels in the microfluidic device.[447] Reprinted with permission from the American Chemical Society.

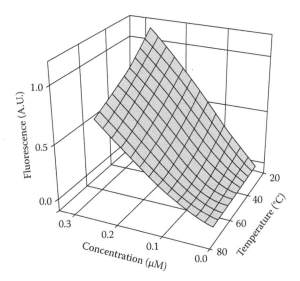

FIGURE 3.26   Three-dimensional plot of fluorescence intensity of carboxyfluorescein dye molecules in aqueous solution as a function of their concentration (0.00715 to 0.266 μM) and temperature (28 to 74°C). The plot was mapped over 110 data points (excluded for clarity) gained from 11 temperature measurements across 10 microchannels.[447] Reprinted with permission from the American Chemical Society.

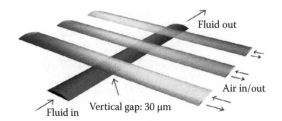

FIGURE 3.27   An elastomeric peristaltic pump using three valve control channels. The fluidic channel is 100 μm wide and 10 μm high. Dimension of microvalve control channels = 100 × 100 × 10 μm; applied air pressure = 50 kPa. Peristalsis was typically actuated by the pattern 101, 100, 110, 010, 011, 001, where 0 and 1 indicate valve open and valve closed, respectively.[167] Reprinted with permission from the American Association for the Advancement of Science.

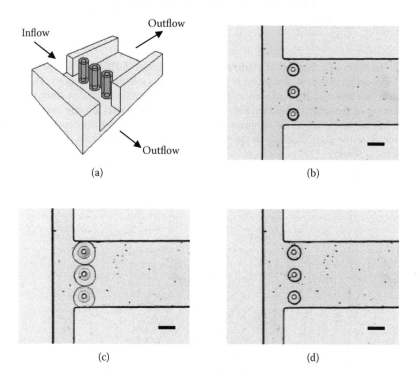

FIGURE 3.28   Prefabricated posts in a microchannel serve as supports for the hydrogels, improving stability during volume changes. (a) A diagram of the hydrogel jackets around the posts. (b) The actual device after polymerization of the hydrogel. (c) The hydrogel jackets block the side channel branch in their expanded state. (d) The contracted hydrogels allow fluid to flow down the side branch. Scale bars, 300 μm.[221] Reprinted with permission from Macmillan Magazines Ltd.

was formed around prefabricated posts in the channel. Since the thickness of hydrogel is reduced, the response time to chemical stimuli was also reduced to 8 s (from 130 s). When acrylic acid (a) was used in the hydrogel, high pH caused swelling and low pH caused contraction. On the other hand, when 2-(dimethylamino)ethyl methacrylate (b) was used instead, high pH caused contraction of the hydrogel. This property was exploited to produce two hydrogel gating valves. In this case, only one gate (A) will contract and open at pH 4.7 (lower pH). On the other hand, at pH 6.7 (higher pH), only gate B will contract and open.[221] A detailed study of the physiochemical properties of these hydrogel polymers has been published.[451]

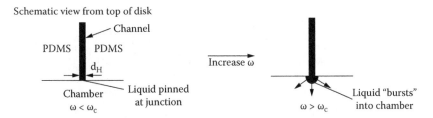

FIGURE 3.29 A schematic view from above the disk of a passive capillary burst valve. A liquid flows in a channel or capillary and is pinned at the discontinuity where the channel meets a chamber or a wider channel. Sufficient fluidic pressure must be exerted by the centrifugal pump to overcome the pressure of curved liquid surfaces and to wet the walls of the chamber with liquid. This pressure is achieved at a characteristic rate of rotation or "burst" frequency, ω_c, above which the liquid exits the channel and enters the chamber. ω_c depends on the hydraulic diameter (d_H) of the capillary and the amount of liquid in the channel, and therefore provides a means of gating the flow of liquid.[1042] Reprinted with permission from the American Chemical Society.

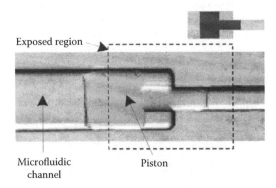

FIGURE 3.30 Photopolymerized piston with cross-sectional area change. The widths of the channels are 150 and 50 μm, and the channels are 25 μm deep. The piston was first UV polymerized in the outlined region and then displaced slightly to the left to show the molded sealing surface. The inset shows the schematic of the piston.[456] Reprinted with permission from the American Chemical Society.

A hydrogel sensor/actuator was constructed in a PDMS chip for creating a flow with regulated pH.[222] The pH-sensitive hydrogel valve was made to output a combined flow of constant pH 7 from an input flow of pH 2 using a compensating flow of pH 12.[222] When the flow of pH 2 is greater, the hydrogel sensor/actuator will shrink and allow a greater flow of pH 12 (than the pH 2 flow) to bring down the pH of the output flow. On the other hand, when the flow pH 2 is lower, the hydrogel will expand and constrict the flow of pH 12. The transition pH point of hydrogel was 5.3, which can be changed to a different value by employing a different polymer composition to synthesize the hydrogel.[452]

Passive capillary burst valves were fabricated on a plastic disk in which centrifugal liquid pumping was used to transport fluids, as shown in Figure 3.29. When the angular velocity (ω) is less than the critical value (ω_c), the liquid cannot enter the big chamber. When $\omega > \omega_c$, the liquid then bursts into the chamber. A wider channel would require less force to burst through and hence a lower rotation rate, or a lower centrifugal force for the valve to burst open.[226,453,454,1042]

Nonstick photopolymer formed inside a glass microfluidic channel was used as a mobile piston for flow control (see Figure 3.30). Sealing against high pressure (>30 MPa or 4,500 psi) could be achieved by the piston. The polymer was compatible with organic solvents. Actuation time less than 33 ms was observed. The photopolymer was formed by irradiating an unmasked section filled with the monomer (trifluoroethyl acrylate/1,3-butanediol diacrylate in a 1:1 ratio) and photoinitiator (2,2'-azobisisobutyronitrile) using UV (355 nm). Using this method, a check valve (one-way flow)

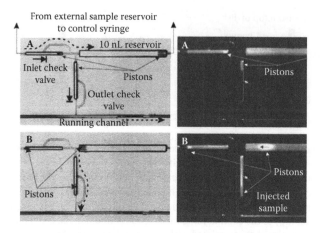

FIGURE 3.31 A 10 nl pipette with two check valves. Check valves are placed at sample inlet and pipette outlet. The outlet of the pipette issues into a pressurized channel at the bottom of the images, flowing from left to right. Dashed lines indicate sample flow. A piston/cylinder reservoir is placed between the sample inlet and a control inlet. The left images are optical images of the pistons; the right images are fluorescence images; only the sample contains dye. Top (a, a'): Suction is applied at the control reservoir; in response, the piston in the reservoir moves to the right, drawing sample through the open inlet check valve, until the reservoir near the control inlet is filled. The check valve at the pipette outlet remains closed. Bottom (b, b'): Pressure is applied at the control inlet; the check valve at the sample inlet closes, the valve at the outlet opens, and the fluid in the reservoir is injected into the channel.[456] Reprinted with permission from the American Chemical Society.

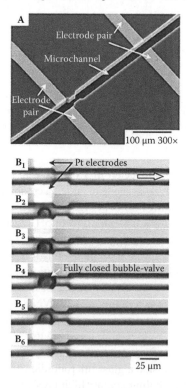

FIGURE 3.32 (a) SEM micrograph of an electrochemical bubble valve chip consisting of a microchannel and a pair of electrodes. (b$_{1-6}$) Optical micrographs showing bubble inflation and deflation. The flow direction is indicated by an arrow. The dark edges near the channel are an optical artifact.[458] Reprinted with permission from the American Chemical Society.

and a diverter valve (like an exclusive OR logic gate) have been fabricated. In addition, a 10 nl pipette was fabricated (see Figure 3.31).[455,456] In a similar way, a high-pressure picoliter injector was constructed to conduct HD injection in a fused silica chip.[457]

Electrochemically generated bubbles were also used as microfluidic valves (see Figure 3.32). The valves closed when the bubbles inflated, and vice versa.[458]

A capillary retention valve was achieved at a constriction that had the highest capillary pressure and hence pinned the interfacial meniscus of the liquid, thus confining the liquid (see Figure 3.33).[459]

Flow switching in a PDMS chip can be achieved by deflection from two side channels (see Figure 3.34).[359] When there is zero hydrostatic pressure, the two streams of fluorescent beads flow equally to two channels. When the hydrostatic pressure from the bottom channel is greater (i.e., 2 mm liquid level difference or 20 Pa pressure), the stream is deflected upward. In a special design of channel shown in Figure 3.35, the liquid flow into the side loops can only occur when the flow is from left to right, but not vice versa. This flow rectifier works because the momentum of the beads naturally took them into the side channel only when the flow is from left to right.[395]

Prefocused flow switching for M inputs and N outputs was achieved on a quartz chip. In Figure 3.36, six input channels and six output channels are illustrated. The sample from A_2, which was focused by streams coming from A_1 and A_3, was directed to channel 1. Similarly, the sample from B_2, after being focused by streams from B_1 and B_3, was directed to channel 2. Figure 3.37 illustrates the situations of (a) no focusing, in which the sample also enters adjacent output channels, and (b) with focusing, in which a narrow, but clean sample stream flows to the output channel.[263] A five-way multiposition valve with five outlet parts for flow switching has been constructed on a Si-Pyrex chip.[307]

Thermal gelation of methylcellulose has been used for valving in a Y-channel to sort fluorescent beads.[460] Thermally responsive monolithic polymer plugs were used as a flow control valve in a glass chip. The polymer plugs were formed by poly(N-isopropylacrylamide) cross-linked with methylene-bisacrylamide. The volume phase transition of the polymer occurs at 32°C. Thermoelectric elements were used to swell and close the valve at 17°C (due to polymer solvation) and shrink and open the valve at 57°C (due to polymer desolvation). The valve closed in ~5 s and opened in 3 s.[341] A valve was also created using a cold finger that employed an ice plug formed in the channel to create the off-state. When a heater was turned on to melt the ice, the channel was opened, creating the on-state.[461] Paraffin wax (and other waxes) was used as a meltable piston for valving on a Si-glass chip[98] or PC chip.[462] A resistive heater melts the wax and moves it to close a Y-junction for a PCR study. The wax is melted again and sucked out to open the valve. Valve actuation time can be within a few tens of milliseconds.[98]

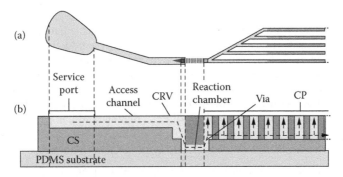

FIGURE 3.33 Geometry and hydrodynamic characteristics of the microfluidic capillary system (CS). (a) Top view of a CS. (b) The flow of liquid (arrows) is superimposed on the cross section (not to scale) of the CS. CRV, capillary retention valve; CP, capillary pump.[459] Reprinted with permission from the American Chemical Society.

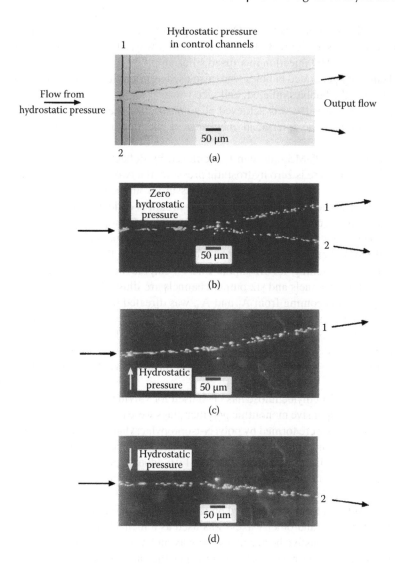

FIGURE 3.34 Actuation of the fluid deflection switch by hydrostatic pressure. (a) Optical micrograph of the fluid deflection switch. (b) Fluorescence image of the flow of beads when the levels of liquid in the reservoirs of the two control channels were the same. Since there was no net hydrostatic pressure in the control channels, the input flow of liquid was not deflected, and the flow partitioned equally into the two output channels. (c) About 50 μl of liquid was added to control reservoir 2 to produce a net hydrostatic pressure in the control channels (shown by a white arrow). As a result, the beads were deflected into output channel 1. (d) Fluorescence image of the flow of beads when the direction of the hydrostatic flow in the control channels was reversed (shown by white arrow) by removing about 100 μl of liquid from reservoir 2 and adding it to the reservoir of control channel 1. The pressure difference in the two control channels that resulted from the difference in height of the liquid in the reservoirs was about 20 Pa.[359] Reprinted with permission from the Institute of Physics Publishing.

A magnetic microvalve was constructed by electroplating Ni/Fe alloy on a Si membrane. Magnetic actuation of the normally open valve was achieved by a Cu coil.[463] Movement of ferrofluid (controlled by a magnet) was also employed as a valve control.[464]

Bidirectional pumping was achieved in a single pneumatic structure in a PMMA chip. A liquid drop can be switched from one channel to another.[465]

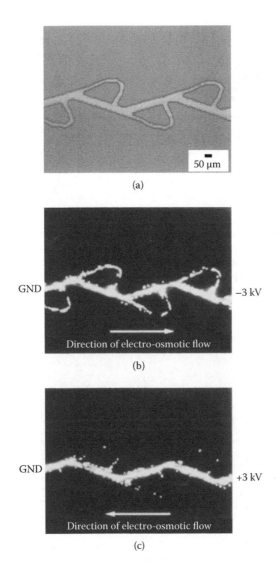

FIGURE 3.35 (a) Optical micrograph of a side channel fluidic controller. The light areas are PDMS channels. (b) Fluorescence image of fluorescent beads when an electric potential was applied that gave rise to EOF from left to right. The momentum of the beads carried them into the side channels. (c) Fluorescence image of the beads when the polarity of the applied voltage was reversed. EOF was from right to left. It was shown that the momentum of the beads in the central channel did not take them into the side channels.[359] Reprinted with permission from the Institute of Physics Publishing.

3.2.5 FLUID MIXING

The laminar flow is prevalent in the microfluidic channel. Solution mixing, which has mainly been accomplished by diffusion, is slow, even in the presence of pillar structures.[739] Other than diffusive mixing, various approaches employed to promote the mixing efficiency are described in subsequent sections.

3.2.5.1 Diffusive Mixing

A mixer was constructed on Si based on distributive mixing (see Figure 3.38).[466] Two liquid streams were split into sixteen ministreams to enhance diffusive mixing. Thereafter, the

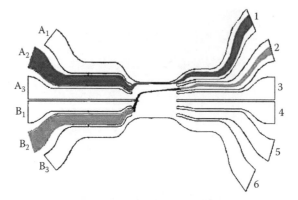

FIGURE 3.36 Schematic representation of a micromachined prefocused 2 × 6 flow switch with six inlet ports and six outlet ports. The flow switch integrates two important microfluidic phenomena, including hydrodynamic focusing and valveless flow switching. The prefocused samples can be injected into desired outlet ports precisely.[263] Reprinted with permission from the Institute of Physics Publishing.

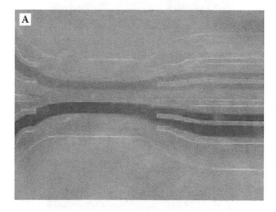

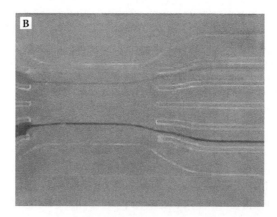

FIGURE 3.37 Sample flows (a) without the prefocusing function and (b) with the prefocusing function. It can clearly be seen that smearing among outlet channels can be avoided when the prefocusing function has been employed.[263] Reprinted with permission from the Institute of Physics Publishing.

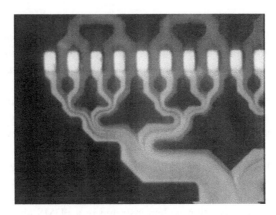

FIGURE 3.38 A micromixer based on distributive mixing. Flow visualization using fluorescein and rhodamine B at a total flow rate of 50 ml/min. Only ten out of sixteen ministreams are shown.[466] Reprinted with permission from the Royal Society of Chemistry.

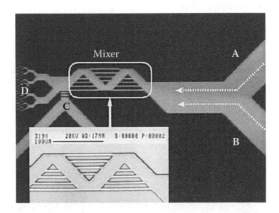

FIGURE 3.39 Photomicrograph of the microfabricated mixer based on eddy diffusion. This mixer is about 100×200 μm wide and 10 μm in depth. Effect of mixing is evaluated by bringing in two fluids from channels A and B, and flowing to D. Channel C is not used in this work. The inset shows the SEM image of the mixer.[493] Reprinted with permission from the American Chemical Society.

ministreams were recombined. This mixer has been used to study chemically induced conformational change of proteins,[467] and chemiluminescent reaction catalyzed by Cr^{3+}.[468] In another report, a similar mixer with sixteen channels was built.[280,469] Mixing was also carried out in a PET chip consisting of a distributor and a dilutor to produce sixteen concentrations from two inputs.

A micromixer (100 pl) was constructed on a quartz chip with multiple small channels (5 μm wide) intersecting 45° with the main channels (27 μm wide) (see Figure 3.39). EOF flow of two solutions containing different concentrations of fluorescein was initiated to achieve mixing based on the principle of eddy diffusion after splitting in the smaller channels.[493]

3.2.5.2 Chaotic Advection

To facilitate mixing, mixers based on chaotic advection in special microstructures were constructed. For instance, in a PDMS fluidic mixer, small chevron-shaped indentations, which were not centered, were constructed in the channel. Such an arrangement forced the fluid to recirculate in order to achieve a mixing effect.[184]

Mixing in a PMMA T-junction was achieved by chaotic advection in the channel with slanted grooves created by laser photoablation (see Figure 3.40).[193,257,470] Mixing could also be achieved in a PDMS microchannel using square grooves in the channel bottom (see Figure 3.41).[471]

A Si micromixer was constructed with seven vertical pillars (10 μm diameter) arranged perpendicular to the flow in a staggered fashion within a 450 pl mixing chamber, as shown in Figure 3.42. Turbulent mixing of sodium azide and horse heart myoglobin was achieved in 20 μs. This provided the fast mixing essential for a downstream freeze-quench procedure to trap metastable intermediates.[472]

A mixer was constructed by porous hydrogel formed at a Y-junction. It was found that the mixing efficiency was higher for a more porous structure.[496]

Mixing was enhanced in a PDMS-glass chip in which polybrene was used to produce a staggered pattern of positively charged surface. It was found that the generation of flow circulation at the positively charged polybrene surface and the negatively charged glass surface enhanced solution mixing.[473]

Passive mixing by chaotic advection is demonstrated in a PDMS chip consisting of a winding channel (see Figure 3.43). An essential component is a water-immiscible solvent, perfluordecaline (PFD). As shown in Figure 3.43b, with no PFD, mixing of a colorless stream (0.2 mM KNO_3) and a red stream (Iron(III) thiocyanate formed by mixing 0.067 M Fe $(NO_3)_3$ and 0.2 M KSCN) did

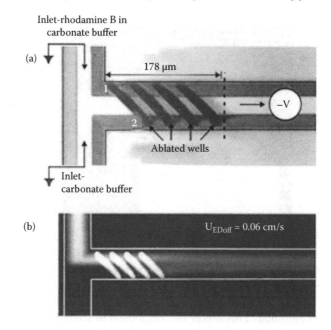

FIGURE 3.40   (a) Configuration of the experimental setup and white light microscopy image of an imprinted T-channel with a series of ablated wells. (b) Fluorescence images of electroosmotic flow past the mixer at flow rates of 0.06 cm/s.[193] Reprinted with permission from the American Chemical Society.

FIGURE 3.41   Optical micrograph taken from above of a stream of black dye flowing in a microchannel containing square grooves in the bottom of the channel. A narrow stream of black dye in water is injected alongside a broad stream of clear water (flow rate of the clear stream is twenty times that of the dyed one). The average flow speed in the channel is 1 cm/s (Reynolds number Re ~ 1).[471] Reprinted with permission from the American Chemical Society.

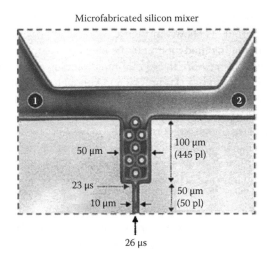

Microfabricated silicon mixer

FIGURE 3.42 Photograph of the microfluidic silicon mixer constructed with seven vertical pillars.[472] Reprinted with permission from the American Chemical Society.

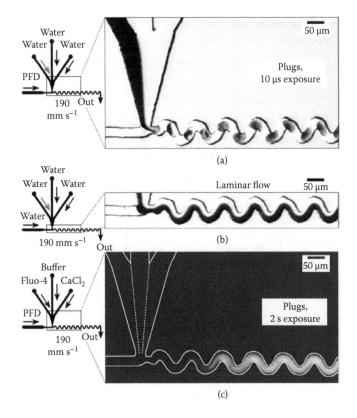

FIGURE 3.43 (a, b) Microphotographs (10 μs exposure) illustrating rapid mixing inside plugs (a) and negligible mixing in a laminar flow (b) moving through winding channels at the same total flow velocity. (c) A false-color microphotograph (2 s exposure, individual plugs are invisible) showing time-averaged fluorescence arising from rapid mixing inside plugs of solutions of Fluo-4 (54 μM) and $CaCl_2$ (70 μM) in aqueous sodium morpholine propanesulfonate buffer (20 mM, pH 7.2); this buffer was also used as the middle aqueous stream. All channels were 45 μm deep; inlet channels were 50 μm and winding channels 28 μm wide; Re ~ 5.3 (water), ~ 2.0 (PFD).[474] Reprinted with permission from Wiley-VCH Verlag.

not occur, and laminar flow was apparent. In the presence of PFD (a 10:1 mixture of PFD and $C_6F_{11}C_2H_4OH$) mixing occurred, as shown in Figure 3.43a. The principle of chaotic advection is based on time-periodic, alternating fluid motion in the plug relative to the channel wall. Such a fast mixing (2 ms) was employed to measure the rate of rapid binding ($k_{on} = 7.1 \times 10^8$ M^{-1} s^{-1}) of Ca^{2+} ions to the calcium-sensitive dye fluo-4 (Figure 3.43c). Although the spatially resolved fluorescent image was acquired over 2 s, it was noted that mixing of CaCl$_2$ and fluo-4 in the plugs was achieved within a short distance (~500 μm), corresponding to mixing in ~2 ms under a flow of 190 mm/s.[474] Later developments in this area are also reported.[475,476]

A 3D serpentine microchannel was fabricated on a Si-glass chip to enhance mixing by chaotic advection (see Figure 3.44). It was found that mixing in the 3D channel was faster and was more uniform than in either a "square wave" channel or straight channel.[477]

3.2.5.3 Oscillating Flow

Bidirectional flow, which occurs at a funnel-shaped channel because of the interaction of EOF and HDF, creates a mixing effect.[478,479] Such a bidirectional flow has also been used to trap particles (see Chapter 8, Section 8.1.5).

An active mixer based on an oscillating EOF induced by sinusoidal voltage (~100 Hz, 100 V/mm) was devised and modeled for mixing of fluorescein with electrolyte solutions. This is termed electrokinetic instability micromixing, which is essentially a flow fluctuation phenomenon created by rapidly reversing the flow. Various microchips materials (PDMS, PMMA, and glass) and various electrolytes (borate, HEPES buffers) have been used to evaluate this method of micromixing.[480]

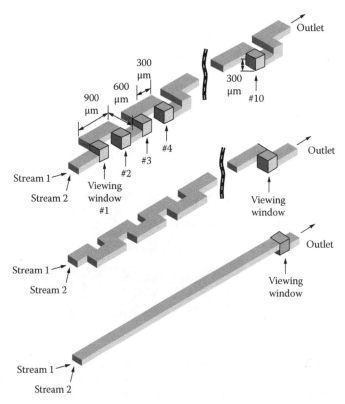

FIGURE 3.44 Top: Schematic of the three-dimensional serpentine channel. "Viewing windows" in the channel are labeled 1 to 10. Middle: Schematic of a square wave channel. Bottom: Schematic of a straight channel.[477] Reprinted with permission from the Institute of Physics Publishing.

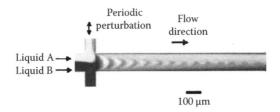

Periodic perturbation

Flow direction

Liquid A →
Liquid B →

100 µm

FIGURE 3.45 Experimental results of a mixer constructed with one pair of side channels. The pressure perturbations created from the side channels induce lobe-like distortions of the interface and facilitate rapid mixing.[481] Reprinted with permission from the Institute of Physics Publishing.

A micromixer in which the fluid can be stirred by periodically pumping through the side channels is shown in Figure 3.45.[481] The periodic perturbation applied via the side channel allows liquids A and B to be mixed. Other mixers based on oscillating pressure-pumped flow have also been reported.[482,483] Two droplets (600 pl) were merged and mixed by a push–pull (shuttling) method in a PDMS device consisting of a hydrophobic microcapillary vent (HMCV).[364]

Self-circulation in a circular chamber was used for mixing. It was found by simulation and flow visualization that mixing was more effective in a hollow chamber than an annular one.[484]

A rotary mixer was constructed on PDMS using multilayer lithography.[485] Peristaltic action through pneumatic control causes rapid binding of neutravidin-coated fluorescent beads on biotinylated spots in the PDMS device. Using this mixing action, more than 80% of the beads were attached within 4 min. But without the rotary mixing, it took 4 h for the binding to be visualized fluorescently due to slow diffusion of the beads ($D = 2.5 \times 10^{-9}$ cm²/s).

3.2.5.4 Acoustic Mixing

Mixing can be achieved using ultrasonic acoustic waves. For instance, a laser-cut acrylic laminated chip, to which bulk piezoelectric transducers (PZT-4) were bonded, was shown to have mixing action using 10 µm polystyrene beads. The transducer had a half-wavelength resonance thickness and exhibited a primary resonance peak at 365 kHz. Bead mixing was achieved for 1, 3, 5, 6, and 10 µm polystyrene beads and glass beads.[486] A PZT disk was also used to enhance ultrasonic mixing in a PC chip.[462]

Another active micromixer based on ultrasonic vibration of a thin Si membrane (0.15 mm) is shown in Figure 3.46. The membrane was actuated by a bulk PZT transducer powered by AC voltage (48 kHz, 150 Vpp). The Si wafer was bonded to a Pyrex glass plate consisting of an etched chamber. The PZT can generate a great force, but its mechanical displacement is small, so high frequency should be used to compensate for this. The laminar flow of dyed ethanol (115 µl/min) and water (100 µl/min) was mixed effectively when the PZT was excited (see Figure 3.47).[327] Unfortunately, considerable heat was generated from the large-amplitude ultrasonic vibration, which is undesirable to study temperature-sensitive fluids.[487]

In another approach, ultrasonic acoustic flexural plate waves (or Lamb waves) were used to generate fluid mixing motion of microspheres in water in a Si device. A 10 µm thick piezoelectric ZnO film was deposited on the backside of a Si wafer (see Figure 3.48). This thickness is an odd multiple of the acoustic half wavelength. The Al electrode (0.5 µm thick) patterned on the ZnO film consists of concentric circles, serving to focus the acoustic energy at a few hundred micrometers above the liquid–air interface. At the fundamental frequency (240 MHz) and second harmonic frequency (480 MHz), greater fluid motion was generated, as compared to other frequencies. The heat generated in the mixer is very low compared to the PZT-actuated mixer.[488]

In another report, an ultrasonic mixer was constructed on a PDMS-quartz chip using a ZnO film. The ZnO film (8 µm) was deposited on the quartz plate with a patterned Au electrode (300 nm) for

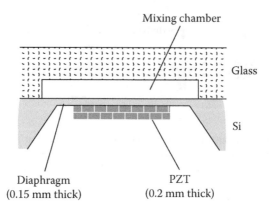

Mixing chamber

Glass

Si

Diaphragm
(0.15 mm thick)

PZT
(0.2 mm thick)

FIGURE 3.46 Schematic drawing of the cross section of a mixer. A glass plate was etched with channels. It was anodically bonded with a Si wafer consisting of the oscillating diaphragm to which a PZT disk was adhered.[327] Reprinted with permission from Wiley-VCH Verlag.

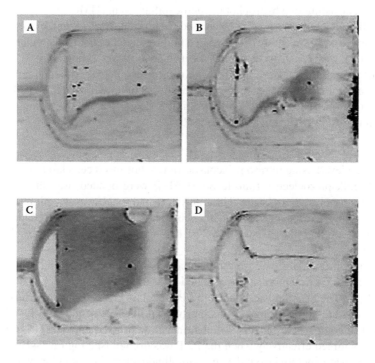

FIGURE 3.47 The ultrasonic mixing process. (a) Videocamera scene at standby state. Ethanol and water were in laminar flow. Limited diffusion occurred at the interface of the ethanol and water flows. (b) Ultrasonic vibration ON. The laminar flow changed and turbulence occurred. (c) After 2 s of ultrasonic vibration, ethanol was mixed well with water. (d) Ultrasonic vibration OFF. The laminar flow resumed.[327] Reprinted with permission from Wiley-VCH Verlag.

excitation. Operation frequency was 450 MHz. No heating or bubble formation was observed with a power level of 15 dBm (or 30 mW at 1.2 V_{rms}).[489]

It should be noted that at some resonance frequencies (or their harmonics), a standing wave will be established in the chamber when the two chamber walls are vertical and parallel to each other. This results in stagnant zones at the nodes of the standing wave, in which no mixing occurs.[84]

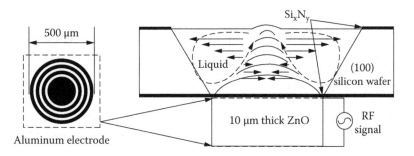

FIGURE 3.48 Cross-sectional view (along with typical liquid flow directions) of the liquid mixer microfabricated on a silicon wafer. Also shown on the left is the electrode pattern of the transducer to form the acoustic wave sources for a constructive wave interference in the liquid.[488] Reprinted with permission from Elsevier Science.

3.2.5.5 Other Mixing Methods

Turbulent flow in a Si-Pyrex chip was used for mixing (see Figure 3.49). A higher flow (0.75 ml/s) resulted in complete mixing, which was not achieved in a lower flow (0.2 ml/s). The volume of the mixer is 0.27 μl (in a 2.66 mm long channel). The higher flow created a turbulent flow, which assisted mixing.[495]

A small magnetic stirring bar (2 mm diameter × 7 mm) was used to enhance mixing in a PDMS-glass chamber (50 μl). This led to a three- to fourfold increase in the DNA hybridization efficiency.[490]

Fast mixing of reagents in a PC microchannel was achieved by thermally triggering individual liposomes containing these reagents, instead of by flowing them in channels. The permeability of liposomes near the gel-to-liquid phase transition temperature was greatly increased through the modulation of temperature, leading to release of reagents and their mixing.[491]

Other micromixers based on various principles have also been constructed. These principles include vortex,[492] eddy diffusion,[493–501,654,955] rotary stirring,[502] turbulence,[495,503] EK instability,[504–506] chaotic advection,[248,507–513] magnetic stirring,[514] bubble-induced acoustic mixing,[515] and piezoelectric actuation.[516,517]

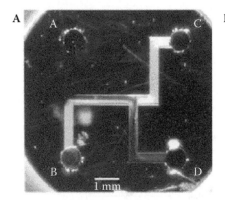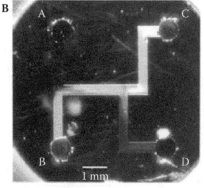

FIGURE 3.49 (See color insert following page 144.) Visualization of the mixing events in the 2.66 mm long silicon mixer manufactured using deep reactive ion etching. An acid–base indicator, bromothymol blue (inlet A, neutral form, green), is mixed with HCl (inlet B). The indicator changes color to yellow (acidic form) and is subsequently mixed with NaOH (inlet C), resulting in a blue color (basic form) before the mixture leaves the chip (outlet D). (A) Flow velocity of 0.2 ml/s results in incomplete mixing. (B) Flow velocity of 0.75 ml/s results in complete mixing.[495] Reprinted with permission from the American Chemical Society.

3.2.6 Liquid Dispensing

Electrowetting-on-dielectric (EWOD) was developed for moving droplets. In EWOD, the local wettability of a hydrophobic, dielectric layer is reversibly changed by applying potentials to the electrodes beneath the layer. Liquid droplets can be made to travel from one location to another for mixing and reaction. In the EWOD device shown in Figure 3.50, the bottom plate was formed from a quartz chip patterned with poly-Si electrodes insulated with SiO_2 and coated with hydrophobic Teflon-AF. The top electrode was a contiguous electrode that was an indium-tin oxide (ITO)-coated glass chip. Aqueous droplets (0.5 μl) were sandwiched between the two plates and were moved by applying AC potentials (1 kHz, 75 V_{rms}).[518]

The hydrophobic passive valve was constructed at the junction of a narrow side channel and a wide main channel on a PDMS chip for liquid droplet dispensing and mixing (see Figure 3.51). First, liquids A and B were introduced from the sample channels. Second, air was used to remove liquids

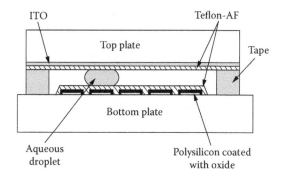

FIGURE 3.50   Side view of an EWOD device (not to scale). The bottom plate served as the base for a pattern of polysilicon EWOD electrodes (1 mm², 4 μm gap) buried under the thermal oxide. A top plate was formed from ITO on glass; both plates were coated with Teflon-AF. The plates were joined with double-sided adhesive tape as a spacer.[518] Reprinted with permission from the American Chemical Society.

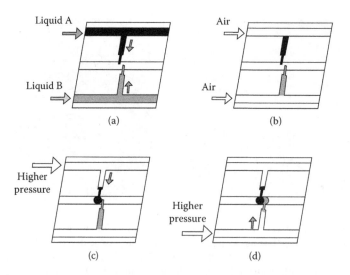

FIGURE 3.51   Schematic diagrams of droplet operation; (a) introduction of liquid A and liquid B, (b) introduction of air, dispensing droplets with certain volumes (3.5 or 20 nl) in metering channels, (c) injection of liquid A into the mixing channel by applying a higher pressure, and (d) injection of liquid B and mixing.[260] Reprinted with permission from the American Chemical Society.

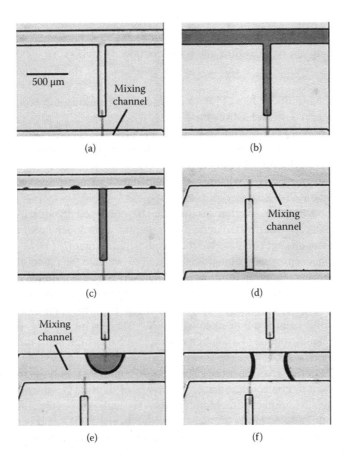

FIGURE 3.52 Photographs of liquid dispensing and mixing; (a) before the introduction of liquid, (b) introduction of dye solution, (c) dispensing of dye solution by introducing air from the upper left inlet, (d) dispensing of reagent solution by introducing air from the lower left inlet, (e) injection of dye solution, and (f) injection of reagent solution, mixing, and reaction. Dispensed volume was 20 nl for each liquid. Hydrochloric acid (0.1 N) was used for the reagent, and the color of the dye solution was bleached within a few seconds.[260] Reprinted with permission from the American Chemical Society.

in the sample channels, leaving the liquid only in the metering side channel. Third, air pressure was increased for the liquids to burst through the hydrophobic passive valve into the main channel, and to be mixed together. The images showing the sequences of events are depicted in Figure 3.52. Figure 3.52f shows the situation after the reaction. Up to fifty such mixing junctions have been constructed on a circular PDMS chip so that dispensing and mixing at these junctions occur simultaneously by only two air pressure control channels. Mixing of glucose oxidase, peroxidase, and mutarotase (liquid A) and glucose, phenol, and 4-aminoantipyrine (liquid B) was achieved at 37°C to colorimetrically detect the amount of red quinone dye formed at different glucose concentrations (2 to 10 mM).[260]

Electrochemical generation of gas bubbles has also been employed for dosing precise nanoliter amounts of liquid.[391]

3.3 PROBLEM SETS

1. Distinguish between laminar flow and turbulent flow. (4 marks)
2. Distinguish between passive and active micromixers. List three examples in each case. (6 marks)
3. Distinguish the use of ZnO and PZT in constructing the ultrasonic acoustic wave micromixer (see Figures 3.46 and 3.48). (4 marks)
4. Describe the principle of mixing in a single channel with slanted wells fabricated at an angle to the liquid flow (see Figure 3.40). Will the slanted wells constructed perpendicular to the flow also work? Why?[193] (4 marks)
5. In the ultrasonic micromixer, what was the method used to measure acoustic displacement generated by the PZT transducer?[487] (2 marks)
6. A micromixer based on solution mixing can be enhanced by distributive mixing. In Figure 3.38, a stream was divided into sixteen streams. How much faster can the mixing in these sixteen streams be, compared to the unsplit single stream?[193,466] (2 marks)
7. A fluorescent quenching experiment was devised to evaluate the rate of mixing of fluorescein and a quencher. Name an example of a quencher. Why did a simple mixing method of fluorescein and water not work to evaluate the mixing rate?[466] (3 marks)
8. For a small molecule with a diffusion coefficient of 10^{-6} cm^2/s, how far would it diffuse after flowing over 1 cm at a flow rate of 10 cm/s?[198,163] (2 marks)
9. Calculate the liquid volume of a solution reservoir of 5 mm high and 2 mm i.d. mounted on a microchip.[698] (2 marks)
10. A native PETG chip gives an EOF of $+4.3 \times 10^4$ cm^2/(V.s). Polyelectrolyte multilayers (PEMs) are used to control the surface chemistry of the PETG substrate using alternate layers of poly(allylamine hydrochloride) (PAAH) and poly(styrene sulfonate sodium) (PSS). When three PEM layers are used, the EOF of the PETG chip becomes -1.8×10^4 cm^2/(V.s). Draw a diagram of the chip to indicate the arrangement of the PETG, PAAH, and PSS layers. (2 marks)
11. In an electrohydrodynamic mixer, a DC electric field (400 kV/cm) is applied on two electrodes across a channel to induce secondary transverse flow for solution mixing. Explain why there is no problem of electrolysis. (2 marks)
12. Describe, with a labeled diagram, the origin of EOF in glass microchannels. (2 marks)
13. Laminar flow is prevalent in microfluidic channels. Describe one advantage and one disadvantage of such a physical phenomenon. (2 marks)
14. Describe three methods to generate liquid flows within microchannels. (3 marks)
15. Explain why a coating that is applied to the chemical surface can prevent sample adsorption, and can suppress EOF. (2 marks)

4 Sample Introduction

In most cases, sample introduction on-chip is achieved using electrokinetic (EK) flow.[3] Two important EK injection modes, namely, pinched injection and gated injection, have been developed. Furthermore, some alternative injection methods are described.

4.1 ELECTROKINETIC INJECTION

For EK injection, a cross-injector and a double-T injector have been constructed. The injection and separation modes for a cross-injector vs. double-T injectors (with the 100 and 250 μm "sampling loop") were visualized as shown in Figure 4.1.[620]

The aforementioned injection of a solution plug is termed plug injection. Another sample introduction mode is stack injection. Stack and plug injections used in DNA separation have been compared (see Figure 4.2). Under similar conditions, the plug injection produced a better resolution, whereas the stack injection produced a higher sensitivity (see Chapter 6, Section 6.2 for more on capillary electrophoresis (CE) separation).[315]

With EK injection, reasonable reproducibility (relative standard deviations [RSDs] in migration time and peak area are 0.1 and 2%, respectively) in eleven successive injections/separations has been reported (see Figure 4.3). Here, a double-T injector consisting of a sampling loop of 150 μm long was used.[340] In a later report, repetitive injections and separations also resulted in good migration time RSD (0.06%) and peak height RSD (1.7%).[547]

It was found that a 250 μm double-T injector increased in the peak signal compared to a straight cross-injector.[548,620] Although a fivefold increase was expected based on the intersection volume calculation of a 250 μm injector vs. a cross-injector, the discrepancy was likely to be caused by the backflow of analyte upon applying the push-back voltage (for preventing leakage).[620] Another way is to create the cross-injector (20 μm deep) by aligning the top and bottom channel plates (each 10 μm deep) (see Figure 4.4).[314]

In EK injection, sample introduction is usually biased. However, the sample can be introduced to a microchip in a nonbiased manner, as long as the electric voltage was applied for a sufficient time so that the slowest migrating component has passed the intersection and entered the analyte waste channel. Accordingly, the sample composition at the intersection is representative of that in the original sample, though it is not the case somewhere downstream in the waste channel.[107,340]

EK injection bias still exists in pinched injection, with neutral species injected in a greater amount than anionic species. However, the orthogonal nature of the loading and injecting steps reduced this bias, although it did not completely eliminate it.[549]

In a T injector, which eliminates the need for a waste channel and reservoir, biased EK injection may result in loading higher-mobility analytes in a larger amount at the expense of the lower-mobility ones. A delayed back biasing (for 25 s) was useful that allowed sufficient loading of higher molecular weight (low mobility in gel) DNA fragments, before a push-back voltage was applied.[980]

According to another report, when DNA fragments are loaded from an aqueous reservoir to a gel-filled channel, a differential concentration effect occurs. First, the fragments are concentrated in the gel phase because of lower electrical mobilities of the fragments in the gel relative to the aqueous phase. This concentration is biased toward larger DNA fragments. Such bias is favorable as the amount of larger fragments is usually smaller than smaller fragments in a DNA sequencing mixture.[265]

A wide sample exchange channel (1 mm wide, 200 μm deep) was interfaced to the shallow EK sample introduction channels (36 μm wide, 10 μm deep) in a glass chip. This allowed fast sample loading

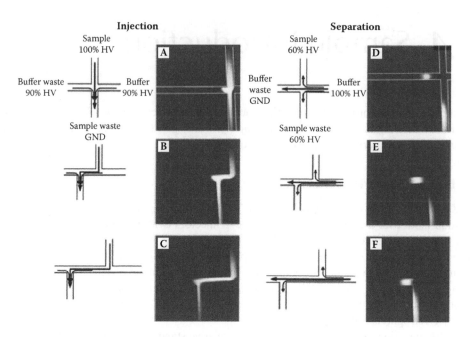

FIGURE 4.1 Fluorescence micrographs of injection and separation of a marker (0.4 mM fluorescein) performed on a glass microchip, using different injector configurations. Injections and separations are shown in (a–c) and (d–f), respectively. Injector configurations: Straight cross (a, d), 100 μm offset double-T (b, e), and 250 μm offset double-T (c, f). The illustrations to the left of the photographs show the injector configurations and applied voltages during injection and separation modes.[620] Reprinted with permission from the American Chemical Society.

and replacement into a CE chip. This automated procedure was achieved by hydrodynamic flow (HDF), without disturbing the liquids within the shallower and narrower sample introduction channels.[117]

4.1.1 PINCHED INJECTION

In plug injection, there is sample leakage of the sample plug around the intersection. The leakage has been attributed to diffusive and convective phenomena.[550] This leakage can be reduced by using the pinched mode in which the buffers from two adjacent channels are flowed in to shape the plug (see Figure 4.5).[136] For clarity, the plug injection without using the pinching voltage is thus called floating injection. It was found that an increase in the pinching voltage resulted in fewer injected materials, whereas a decrease in the pinching voltage caused less peak symmetry.[551]

The temporal stability of the injection plug size is quite good in the pinched mode regardless of the loading time. However, in the floating mode, the plug size continually increased with the loading time. Reproducibility for five successive injections gives 1.7% RSD in the peak area for the pinched injection, compared to 2.7% RSD for the floating injection.[136]

In one report, even though pinched injection was used, the actual injected amount did depend on loading time if the ionic strengths of the sample and run buffers did not match.[552] It was also found that unless the sample is strongly pinched, pinched injection will inject different volumes depending on the injecting voltage.[71]

Even if pinched injection was used, baseline drift of repetitive pinched injections was noted. This was attributed to the meniscus surface tension (Laplace pressure) effect, even though other possible effects (evaporation, buffer depletion due to electrolysis, siphoning, Joule-heat-induced viscosity change) are minimized.[553]

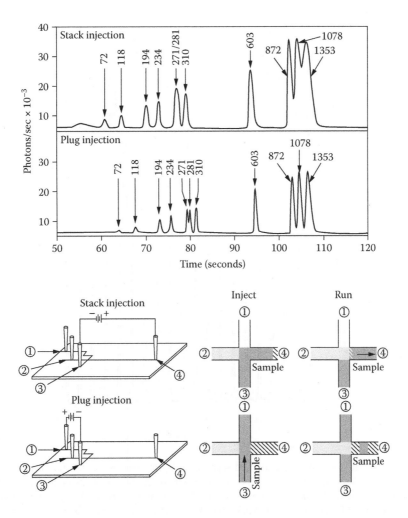

FIGURE 4.2 (a) Schematic diagram of the stack injection and plug injection methods. The hatched regions indicate the sieving gel medium in the separation channel. (b) Electropherogram comparing the stack and plug injection methods. A sample containing φX174 *Hae*III fragments at 10 ng/μl was injected for 1 s in each experiment. The buffer consisted of the standard Tris/acetate/EDTA/hydroxyethylcellulose (TAE/HEC) sieving medium with 1 μM TO dye. A signal of eight thousand photons per second over background corresponds to 100 pg of DNA per μl in the separation channel.[315] Reprinted with permission from the National Academy of Sciences.

Sample leakage can cause peak tailing.[102] To avoid this, the buffer is pushed back into the analyte channel and analyte waste channel by applying a push-back voltage.[136,324] This can be applied for a short duration (~120 ms), rather than continuously, to provide a clean cut of the injection plug.[551] Sample leakage during preinjection and separation can also be avoided by putting resistors (at least 30 MΩ) between the sample waste and ground.[365,366] In the case of repetitive injection/separation, the loading time for a subsequent injection should be long enough to compensate for the location of the sample front pushed back in the previous separation.[547]

The floating injection (with no push-back voltage) was successful without sample leakage in some examples of the separations of DNA[554,925] and proteins,[555] possibly because of slow molecular diffusion in the viscous gel separation media.

In non-gel solution, if the sample channel width was narrower than the separation channel width (by fivefold), floating injection did not result in sample leakage, and no pinched injection

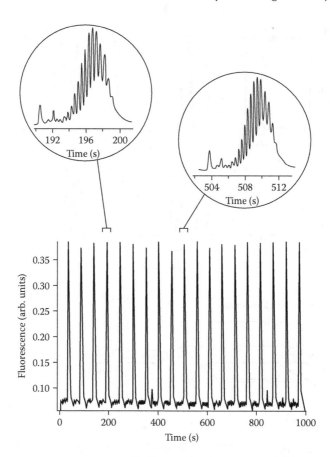

FIGURE 4.3 Repetitive sample injection and separation. Cycle conditions: Injection time, 5 s; dead time, 1 s; separation time, 45 s; dead time, 1 s. The insets show two separation events on an expanded timescale. Sample: Fluorescein-labeled phosphorothioate oligonucleotide mixture, poly(dT)$_{10-25}$. Separation conditions: Buffer, 100 mM Tris, 100 mM boric acid, 2 mM EDTA, 7 M urea, pH 8.5; electric field strength, 2,300 V/cm; separation length, 3.8 cm.[547] Reprinted with permission from the American Chemical Society.

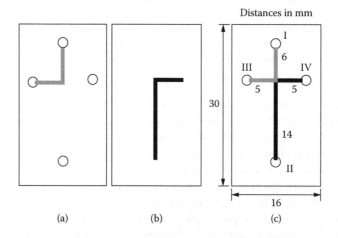

FIGURE 4.4 Slides (a) and (b) are composed of HF etched channels with powder-blasted access holes in (a). The bonding of the two slides gives a cross-shaped channel layout (c) of injection volume of 27 pl.[314] Reprinted with permission from the American Chemical Society.

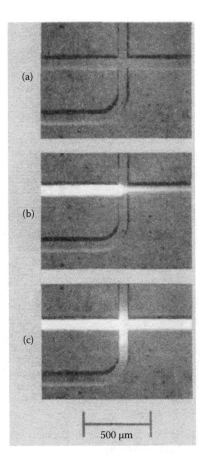

FIGURE 4.5 Images of sample injection at a cross-injector in a glass microchip; (a) no fluorescent analyte, (b) pinched injection of rhodamine B, (c) floating injection of rhodamine B.[136] Reprinted with permission from the American Chemical Society.

was necessary. Moreover, a push-back voltage was not necessary during separation. In addition, a T injector, rather than a cross-injector, was sufficient to perform floating injection without leakage, but the resolution obtained using the T injector was inferior to the cross-injector.[556] If T injectors are used, the number of reservoirs in a multichannel (S) system is reduced to S + 2, which is the real theoretical limit,[556] rather than S + 3.[557]

Various injector geometries (simple cross, double-T, triple-T) were investigated on a glass microchip. The triple-T injector allowed for a selection of different injection volumes. For instance, a triple-T injector allowed injection of three different volumes depending on whether a cross, double-T, or triple-T configuration was used (see Figure 4.6).[529] Pinched injection has been used consecutively to inject two samples into the same separation channel (see Figure 4.7).[557]

It was found that during pinched injection mixing occurred at the sample plug[655] if there were two additional channels shaping the sample plug.[578] To create sample plugs of improved plug width and symmetry, a six-port injector with two additional channels was designed.[558,559] A recent report has provided the simulation results for comparison with the experimental observation obtained in a six-port injector (see Figure 4.8).[109]

Pinched injection of APTS-derivatized sugar isomers in a PMMA chip is shown in Figure 4.9. A field-amplified stacking effect is apparent, as shown in the timed events (see Chapter 5, Section 5.1 for more on stacking). In this pinched injection mode (termed default), even though both B and BW are at GND, the fact that the BW channel is longer than the B channel results in a larger

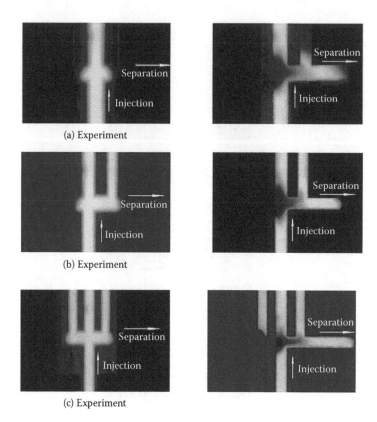

(a) Experiment

(b) Experiment

(c) Experiment

FIGURE 4.6   Sample loading and injection at (a) cross, (b) double-T, and (c) triple-T injector.[529] Reprinted with permission from the American Chemical Society.

amount of sample protruding into the BW channel during loading. To enhance the stacking effect, a larger amount of sample protruding into the B channel is achieved by applying a negative potential (–540 V) to BW and +80 V to B, and this pinched injection mode is known as asymmetric (see Figure 4.10b). For comparison, the case when equal amounts of sample protrude into both B and BW channels is termed symmetric and is achieved by applying –540 V to BW and B at GND (see Figure 4.10a). The resulting electropherograms for CE separation of sugar isomers indicate an enhanced signal (or stacking effect) in the case of asymmetric pinched injection (see Figure 4.11).[560] In another report, depending on the conductivity difference between the sample and run buffers, three modes of pinched injection have been proposed and compared. These are the destacking, usual, and stacking modes.[561]

Another way to implement pinched injection is to apply the pinching voltage directly from the sample loading voltage (A) (see Figure 4.12). A gold electrode joins some downstream position (A') in the sample loading channel (hence a lower voltage) to some positions in the buffer channel (C') and buffer waste channel (D'). Note that there is no fluidic connection between A' to C' and to D'.[562]

Usually EK injection is conducted with three to four ports on a chip. However, EK injection has also been achieved with only two ports, as shown in the cross section diagram in Figure 4.13. Figure 4.13a shows that the sample first goes down the left reservoir and then goes up the right reservoir, and fills up the common region in the channel underneath both reservoirs. Subsequent procedures completed the injection to the left. This procedure is necessary in synchronized cyclic capillary electrophoresis (SCCE) for dsDNA separation (see Chapter 6, Section 6.2.6 for more on SCCE). Moreover, this procedure will save the number of reservoirs in a multichannel chip.[311]

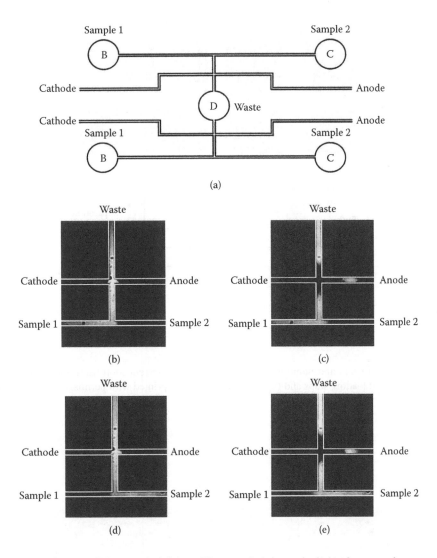

FIGURE 4.7 (a) Layout of the sample injector. The sample injector includes four sample reservoirs, two separation columns, and one common waste reservoir. (b–e) Fluorescence images illustrating the operation of the injector with fluorescein as the sample.[557] Reprinted with permission from the National Academy of Sciences.

In addition to the use of pinched injection for liquid samples, the method is also used for injection of an air sample, as pinched by He gas. Seven injections of air (N_2) were made into a He carrier flow for GC analysis, as shown in Figure 4.14. N_2 was detected by a capillary OED (by the N_2 band at 337 nm).[563]

4.1.2 GATED INJECTION

For continuous sample introduction, gated injection was adopted. With EK flow, the analyte continually flowed in parallel with a separation buffer to the analyte waste reservoir (see Figure 4.15). Injection of the sample analyte was achieved by interrupting the flow of the buffer for a short time (known as the injection time) so that the analyte stream was injected. This scheme was achieved

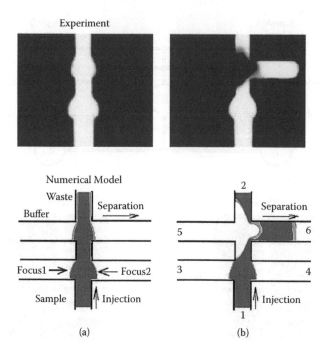

FIGURE 4.8 Experiments and numerical modeling of sample introduction using the double cross-injection system during (a) loading steps and (b) dispensing steps. Reprinted with permission from the American Chemical Society.[109]

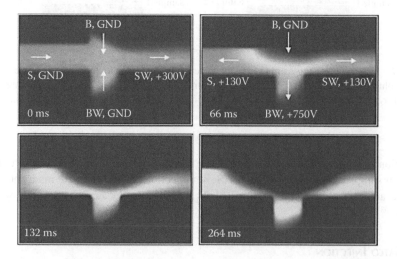

FIGURE 4.9 Fluorescence images of the sample plug formation during the default pinched injection with field-amplified stacking. Arrows indicate the direction of flow in the microchip channels. In loading phase (0 ms), pinching potentials at B, and BW wells were at ground (GND). In separation phase, potentials at 132 and 264 ms were the same as potentials at 66 ms. Buffer: 0.5% methylcellulose in 20 mM phosphate-KOH buffer (pH 6.66).[560] Reprinted with permission from the American Chemical Society.

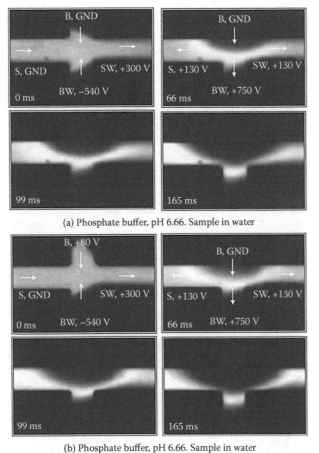

(a) Phosphate buffer, pH 6.66. Sample in water

(b) Phosphate buffer, pH 6.66. Sample in water

FIGURE 4.10 Manipulation of the injected sample plug formation by adjusting the pinching voltages. (a) Symmetric pinched injection. In loading phase, field strengths at B and BW channels were adjusted to be equal to 171 V/cm, using B at ground and BW at –540 V. (b) Asymmetric pinched injection. Field strengths at B and BW channels were adjusted to be equal to 171 V/cm, using B at +80 V and BW at –540 V during downloading. Other conditions are unhighlighted, as in Figure 4.9.[560] Reprinted with permission from the American Chemical Society.

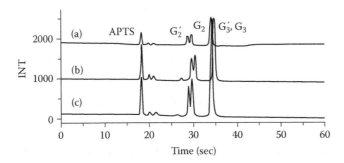

FIGURE 4.11 Microchip electropherograms of APTS-G'_2 -G'_2, -G_3, and -G'_3 (2.1×10^{-7} M in water). (a) Symmetric pinched injection; (b) default pinched injection; (c) asymmetric pinched injection. Conditions: 0.5% methylcellulose in 20 mM phosphate buffer (pH 6.66). Field strength: 300 V/cm. Sample: 2.1×10^{-8} M APTS-G_2, -G'_2, -G_3, and -G'_3 in water. Note G2 is maltose, G'_2 is cellobiose, G_3 is matriose, G'_3 is panose.[560] Reprinted with permission from the American Chemical Society.

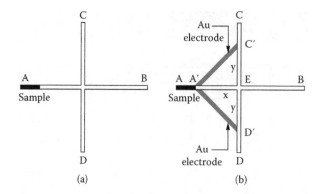

FIGURE 4.12 Schematic drawings of CE chips with added microelectrodes to achieve the pinching effect: (a) The conventional cross-type CE microchip, (b) two Au electrodes of equal lengths were added ($x = y = 0.5$ mm).[562] Reprinted with permission from the Royal Society of Chemistry.

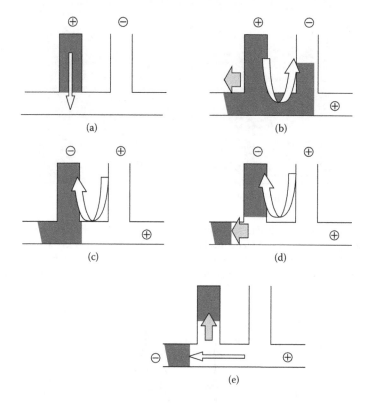

FIGURE 4.13 Scheme of electrokinetic injection achieved by only two reservoirs.[311] Reprinted with permission from the Royal Society of Chemistry.

by four reservoirs (without considering the reagent reservoir) and two power supplies.[317] Gated injection has also been achieved using one power supply and three solution reservoirs.[564]

This gated injection method has allowed the sample loading to be achieved in a continuous manner, whereas a pinched injection mode cannot.[317] Therefore, the gated injection has also been employed for two-dimensional (2D) separation: open-channel electrochromatography/capillary electrophoresis (OCEC/CE)[333,666] or micellar electrokinetic capillary chromatography/CE (MECC/CE)[565] (see Chapter 6, Section 6.4 for more on 2D separation).

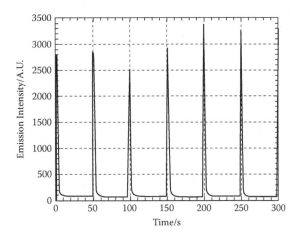

FIGURE 4.14　Seven repeated injections (spaced 50 s) of air into the helium carrier flow using a chip with a cross-injector for GC analysis. Detection with capillary plasma detector using the 337 nm N2 band.[563] Reprinted with permission from the Royal Society of Chemistry.

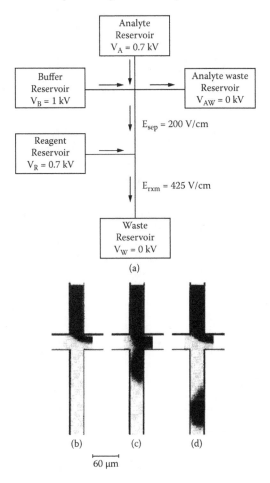

FIGURE 4.15　Schematic diagram of flow pattern for a gated injector. Charge-coupled device (CCD) images of the gated injection using rhodamine B (b) prior to injection, (c) during injection, and (d) after injection into separation column with $E = 200$ V/cm.[317] Reprinted with permission from the American Chemical Society.

An OCEC/CE separation has been achieved on a glass chip, shown in Figure 4.16. The sample injection scheme is shown in Figure 4.17, and the timed events are shown in Table 4.1. For the first-dimensional OCEC separation, the initial injection (a → b) was performed for 0.5 s. Subsequently, serial sampled injections (a → c) were conducted for the second-dimensional CE separation (for 0.2 s every 3.2 s). The relay-controlled gated injection for OCEC/CE separation has RSD values of 1.9% in migration time and 5.5% in peak area for rhodamine B (10 μM) during two hundred serial injections. The plate height remained below 2 μm as long as the concentration of rhodamine B was below 100 μM.[333] When faster-switching power supplies were used, the sampling time for the second dimension was reduced to 20 ms,[565] compared to 0.2 s, as mentioned above.[333,666] This leads to a faster sampling rate of 1 s/sample,[565] compared to 3 s/sample in previous reports.[333,666]

However, the gated injection mode has two problems: (1) the injection plug length increases with the injection time, and (2) the plug length is longer for faster-migrating species, leading to a biased injection.[317] Moreover, because of the turn at the injector, there was an additional sampling bias due to transradial electrokinetic selection (TREKS); i.e., the faster-moving molecule traced out a larger turning radius than the slower-moving molecules.[566]

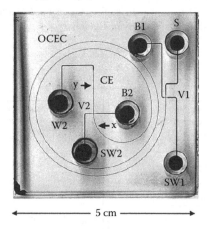

FIGURE 4.16 Image of the glass microchip used for 2D chemical separations. The separation channel for the OCEC (first dimension) extends from the first valve V1 to the second valve V2. The CE (second dimension) extends from the second valve V2 to the detection point y. Reservoirs for sample (S), buffers 1 and 2 (B1, B2), sample wastes 1 and 2 (SW1, SW2), and waste (W2) are positioned at the terminals of each channel. The arrows indicate the detection points in the OCEC channel (x) and CE channel (y).[333] Reprinted with permission from the American Chemical Society.

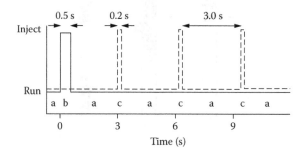

FIGURE 4.17 Timing diagram for making the initial 0.5 s injection into the OCEC channel (solid line) and the subsequent 0.2 s injections into the CE channel with a cycle time of 3.2 s (dashed line).[333] Reprinted with permission from the American Chemical Society.

TABLE 4.1
Sample Injection Scheme for OCEC/CE in Figures 4.16 and 4.17

	Electric Voltages (kV)					
	S	B1	SW1	B2	SW2	W2
a	9.5	10.0	5.0	3.0	2.5	0.0
b (0.5 s)	9.5	8	5.0	3.0	2.5	0.0
c (0.2 s)	9.5	10.0	5.0	Open	Open	0.0

4.2 HYDRODYNAMIC INJECTION

To avoid sampling bias that is inevitable in EK injection, hydrodynamic injection (due to liquid pressure difference) is used. This can be implemented even after EK loading (see Figure 4.18). As soon as the sample was EK loaded up to the intersection junction, the voltage was turned off for hydrodynamic injection to occur. The sample plug was truncated when voltage was turned on again. If a positive voltage was applied, the injection procedure worked only for cations and slowly migrating anions (because of the need of EK loading). For the injection of other anions (e.g., Cl^-, $Cr_2O_7^{2-}$), negative voltages should be applied.[746]

A similar HD injection procedure was also developed for CEC separation of 3-fluorescein isothiocyanate (FITC)-labeled peptides (see Figure 4.19). This HD injection has resulted in less biased amounts of the faster-moving components (see Figure 4.19b), compared to the case when EK injection was used (see Figure 4.19a).[566]

Another variation is using a continuous HD flow for both loading and injection, but using an EK gating flow (see Figure 4.20). As long as the gated voltage is greater than a critical value, the HD flow is prevented from injecting downstream until needed. Injection is implemented by interrupting the EK gating flow for a short time.[567] In a subsequent report, the structure shown in Figure 4.20 was used to carry out full HD injection. The flow-through inlet–outlet channel allows for uninterrupted sampling.[568]

Pure HDF injection can be achieved for sample introduction by using vacuum suction. The sample was sucked through an inserted capillary into a short section of microchannels between three ports. Different amounts could be selected by filling different sections of the microchip.[813] Pressure injection of a DNA sample was also achieved via a transfer capillary sequentially to a five-channel microchip. The use of one capillary allowed for automated sampling necessary for continuous monitoring of an enzymatic DNA restriction digestion experiment.[320]

A microchip was constructed to sequentially load and inject samples from a ninety-six-well plate for electrospray ionization–mass spectrometry (ESI-MS) analysis. As shown in Figure 4.21, a sampling capillary is inserted in the guide tube (P) on the left side by applying a vacuum at the sample loading port (Q). The injection sequence is also shown in Figure 4.22. After loading the sample, a short vacuum pulse was applied to the sample injection port (R) to fill the sample loop (Figure 4.22b). Then the vacuum turned off, and the voltage was applied across the serpentine separation channel (i.e., between T and the liquid junction) for CE separation and subsequent ESI-MS analysis (Figure 4.22c). After analysis, all the channels are washed by applying positive pressure on the liquid reservoir (T, U) and negative pressure on the liquid junction (Figure 4.22d). The microchip is ready for the analysis of another sample on the microwell plate.[296]

Like gated EK injection, gated HD injection was achieved (see Figure 4.23). Sample injection volumes from 0.5 to 10 nl were made in real time.[569] HD injection conducted in the pinched and gated modes was achieved on a PET chip using three syringe pumps and an eight-port rotary valve.[570]

HD injection can be achieved using a pressure pulse (2.5 to 3 kPa) generated by the electromagnetic actuation of a flexible poly(dimethylsiloxane) (PDMS) membrane placed on the sample reservoir (see Figure 4.24). Sample plug volume was easily increased by increasing the pulse duration. Nine consecutive injections for pulse durations of 0.3, 0.35, and 0.4 s are shown in Figure 4.25.

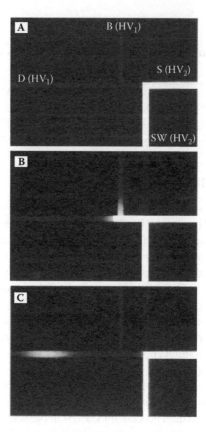

FIGURE 4.18 Fluorescent micrographs of the EK loading and HD injection process. (a) EK loading is achieved when the high-voltage sources HV1 and HV2 are switched on. The sample is transported from the sample reservoir (S) to the sample waste (SW), and run buffer from the reservoir (B) to the detector (D). (b) For HD injection, both HV sources are switched off for ~3 s. Sample flows in due to hydrodynamic flow. (c) The sample plug is introduced into the separation channel by switching back on the HV sources as in (a).[746] Reprinted with permission from the American Chemical Society.

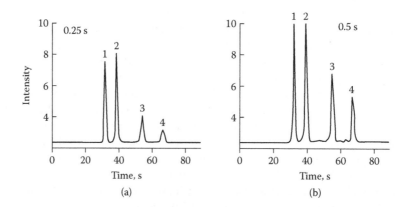

FIGURE 4.19 Microchip separation of an FITC-labeled synthetic peptide mixture following (a) electrokinetic injection (0.25 s) and (b) diffusion-based injection (0.5 s). Mobile phase: 1 mM carbonate buffer (pH = 9.0); separation electric field: 300 V/cm; 1, FITC-Gly-Phe-Glu-Lys-OH; 2, FITC-Gly-Phe-Glu-Lys(FITC)-OH; 3, FITC; 4, FITC-Gly-Tyr-OH. Analyte concentrations: $1 + 2 = 4 = 50 \mu M$.[566] Reprinted with permission from the American Chemical Society.

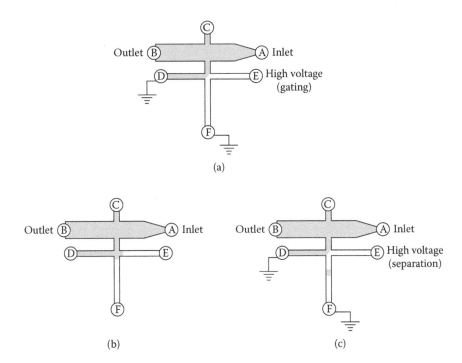

FIGURE 4.20 HD loading, HD injection using EK gating. (a) Gating voltage on, no sample loaded; (b) gating voltage off, sample loaded; (c) gating voltage back on, sample injected.[567] Reprinted with permission from the American Chemical Society.

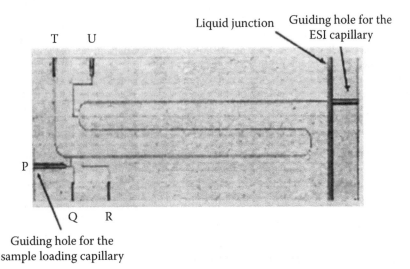

FIGURE 4.21 Photograph of the glass microchip (5 × 2 cm) used for sample injection, separation, and interfacing into the MS system. To minimize the diffusion loss of the sample during separation, the connection between the side channels (leading from Q, R, T, U) and the serpentine separation channel (75 μm deep) was etched to 25 μm (one-ninth of the cross section area of the separation channel).[296] Reprinted with permission from the American Chemical Society.

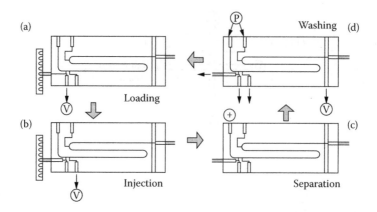

FIGURE 4.22 Sequence of the operation cycle of the microchip: A, loading; B, injection; C, separation; D, washing. V, vacuum; P, pressure; +, separation voltage.[296] Reprinted with permission from the American Chemical Society.

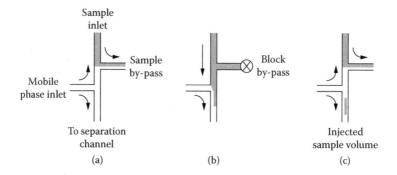

FIGURE 4.23 Sample injection sequence on a PDMS chip. (a) The default state of the analyzer. (b) A sample injection being loaded by interrupting the flow out of the sample bypass, using a valve created in-house. The sample stream backs up and begins flowing down the main channel. (c) Reinstating the flow out of the bypass, the injection pulse is terminated, and returns the system to its default conditions. This technique has been used to make injection pulses as small as 0.5 mm in length (0.5 nl volume).[569] Reprinted with permission from Elsevier Science.

Reproducibilities of 1.6, 1.1, and 0.4%, respectively, are demonstrated. Resolution was enhanced when the pulse duration was longer, as shown by the separation of calcein derivatives.[314]

Another HD injection on a Si-glass chip similar to EK injection shown previously in Figure 4.13 was reported.[571] HD injection has also been used in various reports.[572–575]

4.3   OTHER SAMPLE INJECTION METHODS

An optically gated injection was demonstrated for the CZE separation of four amino acids labeled with 4-chloro-7-nitrobenzofurazan (NBD-F) in a one-channel chip[576] or a four-channel chip.[577] The gating beam was used to continuously photobleach the sample, except for a short time during injection, by interrupting the beam (100 to 600 ms) using an electronic shutter. With only a sample reservoir and a waste reservoir, the sample continuously flowed electrokinetically. Six consecutive separations of the same sample mixture have been accomplished in under 30 s.[576,577]

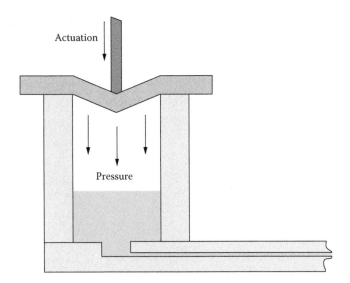

FIGURE 4.24 Schematic view of the mechanical actuation of the PDMS membrane on the sample reservoir for pressure pulse injection.[314] Reprinted with permission from the American Chemical Society.

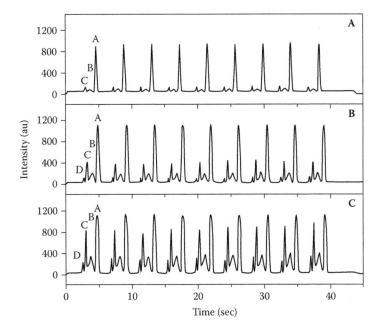

FIGURE 4.25 Sequences of nine consecutive pressure pulse injections with 4 s delay using a mixture of 38 μM calcein and 19 μM fluorescein (a). Species B, C, and D are derived from calcein. The durations of pressure pulse are (a) 0.3, (b) 0.35, and (c) 0.4 s.[314] Reprinted with permission from the American Chemical Society.

4.4 PROBLEM SETS

1. What advantages does HD injection provide over EK injection for sample introduction?[314] (2 marks)
2. Describe three operation modes of EK injection. (3 marks)
3. Describe two types of HD injections with different loading methods. (2 marks)

4. In sample introduction for capillary gel electrophoresis (CGE) separation of oligonucle-otides, a hydrodynamic method cannot be used, and an electrokinetic method must be used. Why? (2 marks)

5 Sample Preconcentration

Several sample preconcentration strategies have been devised to enrich dilute samples. The microfluidic chip has the advantage to integrate various microstructures to achieve sample preconcentration by stacking, extraction, or other methods.

5.1 SAMPLE STACKING

Sample stacking, which is based on a lower electrical conductivity in the sample buffer relative to the run buffer, has been achieved on-chip for sample preconcentration.[578–581] A tenfold enhancement in detection signal was reported for the samples prepared in water, compared to those prepared in the run buffer.[560]

Although stack injection has been employed previously,[316,582] the benefit of stacking for sample preconcentration was only studied in detail later.[346] With the sample buffer (0.5 mM) at a tenfold lower conductivity than the separation buffer (5 mM), simple EK stacking using the gated injection was observed. This was applied to the separation of dansylated amino acids (dansyl-lysine, didansyl-lysine, dansyl-isoleucine, and didansyl-isoleucine).[346]

It was found that sample stacking did not improve further if the dilution factor in the sample buffer (relative to the run buffer) was more than 350-fold. This is because of a loss of resolution due to parabolic flow formed from the difference in the electroosmotic flow (EOF) between the sample buffer (faster EOF) and the run buffer (slower EOF).[578]

Owing to the EK bias in favor of faster-migrating species in stack injection, the signal enhancement ranged from 31 to 8. However, stack injection produced less resolution, i.e., fewer plate numbers (N), compared to those obtained in pinched injection. In addition, relative standard deviations (RSDs; $n = 6$) in peak areas for stack, nonstacked, and pinched injection are 2.1, 1.4, and 0.75%, respectively.[346]

Stacking of a neutral analyte can also be achieved, but only when a high-salt (NaCl) sample buffer is employed. The co-ion (Cl$^-$) should be present in the sample buffer at a concentration sufficiently higher than that of the electrokinetic vector ion (cholate). This causes the formation of a pseudo-steady-state co-ion boundary that forces an increased concentration of cholate near the boundary, leading to the stacking effect.[583]

Full-column stacking with subsequent sample matrix removed by polarity reversal was achieved using a preconcentration channel coupled to the separation channel (see Figure 5.1). First, the preconcentration column is completely filled with the sample by applying a negative voltage at the stacking waste reservoir and ground at the sample reservoir. Then the electric field is reversed to remove the sample matrix toward the sample waste reservoir. During this step, the anions are dragged slowly and hence stacked. Finally, the stacked sample plug leaves the preconcentration channel and enters the separation channel (as monitored by current measurement), and capillary electrophoresis (CE) separation starts. This results in a 65-fold preconcentration when the sample buffer is 1/160-fold as diluted as the run buffer.[578]

To enhance the preconcentration factor in field-amplified sample stacking, the conductivity gradient boundaries should be stabilized. To achieve this, a photoinitiated porous polymer structure was formed near a double-T injector in a glass chip, as shown in Figure 5.2. The main function of the porous polymer is to maintain the boundary between the high- and low-conductivity buffers previously loaded hydrodynamically (see Figure 5.2a and b). Therefore, during EK injection (see Figure 5.2c), the boundary is not disturbed. Separation of fluorescein and BODIPY is subsequently carried out (see Figure 5.2d). In this manner, a signal enhancement factor of 1,110 was obtained (see

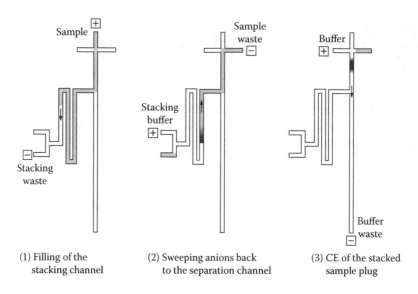

FIGURE 5.1   The three process steps for full-column sample stacking in a coupled column configuration fabricated on a Pyrex chip. A sample of 100 nM of FITC-Arg and FITC-Cly was used.[578] Reprinted with permission from Wiley-VCH Verlag.

Figure 5.3).[584] This enhancement is a factor of 10 over previously reported on-chip field-amplified sample stacking.[578,585]

Another example of field-amplified sample stacking, which occurs during pinched injection, has been described in Chapter 4, Section 4.1.1.

5.2   EXTRACTION

Both liquid-phase and solid-phase extractions have been used for sample preconcentration or cleanup. For instance, an octadecyltricholorosilane (OTS)-coated glass channel was employed for liquid-phase extraction, leading to an enrichment of samples (the neutral coumarin dye C460). It was found that the dye (8.7 nM) was enriched by eightyfold (in 160 s). Acetonitrile (15%) was used for the loading and enrichment in the liquid-phase extraction procedure. Subsequently, 60% acetonitrile was used for elution.[586]

Sample preconcentration was also achieved based on solid-phase extraction. This was carried out on ODS-coated silica beads (of diameter 1.5 to 4 μm) trapped in a cavity (10 μm deep) bound by two weir-type structures (9 μm high) (see Figure 5.4). Concentration enhancement up to five hundred-fold was demonstrated using a nonpolar analyte (BODIPY) eluted by acetonitrile.[587,639]

In another report, an on-chip chamber containing a capture matrix was used for sample cleanup. A 2 μl unpurified sample is concentrated into a smaller volume (~10 nl), and this yields a volumetric concentration factor of ~200. Longer residence time can be achieved by using a doubly tapered chamber. Here, the electric field is tenfold lower within the chamber than within the tapered channel. So, a high-field (fast) sample introduction is followed by a low-field (slow) sample flow past the capture matrix, increasing the residence time so that the binding kinetics are dominant over electromigration.[971]

To provide a stable and reproducible solid phase, the silica beads that were packed into glass microchips should be immobilized. This was achieved by using a sol-gel (from tetraethoxysilane [TEOS]) to act as the interparticle glue. The immobilized beads were employed to purify DNA from human whole blood and bacterial samples.[331]

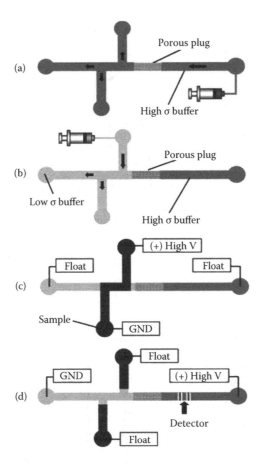

FIGURE 5.2 Schematic of a CE assay protocol with field-amplified sample stacking. (a) High-conductivity buffer is injected from the east reservoir, through the porous structure, and into all channels. This first step takes longest (about 2 min) because of the high fluidic resistance of the porous plug. Arrows show the direction of pressure-driven flow. (b) Low-conductivity buffer is introduced from the west reservoir at a flow rate of approximately 0.1 µl/min for 0.5 min. Here, the porous structure provides high fluidic resistance, which minimizes the mixing of two buffers at the upstream plug–buffer interface. (c) Sample is then electrokinetically loaded into the double-T injector. Negatively charged sample ions electromigrate from the south reservoir (grounded) to the north reservoir (1 kV). (d) Stacking, separation, and detection of samples in the separation channel. The separated sample bands are depicted in the figure as two rectangles near the detector.[584] Reprinted with permission from Wiley-VCH Verlag.

Other than beads, porous polymer monoliths, which were photopolymerized in a COC chip, were used for solid-phase extraction. It is known that priming polymeric surfaces is not as simple as priming silica surfaces, which use a common surface primer agent, 3-(trimethyoxysilyl)propyl methacrylate (TMPM). Therefore, the grafting method as initiated by UV should be used to attach the polymer monoliths.[588] A similar strategy was used for sample preconcentration of polyaromatic hydrocarbon (PAHs; e.g., pyrene). Pyrene (900 nM) was first concentrated by four-hundred-fold in 24% acetonitrile (ACN) before switching to 56% ACN for CEC separation (see Figure 5.5).[148]

A monolithic hydrophobic polymer formed by photoinitiated polymerization for on-chip solid-phase extraction is shown in Figure 5.6. The polymer mixture includes butyl methacrylate (BMA) and ethylene dimethacrylate (EDMA), with the pore size controlled by the composition of the hexane–methanol porogenic mixture. The degree of preconcentration depends on the flow rate, as shown in the preconcentration of green fluorescent protein (GFP) at three flow rates (see Figure 5.7).

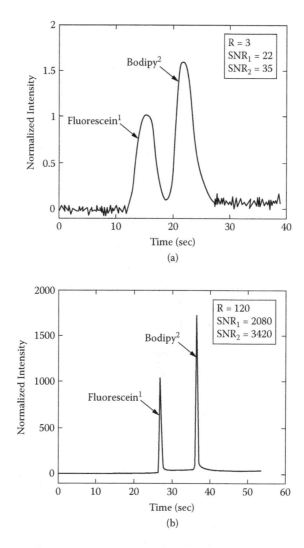

FIGURE 5.3 Electropherograms of fluorescein and BODIPY separations (a) without and (b) with field-amplified sample stacking. Fluorescence signal is normalized with exposure time in both plots. The position of the detector is 10 mm from the downstream channel intersection of the chip. The signal increase is 1,100-fold for the stacked case, and resolution increases from 3 to 120.[584] Reprinted with permission from Wiley-VCH Verlag.

The factors of preconcentration were 355, 756, and 1,002 for the flow rates of 3, 1.03, and 0.53 μL/min, respectively.[342]

Sample preconcentration was also achieved for gaseous samples. A flow-through Pyrex chip consisting of the silica-based solid absorbent was used for preconcentration of the benzene, toluene, and xylene (BTX) gaseous mixture. A thin-film heater was used to desorb the adsorbed gas molecules, which flowed downstream for UV absorbance detection. An LOD of 1 ppm (toluene) was achieved, compared to 100 ppm without preconcentration.[131,715] With an additional air-cooled cold-trap channel located after the adsorbent region, the desorbed molecules were prevented from being diluted before reaching the detection cell, and a further improvement of LOD of toluene to 0.05 ppm was achieved.[589]

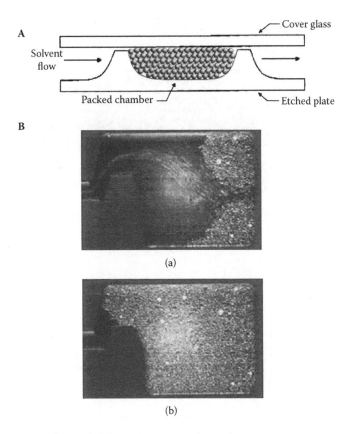

A

Cover glass

Solvent flow

Packed chamber — Etched plate

B

(a)

(b)

FIGURE 5.4 (A) Drawing of cross section of a packed chamber, showing weir heights in relation to channel depth and particle size. (B) Images of the chamber (a) at an intial stage of EK packing and (b) after it is completely filled with beads.[639] Reprinted with permission from the American Chemical Society.

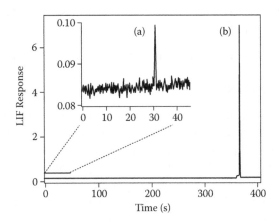

FIGURE 5.5 Chromatograms of (a) 1 s injection without sample preconcentration and (b) 320 s injection with concentration. The sample was 900 nM pyrene.[148] Reprinted with permission from the American Chemical Society.

FIGURE 5.6   SEM image of the monolithic polymer in a borofloat glass chip.[342] Reprinted with permission from the American Chemical Society.

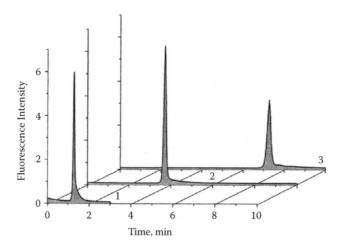

FIGURE 5.7   Elution of green fluorescent protein from the hydrophobic monolithic concentrator. Conditions: Loading, 200 µl of 18.5 nmol/L protein solution in 8 mmol/L Tris-HCl buffer (pH 8) containing 0.95 mol/L ammonium sulfate; flow rate, 3 µl/min; elution with 1:1 acetonitrile/water at a flow rate of 3 (1), 1.03 (2), and 0.53 µl/min (3).[342] Reprinted with permission from the American Chemical Society.

5.3   POROUS MEMBRANE

A porous membrane structure (made of sodium silicate) was fabricated to preconcentrate large molecules such as DNA. This structure (of width 3 to 12 µm) allows electric current to flow, while preventing large molecules from passing through (see Figure 5.8). This is a physical stacking method based on a reduction in the sample volume by removing small solvent molecules.[590] With this method, preconcentration of PCR products (199 bp) has been achieved (see Figure 5.9). The product was obtained in twenty PCR cycles with a starting DNA template copy number of 15 in an injection volume of 80 pl. In fact, as few as ten PCR cycles, with a short analysis time of less than 20 min, can be achieved.[925] More details on PCR are described in Chapter 9, Section 9.2.1.

Another common sample preconcentration method is dialysis, which serves to remove small molecules. For instance, affinity dialysis and preconcentration of aflatoxins were achieved in a copolyester chip (see Figure 5.10). After affinity binding to the aflatoxin B_1 antibody, various aflatoxins (B_1, B_2, G_1, G_2, G_{2a}) in a sample were retained, while the other small molecules passed through a poly(vinylidene fluoride) (PVDF) dialysis membrane. Thereafter, the sample solution was

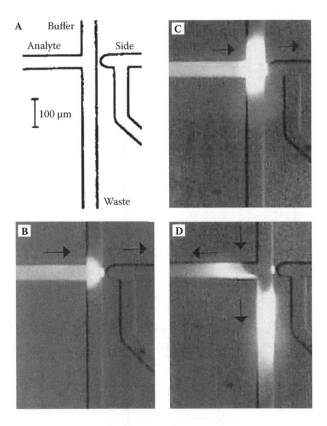

FIGURE 5.8 Schematic of the injection tee and porous membrane structure is shown in (a). Porous membrane region width is 7 μm. CCD images of analyte concentrated for (b) 2 min and (c) 3 min. Injection of concentrated analyte plug is depicted in (d). All channels are filled with 3% LPA in 1× TBE buffer. DNA sample: 25 μg/mL φX174-*Hae*III digest with 6.0 μM TO-PRO dye added.[590] Reprinted with permission from the American Chemical Society.

exposed to a countercurrent flow of dry air, leading to water evaporation and analyte concentration. The concentrated and desalted sample was used in subsequent MS analysis.[821] More details for MS analysis are described in Chapter 7, Section 7.3.

Countercurrent dialysis was achieved through a dialysis membrane sandwiched between a laser-ablated polycarbonate (PC) sheet for sample flow (0.5 μl/min) and a polyimide sheet for buffer flow (100 μl/min) (see Figure 5.11). This device produced a desalted sample (horse heart cytochrome *c*) for subsequent MS analysis (see Figure 5.12).[811]

Microdialysis was achieved in a fused silica chip with *in situ* photopatterned porous membrane, as shown in Figure 5.13. Phase-separation polymerization of the membrane (7 to 50 μm thick) was formed between posts. The posts maximize the mechanical strength of the membrane so that it can withstand a pressure drop of ~1 bar. A low molecular weight cutoff (MWCO) membrane, which can be formed by using less organic solvent, 2-methoxyethanol, appears to be more transparent (see Figure 5.13). This low MWCO membrane can be used to dialyze away low MW molecules, such as rhodamine 560, but not fluorescently labeled proteins (insulin, BSA, anti-biotin, and lactalbumin). A high MWCO membrane, which was formed by more organic solvent, allows diffusion of lactalbumin.[347]

A nanocapillary array membrane was sandwiched between two PDMS layers to achieve preconcentration (see Figure 5.14).[591] The array is a PC nuclear-track etched (PCTE) membrane. The membrane is ~10 μm thick and consists of 200 nm pores that are coated with poly(vinylpyrrolidone) (PVP). Before use, the array was conditioned for at least 5 min by applying ~10 V to induce a buffer flow through the pores. Gated injection of fluorescein isothiocyanate (FITC)-labeled amino acids

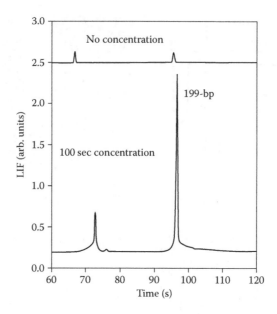

FIGURE 5.9 Electrophoretic analysis without and with 100 s preconcentration of PCR product (199 bp), which was obtained after twenty PCR cycles.[925] Reprinted with permission from the American Chemical Society.

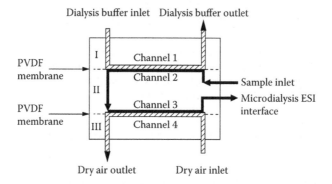

FIGURE 5.10 Side-view schematic of miniaturized affinity dialysis and preconcentration system. I, II, and III indicate top, middle, and bottom imprinted copolyester pieces, respectively. Piece II is imprinted on both sides. Two PVDF membranes separate the copolyester channels.[821] Reprinted with permission from the American Chemical Society.

was achieved by biasing both ports (P, Q) at the upper injection channel vs. the waste port (S) at the bottom separation channel. Separation occurs when R is biased vs. S in the separation channel. Loading and injection are shown in Figure 5.15. The volume at the gating intersection (75 pl) defines the smallest sample plug volume injected. However, the volume can be increased by increasing the injection time (0.3 to 7.5 s). For instance, an injection volume of ~200 pl is obtained by a 0.5 s injection. Reproducible injection and separation are observed, as shown in Figure 5.16.

The injection process is also monitored graphically (Figure 5.17). The detection point was located downstream of the separation channel (see Figure 5.14 for notations). Upon application of +40 V at S, buffer flows back from S and dilutes the fluorescence background, showing a negative peak around 50 s (see Figure 5.17, right). Upon application of –40 V at S, fluorescein flows to S, showing a positive peak around 150 s. This phenomenon is similar in the case of using a neutral

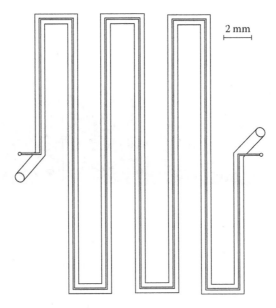

2 mm

FIGURE 5.11 Schematic representation of a dialysis chip that is formed by sandwiching a dialysis membrane (MW cutoff 8,000) between the sample channel (160 μm wide) and buffer channel (500 μm wide) after alignment of the microchannel device.[811] Reprinted with permission from the American Chemical Society.

dye (BODIPY), indicating the microfluidic transport in the microfluidic PDMS channels (with negatively charged surface) is dominant over nanofluidic transport in the nanocapillary array (with positively charged surface). On the other hand, if the nanopore size decreased to 15 nm, the flow of fluorescein occurred when S was biased at +40 V. This suggests a dominant nanofluidic transport due to a greater voltage drop across the smaller nanopore. A modeling study shows a ~25% voltage drop across the nanocapillary array consisting of 15 nm pores, compared with <2% for 200 nm pores.[592] Such a nanocapillary array membrane has also been used for fluidic switching.[593]

Preconcentration of anionic FITC-labeled peptides (human necrofibrin, melanocyte-stimulating hormone (MSH)-release inhibiting factor) was also carried out with the nanocapillary array membrane. It was found that negatively charged fluorescein was concentrated by three-hundred-fold, whereas positively charged R6G was not retained at the membrane and no concentration was observed. This is consistent with the fact that the diffuse layer on the interior of the nanocapillaries is negatively charged, which repels and retards the flow of fluorescein, leading to its concentration in front of the membrane.[358]

Preconcentration of proteins was achieved by a nanoporous membrane formed in a fused silica chip. Subsequent CE separation was conducted to achieve a concentration factor of 130 for BSA and 160 for phosphorylase b.[594]

5.4 OTHER PRECONCENTRATION METHODS

One preconcentration technique is based on sample volume reduction due to flow confinement. A sample flow through a channel (20 μm deep) can be confined by a perpendicular makeup flow into a thin layer (of thickness 2 to 6 μm) above the sensing area (see Figure 5.18).[694] The fabrication of the PDMS three dimensional (3D) microchip was previously described.[181] This 3D microfluidic confinement reduces the sample volume, increases the flow velocity, and benefits any mass-transport-limited processes, such as heterogeneous immunoassays. The degree of confinement and hence preconcentration can be adjusted through the volume flow rate of the confining flow.[595,694] For the 156 μl/min whole channel experiment and the corresponding 25:1 confinement experiment, the consumed sample volume reduced from 1.9 ml

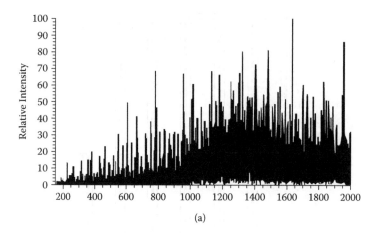

(a)

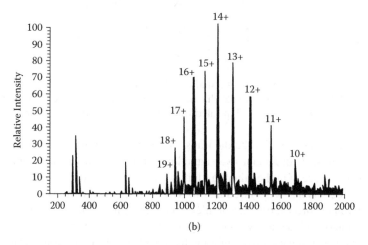

(b)

FIGURE 5.12 ESI-MS of 5 µM horse heart myoglobin in 500 mM NaCl, 100 mM Tris, and 10 mM EDTA by direct infusion. (b) ESI-MS of previous myoglobin sample after desalting, which was achieved by on-line microdialysis using 10 mM NH$_4$OAc and 1% acetic acid as dialysis buffer.[811] Reprinted with permission from the American Chemical Society.

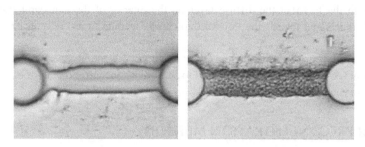

FIGURE 5.13 Left: Low MWCO membrane (deionized water:2-methoxyethanol = 3.7:1). Right: High MWCO membrane (deionized water:2-methoxyethanol = 0.34:1). The post diameter is ~50 µm. For phase-separation polymerization, the monomer is 2-(N-3-sulfopropyl-N,N-dimethylammonium) ethyl methacrylate, the cross-linker is methylene bisacrylamide, and the plotoinitiator is 2,2'-azobis(2-methylpropanimidamide dihydrochloride). To prevent unwanted polymerization that may occur by heat and molecular diffusion outside the UV-irradiated region, a polymerization inhibitor, hydroquinone, is also added. To facilitate covalent attachment of the porous membrane to the silica surface, it is first coated with 3-(trimethoxysilyl)propylacrylate.[347] Reprinted with permission from the American Chemical Society.

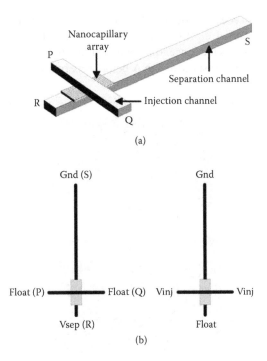

FIGURE 5.14 (a) Schematic of a three-dimensional gated-injection separation device consisting of two crossed microfluidic channels with a PCTE membrane interconnect. (b) Electrical bias configurations for active electrokinetic injection control. Right: Nanocapillary array gated injections. Left: Main channel separations.[591] Reprinted with permission from the American Chemical Society

to 80 μl (i.e., a twenty-four-fold reduction), but the reaction completion time was very similar, presumably due to the diffusion loss of the sample into the confinement flow. This could be alleviated by using a high-density aqueous phase (e.g., glycerol solution) or an organic phase as the confining flow.[694]

An H-filter was constructed to remove blood cells (see Figure 5.19). Subsequent off-chip high-performance liquid chromatography (HPLC) analysis was conducted to determine the amount of an antibiotic, cephalosporin cephradine (0.2 to 100 μg/ml), in blood.[596]

Another preconcentration method is based on temperature gradient focusing (TGF) (see Figure 5.20). The focusing effect is caused by the balance of the electrophoretic flow against the bulk EOF flow. Since electrophoretic flow depends on the temperature gradient (cf. pH gradient in IEF; see Chapter 10, Section 10.2), by using an appropriate buffer, a balance is attained at a certain temperature. TGF has been demonstrated with various ionic species (Oregon Green 488 carboxylic acid, Cascade Blue hydrazide, FQ-labeled aspartic acid, CBQCA-labeled serine and tyrosine, GFP, TAMRA-labeled 20-mer oligonucleotide, and fluorescently labeled PS particles; see Figure 5.20). This focusing effect leads to a preconcentration factor of 10,000 or greater (from 8 nM to 90 μM for Oregon Green).[597]

The temperature gradient can also be achieved at the junction of two microchannels of two different cross-sectional areas in the presence of an electric field. Since there is a higher current density in the narrower channel than in the wider channel, it is hotter in the narrower channel.[597]

Although TGF can be applied to a capillary, this method has not been first demonstrated in the capillary format, but in the microfluidic format. This is mainly because of the ease of temperature gradient implementation on the planar chip surface and the ease of visualization through the optically transparent planar glass substrate. TGF appears to complement common preconcentration strategies, which suffer from various disadvantages. For instance, in conventional methods, multiple buffers (for stacking) are required, and reproducible surface modification (for extraction) is usually needed. Moreover, in dialysis, preconcentration is applied only to large molecules or particles; in

IEF, pI values may not be accessible by the pH gradient and many proteins at their pI values are insoluble; in the stacking method, electrode products are generated. However, one limitation of TGF is that it cannot be used to focus (or concentrate) any neutral analyte with zero electrophoretic flow.

A pressurization technique for stacking DNA before separation will be described in Chapter 9, Section 9.4.3.

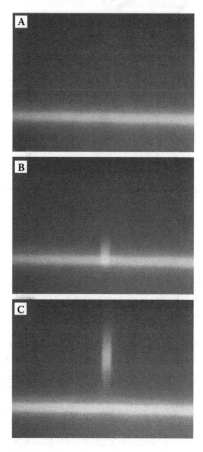

FIGURE 5.15 Fluorescence image series for gated injection of an analyte (5 µM fluorescein in 5 mM phosphate buffer, pH 7.8) band across the PCTE membrane consisting of 200 nm diameter capillaries: (a) before injection with separation voltage bias (400 V) applied, (b) during injection voltage bias (400 V, 0.5 s), and (c) after switching back to separation voltage bias (~1 s after injection). Images were captured from a video recording of a 100-repetitive-injection experiment. Fluorescence from the injection channel (horizontal) appears with artifacts associated with being out of focus as a result of being above the focal zone of the separation channel (vertical).[591] Reprinted with permission from the American Chemical Society.

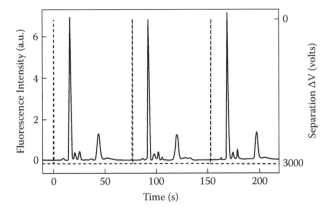

FIGURE 5.16 Reproducibility for the three consecutive separations achieved after gated injection through the PCTE membrane. Separation of FITC-labeled arginine (2.5 µM, first major peak) and tryptophan (7.5 µM, second major peak) was achieved in 5 mM phosphate buffer (pH 7.8). Fluorescence intensity (left axis) in the detection region is depicted as the solid line. Detection was performed 2.0 cm downstream of the gated injection area. The applied separation bias (V_{sep} or ΔV), which is either 0 or 3 kV, is depicted as the dashed line. Injection bias (V_{inj}) was applied concurrent with the separation bias being switched to 0 V, thus marking the 0.3 s injection times. The two smaller peaks at intermediate retention times are identified as impurities associated with FITC labeling.[591] Reprinted with permission from the American Chemical Society.

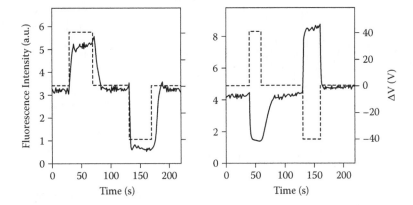

FIGURE 5.17 Migration of a probe across the PCTE membrane consisting of capillary arrays with 15 nm (left) and 200 nm (right) pore diameters connecting two 100 µm wide PDMS channels. Fluorescence intensity (solid line, left axis) and applied bias, ΔV (dashed line, right axis), are plotted as a function of time to monitor the transport of 0.17 µM fluorescein in 5 mM (pH 8) phosphate buffer.[592] Reprinted with permission from the American Chemical Society.

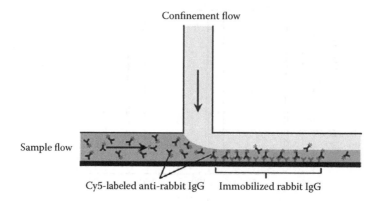

FIGURE 5.18   Illustration of the flow confinement concept. In microchannels, a sample flow is joined with a confinement flow (e.g., water or sample medium) in a perpendicular orientation. Under laminar flow conditions, no mixing occurs. The sample flow is confined into a thin layer of higher velocity. For immunoassay application, rabbit IgG is immobilized on a planar waveguide and Cy5-labeled anti-rabbit IgG is introduced as analyte in the sample flow.[694] Reprinted with permission from the American Chemical Society.

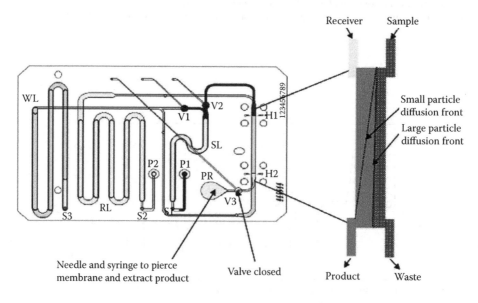

FIGURE 5.19   The H-filter cartridge was constructed on a polyester chip for blood cell removal. The inset illustrates the diffusive mass transfer. First, to prime the device: (1) The pneumetic valve V2 was closed and blood sample placed in P1 was pumped (via S1) to fill SL up to V2, and the receiver solution placed in P2 was pumped (via S2) to fill RL up to H1. (2) Then V2 was opened (V1 is closed), the blood sample was pumped to reach H1 (the start of the H-filter; see inset), receiver solution was pumped from H1 to H2, and the waste loop (WL) was backfilled (via S3) to H2. Second, to start the process of solution flow and diffusive mass transfer, solution pushing (via S1 and S2) and pulling (via S3) was carried out. The products went to PR and blood cells (which diffused to a lesser extent) went straight to WL.[596] Reprinted with permission from Elsevier Science.

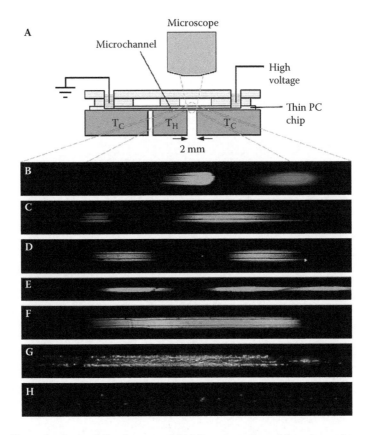

FIGURE 5.20 **(See color insert following page 144.)** Demonstration of focusing and separation of a variety of different analytes using TGF on a PC chip. (a) Schematic drawing of the apparatus. The temperature gradient (~25°C/min) was achieved by two end cool Cu blocks (T_C = 10°C using cold water) and one middle hot Cu block (T_H = 80°C using an embedded resistive heater). For the study of GFP, the hot temperature was lowered to 60°C to avoid denaturing of the protein. (b–g) Fluorescence images of focusing zone for various analytes. Magnification is constant for all images. The total length of each image is 1.9 mm. (b) Oregon Green 488 carboxylic acid and Cascade Blue hydrazide. (c) The two products resulting from labeling of aspartic acid with FQ. (d) Mixture of CBQCA-labeled serine and tyrosine. (e) Green fluorescence protein. (f) Fluorescein- and TAMRA-labeled DNA. (g) Six-millimeter-diameter fluorescently labeled polystyrene particles. (h) Same as (g), but after the channel was rinsed and refilled with the unfocused particle solution.[597] Reprinted with permission from the American Chemical Society.

5.5   PROBLEM SETS

1. Sample stacking is employed in CE separations. In this method, the differences in two types of physiochemical properties between the sample buffer and run buffer are exploited. What are these two properties? (2 marks)

2. Describe two methods to preconcentrate a liquid sample by removing the small molecules in the sample. (2 marks)

FIGURE 3.20

3.3 PROBLEMS

6 Separation

6.1 GAS CHROMATOGRAPHY (GC)

The first miniaturized chip was fabricated on a Si wafer for gas chromatography (GC; see Figure 6.1).[1] Isotropic etch on a 200 μm thick Si <100> wafer (5 cm diameter) produced a spiral GC channel (200 μm wide, 30 μm deep, 1.5 m long). A relatively short capillary length (1.5 m) with an OV-101 stationary phase was used for GC analysis of hydrocarbons. An etched Si miniature valve with a Ni diaphragm (activated by a solenoid plunger) was used for injections. The separation was completed in less than 10 s. However, the plate numbers only range from 385 to 2,300, and n-hexane and chloroform were not resolved. The GC chip was incorporated with a thermal conductivity detector (TCD), which was separately fabricated on another Si chip. It consisted of an etched cavity in order to reduce thermal mass and to shorten the thermal time constant to about 1 ms. The cavity was insulated by thermal SiO_2 (100 nm), sputtered with Pyrex glass (1.5 μm), and patterned with a Ni thin-film (100 nm) resistor.[1]

GC analysis was also carried out on a Si-glass chip using an off-chip flame ionization detector (FID). Figure 6.2 shows the Golay plots (cf. Van Deemter plots) of the n-C_9 peak when both the native and oxidized Si surfaces were used. H_{min} is lower in native Si because of the less polar stationary phase. Figure 6.2 also shows the Golay plots when air and H_2 were used as the carrier gas. As the viscosity of air is greater than H_2, the μ_{opt} obtained when air is used is lower. Isothermal GC separation of a twenty-component hydrocarbon mixture is shown in Figure 6.3. To improve resolution, temperature programming was employed, with the results shown in Figure 6.4.[598] GC separations of alkanes[599–602] and methyl esters[603] have also been performed on chip.

6.2 CAPILLARY ELECTROPHORESIS (CE)

6.2.1 FREE-SOLUTION CAPILLARY ELECTROPHORESIS (FSCE)

Successful liquid-phase separation on-chip was first carried out in CE, because electroosmotic flow (EOF) pumping can be easily achieved in the microscale. For instance, six fluorescein-labeled amino acids are separated by CE on a Pyrex glass chip (10 μm deep and 30 μm wide channel) (see Figure 6.5). Separation was achieved in a very short time of about 15 s.[324] Similar CE separation of calcein and fluorescein was also reported.[582] Separation of a binary mixture of rhodamine B and dichlorofluorescein was even achieved in only 0.8 ms using a short separation length of 200 μm.[604]

An index of separation speed, N/t, was defined, where N is the number of theoretical plates and t is the migration time.[137,340] The theoretical upper limit of 8,500/s for Gln–fluorescein isothiocyanate (FITC) (H = 0.3 μm) is close to the experimental value of 8,300/s.

As in conventional CE, the plate height, H, is given by Equation 6.1:[136,340,547,582]

$$H = H_{diff} + H_{inj} + H_{det} \tag{6.1}$$

The parameters H_{diff}, H_{inj}, and H_{det} denote the plate heights due to longitudinal or axial diffusion, injector length, and detector length, respectively. H_{inj} usually accounts for about 50% of the total H.[547] These three parameters are given in Equations 6.2 to 6.4:[102,340,547,550]

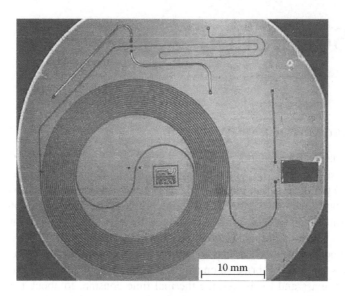

FIGURE 6.1 Photograph of a gas chromatograph integrated on silicon.[1] Reprinted with permission from IEEE.

$$H_{diff} = \frac{2Dt_m}{L} + \frac{2D(t_{inj} + t_{dl})}{L} \tag{6.2}$$

$$H_{inj} = \frac{L_{inj}^2}{12L} + \frac{2D(t_{inj} + t_{dl})}{L} \tag{6.3}$$

$$H_{det} = \frac{L_{det}^2}{12L} + \frac{(\mu E\tau)^2}{L} \tag{6.4}$$

where t_m is migration time; D, diffusion coefficient; t_{dl}, delay time between loading and separation; t_{inj}, injection time; L, distance from the injection to the detection points; L_{inj}, injector length; L_{det}, width of detector window (equals the pinhole size divided by the microscopic magnification); μ, electrical mobility; E, electric field strength; and τ, time constant of the detection system.

To increase the separation channel length within the compact size on a chip, a serpentine channel (165 mm long) within an area of 10 × 10 mm was designed (see Figure 6.6). Unfortunately, band distortion around the channel turns (of radius 0.16 mm) is substantial (see Figure 6.7). This is caused by the slower migration speed of a molecule traveling on the outer track of the channel, compared to the inner track, due to longer migration distance and lower electric field strength. The additional contribution to band broadening is termed H_{geo}, given by Equation 6.5:[136]

$$H_{geo} = \frac{n(\omega\theta)^2}{12L} \tag{6.5}$$

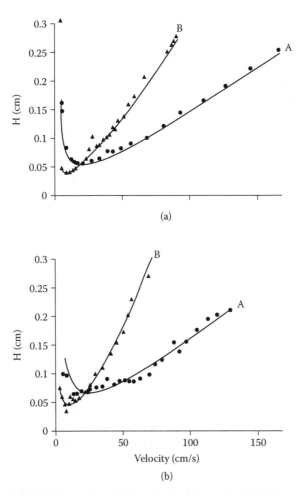

FIGURE 6.2 Height equivalent to a theoretical plate (H) of a GC-separated compound vs. average carrier gas velocity for a 3 m long silicon channel coated with nonpolar dimethyl polysiloxane stationary phase: (a) native silicon surface, (b) oxidized silicon surface. Plots A are for hydrogen carrier gas, and plots B are for air carrier gas.[598] Reprinted with permission from the American Chemical Society.

where n is the number of turns, Γc is the turn radius, and θ is the turn angle. For a channel consisting of one 90° turn and four 180° turns, H_{geo} is calculated to be 0.85 μm.[136] With this additional term, Equation 6.1 is modified to give Equation 6.6:

$$H = \underset{0.23\,\mu m}{H_{inj}} + \underset{0.25\,\mu m}{H_{det}} + \underset{0.85\,\mu m}{H_{geo}} + H_{diff} \tag{6.6}$$

It is found that H_{geo} is a significant portion of the solute-independent plate heights (sum of first three terms); see Equation 6.6.[136]

During CE separation, there is Joule heating due to the flow of electric current. This effect has been studied by examining the nonlinearity in a plot of current vs. E.[107,352] CE current could be determined by measuring the voltage drop across a 100 kΩ resistor (typically less than 1/1,000 the microchannel resistance).[193] Deviation from linearity appeared when E was greater than 720 V/cm. At this value, the electrical power was around 2.3 to 3.8 W/m. This value was a twofold improvement over 1 W/m typically found in fused silica capillary.[102]

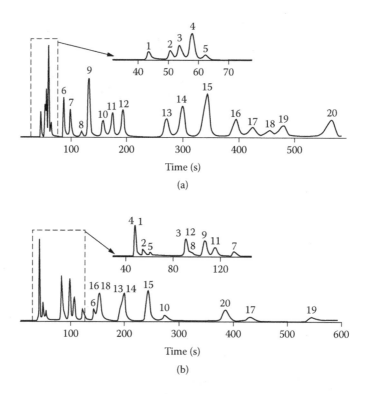

FIGURE 6.3 Isothermal gas chromatograms at 22°C of the twenty-component hydrocarbon mixture with air as carrier gas using channels coated with (a) the nonpolar and (b) the moderately polar stationary phases. For (a), the native silicon column was used, and for (b), the oxidized silicon column was used. The insets show the early-eluting peaks on expanded timescales.[598] Reprinted with permission from the American Chemical Society.

The additional plate height contribution due to Joule heating is given by Equation 6.7:[352]

$$H_{\text{Joule}} = \frac{7 \times 10^{-9}\,\mu E^5 d^6 \lambda^2 C^2}{D\kappa^2} \tag{6.7}$$

where d is the channel depth, λ is the molar conductivity of the buffer, C is the buffer concentration, and κ is the thermal conductivity of the buffer.

Using Equation 6.7, the calculated H_{Joule} value is insignificant compared to the observed value, even though the power becomes greater than 1 W/m when E is greater than 580 V/cm.[352] The discrepancy may be caused by higher H_{Joule} values in narrower adjacent sample channels, causing the flow of heated fluid into the separation channel.[193,339]

Capillary zone electrophoresis (CZE) separation of metal ions (Zn, Cd, Al), which were 8-hydroxyguinline-5-sulfonic acid (HQS) derivatized, was achieved in a microchip. To enhance separation resolution without using an excessively long channel or high voltage, EOF was eliminated by coating the channel wall with polyacrylamide, and separation was carried out in the cathodic mode. The use of a fused quartz substrate resulted in a reduction in fluorescent background, leading to a limit of detection (LD) of 30 to 57 ppb.[352]

CZE separation of sugar isomers in a poly(methylmethacrylate) (PMMA) chip was enhanced by using a phosphate-based, rather than a borate-based buffer. The resolution enhancement may be caused by a rapid carbohydrate–phosphate complex formation process, compared to a slow carbohydrate-borate complex formation process. To reduce adsorption of the labeled sugar isomers to

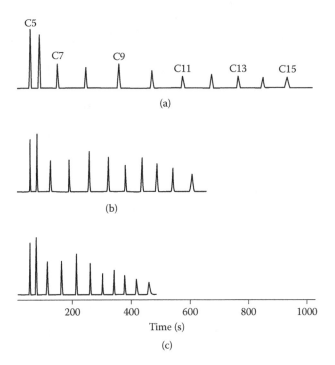

FIGURE 6.4 Gas chromatograms of a mixture containing C5–C15 normal alkanes using temperature-programming rates of (a) 10, (b) 20, and (c) 30°C/min. The silicon column was native nonpolar, and the carrier gas was air. In all cases, the initial column temperature was 30°C, and the temperature program was initiated at the time of injection.[598] Reprinted with permission from the American Chemical Society.

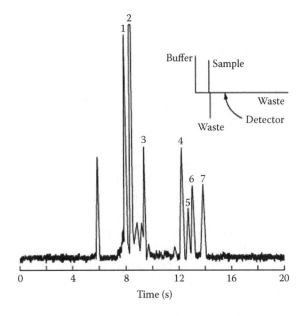

FIGURE 6.5 Electropherogram of the CE separation of six FITC-labeled amino acids (10 µM) in pH 9.0 buffer. A potential of 2,330 V was applied between the injection and detection points, which are 2.2 cm apart, and a potential was applied to the side channels to reduce leakage of the sample. The peaks were: 1, Arg; 2, FITC hydrolysis product; 3, Gln; 4, Phe; 5, Asn; 6, Ser; 7, Gly. The inset shows the approximate layout of the device, with a buffer-to-waste distance of 10.6 cm. Reprinted with permission from D.J. Harrison.

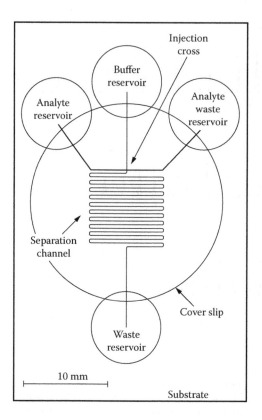

FIGURE 6.6 Schematic of a serpentine channel on a microchip. The large circle represents the coverslip and the smaller circles represent the reservoirs.[136] Reprinted with permission from the American Chemical Society.

the PMMA surface, 0.5% methylcellulose (viscosity of 2% aqueous solution at 20°C, 4,000 cP) was added to the run buffer.[560]

To resist protein adsorption during separation, poly(dimethylsiloxane) (PDMS) chips were modified with hydrophilic polyacrylamide through atom-transfer radical polymerization. This method has shortened CE analysis time for the separation of lysozyme and cytochrome *c*. Migration of TRITC-labeled bovine serum albumin (BSA) at various separation voltages was shown in Figure 6.8. When the electropherograms were replotted on a scale normalized for voltage, the peaks nearly lined up, as expected (see Figure 6.9). A more quantitative assessment of the contribution of adsorption/desorption (or mass transfer) to zone width can be achieved by examining the Van Deemter plot (plate height vs. migration velocity), as shown in Figure 6.10. A flat line indicates the absence of the mass transfer term. It also indicates that the diffusion term is negligible, as the diffusion coefficient of BSA is 10^{-6} cm^2/s or less. If conventional polyacrylamide polymerization is used, the capillary electrophoresis results indicate substantial adsorption. This is because the conventional method gives a film of low density, allowing irreversible entanglement of protein and polymer chains.[174]

FSCE separation of a negatively charged bovine anhydrase II ladder was achieved in an oxidized PDMS channel, where the negative charge of the oxidized surface minimized protein adsorption through coulombic repulsion. Separation of positively charged proteins was also achieved after the attachment of a cationic polymer to the PDMS surface.[1033]

Fast chiral separations were carried out in a quartz chip using a linear imaging UV detector. Figure 6.11 shows the chiral separation of a tocainide derivative (an antiarrhythmic drug). UV imaging with a diode-array detector located along the 25 mm long separation channel reveals the separation between the two enantiomers. Another chiral separation of pseudoephedrine was achieved in

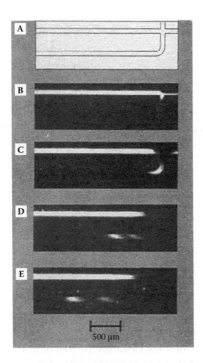

FIGURE 6.7   (a) Schematic of the injection cross region imaged; charge-coupled device (CCD) images obtained during (b) pinched sample loading and (c–e) separations at 1, 2, and 3 s, respectively, after switching to the separation mode using an electric field strength of 150 V/cm. Rhodamine B migrated faster, and sulfo-rhodamine migrated later.[136] Reprinted with permission from the American Chemical Society.

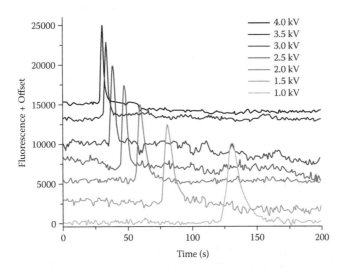

FIGURE 6.8   Electropherograms of TRITC-labeled bovine serum albumin in the PDMS microchip, surface-modified by polyacrylamide, at varying applied voltages over 4.5 cm and a separation length of 3.5 cm. The electropherograms are artificially offset for display. The PDMS surface was first oxidized in an ozone plasma, then treated with 1-trichlorosilyl-2-(m,p-chloromethylphenyl)ethane. Subsequently, a solution of acrylamide/Cu(I) Cl/Cu(II) Cl/tris[2-dimethylamino)ethyl]amine (100:1:0.1:1) was used to perform polymerization of acrylamide. It is essential that generation of the hydrophilic polyacrylamide only occurs in the channel, but not on the sealing region; otherwise, leakage will be worsened by the hydrophilicity of the surfaces.[174] Reprinted with permission from the American Chemical Society.

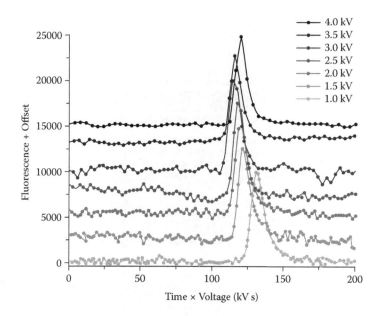

FIGURE 6.9   The same electropherograms as in Figure 6.8, except that these are plotted vs. voltage × time to make the peaks nearly coincide, allowing easier comparison of peak widths and shapes. The individual data points are shown to clarify the starting and ending of the peak.[174] Reprinted with permission from the American Chemical Society.

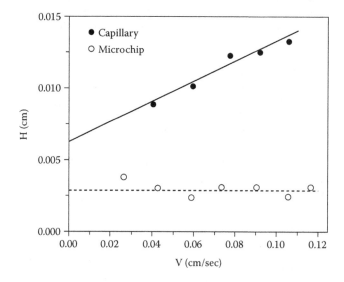

FIGURE 6.10   Van Deemter plots for capillary electrophoresis (Σ) and microchip electrophoresis (o), where plate height (H) of TRITC-labeled bovine serum albumin is plotted vs. migration velocity (v).[174] Reprinted with permission from the American Chemical Society

13 s. The compound was found to consist of only 2% excess of the (–)-enantiomer, as depicted in Figure 6.12.[605] A protein separation was also reported.[910]

Free-solution (gel-free) CE separation has been used in the detection of a target gene sequence amplified using the CPT reaction (see Chapter 9, Section 9.2.1.7 for details). The intact chimeric oligonucleotide probe (fluorescein labeled at the 5' end and biotin labeled at the 3' end) was separated

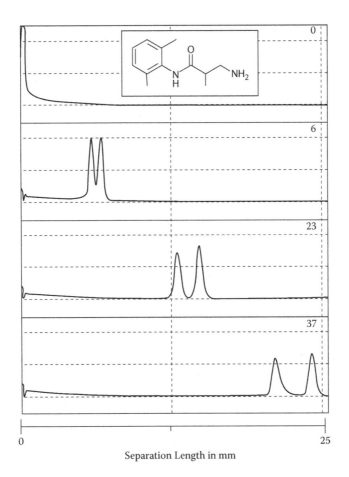

FIGURE 6.11   Real-time migration patterns (0 to 37 s) for the chiral separation of a tocainide derivative (2 mg/ml). Electrolyte, 5% highly sulfated-γ-CD, 25 mM triethylammonium phosphate buffer, pH 2.5; detection, UV at 200 nm.[605] Reprinted with permission from Wiley-VCH Verlag.

from a cleaved probe (only fluorescein labeled at the 5' end). Owing to the presence of biotin in the intact probe, it migrated later than the cleaved probe, even in the absence of a gel sieving matrix.[606]

CE separation usually requires high voltage. However, application of voltage >30 kV was not feasible because of arcing between the Pt electrodes, even though they were wrapped in Tygon tubings for insulation.[339] Lower voltage can be achieved by using a moving electric field on a series of electrodes constructed along the separation channel.[607]

In one report, open-access channel electrophoresis was achieved on a glass channel plate, with no cover plate.[792]

Other CE separations are tabulated in the appendix.

6.2.2   Capillary Gel Electrophoresis (CGE)

Capillary gel electrophoresis has mostly been associated with DNA analysis, which is also covered in Chapter 9, Section 9.4.

The first miniaturized CGE separation was achieved in a 12 μm deep and 50 μm wide glass channel. A non-cross-linked sieving medium (10% T/0% C polyacrylamide) was used to separate single-stranded antisense oligonucleotides (10 to 25 b). Separation was achieved at a separation speed of 5,000 plates/s

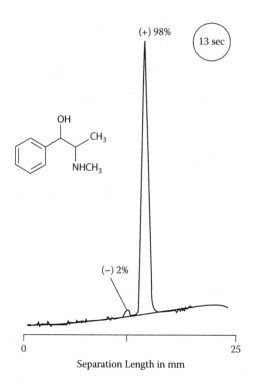

FIGURE 6.12 Optical purity determination of pseudoephedrine. Conditions are as in Figure 6.11.[605] Reprinted with permission from Wiley-VCH Verlag.

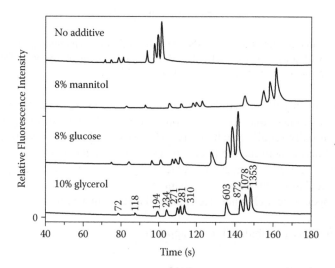

FIGURE 6.13 Electropherograms of ΦX174 *Hae*III restriction fragments on the electrophoresis microchip using 2% HPMC-50/1 × TBE solution without or with certain amounts of polyhydroxy buffer additives. Numbers on the bottom electropherogram correspond to the size of DNA fragments in bp. Applied field strength was 300 V/cm.[613] Reprinted with permission from Wiley-VCH Verlag.

and plate height of 0.2 μm within 45 s! Separation efficiency of two-thirds of the theoretical limit (N = 330,000) was achieved, suggesting band broadening caused by Joule heating was likely to be small.[547]

CGE separation for oligonucleotide DNA fragments (ΦX174 HaeIII) has been achieved in about 120 s in a microfabricated glass channel of 8 μm depth and 50 μm width. A sieving matrix of TAE buffer and 0.75% (w/v) HEC was used for separation, and an intercalating dye (1 μM TO or 0.1 μM TO6) was used for laser-induced fluorescence (LIF) detection (λ_{ex} = 488 nm, λ_{em} = 530 nm). An increase in the electric field strength from 100 to 180 V/cm shortens the separation time from 400 s to 200 s, with only a slight loss in resolution.[316]

Another CGE separation was also performed in a 50 μm wide and 20 μm deep channel cast on PDMS. The heat dissipation of PDMS (~0.3 W/m) is less efficient than fused silica (1 W/m), but it is sufficient for most applications with E of about 100 to 1,000 V/cm.[159]

Various sieving matrices for DNA fragment separation in a glass chip were evaluated.[611, 616] It was found the performance of HEC was comparable to that of PDMA, but was superior to that of polyacrylamide and polyethylene oxide (PEO).[611] Hydroxylpropylcellulose (HPC), instead of HEC or HPMC, was used as a sieving matrix to achieve better DNA separations. In addition, the lower viscosity of HPC allows for more concentrated solution to be used without filling problems. CGE separations of DNA for analyzing hepatitis C virus,[617] and mutations associated with Duchenne muscular dystrophy (DMD)[612] were demonstrated.

CGE separation of dsDNA fragments was conducted on a PMMA chip (see Figure 6.13). Separation using a low-viscosity hydroxypropylmethylcellulose (HPMC) was enhanced by using polyhydroxyl compounds (e.g., mannitol, glucose, and glycerol) as buffer additives. Apparently, the mannitol–tetraborate complex might form a bridge over HPMC chains and modify the sieving behavior of the polymer solution.[613]

CGE separation of dsDNA was performed in a PDMS chip[183] and a PMMA chip,[614] which used a sieving medium containing Pluronic (F127), a thermoreversible gel.

When a cross-linked gel is used for CGE separation, high E field or long separation length was not required, unlike in non-cross-linked gel systems. However, highly viscous cross-linking gel is very difficult to fill into microchannels. Therefore, a cross-linked polyacrylamide gel was photopolymerized in situ within microchannels. The use of cross-linked gel also allowed sample compaction and injection (see Figure 6.14). Sample compaction was accomplished using electrophoretic injection through the cross-linked gel (see Figure 6.14c). As shown in Figure 6.14d, CGE separation was performed in less than 15 min over a short length of 0.18 cm![615]

CGE separation of dsDNA fragments was performed on a PMMA chip using a near-IR dye (TOPRO-3). This allows the LIF detection of fragments (603 to 1,353 bp) as low as 0.1 μg/mL. DNA sizing of hepatitis C virus (HCV) amplicon was demonstrated.[149]

CGE separation of ssDNA was performed on a Si-glass chip filled with cross-linked polyacrylamide gel. Separation was carried out at a denaturing temperature of 50°C, as provided by the Pt thin-film heater fabricated along the separation channel, as shown in Figure 6.15. With the use of a cross-linked gel, separation can be achieved in a short distance. To improve resolution, the width of the sample plug was reduced by an electrode-defined injection scheme using an array of on-chip electrodes. An E field was applied in the sample loading region, which caused the DNA to migrate toward and be collected at the capture electrode located just outside the gel interface.[95]

Other CGE separations of DNA are included in Chapter 9 (Section 9.4) and the appendix.

CGE separation has also been used for resolve proteins.[555,609,618,619] For instance, sodium dodecyl sulfate–polyacrylamide gel electrophoresis (SDS-PAGE) of various proteins was carried out on a glass chip. The separation was ~100 times faster than slab-gel SDS-PAGE.[555] SDS-PAGE of several cytokines (FITC-labeled interferon gamma [IFN-γ], interleukin 2 [IL-2], and insulin growth factor [IGF]) was performed on a glass chip with UV-initiated photopatterned polyacrylamide.[618] In one report, pressurization was applied prior to CGE separation of proteins in a PMMA chip. This approach dramatically reduced migration time, but still retained a plug flow profile without compro-

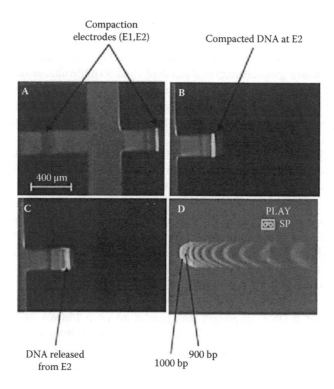

FIGURE 6.14   Video sequence depicting the separations using electrode-defined sample compaction/injection and photopolymerized polyacrylamide gel sieving matrix. In (a) and (b), a fluorescently labeled 100 bp ladder DNA was compacted at a 50 mm electrode by applying E of ~12 V/cm. (c) Compacted sample was released by switching the electric field to two electrodes spanning the gel matrix. (d) Separation was initiated from left to right at $E = 20$ V/cm. Complete resolution of all fragments was achieved.[615] Reprinted with permission from Wiley-VCH Verlag.

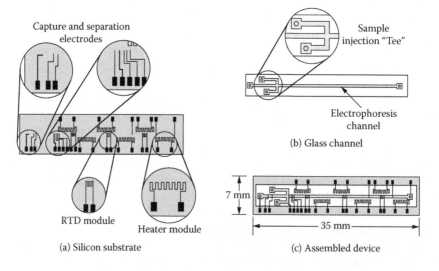

FIGURE 6.15   Schematic drawings (top view) depicting (a) the capture and separation electrode, heater, and resistive temperature detector (RTD) module on the silicon substrate; (b) the electrophoresis channel layout on the glass substrate; and (c) the assembled device consisting of the glass channel bonded on top of the silicon substrate.[95] Reprinted with permission from Wiley-VCH Verlag.

mising resolution. Sample plugs were first loaded and injected electrokinetically. Then pressurization was applied and it was employed before but not during subsequent CGE separation.[619]

6.2.3 MICELLAR ELECTROKINETIC CAPILLARY CHROMATOGRAPHY (MECC)

In MECC separation, the elution range (or window) is determined by t_0/t_m, which is given in Equation 6.8:[107]

$$k' = \frac{t_R - t_0}{t_0 \left(1 - \dfrac{t_R}{t_m} \right)} \tag{6.8}$$

where k' is the retention factor, and t_0, t_r, and t_m are migration times of neutral solutes, solutes of interest, and micelles, respectively. In this mode of chromatography, sorption/desorption kinetics and polydispersity of micelles decrease the separation efficiency. These effects are summarized in Equation 6.9:[107]

$$H = H_{ec} + H_{diff} + H_{mc} + H_{ep} \tag{6.9}$$

where H_{ec} is the extra-column band broadening due to finite size of injection and detection—still the dominant factor; H_{mc}, sorption/desorption kinetics—small for fast hydrophobic interactions between neutral solute and micelles, but large for slow ionic interactions between charged solutes and micelles; and H_{ep}, electrophoretic dispersion due to polydispersity in micelle sizes (as large as 20% RSD)—reduced if the micelle concentration is larger, but will cause excessive Joule heating. Both H_{mc} and H_{ep} are larger for more retained solute with high k'.[107]

MECC separation of three neutral coumarin dyes was achieved on a glass chip with fluorescent detection.[107] The separation efficiency of the coumarin dyes was found to be better in MECC[107] than in CEC (see Section 6.2.5),[112] when a high E was used in MECC.[107]

Direct comparison of MECC performed in a microchip and in a fused silica capillary has been made. It was found that separation efficiency for FITC-Ser (10 µM) was higher in the chip than in the capillary. This is because (1) the channel cross section is smaller in the chip (40 × 10 µm) than in the capillary (75 µm i.d.), and (2) the E value is higher in the chip (1175 V/cm) than in the capillary (215 V/cm).[551]

The MECC separation of explosives was achieved, except that three isomers of nitrotoluenes cannot be resolved.[620] A peak height RSD was 1.7 to 3.8% for TNB, DNB, TNT, tetryl, 2,4-DNT, 2,6-DNT, and 2-amino-4,6-DNT. But the linear ranges for TNB, DNB, TNT, and tetryl were only 1 to 5 ppm! This narrow linear range, which is caused by the indirect LIF detection based on fluorescent quenching, may be sufficient for screening, but is certainly not useful for quantitation.[620]

Another buffer additive, 18-crown-6-ether, which complexed with K^+, was added to the CE run buffer to separate K^+ from NH_4^+. This method was applied to separate a mixture containing seven explosive-related cations and anions in a PMMA chip. NH_4^+, K^+, and Na^+ are preexplosive inorganic cations; NH_4^+, $CH_3NH_3^+$, K^+, and Na^+ are postexplosive cations; and Cl^-, NO_3^-, and ClO_4^- are postexplosive anions.[621]

Nonaqueous separation of explosives was achieved in a glass chip. Colorimetric detection was based on the formation of a red-colored derivative, which was formed between the nitroaromatic compounds and hydroxide or methoxide ions. This formation was more favored in basic CH_3CN solution than in basic CH_3OH or aqueous solutions. Figure 6.16 shows the need of surfactant to enhance separation of three explosives. Both negatively charged SDS (hydrophobic interaction) and positively charged cetyltrimethylammonium bromide (CTAB; electrostatic interaction) produced similar separation enhancement and resulted in the same migration order.[622]

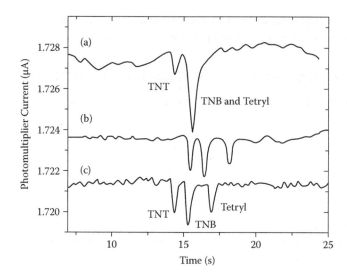

FIGURE 6.16 Effect of surfactant type on the microchip separation of 2 mg/L TNT, 1 mg/L TNB, and 2 mg/L tetryl. The separation buffer contained MeCN/MeOH (87.5/12.5 [v/v]), 2.5 mM NaOH, and with (a) no surfactant, (b) 0.5 mM CTAB, (c) 1.0 mM SDS. Applied separation field strength, ~506 V/cm, using a 1 s floating injection. Colorimetric detection was achieved using a green light-emitting diode (LED) source (505 nm).[622] Reprinted with permission from Elsevier Science.

Surfactants, such as SDS[551] and Tween-20,[330] have been employed in uncoated glass chips for determination of theophylline in human serum without the severe adsorption problem of the protein (anti-theophylline antibody). The separation was achieved in fifty-fold shorter analysis time than a competitive immunoassay first achieved on a fused silica capillary.[551]

The MECC separation of nineteen naturally occurring amino acids has been achieved with an impressive average N value of 280,000, mainly because of the use of a 25 cm long channel. The dispersion introduced by the turn geometry was reduced by using a spiral channel, instead of a serpentine channel. The spiral channel with turns of much larger radius of curvature, r_c,[339] was akin to that first introduced in a GC chip.[1]

The MECC separation of amino acids using a more conducting buffer (2,370 μS) generated fewer N (average 280,000) than a CZE separation of dichlorofluorescein (DCF) and contaminants (N = 1,100,000) using a less conducting buffer (438 μS). This is because more Joule heating is generated in MECC using a more conducting buffer.[339] The separation was achieved in only 165 s, with a resolution greater than 1.2 when 10% (v/v) propanol, instead of 20% (v/v) methanol, was used as the organic modifier.[339]

MECC of primary amines from lysate of neuron cells (CATH.a) was achieved in a glass chip coated with polyacrylamide.[623] MECC separation of biogenic amines (in soy sauce) was also carried out in a glass chip using LIF detection.[624]

The MECC separation of organophosphate nerve agents was achieved using amperometric detection. Studies show that the use of MES buffer (pH 5.0) and 7.5 mM SDS provides the best separation performance (short analysis time and adequate resolution).[625]

MECC separation with gradient elution was also reported. This was achieved using different compositions of MeOH or CH_3CN on a glass chip.[626]

On-chip chiral separation of enantiomers of adrenaline, noradrenaline, and dopamine was achieved using a mixture of carboxymethyl-β-cyclodextrin (CMCD) and a polyamidoamine dendrimer (Starburst).[120] On-chip chiral separation of enantiomers of homovanillic acid, DOPA, cDOPA, methoxytyramine (MT), metanephrine, and normetanephrine was achieved using a mixture of 18-crown-6-ether and carboxymethyl-β-cyclodextrin.[120]

Chiral separation of FITC-labeled amino acid enantiomers was performed on a glass chip using fluorescent detection. Analysis time ranged from 75 s for the most basic amino acids to 160 s for the most acidic ones. γ-CD was used as the chiral selector.[627] Chiral separation of amino acids in extraterrestrial samples or meteorites were also performed.[610,628]

There are other MECC applications, which are given in the appendix.[629]

6.2.4 ISOTACHOPHORESIS (ITP)

Isotachophoresis, which is another CE method, has been used either as a separation method or as a sample preconcentration method. For instance, ITP alone[630] and ITP preconcentration before CZE[630,631] have been achieved on a PMMA chip (see Figure 6.17). With valves (V_1, V_2, V_3, V_4), appropriate solutions are filled to four consecutive columns on a chip in the following order: (1) P_2 to W for filling LE to SC2 (and SC1), (2) P_1 to W for filling SC1, (3) P_T to W for filling TE to S_T (and S), and (4) P_s to W for filling S.[631]

When a second ITP was cascaded with the first ITP, known as concentration-cascade ITP, SC2 was filled with LE (as in SC1). This concentration-cascade ITP was achieved for fourteen anions (200 μM) within 600 s (see Figure 6.18). When ITP was followed by CZE (ITP-CZE), SC2 was filled with a background electrolyte (not necessarily with the greatest mobility). ITP-CZE has been achieved for nitrate, fluoride, and phosphate (10 μM) in the presence of high contents of sulfate (800 μM) and chloride (600 μM).[631] Subsequent work on the reproducibility of this on-chip ITP separation has been reported.[632] When ITP is performed at pH of greater than 7, there is a common problem of carbonate contamination caused by dissolved atmospheric CO_2.[633]

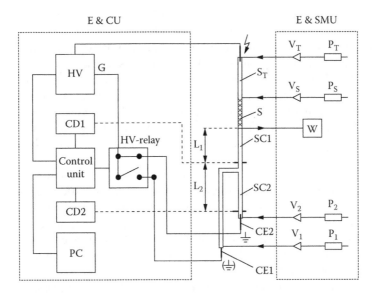

FIGURE 6.17 Block scheme of the ITP: E & CU, electronic and control unit; HV, high-voltage power supply; CD1, CD2, platinum conductivity detectors for the first and second separation channels, respectively; HV-relay, a high-voltage relay switching the direction of the driving current in the separation compartment; S_T, a high-voltage (terminating) channel; S, sample injection channel; SC1, the first separation; SC2, the second separation; CE1, CE2, counter electrodes for the first and second separation channels, respectively; L_1, L_2, separation paths for the ITP measurements of the migration velocities on the chip; E & SMU, electrolyte and sample management unit; V_1, V_2, V_T, needle valves for the inlets of the separation and terminating channels; V_S, a pinch valve for the inlet of the sample injection channel; W, waste container; P_1/P_2, P_S, P_T, syringes for filling the separation, sample injection, and terminating channels via FEP tubings.[631] Reprinted with permission from the American Chemical Society.

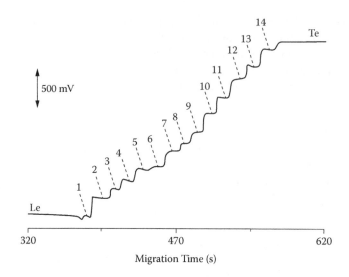

FIGURE 6.18   An isotachopherogram from the separation of a fourteen-component model mixture of anions using concentration-cascade ITP. CD2 was used to monitor the separation. The injected sample contained the anions at 200 μM concentrations. Zone assignments: LE, leading anion (chloride); 1, chlorate; 2, methanesulfonate; 3, dichloroacetate; 4, phosphate; 5, citrate; 6, isocitrate; 7, glucuronate; 8, bromopropionate; 9, succinate; 10, glutarate; 11, acetate; 12, suberate; 13, propionate; 14, valerate; TE, terminating anion (capronate).[631] Reprinted with permission from the American Chemical Society.

In another report, transient ITP followed by CZE (tITP-ZE) was achieved on a PMMA chip. The stacked zone of fluorescein was visualized in Figure 6.19.[634]

Back-transient ITP was carried out for temporary stacking of samples in an acrylic chip. This stacking works when there is EOF suppression (i.e., the background electrolyte contains 25 mM HEPES, pH 7.4, 1% PEO). However, this stacking effect was not satisfactory in the usual pinched injection with push-back voltage (p + pb) during separation, as shown in Figure 6.20. This is because of the loss of sample during the push-back step. To restore the stacking effect, the floating injection (float) must be used. The stacking effect is further enhanced when the sample conductivity is higher than the background electrolyte conductivity (i.e., by adding 25 mM NaCl into the fluorescein sample) (see Figure 6.20).[635]

In another report, ITP followed by CGE was employed for separation of dsDNA or a PMMA chip. ITP was employed to concentrate dsDNA, even in a high-salt buffer (i.e., >60 mM Cl$^-$), and no desalting step was required. This was because Cl$^-$ was used as the leading electrolyte with its net mobility higher than the mobilities of the dsDNA fragments, which in turn were higher than that of the terminating electrolyte (e.g., HEPES). A forty-fold signal enhancement, leading to an LOD of 9 fg/μl (72 to 1,353 bp), was obtained. Figure 6.21 shows the CGE results with or without the prior ITP preconcentration process.[636]

ITP was also used to concentrate selenoamino acids (methionine, ethionine, cysteine) prior to CE separation. The transfer of the analytes was carried out at a bifurcation junction as shown in Figure 6.22. As shown by the two colored dyes in Figure 6.22a, the ITP zones reached the junction. Then the zones moved pass the junction to the right CE channel (Figure 6.22b), and continued to separate further as they moved downstream (Figure 6.22c).[637]

In one report, bidirectional ITP was achieved on a PMMA chip. A common terminating electrolyte (TE) was employed to achieve simultaneous cationic and anionic separations. Without a complex injector design, sample introduction was achieved hydrodynamically for ITP separation.[638]

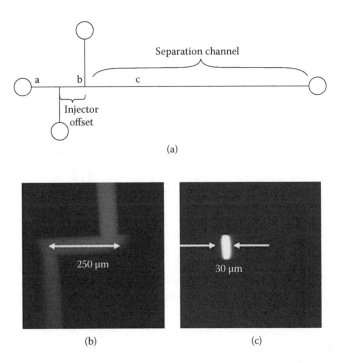

(a)

(b)   (c)

FIGURE 6.19 Fluorescence CCD images of tITP-ZE separation during injection (b) and after ITP concentration (c). (a) The general microfluidic channel configuration. Panels (b) and (c) show results obtained with a 250 μm injector. Conditions: Sample was 1 mM fluorescein. Leading electrolyte: 25 mM Tris+, 25 mM Cl−. Trailing electrolyte: 25 mM Tris+, 25 mM TAPS−. Injection field: 300 V/cm, current ~18 μA. Separation field: 200 V/cm, current ~10 μA.[634] Reprinted with permission from the American Chemical Society.

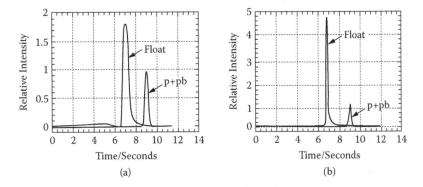

(a)   (b)

FIGURE 6.20 Experimental electropherograms of fluorescein under floating (float) and pinch-and-pull-back (p + pb) conditions with (a) no NaCl and (b) 25 mM NaCl added to the sample.[635] Reprinted with permission from the Royal Society of Chemistry.

6.2.5 CAPILLARY ELECTROCHROMATOGRAPHY (CEC)

Electrochromatography was first demonstrated in the open-tubular microfluidic channel coated with octadecylsilane (ODS) as the chromatographic stationary phase. EOF was used as the pumping system within the 5.6 μm deep and 66 μm wide channels. Plate heights of 5.0 to 44.8 μm were achieved for three coumarin dyes.[112] Open-channel CEC separation was achieved using a stationary phase of octadecyltrimethoxysilane.[148] CEC separation of some polyaromatic hydrocarbon (PAHs)

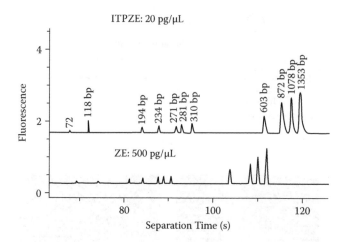

FIGURE 6.21 Comparison of resolution and detection sensitivity of ITPZE and ZE separations. Sample, dsDNA ΦX174/*Hae*III in 0.1 × PCR buffer. The ZE sample has a 25× higher total DNA concentration than the ITPZE sample. The average signal enhancement from ITP is forty-fold.[636] Reprinted with permission from Wiley-VCH Verlag.

has also been carried out on a Pyrex chip with a thin film of C8 stationary phase formed using a sol-gel process.[338]

In CEC, there is an additional factor in band broadening due to mass transfer, H_m. The Van Deemter equation is given in Equation 6.10:[112]

$$H = H_{inj} + H_{det} + H_{geo} + H_{diff} + H_m$$

$$0.13\mu m \quad 0.014\mu m \quad 0.53\mu m$$

(6.10)

H_{geo}, which is the greatest contribution among the first three factors, can be reduced by decreasing the channel width (at the expense of detector path length), decreasing the number of turns, and decreasing the turn angle (see Equation 6.5). The plate heights as contributed by Joule heating and mass transfer for the thin stationary phase film were neglected. Other phenomena, such as electric field effects or eddy flow, though conceivable, were claimed to be not observed.[112]

Upon curve fitting of experimental data to the Van Deemter equation ($H = A + B/v + C_m v$), the fitted C_m values are two orders of magnitude higher than the calculated C_m values (based on channel height and diffusion coefficient of the analyte in aqueous solution). This discrepancy has been attributed to the trapezoidal geometry of the channel cross section, leading to more broadening near the triangular cross section than the middle rectangular cross section.[112] In this connection, dispersion (due to transverse diffusion) in microchannels was studied with reference to the isotopically etched cross section.[118]

CEC separation has been achieved on a reverse-phase chromatography bed, which is a 200 μm bed packed with ODS-coated silica beads (see Figure 5.4 in Chapter 5). Separation of a mixture of BODIPY and fluorescein was achieved in less than 20 s (see Figure 6.23). Five bioactive peptides (papain inhibitors, proctobin, opioid fragments 90 to 95, ileangiotensin III, and angiotensin III) were also separated in this manner.[639] A similar method has been used to separate two coumarin dyes or to separate Alexafluor-labeled angiotensin from the excess dye.[587,640] CEC separation of four FITC-labeled synthetic peptides was achieved on a PDMS chip packed with beads.[566,641] Open-channel CEC (with coupled solid-phase extraction) was also performed for PAHs.[642]

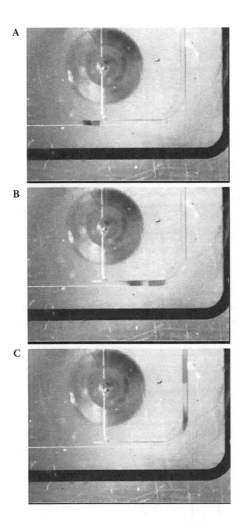

FIGURE 6.22 Section from the ITP preconcentration to the electrophoretic separation at the bifurcation block on a PMMA chip. (a) Stack of zones at the end of the enrichment. (b, c) Transition from the isotachophoretic migration to zone electrophoretic movement.[637] Reprinted with permission from Wiley-VCH Verlag.

Solvent programming with a stepwise or linear gradient was used to enhance CEC separation of four PAHs (Figure 6.24). This was achieved using aqueous acetonitrile of two concentrations delivered from two inlet reservoirs at different mixing ratios, leading to the CH_3CN concentrations in the range of 24 to 56%.[148, 643]

To increase the surface area of the stationary phase for CEC separation, a collocated monolithic support structure (COMOSS) was constructed in a Si chip. A polystyrene–sulfonic acid stationary phase was then immobilized.[349] Design of the COMOSS required that the combined cross-sectional area at the column head be the same at any point in the inlet distributor.[644] A study for the reduction of band broadening in COMOSS was also reported.[645]

The COMOSS has also been fabricated on a PDMS chip for CEC separation of FITC-labeled peptides (Figure 6.25). However, in the CEC separation of a mixture of rhodamine and fluorescein, a broad rhodamine peak was obtained, but fluorescein did not have this problem. This was possibly because the neutral rhodamine had diffused into the PDMS substrate, as illustrated in the fluorescent image in Figure 6.26.[360] In another report, CEC separation of a peptide mixture was performed on a PDMS chip after cerium(IV)-catalyzed polymerization of the stationary phase within the channels.[646]

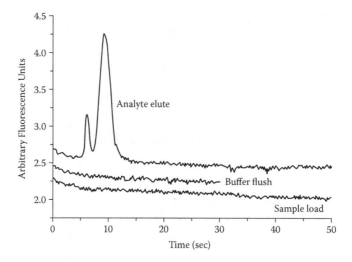

FIGURE 6.23   Electrochromatogram of fluorescein (first) and BODIPY (second), showing different steps of the separation: step 1, 100 s loading; step 2, 30 s buffer flush; step 3, an isocratic elution from the 200 μm long column with a 30% acetonitrile/70% 50 mM ammonium acetate mobile phase.[639] Reprinted with permission from the American Chemical Society.

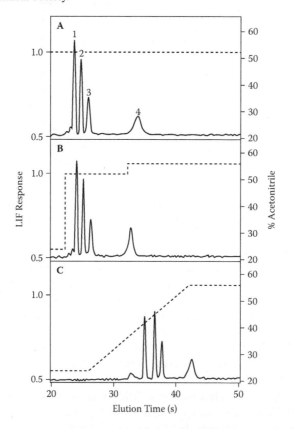

FIGURE 6.24   (a) Isocratic, (b) step gradient, and (c) linear gradient used in the CEC separations of (1) anthracene, (2) pyrene, (3) 1,2-benzofluorene, and (4) benzo[a]pyrene. The injection time was 20 s. The concentrations of analytes 1 to 4 were 2.8, 0.9, 5.8, and 5.0 M, respectively. The dashed lines show the gradient profiles as seen at the detector.[148] Reprinted with permission from the American Chemical Society.

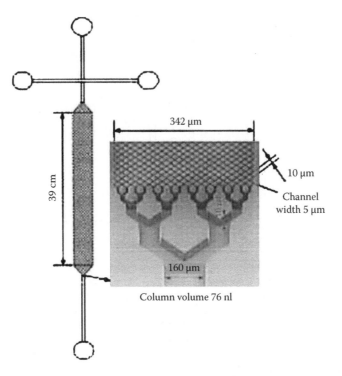

Column volume 76 nl

FIGURE 6.25 Scheme of a COMOSS column incorporated in a PDMS chip for CEC separation. A reverse-phase coating (poly[styrenesulfonic acid]) was bonded to the COMOSS column after silanization treatment of the PDMS surface. Since the usual solvent (toluene) for silanization cannot be used in PDMS, the surfactant, SDS, was used to help dissolve the silanes.[360] Reprinted with permission from Wiley-VCH Verlag.

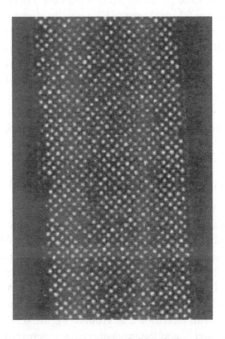

FIGURE 6.26 Fluorescence image of unmodified PDMS COMOSS column (10×) following incubation with rhodamine 110 for 2 days.[360] Reprinted with permission from Wiley-VCH Verlag.

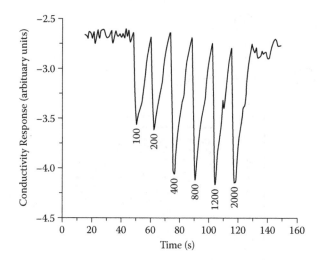

FIGURE 6.27   CEC separation of a double-stranded DNA ladder (400 ng/ml) in a PMMA device. The ladder consisted of 100, 200, 400, 800, 1,200, and 2,000 bp fragments. The PMMA microchannel was modified by chemically attaching a C18 phase to its surface. The mobile phase used for this separation was 25% acetonitrile and 75% aqueous phase containing 50 mM TEAA (ion-pairing agent, pH 7.4). Detection was accomplished using indirect, contact conductivity detection. The conductivity cell was operated at 5.0 kHz and a pulse amplitude of ±0.5 V. The field strength used for the separation was 100 V/cm.[208] Reprinted with permission from the American Chemical Society.

A UV-initiated acrylate-based negatively charged porous polymer monolith was also used as a stationary phase for CEC separation of NDA-labeled peptides (0.1 to 1 µM) in a glass chip. The monolith was cured in a short time (10 to 20 min) within the microchannel without the use of a retaining frit. The monolith was electrokinetic (EK) conditioned to remove residual monomeric materials prior to use. Thiourea (2.5 mM) was used as an unretained marker.[647] CEC separation of ten (out of thirteen) PAHs (10 mM) was performed on a fused silica chip containing UV-initiated polymerization of a sulfonate group-containing monolith. LIF detection was achieved with excitation at 257 nm, with possible fluorescent quenching problems due to dissolved O_2.[1152]

CEC was also applied to separate oligonucleotides. Here, an ion-pair reagent was used to enhance separation. Figure 6.27 shows the results of the separation of a DNA ladder (100 to 1,000 bp) using conductivity detection. No separation occurred if the PMMA channel was not modified with a stationary phase.[208]

6.2.6   SYNCHRONIZED CYCLIC CAPILLARY ELECTROPHORESIS (SCCE)

In high-performance liquid chromatographic (HPLC), column switching has been employed in order to increase the separation column length, and hence the N value. However, because of instrumentation limitations, this advantage has not been exploited in CE. With micromachining on glass, a complex channel system can then be constructed on a chip, leading to synchronized cyclic capillary electrophoresis (SCCE).[648,649] With the use of 4 HV power supplies and 11 HV relays, switching between four columns for CE has been achieved (see Figure 6.28). A sample was loaded between electrode 0 and electrode 1 (Figure 6.28a). Then the separation started in phase 1 when a voltage U was applied at electrode 8 with electrode 5 at ground (Figure 6.28b). The three groups of components were numbered 1, 2, and 3, and the procedure was synchronized to the migration speed of group 2. Separation continued in phase 1 until group 2 reached the detector (Figure 6.28c). Then when voltage U was applied at electrode 2 with electrode 7 at ground, group 2 continued to move in phase 2 (Figure 6.28d). Then voltage U was applied at electrode 5, with electrode 9 at ground, and

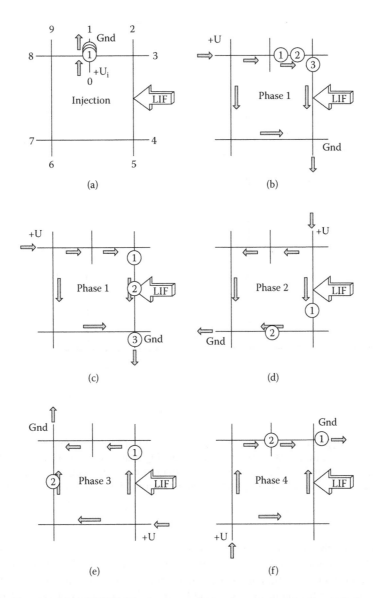

FIGURE 6.28 The principle of SCCE. Three groups of samples are given by the circled numbers 1, 2, and 3. The voltage-switching procedure is synchronized to group 2. (a) Injection phase, (b) during phase 1, (c) at the end of phase 1, (d) phase 2, (e) phase 3, (f) end of the cycle.[648] Reprinted with permission from Elsevier Science.

group 2 continued to move in phase 3 (Figure 6.28e). Then voltage U was applied at electrode 6, with electrode 3 at ground, and group 2 continued to move in phase 4, which was the end of cycle 1 (Figure 6.28f). During cycle 2, phase 1 migration would bring the group 2 to the detector a second time. The migration of several components, synchronized to the group of components A and B for two cycles, is shown in Figure 6.29. An enhancement of resolution (R_s) from 1.7 to 3.0 has been achieved for the CZE separation between two major peaks in an FITC sample.[648] Further development of SCCE was demonstrated in the separations of amino acids[551,650] and oligonucleotides.[650] A similar approach has been used for separation of dsDNA on a Si chip (coated with thermal SiO_2 for insulation) and only low voltage (up to 10 V) was used.[306,311]

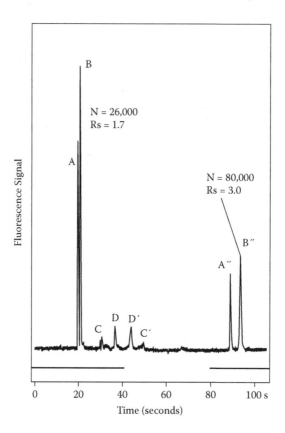

FIGURE 6.29 Electropherogram obtained using SCCE for an FITC sample and its degradation products. The synchronization speed was adjusted to the first two eluting peaks, A and B. Time intervals with forward motion at the detector are indicated by the black bars. After one cycle, the components A and B appear again (80,000 theoretical plates, resolution 3.0).[648] Reprinted with permission from Elsevier Science.

A new layout in SCCE was reported. This permits the investigation of sample behavior at corners and T-junctions. Improved resolutions are found for FITC–amino acid in water as well as FITC-compounds in human urine when more cycles are run. However, fewer components remained at subsequent cycles. In addition, signal intensity decreased but peak area also decreased, probably because photobleaching occurred and materials were lost around the T-junction.[551]

Although SCCE has the advantage of increasing column length by cycling, the method suffers from some drawbacks. These include (1) complicated determination of synchronization time, (2) inherent complexity in peak assignments of complex mixtures, (3) excess analyte dispersion generated in the corners (turn geometry), (4) loss of the analyte at the channel junction, which can be improved by junction design, and (5) a continually decreasing migration window as the separation progresses.[276,339,551,648]

Another variation of SCCE is field-inversion electrophoresis. A sample (e.g., DNA) was passed back and forth within a short length of a microchannel. Using an alternating potential (600 V of several Hz) applied over an 8mm channel, this method can provide a resolution similar to that of a significantly longer microchannel. The inverting field serves to disrupt DNA alignment (reptation) with the electric field, and hence recovers the size-dependent separation.[651]

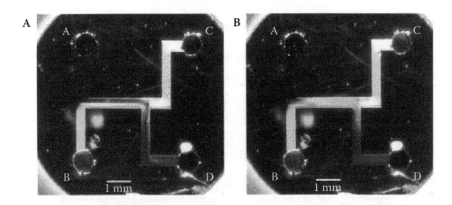

FIGURE 3.49 Visualization of the mixing events in the 2.66 mm long silicon mixer manufactured using deep reactive ion etching. An acid-base indicator, bromothymol blue (inlet A, neutral form, green), is mixed with HCl (inlet B). The indicator changes color to yellow (acidic form) and is subsequently mixed with NaOH (inlet C), resulting in a blue color (basic form) before the mixture leaves the chip (outlet D). (A) Flow velocity of 0.2 ml/s results in incomplete mixing. (B) Flow velocity of 0.75 ml/s results in complete mixing.[495] Reprinted with permission from the American Chemical Society.

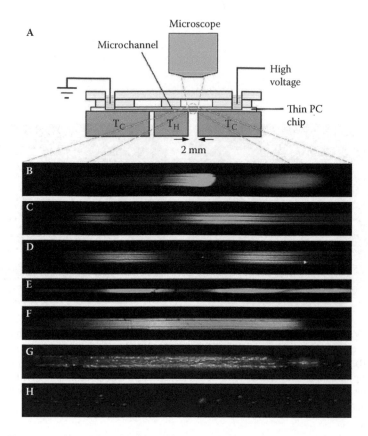

FIGURE 5.20 Demonstration of focusing and separation of a variety of different analytes using TGF on a PC chip. (a) Schematic drawing of the apparatus. The temperature gradient (~ 25°C/min) was achieved by two end cool Cu blocks (T_C = 10°C using cold water) and one middle hot Cu block (T_H = 80°C using an embedded resistive heater). For the study of GFP, the hot temperature was lowered to 60°C to avoid denaturing of the protein. (b–g) Fluorescence images of focusing zone for various analytes. Magnification is constant for all images. The total length of each image is 1.9 mm. (b) Oregon Green 488 carboxylic acid and Cascade Blue hydrazide. (c) The two products resulting from labeling of aspartic acid with FQ. (d) Mixture of CBQCA-labeled serine and tyrosine. (e) Green fluorescence protein. (f) Fluorescein- and TAMRA-labeled DNA. (g) Six-millimeter-diameter fluorescently labeled polystyrene particles. (h) Same as (g), but after the channel was rinsed and refilled with the unfocused particle solution.[597] Reprinted with permission from the American Chemical Society.

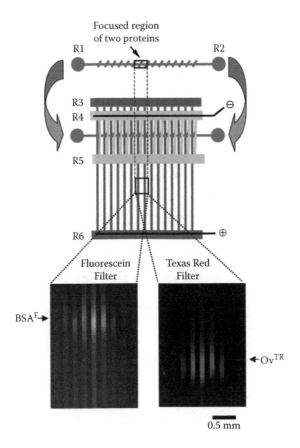

FIGURE 6.43 A sketch of the second-dimensional electrophoresis channels aligned on the first-dimensional channel previously shown in Figure 6.42. The long red line represents the IEF channel, the parallel green lines represent the channels in the top PDMS slab, and the parallel blue lines represent the channels in the bottom PDMS slab. Below the sketch are the fluorescence images of a section of the electrophoresis channels containing the separated proteins (BSA^F, Ovalbumin^{TR}) that overlapped in the IEF separation. The image on the left was obtained using a filter that isolated fluorescence from fluorescein (F), and the image on the right was obtained using a filter that isolated fluorescence from Texas Red (TR). Both images were obtained for the same region of the capillary array where the overlapping peaks of BSA^F and Ovalbumin^{TR} migrated. The weaker fluorescence signal in the lower portion of the fluorescein image may be due to impurities in the BSA^F sample.[182] Reprinted with permission from the American Chemical Society.

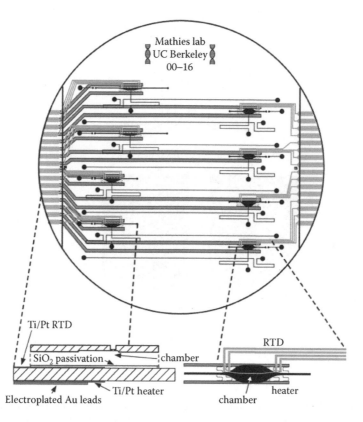

FIGURE 9.4 Overview of the PCR-CE mask design including microfabricated heaters (pink) and green-colored resistance temperature detectors (RTDs). Bottom right: An expanded top view of the chamber area, showing the chamber, the four-lead RTD, and the heater. The leads for the heater should be electroplated with gold (see also Chapter 2, Section 2.6). The four-lead RTD provides more accurate temperature measurement than the two-lead RTD, since the four-lead device separates the measurement due to Joule heating in the device from that due to the actual temperature change within the PCR chamber. Bottom left: An expanded view of a partial cross section of the device, showing the two glass layers and their relative alignment.[282] Reprinted with permission from the Royal Society of Chemistry.

6.2.7 Free-Flow Electrophoresis (FFE)

Another separation method is FFE. The separation depends on the deflection angle of a sample flow stream of flow speed v due to an applied perpendicular E field (Figure 6.30). The deflection angle is greater if v decreases, E increases, or μ (electrical mobility) increases. This gives rise to possible fraction collection at the outlet array (70 μm wide, 50 μm deep, 5 mm long).[89]

Isolation of the separation bed from the two electrode-containing side beds was achieved by two arrays of narrow grooves (12 μm wide, 10 μm deep, and 1 mm long) (see Figure 6.31). While the side-bed flow should be high enough (>15 μl/min) for effective removal of electrolytic gas bubbles, the flow should not be too high (<50 μl/min) to avoid increasing the separation-bed flow rate and reducing the residence time. This is because the residence time should be maximized to increase sample deflection and to enhance separation resolution.[89] Greater resolution was obtained when a greater potential difference across the separation bed was applied by increasing voltage and increasing side-bed buffer conductivity. Separation of a neutral, monoanion, and dianion amino acid was demonstrated (see Figure 6.32).[89] FFE was also conducted on a Si chip (with a 25 μl bed) for separation of various proteins (HSA, bradykinin, ribonuclease A) or tryptic digests of mellitin and cytochrome c.[652]

In another report, FFE separation of two FITC-labeled amino acids was carried out on a PDMS-glass chip consisting of micrometer-sized posts. Separation was achieved in 75 ms using a sample flow rate of 2 nl/s. The presence of these posts resulted in a very small separation-bed volume, which allowed for the application of higher electric field, higher separation speed, and higher resolution.[653]

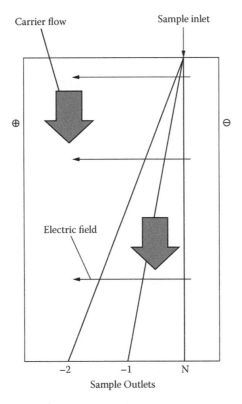

FIGURE 6.30 Schematic diagram illustrating the FFE concept. A sample mixture flows vertically down while a transverse electric field is applied. N, –1, and –2 denote the positions at which a neutral, a monoanion, and a dianion, respectively, are expected to exit the system.[89] Reprinted with permission from the American Chemical Society.

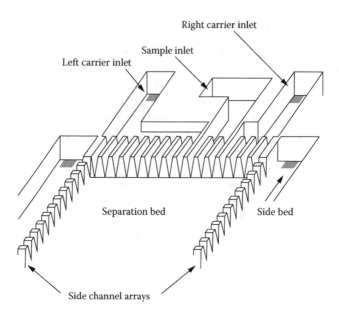

FIGURE 6.31 Three-dimensional view of the inlet region of the silicon FFE device. To avoid the dielectric breakdown of Si, it has been deposited with a composite layer of SiO_2 and Si_3N_4.[89] Reprinted with permission from the American Chemical Society.

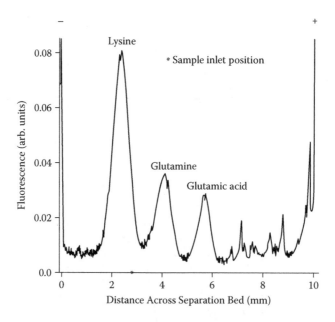

FIGURE 6.32 FFE separation profile of rhodamine-6-isothiocyanate-labeled lysine, glutamine, and glutamic acid obtained 2.4 cm along the separation-bed length. A voltage of 50 V (current 7.5 mA) was applied. The sample (0.05 mM-labeled amino acids based on the concentration of rhodamine-6-isothiocyanate) was dissolved in the carrier buffer, pH 7 phosphate buffer (0.04 M Na_2HPO_4/0.03 M KH_2PO_4; $\kappa = 9.5 \times 10^{-3}$ Ω^{-1} cm^{-1}). The sample and carrier buffer flow rates were 0.2 and 5 pl/min, respectively. The side-bed buffer was pH 7 phosphate buffer containing 0.25 M Na_2SO_4 ($\kappa = 35.5 \times 10^{-3}$ Ω^{-1} cm^{-1}), with a constant flow rate of 50 pl/min. Residence time to the point of detection was 73 s.[89] Reprinted with permission from the American Chemical Society.

6.2.8 DERIVATIZATIONS FOR CE FOR SEPARATIONS

Derivatizations have been employed to enhance detection in CE separations. Derivatization can be achieved by either postcolumn or precolumn means.

6.2.8.1 Precolumn Derivatization

LIF detection (λ_{ex} = 351.1 nm, λ_{em} = 440 nm) of two amino acids (0.58 mM glycine, 0.48 mM arginine) was achieved by precolumn derivatization with 5.1 mM o-phthaldialdehyde (OPA).[106] A reaction chamber was constructed "before" the cross-injection (see Figure 6.33). The chamber is wider than the separation channel to allow for a lower electric field and hence longer residence time for the derivatization reaction.[106] This method may be more advantageous than postcolumn derivatization,[317] when the analysis time is faster than the product reaction time ($t_{1/2}$ for OPA ~4 s[106]).

Precolumn OPA derivatization was also employed to analyze biogenic amines prior to MECC separation on a PDMS chip.[654] Precolumn OPA derivatization and MECC were also performed on a glass chip to analyze amino acids. Usually, OPA was used for fluorescent detection. However, in this report, amperometric detection was used as the OPA derivatives were also electroactive. Voltage (needed for separation) programming was used to decrease the migration time of late migrating species.[655]

Dynamic labeling of a fluorescent dye was employed for the detection of proteins as separated by SDS-PAGE in a glass chip. This labeling method relied on the fact that both the run buffer and sample buffer contained the fluorescent dye Nano Orange. An LOD of 500 ng/ml for BSA was achieved. This is feasible only in the microchip format, but not in the capillary format, thanks to the orthogonal injection mode commonly used in microchips. This mode is biased toward the injection of the slow-migrating protein–SDS complexes, but avoids the injection of fast-migrating species (i.e., free SDS and its micelles).[656]

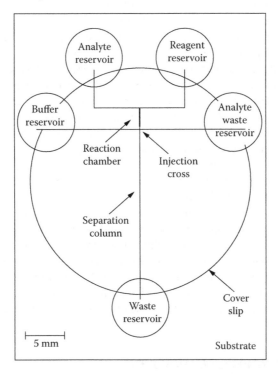

FIGURE 6.33 Schematic of the microchip with integrated precolumn reactor. The reaction chamber is 2 mm long and 96 μm wide (at half depth). The separation column is 15.4 mm long and 31 μm wide. The channels are 6.2 μm deep.[106] Reprinted with permission from the American Chemical Society.

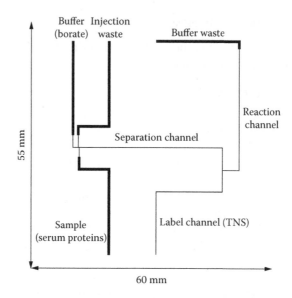

FIGURE 6.34 Schematic diagram of the postseparation labeling design of the chip. Thick lines were 256 μm wide, thin lines were 66 μm wide. Channels were 14 μm deep.[657] Reprinted with permission from Elsevier Science.

6.2.8.2 Postcolumn Derivatization

Postcolumn derivatization was employed to detect the separated proteins[657] or amino acids.[317,658] For instance, 2-toluidinonaphthalene-6-sulfonate (TNS) was used to derivatize serum proteins in the postcolumn format (see Figure 6.34).[657] With the use of fused silica with substantially lower fluorescence background (λ_{ex} = 325 nm), a twofold improvement in signal-to-noise ration (S/N) for phenylalanine detection was achieved.[658]

In another report, after the separation of arginine (2 mM) and glycine (2 mM) in a buffer (20 mM sodium borate, pH 9.2, 2% v/v methanol, 0.5% v/v β-mercaptoethanol), the amino acids were detected after postcolumn derivatization. The separated components were mixed to react with a derivatizing agent consisting of 3.7 mM OPA. Turbulence or band broadening was found in the channel after mixing unless the electric field strength (E) was above a threshold value (840 V/cm). The plate height was found to decrease monotonically with increasing E. Since OPA had a typical half-time of reaction with amino acids of 4 s, the residence time should be long enough, and this was achieved by adjusting E for maximal product formation. It was found that sufficient product detection occurred with E at 1,700 V/cm.[317]

Postseparation labeling was also achieved for separation of four human serum proteins (IgG, transferrin, α1-antitrypsin, and albumin) using 0.2 mM TNS. This is a virtually nonfluorescent reagent that, upon noncovalent association with proteins, produces a fluorescent complex (λ_{ex} = 325 nm, λ_{em} = 450 nm) for detection. TNS was selected among the commonly used labels: OPA (for amine-containing protein) or 8-anilino-1-naphthanesulfonic acid (ANS) (for protein). This postseparation labeling avoids the problems of multiple-site labeling, which is manifested as multiple peaks after separation.[657] Postcolumn noncovalent labeling of protein was also achieved using Nano Orange as the labeling reagent.[659]

6.3 CHROMATOGRAPHIC SEPARATIONS

Hydrodynamic chromatography of fluorescent nanospheres (polystyrene) and macromolecules (dextran) has been achieved on a Si-Pyrex chip. Separation is based on faster movement of larger particles or molecules because they follow the faster fluid density near the center of a channel relative to the channel wall.[336,572]

Open tubular LC using an ODS-coated channel was attempted on a Pyrex chip.[660] LC was also performed on a Si-Pyrex chip coated with C8 stationary phase for analysis of caffeine using UV detection, and of phenol using amperometric detection.[661]

Separation of two proteins (BSA, IgG) by anion-exchange chromatography was demonstrated on a PDMS chip packed with beads.[662]

Affinity chromatography of streptavidin was performed on a polyethylene terephthalate (PET) chip. The microchannel was first filled with the dual-modified latex beads (as shown in Figure 6.35). The biotinylated beads were surface modified with a temperature-sensitive polymer, poly(N-isopropylacrylamide (PNIPAAm; 11 kDa). When the temperature was raised above the lower critical solution temperature (LCST) of PNIPAAm, the beads aggregated and adhered to the channel wall, because of a hydrophilic-to-hydrophobic phase transition. Then streptavidin from a sample solution was captured by these adhered biotinylated beads. Thereafter, when the temperature was reduced below the LCST, the beads dissociated and eluted from the channel wall together with the captured streptavidin.[203]

Shear-driven LC was demonstrated in microchannels, which were formed by printing toner on a transparency film. This channel plate was covered, but not sealed by a glass plate. The shear-driven flow principle is based on the use of channels that are divided into two nonsealed and independently moving plates. By axially sliding one plate past the other, the viscous drag effect establishes a net flow. The flow velocity is equal to one-half of the plate moving velocity, and is independent of the length and depth of the channel and of the physiochemical properties of the fluid. By coating the glass plate with a monolayer of polyethoxysilane, the stationary phase of the shear-driven LC is formed. Separation of two coumarin dyes (1 mM) was achieved. Although the stationary phase (on the glass plate) was quite robust (i.e., good for about sixty runs), the carbon toner channel materials on the transparency were not, and should be replaced after approximately ten runs.[239]

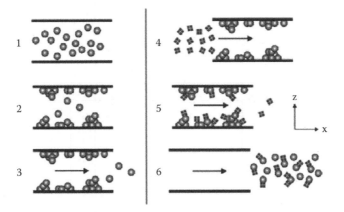

FIGURE 6.35 Schematic of the experimental protocol for streptavidin affinity chromatography. (1) The channel is initially filled at room temperature with a suspension of biotinylated, PNIPAAm-coated beads (100 nm). (2) The temperature in the channel is then raised to 37°C, and the beads aggregate and adhere to the channel walls. (3) Buffer is then pumped through the channel (the presence of flow is indicated in this schematic by an arrow), washing out any unbound beads. (4) A fluorescently labeled streptavidin sample (2.5 µM) is then introduced into the flow stream. (5) Streptavidin binds to the beads, and any unbound streptavidin is washed out of the channel. (6) Finally, the temperature is reduced to room temperature, leading to the breakup of the bead aggregates. Beads, bound to labeled streptavidin, elute from the channel.[203] Reprinted with permission from the American Chemical Society.

6.4   COUPLED SEPARATIONS

6.4.1   FRACTION COLLECTION

Accurate flow switching has been employed for fraction collection of single or a group of oligonucleotides on a glass chip (see Figure 6.36).[663] After CGE separation of a mixture of fluorescently labeled oligonucleotides, $p(dT)_{10-25}$, preselected components were withdrawn by applying a brief 5 s fraction collection pulse. Since the EOF has been suppressed, the oligonucleotide migrated from the negative voltage to the GND (i.e., positive). In a typical procedure, a CGE run occurs for 140 s (phase 1), fraction selection for 5 s (phase 2), removal of other components for 40 s (phase 3), and fraction collection for 65 s (phase 4), and all these steps lead to a total operation time of 250 s. Such a microchip fraction collection system alleviates the problem of dead volume in a conventional system. However, a dilution factor of 3 is still involved based on the calculation of peak band broadening.[663]

Fraction collection of fragments from a 100 bp DNA ladder was also achieved in a PDMS chip[664] or a glass chip. In the latter example, a small reversed field was maintained in the separation column to halt or slow down later migrating DNA in order to assist collection of a DNA fraction.[665] In one report, a single peak from the isoelectric focusing (IEF) separation channel has been isolated and transferred to a subsequent channel by means of microfluidic valve control.[449]

Fraction collection has been achieved using a PC nuclear track etched (PCTE) membrane (with 200 nm nanocapillary pores) sandwiched between two PDMS plates (see Figure 6.37). At the intersection

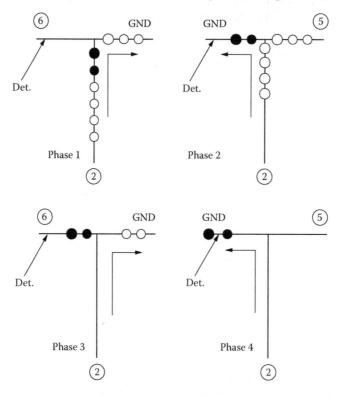

FIGURE 6.36   Various phases implemented for microchip fraction collection at reservoir 6. During the whole sequence, reservoir 2 is kept at a negative high voltage. Phase 1: Application of a ground voltage at reservoir 5 initiates the separation of the injected sample plug toward reservoir 5. Phase 2: When the sample zones of interest (filled circles) arrive at the intersection, reservoir 6 is grounded for a few seconds and reservoir 5 left floating. Phase 3: The remaining zones are removed from the system by switching back to the phase 1 conditions. Phase 4: The extraction fractions are moved past the detector by grounding reservoir 6 again while leaving reservoir 5 flowing.[663] Reprinted with permission from the American Chemical Society.

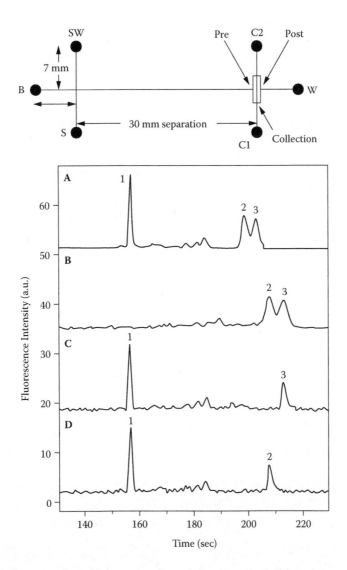

FIGURE 6.37 Separation and selective collection of arginine and glutamate derivatized with FITC. Schematic on the top shows the sampling geometry, i.e., a electrophoresis microchip (S-SW, B-BW) coupled with a molecular gate collection channel (C1-C2). LIF signals were monitored before and after the membrane gate, as indicated with the arrows, with the data shown here recorded from the post-gate detector. The electropherogram shown in (a) was obtained for a separation where no collection was performed. For selective collection experiments (b–d), the gate collection state was activated for a period matching the analyte bandwidth as judged from (a). The electropherograms demonstrate the separation and selective collection of three bands: (1) glutamate, (2) first arginine band, and (3) second arginine band. The difference in elution times between the nongated and gated experiments is from the actuation of the low-voltage gate located prior to the detection area affecting the elution time. The electrical field strength used for separation was 170 V/cm.[592] Reprinted with permission from the American Chemical Society.

of the separation channel (upper) and collection channel (lower), the membrane is located to separate the two channels. First, separation occurred along channel B-W. When the desired component reached the intersection, C1 and C2 were biased (with W floating) to divert the component through the membrane to the collection channel. Figure 6.37a shows the separation of three FITC-labeled amino acids. Figure 6.37b shows the remaining two amino acids after the collection of component 1. Figure 6.37c and d shows the other situations in which components 2 and 3 are collected, respectively. The use of the PCTE membrane has been previously described in Chapter 5, Section 5.3.[592]

In another report, fractionation of DNA samples was achieved on a Si-Pyrex device using on-chip electrodes (see Figure 6.38). The DNA fragments migrate and separate in the main gel-filled separation channel. As the individual band of interest migrates into the intersection formed by the separation channel and side channel, an electric field perpendicular to the direction of the separation field is applied, and the target band can then be attracted to the capture electrode in the side channel out of the gel.[96]

Fluorescent images showing this sequence of events are shown in Figure 6.39, where the capture electrode is located in the lower part of the side channel. Multiple fractions can be captured using a series of capture electrodes in the side channel.[96]

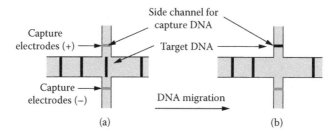

FIGURE 6.38 (a) Schematic operation showing DNA bands migrating through the intersection formed by the separation and side channels. (b) Schematic operation showing the target DNA captured by the electrode in the side channel.[96] Reprinted with permission from Elsevier Science.

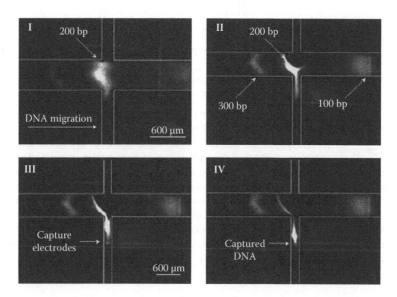

FIGURE 6.39 Sequences for the extraction of the second migrating DNA fragment (200 bp) during separation of a 100 bp ladder in cross-linked polyacrylamide (8% T, $E = 30$ V/cm). Capture electric field is 31 V/cm. Pictures are shown at (I) 0, (II) 100, (III) 200, and (IV) 400 s after extraction begins.[96] Reprinted with permission from Elsevier Science.

6.4.2 TWO-DIMENSIONAL SEPARATIONS

Two-dimensional liquid-phase separation was conducted on-chip. A micellar electrokinetic capillary chromatography/capillary electrophoresis (MECC/CE) was first reported. Here, the first-dimensional MECC separation was followed by a second-dimensional CE separation to separate the tryptic digests of cytochrome c, β-lactalbumin, and ribonuclease A. However, the CE and MECC separations are not completely orthogonal (or uncorrelated), limiting the resolving power of this method.[666]

Thereafter, a two-dimensional (2D) open-channel electrochromatography/capillary electrophoresis (OCEC/CE) was achieved on a glass chip consisting of a spiral column (see Figure 4.16). The separation of a tryptic digest of β-caesin was achieved using a mobile phase of 10 mM sodium borate (for CE) and of 30% v/v acetonitrile (for OCEC).[333] A 2D contour plot was generated by plotting each electropherogram at the corresponding time on the OCEC axis (see Figure 6.40).[333] Here, seventeen peaks from OCEC generated twenty-six spots in the 2D plot.[333] The chip layout and the OCEC/CE operation have been described in Chapter 4, Section 4.1.2.

A 2D channel array has been fabricated on a quartz chip intended for 2D CE application, as shown in Figure 6.41. The first dimension consisted of a single channel of 80 μm wide, whereas the second dimension consisted of five hundred channels of 900 nm wide. In both dimensions, the channel depth was 3 to 7 μm.[150]

In one report, two PDMS chips were physically connected together to achieve 2D separation of fluorescently labeled proteins (BSA, CA, ovalbumin). The first-dimensional IEF separation was carried out along a horizontal channel in one chip (Figure 6.42). After IEF, the chip was aligned to

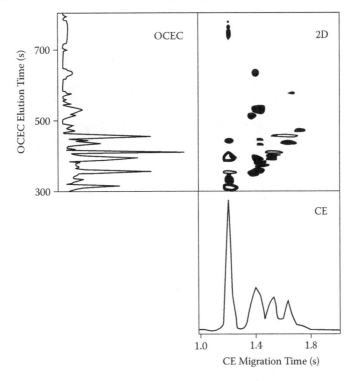

FIGURE 6.40 2D separation of TRITC-labeled tryptic peptides of β-caesin. The projections of the 2D separation into the first dimension (OCEC) and second dimension (CE) are shown to the left and below the 2D contour plot, respectively. The field strengths were 220 V/cm in the OCEC channel and 1,890 V/cm in the CE channel. The buffer was 10 mM sodium borate with 30% (v/v) acetonitrile. The detection point was given as y in Figure 4.16 (Chapter 4) and was 0.8 cm past valve V2 in the CE channel.[333] Reprinted with permission from the American Chemical Society.

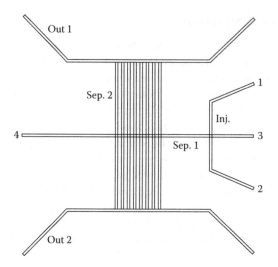

FIGURE 6.41 Layout of the quartz chip for 2D separation. Inj., the injection channel; Sep. 1, the first-dimension separation channel; Sep. 2, the second-dimension separation array of five hundred parallel channels; Out 1 and Out 2, the waste collection channels. The channels cover an area of 16 × 16 mm; the overall chip size is 23 × 23 mm.[150] Reprinted with permission from the Institute of Physics Publishing.

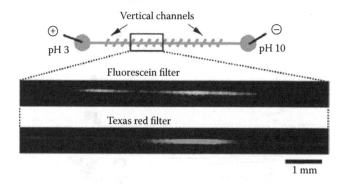

FIGURE 6.42 A sketch of the first-dimensional IEF channel. The slanted lines crossing the long IEF channel represent the vertical short channels used to connect to the array of channels used for separation in the second dimension. Below the diagram are the fluorescence images of a section of the IEF channel containing the focused proteins (CA^F, BSA^F, Ovalbumin^{TR}). The image on top was obtained using a filter that isolated fluorescein (F) fluorescence, and the image below was obtained using a filter that isolated Texas Red (TR) fluorescence. Both images were obtained over the same region of channels.[182] Reprinted with permission from the American Chemical Society.

a second chip consisting of several vertical channels (Figure 6.43), in order to carry out SDS-PAGE separation in the second dimension. Fluorescein-conjugated BSA was seen to be separated from Texas Red-conjugated ovalbumin.[182]

An IEF/CGE separation for proteins has been achieved on a PDMS chip. Microfluidic valves were used to prevent intermixing between the two separation buffers used in IEF and CGE separations. The IEF ampholyte was very sensitive to high buffer concentration, but a small amount of ampholyte in the CGE did not affect its separation resolution.[449]

Another MECC/CE separation was applied to analyze tryptic digest of BSA, ovalbumin, and hemoglobulin (human and bovine). In this work, asymmetric turns were incorporated in the serpentine channel.[565] Faster-switching power supplies were used, leading to a threefold increase in the

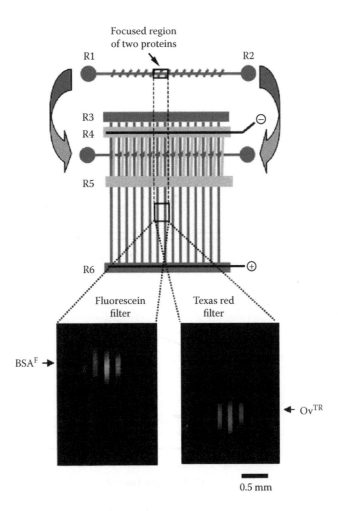

FIGURE 6.43 **(See color insert following page 144.)** A sketch of the second-dimensional electrophoresis channels aligned on the first-dimensional channel previously shown in Figure 6.42. The long red line represents the IEF channel, the parallel green lines represent the channels in the top PDMS slab, and the parallel blue lines represent the channels in the bottom PDMS slab. Below the sketch are the fluorescence images of a section of the electrophoresis channels containing the separated proteins (BSA[F], Ovalbumin[TR]) that overlapped in the IEF separation. The image on the left was obtained using a filter that isolated fluorescence from fluorescein (F), and the image on the right was obtained using a filter that isolated fluorescence from Texas Red (TR). Both images were obtained for the same region of the capillary array where the overlapping peaks of BSA[F] and Ovalbumin[TR] migrated. The weaker fluorescence signal in the lower portion of the fluorescein image may be due to impurities in the BSA[F] sample.[182] Reprinted with permission from the American Chemical Society.

sampling rate into the second dimension. This caused less SDS (from the first-dimension MECC) sampled into the second-dimension channels, resulting in higher separation resolution.[565] Although complete tryptic digestion of ovalbumin should produce thirty-three unique fragments, the 2D MECC/CE results shown in Figure 6.44 produced over fifty spots. This may be due to multiple-labeled peptides by 5-TAMRA (5-carboxytetramethylrhodamine, succinimidyl ester) or incomplete tryptic digestion. Spiking of a known ovalbumin tryptic fragment revealed an increase in the spot intensity as shown by the arrow position in Figure 6.44b, compared to Figure 6.44a.[565]

An IEF/CGE separation was carried out for protein analysis. The PMMA chip consisted of one IEF channel and three hundred CGE channels. The interface is shown in Figure 6.45. Transfer

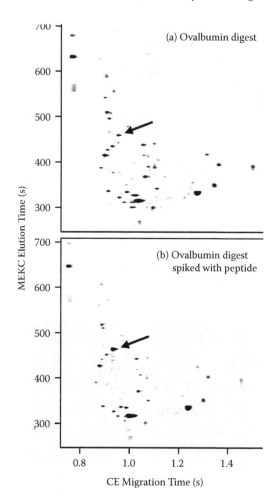

FIGURE 6.44   Two-dimensional separations of (a) an ovalbumin tryptic digest and (b) an ovalbumin tryptic digest spiked with a peptide of the sequence GGLEPINFQTAADQAR. The arrow indicates the identified peptide.[565] Reprinted with permission from the American Chemical Society.

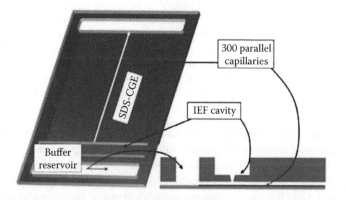

FIGURE 6.45   A PMMA chip for 2D electrophoresis. The credit-card-sized chip (85 × 55 × 2 mm) contains a cavity for the first-dimensional IEF channel and two buffer reservoirs for electrodes, respectively. These buffer reservoirs are physically connected by three hundred parallel second-dimensional channels (50 × 50 μm cross section, 64 mm long) for SDS-CGE. The cavities for IEF and the CGE channels are connected to each other by an opening of 50 μm width.[214] Reprinted with permission from the Royal Society of Chemistry.

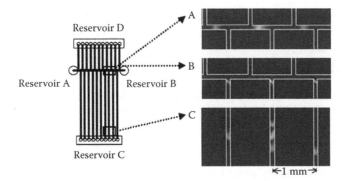

FIGURE 6.46 Fluorescent images of on-chip 2D separation of five model proteins using multiple separation media. The five FITC-conjugated proteins (10 ng/ml) are denatured as well as alkylated in order to maintain their denatured state. (a) Nonnative IEF with focusing order (from left to right) of (i) actin, (ii) bovine serum albumin, ovalbumin, and trypsin inhibitor, and (iii) parvalbumin; (b) electrokinetic transfer of focused proteins; (C) SDS–gel electrophoresis (using PEO gel). Images were captured at 90 s following the initiation of IEF or SDS–gel electrophoresis separations. Images were obtained using either green fluorescence of protein-fluorescein conjugates in IEF or red fluorescence of Sypro Red-labeled proteins during electrokinetic transfer and size-based separation.[667] Reprinted with permission from the American Chemical Society.

of proteins after IEF was accomplished via the 50 μm opening into the three hundred CGE channels.[214]

Another IEF/CGE separation was carried out on a PC chip for protein analysis. The first dimension in one horizontal channel was used to perform denaturing IEF, and the second dimension in an array of ten channels was used for SDS–gel electrophoresis, as shown in Figure 6.46. Instead of sequentially sampling proteins eluted from IEF, they are electrokinetically transferred in a parallel fashion into an array of orthogonal microchannels for CGE. The high resolution in this 2D separation was mostly contributed by IEF under denaturing conditions. Moreover, since the protein analytes were denatured, they were well prepared for rapid and effective formation of SDS–protein complexes, in the second dimension. The proteins were separated into three groups in the first dimension, as shown in Figure 6.46a. The IEF order is actin (BSA, ovalbumin, and trypsin inhibitor) and paralbumin, according to the pI values of 5.20 (4.60, 4.50, 4.55) and 4.10, respectively. These three groups of proteins were then transferred into the CGE channels, as shown in Figure 6.46b. Owing to the use of EK transfer and EOF suppression in the channel, there was negligible transfer of undesirable species (i.e., urea and focused ampholytes) to the second dimension. Subsequent CGE separation of BSA, ovalbumin, and trypsin inhibitor was achieved according to their MW.[667]

An IEF/CGE separation of proteins (GFP, FITC-labeled ovalbumin) was carried out on an acrylic chip. EK mobilization at ~20 μm/s allows repeated injections of the focused bands into the cross-channels for CE separations. Even though the ampholyte entered into the CE channel, normal CE occurred at a single pH of 8.5 and IEF did not occur. This was because there were no acidic and basic boundary conditions at the terminal reservoirs for IEF to proceed.[668]

In another report, 2D separation was applied to resolve DNA duplexes. Three types of DNA duplexes (136 bp homoduplex, 339 bp heteroduplex, and 450 bp heteroduplex) were involved in a single nucleotide polymorphism (SNP) study. First-dimension CGE separated the three types of DNA molecules. Subsequent temperature gradient gel electrophoresis (TGGE), which is an analogous counterpart to denaturing gradient gel electrophoresis, was used to resolve the duplexes to reveal SNP.[669]

Separation of proteins extracted from Jurkat cell lysates was achieved on a microchannel on a PMMA chip. Pressurization has been used prior to CE to improve resolution, as described in Section 6.2.2. An additional merit is that transient size separation occurred during pressurization.

During subsequent CE separation, IEF occurred in the microchannel. This combination of transient size separation and IEF improved resolution, which is akin to 2D separation.[619]

6.5   PROBLEM SETS

1. What neutral molecules have been used as EOF markers?[670,748]
2. In the use of cross-linked gel in CGE separation, there is the bubble formation problem. Explain its origin. (2 marks)
3. In carrying out CGE separation on-chip, why is the channel usually coated? (2 marks)
4. What is the method used to adjust or control the pore size in a cross-linked gel? (2 marks)
5. What is the cross-linking agent used for producing a cross-linked polyacrylamide gel? (2 marks)
6. In a two-dimensional separation for protein analysis, the separation mode in the first dimension is IEF and that in the second one is CGE. Describe briefly the basis of these two CE modes for protein separation. (4 marks)
7. In the method of SDS-PAGE for protein separation, what is the function of SDS, which is an anionic surfactant? (2 marks)
8. In the separation of various explosives using MECC (Figure 7.9), explain why the compounds are detected as negative peaks.[620] (2 marks)
9. In labeling amino acids, peptides, and proteins, FITC can only label primary and secondary, but not tertiary, amino groups. Why? (2 marks)
10. In chromatography or electrophoresis, what does the plate number or number of theoretical plates describe? (2 marks)
11. Write down the general form of the Van Deemter equation. Hence, explain why CE separation is inherently more efficient than chromatography. (3 marks)
12. Cross-linked polymer gels have higher separation resolution than linear polymer gels, but the former was seldom used in microchip CGE. Why? (2 marks)
13. What caused the breakthrough in the early 1990s in developing the first miniaturized liquid separation chip since the GC chip was first reported in 1979? (2 marks)

7 Detection Methods

7.1 OPTICAL DETECTION METHODS

7.1.1 Fluorescence Detection

7.1.1.1 Single-Channel Fluorescence Detection

The most commonly used detection system for microchip applications has been laser-induced fluorescence (LIF) (see Figure 7.1).[159,316,550,670] In LIF, the excitation light is provided by a laser in order to increase the measured fluorescence emission. To reduce the amount of excitation light entering the photodetector, bandpass emission filters are usually used. In some reports, even two 530 nm bandpass filters have been employed.[152,557]

For multiple wavelength detection, several emission filters of different wavelength band pass are used. A four-color LIF detection system that detected at four wavelengths is shown in Figure 7.2. It consisted of three dichroic filters and four photomultiplier tubes (PMT) to detect at four wavelengths.[543]

Optical fibers have been employed as waveguides to confine the excitation beam and fluorescence emission beam. With an excitation optical fiber, the excitation laser beam has been guided along the channel, leading to the so-called longitudinal excitation. As compared to the conventional transverse excitation, in which the excitation beam is perpendicular to the channel, longitudinal excitation has resulted in a twentyfold improvement in the signal-to-noise ration (S/N). This is because the longitudinal excitation beam is narrower than the channel, and so there is less background caused by the light scattered from the channel wall. The use of the optical fibers for longitudinal excitation and emission collection has led to an improved detection limit of fluorescein of 3 nM (or twenty thousand molecules).[671]

In another report, in order to reduce channel wall light scattering, the laser beam size was reduced. This was achieved by reshaping the Gaussian beam shape into a flattop beam shape by using a Keplerian beam reshaper. In this way, a collimated laser beam was reduced to a size of 33 μm, facilitating its use inside a 50 μm wide channel in a glass microchip for LIF detection.[672]

In one report, the sample stream was physically focused so that it was within the illumination region of the laser probe beam. In this way, fluorescence detection of DNA was enhanced.[329] In another report, the effective observation volume was reduced by confinement within nanochannels in a fused silica chip. In this way, fluorescent detection of even single molecules was achieved.[673]

Optical fibers (excitation and emission) have also been integrated on a poly(methylmethacrylate) (PMMA) chip for fluorescent detection. An intercalating near-infrared (IR) dye (TOPRO-5) with an excitation wavelength of 755 nm was used to label DNA fragments. A subattomol detection limit of the labeled DNA fragments was achieved.[674]

Although fluorescence detection does not depend on the optical path length as much as absorbance detection does, an increase in the path length does provide an increased fluorescent intensity. For instance, the measured fluorescence intensity is increased when a greater channel depth is used. This is achieved with a channel geometry in which the depth (20 μm) is twice as large as the width (10 μm). This aspect ratio can be made possible only by using anisotropic Si etch to create a molding master for producing the poly(dimethylsiloxane) (PDMS) replica.[159]

Two-photon fluorescent (TPF) detection, which was initiated by a nonlinear optical absorption process, has been performed on a quartz chip. Since the fluorescent efficiency in TPF is inversely proportional to the excitation beam area, the path length dependence problem in fluorescence is

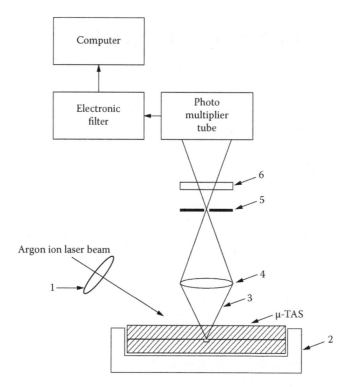

FIGURE 7.1   Schematic of the laser-induced fluorescence detection system. Excitation was provided by an argon ion laser light. It was focused with a lens (1) onto the separation channel in a microchip. The chip was held in place with a Plexiglas holder (2). Fluorescence emission (3) was collected with a microscope objective (4), focused onto a spatial filter (5), emission filter (6), and then detected with a photomultiplier tube.[550] Reprinted with permission from the American Chemical Society.

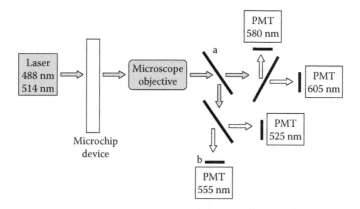

FIGURE 7.2   Four-color optical excitation and detection system for LIF. Excitation was provided by a laser emitting at 488 and 514 nm. (a) Dichroic mirrors; (b) emission bandpass filters. Detection was carried out at four wavelengths (525, 555, 580, and 605 nm).[543] Reprinted with permission from Wiley-VCH Verlag.

significantly reduced. This method is used for analysis of β-naphthylamine (excitation at 580 nm), which is the enzymatic product of leucine aminopeptidase (LAP) acting on the fluorogenic substrate: leucine β-naphthylamide.[675]

Time-resolved fluorescence was used to detect rhodamine 6G (R6G), sulforhodamine 101 (SR101), and rhodamine B (RB). A Ti-sapphire laser (800 nm, 50 fs) was used, but the excitation wavelength has been converted to 400 nm (for R6G) or 532 nm (for SR101 or RB) by an optical parametric amplifier or by second harmonic generation.[676] In another report, fluorescence burst detection was used for detection of single chromophore molecules.[677]

Laser excitation has also been provided by laser diodes. For instance, a red diode laser (635 nm) has been used as the light source for LIF detection.[619,678] Detection of the Cy 5 dye in a 13 μm deep channel in a Pyrex glass chip has produced a limit of detection (LOD) of 20 μM, as optimized using different filter sets. However, with a 20 μm deep channel, the LOD of Cy 5 can be further improved to 9 pM.[678] A laser diode (GaAlAs, 780 nm) was also used for near-IR excitation.[679] A near-IR dye, IRD800, was used to label DNA primers for their LIF detection after separation on a PMMA chip.[380]

Fluorescent excitation can be achieved by nonlaser sources, such as xenon arc lamp or LED. A xenon arc lamp with an inverted microscope was used to measure cellular fluorescence. Fluorescent detection (green emission at 520 nm) was performed together with optical observation (using red light and a long-pass filter to avoid interfering with the fluorescent detection) for long-term monitoring of single-cell measurement. In this case, it is ensured that the cellular fluorescence, but not the artifacts, was measured.[680]

The LED has also been used as the excitation source for fluorescent detection. For instance, two LEDs (red and blue) were used to detect fluorescein isothiocyanate (FITC)-labeled amino acids on a PMMA chip. The blue LED was used for excitation and the red LED for background noise compensation, leading to a fourfold S/N enhancement.[681] Moreover, blue LEDs (470 nm) were used to detect various dyes, such as APTS (for sugar isomers),[560] RNA molecular beacon probes,[337] and YOPRO-1 (for dsDNA).[613] In one report, a ring of twenty-four blue light-emitting diodes (LEDs; 470 nm) has been used for flood illumination.[321]

7.1.1.2 Scanning Detector

For fluorescent detection in multiple channels on microchips, scanning detectors and CCD cameras have been used. For instance, scanning LIF detection was first performed using a galvanoscanner, which sequentially probed sixteen channels[265] and forty-eight channels.[557]

Later, a four-color rotary confocal fluorescent scanner was designed to simultaneously detect 96 channels[977] and 384 channels.[980] The confocal scanner used to detect 384 channels on a microcapillary array electrophoresis (μCAE) chip is shown in Figure 7.3.[980]

Scanning LIF detection was also achieved using the acoustic-optical effect. This method was used to sequentially detect eight channels.[682,683] The acoustic-optical detection (AOD) is based on the phenomenon that a transparent crystal (TeO$_2$), through which an acoustic wave propagates, alters the direction of the laser light passing through it (see Figure 7.4). The degree of refraction is adjusted by changing the frequency of an acoustic wave, using a piezoelectric transducer.[683,684] As shown in Figure 7.5, successive detection on eight channels was achieved. This method offered a much faster scan rate (200 Hz) than a translating stage-based scanner (3.3 Hz scan rate). Three scanning modes, raster (uni- or bidirectional), step, and random addressing, are available. The last one is harder to achieve with a translating stage- or galvanometric-based scanner.[682] One obstacle with AOD scanning is laser attenuation as the laser is deflected. Partial correction can be achieved mathematically, i.e., by the use of a *cosh* function.[684]

Charge-coupled device (CCD) cameras have also been used for multichannel fluorescent detection. However, the use of the CCD detector suffers from the need to read out the entire image content information in order to determine the pixel intensity at the flow channel location. Accordingly, a complementary metal oxide semiconductor (CMOS) imager was used. This method offers direct control over individual pixels, and can provide much faster response times and longer integration times for those desired pixels.[252]

On the other hand, the scanning confocal fluorescent detector has better sensitivity than the CCD detector because the confocal detector provides better spatial filtering of the fluorescent background of the substrate (e.g., glass).[108]

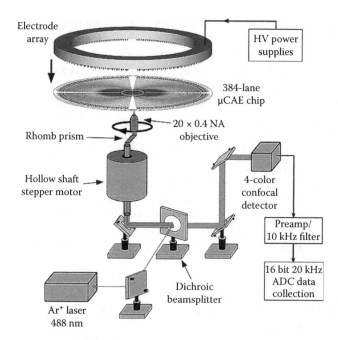

FIGURE 7.3   The rotary confocal fluorescence scanner used to detect μCAE chip separations. Laser excitation at 488 nm (100 mW) is directed up through the hollow shaft of a computer-controlled stepper motor, deflected 1.0 cm off-axis by a rhomb prism, and focused on the electrophoresis lanes by a microscope objective. The stepper motor rotates the rhomb/objective assembly just under the lower surface of the microchip at five revolutions per second. Fluorescence is collected along the same path and spectrally, and is spatially filtered before impinging on the four-color confocal detector.[980] Reprinted with permission from the American Chemical Society.

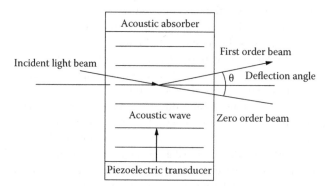

FIGURE 7.4   The acousto-optical detection system. Deflection of the incident laser beam by the acousto-optic crystal. Deflection angle is determined by the frequency of the acoustic wave.[683] Reprinted with permission from John Wiley & Sons.

7.1.1.3   Background Reduction

Reduction of background fluorescence can be achieved by using the fused silica substrate. In this way, the detection limit can be improved.[106] Further reduction in the background could be achieved by optimizing the chip bonding procedure to reduce light scattering centers, by avoiding the curved channel wall in excitation, and by adjusting incidence angle. With these improvements, an LOD of 30 pM fluorescein was achieved.[330]

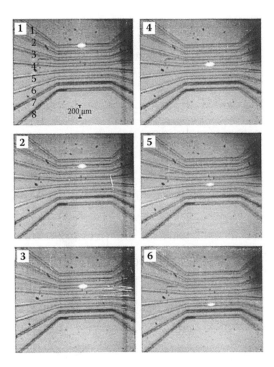

FIGURE 7.5 Acousto-optical-based laser beam scanning of six channels filled with 6-carboxyfluorescein (10^{-5} M). Frames of the laser beam step scanning across the microchannels, captured from the video record.[682] Reprinted with permission from the American Chemical Society.

Another way to reduce the fluorescence background, especially from plastic chips, is to modulate the velocity of a fluorescent analyte. The analyte velocity is modulated by periodic variation (in 7 to 20 Hz) of the separation voltage. Noise rejection is achieved using a lock-in amplifier because only the fluorescent signal but not the background from the chip substrate was modulated. With this method, a decrease in LOD by one order of magnitude has been obtained.[685]

7.1.1.4 Photobleaching Effect

Photobleaching of fluorophores in fluorescent detection can affect quantitative measurement. The extent of photobleaching can be estimated by the photochemical lifetime (τ). This parameter can be determined by measuring the fluorescence signal (F) of the fluorophore as a function of residence time (t). The parameter is obtained from the slope of a plot of ln F vs. t. Using this method, the photochemical lifetime has been determined to be 51 ms for o-phthaldialdehyde (OPA)-arginine and 58 ms for OPA-glycine.[106]

The photobleaching effect can be decreased in LIF detection. This can be achieved by adding various reducing agents, such as 2-mercaptoethanol,[95,323,351,686] ascorbic acid,[1038] or DTT.[687] Photobleaching of an analyte can also be reduced by decreasing its residence time. This can be achieved by increasing the analyte velocity using a narrower detection region, as compared to the main channel (see Figure 7.6).[285]

7.1.1.5 Integrated Fluorescent Detector

An array of circular or elliptical microlenses made of photoresist was constructed on a chip, as shown in Figure 7.7. The microlenses were fabricated on both the bottom glass plate (for focusing the excitation beam) and the top glass plate (for collecting the emission). In addition, an array of entrance apertures was formed around the focusing microlenses to limit the excitation beam;

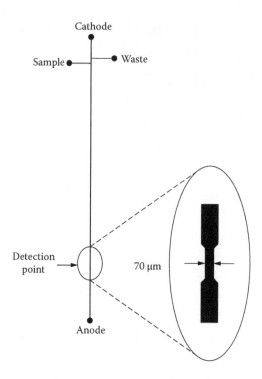

FIGURE 7.6   Reduction of photobleaching using a narrow 70 μm neck at the detection point. This is situated along a straight capillary of 8 cm separation length with a 500 μm offset twin T injector.[285] Reprinted with permission from the American Chemical Society.

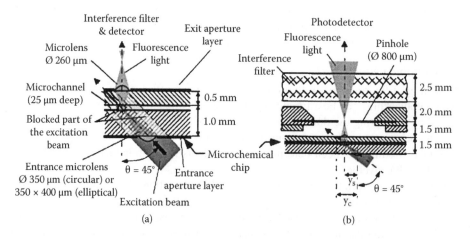

FIGURE 7.7   Cross-sectional view of a microchemical chip integrated with some optical components. (a) The microchemical chip showing the microlens (entrance and exit), the 3,000 Å Cr aperture layers (entrance and exit), and microchannel (25 μm deep); (b) the microchemical chip together with the pinhole (800 μM) and interference filter for detection by the photodetector.[688] Reprinted with permission from the American Chemical Society.

another array of exit apertures was formed around the collecting microlenses to block off any unabsorbed excitation beam from being scattered into the detector. With the reduction in excitation and scattering, the detection limit of Cy5 was found to be reduced to 3.3 nM.[688]

Integration of a lens was also achieved by molding a cylindrical lens together with a PDMS channel. Both the lens and the channel serve as an optical waveguide for fluorescent detection of FITC-labeled albumin (LOD = 0.3 g/L).[689]

Integration of optical filters on the microchip was also reported. This can be achieved by depositing a predetermined thickness of alternate SiO_2 and Si_3N_4 multilayers.[252] An interference filter (515 nm) was also fabricated on Si by depositing forty alternating layers of SiO_2 and TiO_2 thin films, giving a total thickness of ~6 μm.[693] In another report, an optical interference filter was produced by the deposition of the ZnS/YF_3 multilayer.[690]

Integrated fluorescent detection was also achieved by fabricating the interference filter and Si photodiode on a Si substrate on which a parylene channel was built.[691]

In one report, the detection optical fiber (coupled to a blue LED light source) and the microavalanche photodiode were both embedded in a PDMS chip to detect proteins.[692]

Integrated fluorescence detection was also achieved using hydrogenated amorphous silicon positive-intrinsic-negative (PIN) photodiodes fabricated on a glass chip.[690] In another report, the integrated Si photodiode was used for fluorescent detection of DNA, with an LOD of 0.9 ng/μL.[693]

A planar optical waveguide was integrated within a PDMS chip for fluorescent detection. The waveguide consists of a 150 nm thick silicon nitride layer deposited on a 2.1 μm thick SiO_2 buffer layer on a Si substrate. The rabbit IgG was immobilized on the waveguide layer. The evanescent field-based fluorescence detection has been used to detect Cy5-labeled anti-rabbit IgG, which binds to the immobilized rabbit IgG.[694] This evanescent field-based format allows excitation of fluorescent molecules present within ~200 nm of the waveguide surface.[694,695]

A four-layer microchip has been constructed to generate total internal reflection (TIR) and an evanescent field (see Figure 7.8). Surface-adhered Nile red-labeled fluorescent microspheres (1 μm) are excited by the evanescent field for fluorescent measurement. An essential feature on the chip was the micromirror that was constructed by depositing Au/Cr on the slanted wall (54.7° due to anisotropic etch of Si). Operation near the critical angle θ_c ensures strong evanescent intensity.[695]

7.1.2 INDIRECT FLUORESCENT DETECTION

Besides the commonly used direct LIF detection, indirect LIF detection on the microchip has also been reported. This method has been employed to detect explosives in spiked soil samples (see Figure 7.9).[620] In contrast to a capillary-based system, an increase in E from 185 to 370 V/cm for micellar electrokinetic capillary chromatography (MECC) separation did not result in an unstable background fluorescence due to excessive Joule heating. This was probably because of the effective heat dissipation in the glass chip. However, upon multiple injection, it was found that the detection sensitivity decreased, which might be caused by the degradation of the visualizing dye (Cy7).[620] Indirect LIF also allows the detection of unlabeled amino acids.[683]

For other applications using indirect fluorescent detection, see the appendix.

7.1.3 MULTIPLE-POINT FLUORESCENT DETECTION

Multiple-point fluorescent detection has been proposed to enhance detection sensitivity. This method is based on the use of a detector function, such as the Shah function. The time-domain signals were first detected, and they were converted into a frequency-domain plot by Fourier transformation. Therefore, this technique was dubbed Shah convolution Fourier transform (SCOFT) detection. As a comparison, the single-detection point time-domain response is commonly known as the electropherogram.[698,699,701]

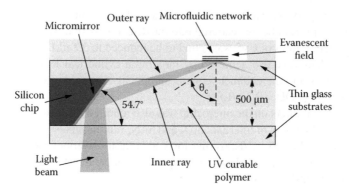

FIGURE 7.8 Schematic cross section of the four-layer TIR chip. The laser beam enters at the bottom of the chip and is redirected upon hitting the micromirror sidewall. Then the beam passes through the thin glass substrate and enters the microchannel. The critical angle (θ_c), which depends on the excitation wavelength, can be fine-tuned by filling the glass cavity (n_1 of glass = 1.526) with a polymer of a different refractive index n_2. Using the UV-curable polymer (n_2 = 1.524), θ_c is determined to be 64.8°. So the reflectance light loss was reduced from 35% to 5%, thus increasing the evanescent field intensity.[695] Reprinted with permission from the Royal Society of Chemistry.

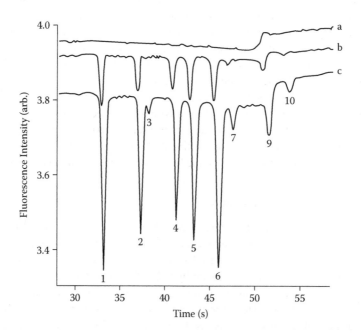

FIGURE 7.9 MECC separation with indirect LIF detection of explosives from spiked soil samples: (a) soil blank, (b) soil containing 1 ppm of each analyte, and (c) soil containing 5 ppm of each analyte. Analytes: (1) TNB, (2) DNB, (3) NB, (4) TNT, (5) tetryl, (6) 2,4-DNT, (7) 2,6-DNT, (9) 2-Am-2,6-DNT, (10) 4-Am-2,6-DNT. Conditions: Run buffer, 50 mM borate, pH 8.5, 50 mM SDS, 5 µM Cy7; separation voltage, 4 kV; separation distance, 65 mm.[620] Reprinted with permission from the American Chemical Society.

The Shah function can be realized physically by using multiple slits. With microfabrication, the separation channel and detection slits (fifty-five) can be fabricated and aligned to each other (see Figure 7.10).[698] A 488 nm Ar+ laser line was expanded to produce a parallel beam illuminating along the separation channel consisting of fifty-five slits. The fluorescent emission was detected by a PMT detector. The time-domain signal for a single component (fluorescein) is shown in Figure 7.11a. The

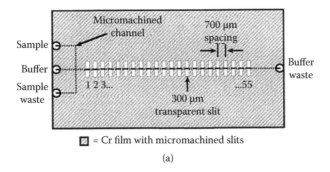

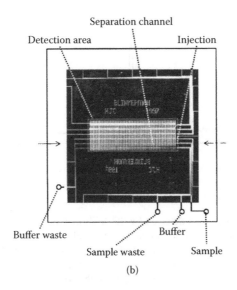

FIGURE 7.10 Micromachined glass electrophoresis chip for SCOFT detection: (a) schematic showing the solution reservoirs, one electrophoresis channel (4.5 cm long, 15 μm deep, 50 μm wide at the top, 20 μm wide at the bottom), and patterned Cr layer with micromachined slits (300 μm wide, 700 μm spaced center to center); (b) scanned image of the actual 75 × 75 mm chip. The six different channel systems can be seen.[698] Reprinted with permission from the American Chemical Society.

frequency-domain signal obtained after Fourier transform is shown in Figure 7.11b. The frequency is related to the migration time in the usual electropherogram format. Using this technique, the baseline drift due to a changing fluorescent background can also be eliminated.[698]

The use of multiple sample injections (up to a maximum of three) was found to enhance S/N; i.e., the S/N is slightly higher than the square root of the number of injected sample plugs. In addition, multipoint detections of a two-component sample[700,701] or four-component sample[699] were also achieved, but the separation resolution was not as good as that obtained from the conventional single-point detection.[701] Besides Fourier transform, wavelet transform was also used in multipoint fluorescent detection to retain some time information in addition to the frequency information.[702]

Instead of realizing the Shah detector function using the multiple slits, multiple excitation points were created. This was achieved by monolithically integrating multiple planar waveguide beam splitters on a fluidic channel. In this way, 128 excitation points were created on the fluid channel for multipoint fluorescent detection.[413]

In contrast to using physical slits or beam splitters, the Shah (or comb) function has been applied after collecting the signal over an unmasked channel (9 mm long) using a cooled CCD detector.[703] Since the detector function was applied after data collection, greater flexibility can occur in the

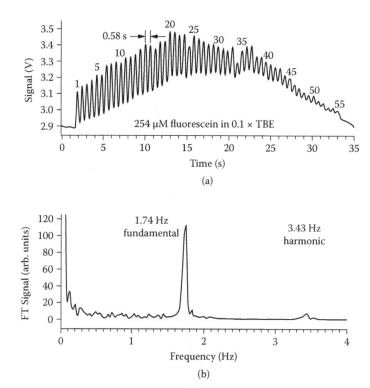

FIGURE 7.11 One-component injection detected by SCOFT. (a) Time-domain data produced when a fluorescein sample is injected down the separation channel: 55 peaks, spaced at 0.58 s, superimposed on the Gaussian distribution of the expanded laser beam. (b) Fourier transformation of part (a). Peaks for 1.74 Hz (fundamental) and 3.43 Hz (second harmonic) can be seen. The frequency of 1.74 Hz was correlated to the migration time of ~0.58 s.[698] Reprinted with permission from the American Chemical Society.

choice of the detector function (Shah, modified Shah, or sine). When the Shah function was used, although the second harmonic peak in the frequency domain was not seen, new artifact peaks appeared. However, when the sine function was used before FT, there was no artifact peak. This data reprocessing capability has provided an advantage over the SCOFT method.[703] A Hadamard transform (HT) was also reported for detection using a CCD camera. An eightfold improvement in S/N was reported for the detection of fluorescein.[704]

However, the sampling frequency of CCD is low (28 Hz), compared to PMT (MHz), limiting the use of CCD for fast-moving analytes or beads.[413]

Another method, cross-correlation chromatography, was also reported for detection sensitivity enhancement (see Figure 7.12). Multiple injections were performed in a continuous but random sequence. A single point detection output was recorded. This output was correlated with the injection profile. This correlation enhanced the detection sensitivity due to the multiplex advantage.[705]

7.1.4 Absorbance Detection

A UV absorbance detector has been achieved on-chip by fabricating a U-shaped detection cell (see Figure 7.13). This cell provides a longer optical path length of 140 μm. An optical fiber was used to launch the excitation light, and a second fiber was used to collect the transmitted light. This longitudinal absorbance method provides an elevenfold increase in absorbance, giving a detection limit of 6 μm for hydrolyzed FITC. This result is better than the theoretical sevenfold increase based on a sevenfold increase in the path length.[671]

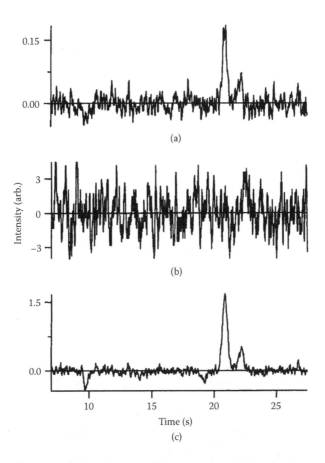

FIGURE 7.12 Electropherogram and correlograms of the separation of dichlorofluorescein (70 pM) and fluorescein (63 pM): (a) an individual seven-bit correlogram, (b) electropherogram from a single 0.25 s gated injection, (c) the averaged twelve-bit correlogram.[705] Reprinted with permission from the American Chemical Society.

In another report, a monolithic optical waveguide was fabricated along a U-shaped cell for the detection of various compounds (see Figure 7.14).[706] This method has been used for absorbance detection at 254 nm[88] or 488 nm.[706]

To increase the absorbance detection sensitivity, both a collimating lens and a detection slit have been used in a PDMS chip to reduce collection of scattered light.[707] An improved S/N ratio was also obtained by using tapered channel waveguides as the collimators as well as the elliptical lens for absorbance detection.[708]

To improve light collection for absorbance detection, a photosensor array was integrated on an acrylic chip with a smooth and transparent surface. This allows absorbance measurement (420 to 1,000 nm) down to 0.0004 AU. Detection of 5 mg/mL blue dextran and 10 mg/ml each of lactic dehydrogenase (39 kDa) and β-galactosidase (116 kDa) has been carried out.[240]

A deeper channel (e.g., 100 μm) can be used to enhance absorbance detection because of the longer optical path length. However, this is feasible only when nonaqueous capillary electrophoresis (CE), which produces low electrical current, is used.[622] An increased path length (720 μm) needed for optical absorbance detection was also achieved by constructing a three-dimensional (3D) fluid path.[709]

An increase in optical path length can also be achieved in a multireflection cell constructed on the microchip. Unlike in a Si chip where the crystal plane <111> could be directly used as the reflective surface,[710] a metal film was needed in a glass chip to create the reflective surface.[711] It was found that an Al film (80 nm thick) provided the best reflectance. Using bromothymol blue,

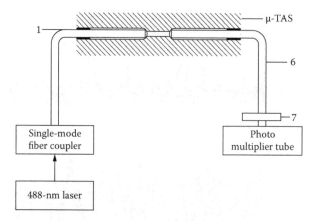

FIGURE 7.13 Schematic of longitudinal absorbance detection on a glass microchip. 1, launch optical fiber; 6, absorbance collection fiber; 7, optical filter. A 3.1 μm core fiber was used for launching and a 7.9 μm core fiber was used for collecting light. Both were single-mode optical fibers that consisted of silica core, 125 μm diameter silica cladding, and 250 μm diameter jacket polyimide coating. The fiber was first stripped of the polyimide coating, and then the silica cladding was etched. The fiber was inserted into a 230 μm wide and 150 μm deep channel. An index matching fluid (1,4-dibromobutane, $n = 1.519$) was also used to fill the channel.[671] Reprinted with permission from the American Chemical Society.

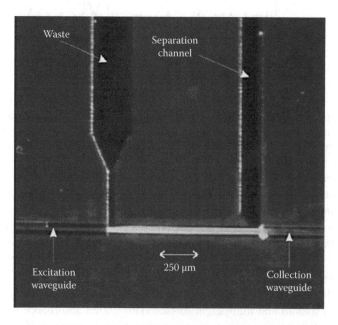

FIGURE 7.14 Micrograph of a U-shaped absorption cell with an optical path length of 1,000 μm on a silica glass microchip. A 250 nM fluorescein solution was used for visualization.[706] Reprinted with permission from Elsevier Science.

a five- to tenfold enhancement in absorbance (633 nm) was achieved. This corresponded to an increased effective path length of 50 to 272 μm, as determined from the channel depth of 10 to 30 μm.[711] Another approach was to employ a liquid-core waveguide in a PDMS chip with a 3D path for absorbance detection. Instead of then using a reflective coating, total internal reflection at a polymer cladding was employed. Since the refractive index of the liquid core (water, $n = 1.33$) was lower than that of the polymer cladding (amorphous Teflon AF, $n = 1.29$), total internal reflection could occur.

This polymer cladding was necessary because the refractive index of PDMS ($n \sim 1.41$) was not lower than that of water ($n = 1.33$). Using this method, the LOD of crystal violet ($\lambda = 590$ nm) was less than 1.3 µM, which was comparable to that obtained in conventional detection using a 1 cm cuvette.[712]

In another approach, optical absorbance detection was achieved in an injection-molded PMMA flow cell consisting of 1,800 pillars (25 µm high) that were used as diffraction elements. The LOD of Nile blue A perchlorate was determined to be 1.2 µM.[713]

Optical absorbance detection was also achieved using a CCD-based spectrophotometer. Two food dyes (FD&C Blue #1 dye and FD&C Red #3 dye) were first separated by open-tubular liquid chromatography and then detected on a PDMS chip. The LODs for the blue and red dyes were determined to be 80 and 200 µM, respectively.[569]

Optical absorbance detection has been achieved based on reflectance measurement. This was achieved on a plastic disk for p-nitrophenol detection (430 nm). The disk contained 1 wt% of titanium white pigment so that the disk was reflective to white light. For enzyme measurements in 48 channels on the chip, a scanning detection method should be used. Instead of using a scanning detector, a stationary detector was used with the disk spinning at 60 to 300 rpm. The disk was in fact spun in order to generate the centrifugal force needed for fluid flow.[1042]

A liquid prism was created on a PDMS chip for detection based on absorption and refractive index shift. The liquid prism was formed by filling a hollow triangular-shaped chamber with a liquid sample. Excitation and emission were arranged at the minimum deviation configuration. At a low concentration of fluorescein (<100 µM), excitation light from an optical fiber was absorbed by the molecule, but there was no shift in the excitation maximum. In this absorption-only mode, the LOD of fluorescein was 6 µM. At higher concentration (i.e., 100 to 1,000 µM), there is an additional shift in the excitation maximum. This leads to a much sharper decrease in the measured intensity, which is more than can be accounted for simply by the absorption effect.[714]

UV absorbance detection was used to analyze gases. For instance, a gaseous mixture of benzene, toluene, and xylene (BTX) was detected using optical fibers along a 20 mm long channel.[131,715]

7.1.5 PLASMA EMISSION DETECTION

A plasma emission detector was constructed on-chip (see Figure 7.15). An image of the helium plasma formed inside the chamber is shown in Figure 7.16.[718] The plasma was generated by atmospheric pressure DC glow discharge using He (99.995% pure).[716,717] Various carbon-containing compounds (hexane, methanol, ethanol, 1-propanol, 1-butanol, 1-pentanol) in a gas chromatography (GC) effluent were directed to the plasma detection chip. The optical emission was measured at 519 nm. For hexane, the detection limit was 10^{-12} g/s (or 800 ppb), but with only two decades of dynamic range.[716]

The microwave-induced plasma (MIP) is the most popular plasma used for conventional GC–optical emission spectroscopy (OES). However, the DC glow discharge plasma has recently received more attention because it can be operated at a low temperature, albeit at a low pressure of 1 to 30 Torr so as to avoid excessive gas heating and arcing. However, in a miniaturized device, the pressure does not have to be very low (e.g., 860 Torr or atmospheric) because of a decrease in the device dimension. Therefore, when the DC glow discharge plasma was produced in the chip, it had an improved lifetime, probably because of the low-temperature operation at atmospheric pressure, minimizing cathodic sputtering.[716] However, the detector signal shows a marked band broadening and peak tailing compared with the FID signal. This is mainly attributed to the dead volumes of the connection between the conventional GC column and the plasma chip detector, thus necessitating the on-chip integration of GC and plasma detector.[716] The LOD of methane was reported to be 10^{-14} g/s with over two decades of linear range.[718] This compares well with the LOD (for carbon) obtained from an FID and from a microwave-induced plasma atomic emission detector (10^{-12} g/s).[716]

Plasma emission detection was also carried out on a glass chip constructed with a main channel and a pair of side channels (75 µm wide and deep). Two 50 µm tungsten wires were inserted through

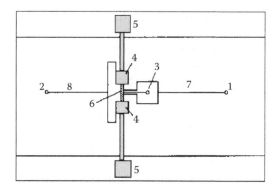

FIGURE 7.15   Schematic of a plasma emission chip consisting of the top and bottom plates. Features of the bottom plate (20 × 30 × 0.5 mm) are: 1, gas inlet; 2, gas outlet; 3, pressure sensor connection; 4, electrodes; 5, electrode connection pads. Etched in the top plate (14 × 30 × 0.5 mm) are: 6, plasma chamber; 7, inlet channel; 8, outlet channel.[718] Reprinted with permission from the Royal Society of Chemistry.

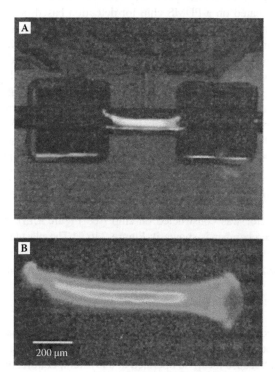

FIGURE 7.16   DC discharge plasma formed in a chamber (1,000 × 350 × 150 μm) at 750 Torr, 500 V, and 60 μA. (A) Original image. (B) False-color image of the same plasma.[718] Reprinted with permission from the Royal Society of Chemistry.

these side channels, and an electrical voltage of 2 kV was applied on them in order to produce emission in the main channel. Off-chip GC effluents (CH_3OH, CH_2Cl_2, $CHCl_3$) were directed into the plasma emission chip for detection. Gas chromatograms using the detection chip (at three wavelengths) and the FID (for comparison) were shown in Figure 7.17. LOD for chlorine was ~8 × 10^{-10} g Cl per second. However, peak tailing was more severe using the plasma emission chip, with respect to FID. This observation was possibly caused by transient fouling of the plasma chamber walls or diffusive loss of analytes into the side electrode channels.[563]

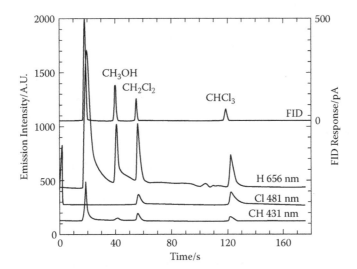

FIGURE 7.17 Gas chromatograms obtained by the plasma chip (emission at three different wavelengths) and by the flame ionization detector (FID) for the separation of methanol, CH_2Cl_2, and $CHCl_3$.[563] Reprinted with permission from the Royal Society of Chemistry.

This plasma detection method was also applied to detect a copper(II) ion solution (0.1 M) in a glass chip. Here, the liquid sample itself was employed as the cathode for the DC glow discharge, using Ar as the carrier gas.[719]

7.1.6 CHEMILUMINESCENCE (CL) DETECTOR

CL detection is based on the optical emission of an excited species formed in a chemical reaction. For instance, an excited species is formed when luminol (or 3-aminophthalhydrazide) reacts with H_2O_2 as catalyzed by various substances, such as metal ions or the peroxidase enzyme. CL detection is achieved as in fluorescent detection, except that no excitation light is needed.

CL detection has been employed in various chemical analyses. For instance, determination of mouse IgG was based on the CL reaction of a horseradish peroxidase (HRP)-goat anti-mouse IgG (HRP-Ab) and the luminol/H_2O_2 system. The amount of HRP-Ab decreased with an increasing amount of mouse IgG (10 to 60 µg/mL). Mouse IgG was then quantified using an internal standard (microperoxidase) after capillary zone electrophoresis (CZE) separation. To enhance light collection in CL detection, an aluminum mirror was deposited on the backside of the microchip.[720]

When luminol is the CL reagent, which works only at high pH (above pH 10), its mobilization by EOF is difficult. This is because at a high pH, the cation population becomes too low to sustain the electrical double layer for EO flow. A way around this is to add surfactant to the system to change the channel wall surface to be positively charged, leading to a reversal of EOF. For instance, CTAB was used to assist the EOF mobilization of luminol for the detection of a metal ion (i.e., Co) in a glass chip. A bonus of the use of surfactant is the finding that the lifetime of the CL emission increases in the presence of surfactants (i.e., CTAB + CTAC). It is also found that N_2 gas, which is a product of the luminol reaction, has caused a fluctuation in the CL intensity. So a reservoir was incorporated into the chip in order to help release the trapped N_2 gas.[284]

Determination of metal ions (Cr^{3+}, Co^{2+}, Cu^{2+}) has been achieved using the metal-catalyzed luminol-peroxide system. The EOF delivery of the CL reagent (H_2O_2, pH 11.7) in the aqueous system can be achieved using a side channel (with luminol present in the separation buffer with a pH of 6.0).[168]

In one report, the model compound, H_2O_2, was detected by CL based on the luminol system using Co^{2+} as a catalyst. Here, photodiodes were integrated onto a Si-glass chip for the CL detection.[310] In another report, H_2O_2 in the range of 0.01 to 1 mM was detected by mixing with a stream of luminol/microbial peroxidase in a Si-Pyrex chip.[307] This CL system was expanded for the detection of glucose by immobilizing glucose oxidase together with microbial peroxidase.[307]

In another report, CL determination of xanthine (1 mM) was achieved. Xanthine oxidase (XOD) first converted xanthine to H_2O_2 and uric acid. Then H_2O_2 reacted with the CL reagents, HRP and luminol, to produce the CL signal.[721]

CL systems other than luminol have also been used, such as the $Ru(bpy)_3^{2+}$ (or TBR) system. Chemical reaction of TBR has been achieved by oxidation using cerium(IV) sulfate[283,722] or lead oxide.[722] The addition of a nonionic surfactant (Triton X-45) strongly enhanced the CL emission.[722] The TBR system has been applied on a glass chip for the determination of alkaloids, such as codeine,[722] or atropine and pethidine.[283]

The CL system, 1,10-phenanthroline/H_2O_2, was used for determination of Cu in water samples in a PMMA chip.[723]

Another CL system, bis[2-(3,6,9,-trioxadecanyloxycarbonyl)-4-nitrophenyl]oxalate (TDPO), together with H_2O_2, has also been used for amino acid analysis.[168,353,724] Dansylated amino acids (e.g., glycine and lysine) were first separated on a quartz chip and then determined using CL detection. The LOD of dansylated lysine was 10 μM. The detection region was near the buffer waste vial where the migrated dansylated amino acids came into contact with the CL reagents at the vial. This configuration, which did not require an additional CL channel, had caused reproducibility problems in the migration time and peak area.[353] Separation and detection of dansylated d/l amino acids (Phe) were also achieved on a PDMS chip using the TDPO/H_2O_2 system. Here, since the CL reagent (TDPO) is dissolved in acetonitrile, TDPO can only be hydrodynamically delivered from a side channel. In addition, the CL reagent should not enter the separation channel; otherwise, the presence of acetonitrile in the separation buffer will alter the electric distribution in the separation channel, leading to reduced separation efficiency.[168]

CL detection can also be achieved using the electrochemical reaction. Therefore, this mode is called electrochemiluminescence (ECL). To date, two systems, $Ru(bpy)_3^{2+}$ (TBR)[725,726,1168] and $Ru(phen)_3^{2+}$ (TPR),[725] have been achieved on-chip.

First, $Ru(bpy)_3^{2+}$ and tripropylamine (TPA) are both oxidized electrochemically to form $Ru(bpy)_3^{3+}$ and TPA$^+$, respectively. Second, TPA$^+$ deprotonates and forms almost immediately TPA-H. Third, TPA-H subsequently reduces $Ru(bpy)_3^{3+}$ to form an excited state of the reduced form, $Ru(bpy)_3^{2+*}$. Fourth, this species relaxes back to $Ru(bpy)_3^{2+}$ and emits at $\lambda_{em} = 610$ nm. Note that TPA was consumed during the reaction, but TBR was recycled.[725] The other ECL system, $Ru(phen)_3^{2+}$, behaves in a similar manner.[725,727]

The ECL reaction can occur at room temperature in aqueous buffered solutions and in the presence of dissolved O_2 or other impurities. To provide the electric field for ECL to occur at the detector region, a pair of connecting floating Pt electrodes (50 μm wide, 50 μm apart, and 100 nm thick) has been used. To improve the adhesion of the Pt layer, a Cr underlayer was used. Unfortunately, Cr was corroded in the presence of Cl^-.[725]

Separation of three amino acids (l-valine, l-alanine, and l-aspartic acid, 300 μM) has been detected using 0.5 mM $Ru(bpy)_3^{2+}$ in the run buffer. The ECL emission intensity can be enhanced by increasing the voltage across the electrodes. However, with the floating electrodes, such an increase could only be achieved by increasing the separation voltage.[725]

ECL detection of $Ru(bpy)_3^{2+}$ (or TBR) was conducted on a Si-glass chip with an ITO anode.[727] Through the transparent ITO anode, orange light (620 nm) was observed and recorded by a detector. It was found by cyclic voltammetry that the oxidation potential was more positive, and the peak current density was less on an ITO anode, compared to the use of a Pt anode. In this work, Au cannot be used as an anode, presumably because of polymerization of TPA at the gold surface.[727]

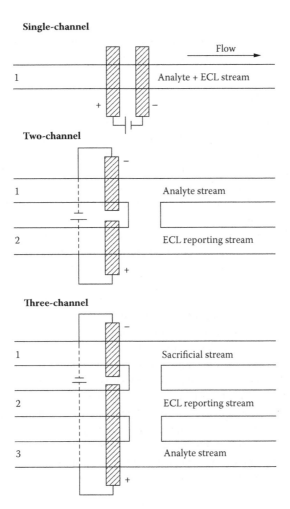

FIGURE 7.18 One-channel, two-channel, and three-channel systems in ECL detection.[728] Reprinted with permission from the American Chemical Society.

To apply the ECL method to detect more general analytes, multichannel systems have been proposed (see Figure 7.18).[274] For instance, a two-channel ECL detection was carried out on a PDMS-glass chip. This allows the detection of some reducible species, such as $Ru(bpy)_3^{3+}$, $Fe(CN)_6^{3-}$, and benzyl viologen (BV^{2+}), based on the reduction reaction not involved in the series of ECL reactions as previously described. These species are reduced at the cathode of one (sensing) channel, with the ECL reagent ($Ru(bpy)_3^{2+}$) electrogenerated at the anode in another (reporting) channel. The two-channel system permits an electrical contact between the ECL-reporting stream and the analyte-sensing stream, while preventing their mixing. If the previous one-channel system (see Figure 7.18) was used, the upper concentration limit of BV^{2+} was 100 µM. This is because when the concentration is above this limit, the products produced at the anode (reporting) and cathode (sensing) overlap in space, resulting in luminescence quenching. This two-channel system allows an increased concentration of BV^{2+} up to 10 mM to be detected, thus increasing the dynamic range.[86]

Other than reducible species, the oxidizable molecule, dopamine, can also be detected by indirect ECL detection, but only with the use of a three-channel system (see Figure 7.18). The first oxidation (reaction a) of dopamine occurred at the anode, and its presence was reported using the second oxidation (reaction b) of TPA by $Ru(bpy)_3^{3+}$, which was electrogenerated at the anode from $Ru(bpy)_3^{2+}$.[728] The cathodic reaction involved a sacrificial species $Ru(NH_3)_6^{3+}$, which was reduced to $Ru(NH_3)_6^{2+}$. In contrast to a previous report on indirect ECL detection,[274] the two oxidation reactions (a and b) were chemically separate, but electrically correlated.[728]

Other CL and ECL applications are given in the appendix.

7.1.7 REFRACTIVE INDEX (RI)

An RI detector was proposed, and this was based on the shift of an interference pattern. When a collimated laser beam passed through a holographic optical element, the beam was divided into two coherent beams. One beam (the probe beam) was directed through the channel, and the second beam (the reference beam) passed through the glass substrate only and acted as a control.[729] The two beams subsequently diverged in the far field and interfered, which generated an interference pattern to be detected by a photodiode array. A change in the RI of the solution in the channel resulted in a lateral shift in the fringe pattern.[729]

A refractive index detector was also constructed on-chip to detect glycerol (743 µM). An improved sensitivity was achieved in the backscatter format, rather than in the forward scatter format, because the probe beam passed the detector channel more than once in the backscatter format. This multiple-pass advantage was enhanced in the unique hemicylindrical shape of the channel. When the channel was irradiated by a He/Ne laser with a beam diameter of 0.8 mm, a series of backscatter fringes was produced. When the solution was changed from water to glycerol, there was a change in the refractive index, and the position of the fringes shifted (191 µM/mRIU).[151]

The change in the refractive index gradient (RIG) between adjacent laminar flow streams is employed as a molecular weight (MW) sensor, as shown in Figure 7.19. A sample was mixed with a mobile phase in a PDMS chip. The two laminar flow streams will be mixed by diffusion and create a concentration gradient. When a laser beam (780 nm single-mode diode laser) is incident upon the channel structure at a detection point, it will be deflected to the extent depending on the concentration gradient established. The deflection angle is detected by a one-dimensional position-sensitive detector (PSD). Such a deflection angle is greater at a second detection point farther downstream (say, 3 cm) than at the first detection point. The ratio of the PSD signals obtained at the two detection points is readily correlated to the analyte diffusion coefficient, and thus to the analyte MW for

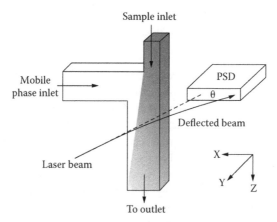

FIGURE 7.19 Illustration of the refractive index gradient (RIG) measurement by the deflection of a diode laser beam on a PSD. The diode laser beam is incident upon the channel structure, orthogonal to both the direction of flow and the concentration gradient. An analyte stream (dark, for illustration purposes only) flows in from the sampling channel and meets with a mobile-phase stream (light). The two streams flow together down the analysis channel as the analyte diffuses into the mobile-phase stream. The laser beam is deflected from a default position (dashed line) by the RIG between the sample and mobile-phase streams. The deflection signal θ is measured.[730] Reprinted with permission from the American Chemical Society.

a given class of compounds. A LOD of 56 ppm of PEG 11840 has been achieved, equivalent to 4.5 × 10^{-6} RI unit.[730]

7.1.8 THERMAL LENS MICROSCOPE (TLM)

Thermal lens microscopy (TLM) is a type of photothermal spectroscopy. TLM depends on the coaxial focusing of the excitation and probe laser beams (see Figure 7.20), which is achieved using the chromatic aberration of a microscopic objective lens.[731] The excitation beam can be provided by a YAG laser (532 nm)[846,1021] or an Ar ion laser (514.5 nm[846] or 488 nm[732]). The probe beam can be provided by a He-Ne laser (632.8 nm).[846,1021] After optical excitation of the analyte molecules, radiationless relaxation of the analytes occurs, and this produces a concave thermal lens (see Figure 7.20). The probe beam, which has a longer wavelength than the excitation beam, is focused just below the center of the thermal lens.[733] The probe beam was detected by a photodiode and the output was recorded as a TLM signal.[732]

The TLM detection method has been applied to detect labeled amino acids. For instance, after separation in a conventional capillary, DABSYL-labeled amino acids were detected in a Pyrex TLM detector chip. The LOD was 4.6 × 10^{-8} M, compared to the LOD of 5.2 × 10^{-6} M obtained by the conventional absorbance method.[343] TLM has also been applied for the analysis of metal ions such as Co,[319,734] Ni,[735] Pb,[736] and Fe,[734] and of organic molecules such as o-toluidine after its oxidation.[737]

The TLM method produces a spatial resolution of 1 µM. Another advantage of TLM is that it is not strongly affected by light scattering (e.g., by the cell membranes in biological systems).[851] It has been demonstrated that the TLM method has a detection limit at the single-molecule level.[425]

7.1.9 RAMAN SCATTERING

Raman scattering has also been employed for detection in microchips. The Raman scattering (520 nm) of water has been used for focusing the laser light at the channel in a glass chip.[285] Herbicides

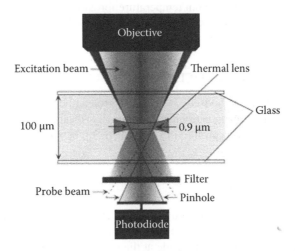

FIGURE 7.20 Schematic illustration of the thermal lens measurement. The excitation beam was focused by the objective lens. After excitation of some analytes, their radiationless relaxation caused the thermal effect and the formation of a concave thermal lens. The probe beam, which was also coaxially focused by the same objective lens, was collected by the photodiode detector defined by the pinhole. Any change in the amount of heat produced by the radiationless relaxation is manifested as the change in the photodiode output as a TLM signal.[733] Reprinted with permission from Elsevier Science.

(diquat or paraquat) have been detected by Raman scattering in a glass chip. They were measured (in the 700 to 1,600/cm range) using a Nd:YVO$_4$ laser (532 nm).[738]

Surface-enhanced resonance Raman scattering (SERRS) has also been achieved using silver colloid aggregates produced *in situ* in the chip. This method was used to detect an azo dye, 5-(2'-methyl-3',5'-dinitrophenylazo)quinolin-8-ol, which is a derivative of the explosive, TNT. With this method, it was possible to detect 10 µl of 10^{-9} M dye (or 10 fmol). This represented a twenty-fold increase in sensitivity over that achieved using a macroflow cell.[739]

7.1.10 SURFACE-PLASMON RESONANCE

Surface-plasmon resonance (SPR) has been used to detect surface-bound chemical species. SPR is achieved to detect the binding events of antibody and antigens in immunoassays. A gold-coated PMMA chip that was sealed by a PDMS channel plate was used. The antibodies were first immobilized on the gold layer. Upon binding with benzo[a]pyrene (BaP) 2-hydroxybiphenyl (HBP), the SPR signal was recorded.[740]

In another report, binding of IgG to anti-IgG immobilized on a Au-coated BK7 chip (sealed by PDMS) was detected by SPR.[741] Lactate was also determined using the SPR method. First, lactate was oxidized by lactate oxidase (LOX) immobilized on an osmium redox polymer to produce H$_2$O$_2$. This molecule was then reduced by HRP immobilized in the same redox polymer. This surface reaction was detected by SPR.[741]

A binding kinetics study of rabbit IgG to protein A was also carried out based on SPR detection. Protein A was immobilized on the Au layer in the channel wall in a PDMS-glass chip.[511] Antibody binding on peptides was also followed by SPR. The peptides were immobilized in cross-PDMS channels on a Au-coated chip.[742]

7.1.11 INFRARED DETECTION

Infrared detection of toluene can be achieved on a microchip. The chip substrate, which should be IR transparent, has been fabricated from CaF$_2$. The microchannels were etched on CaF$_2$ using a saturated Fe(NH$_4$)(SO$_4$)$_2$ solution at room temperature for 24 h. The etched depths were ~18 µm or ~8 µm, when the etchant was stirred or unstirred, respectively. Bonding was achieved using photoresist as an adhesive layer and with heating (135°C, 30 min).[743,744]

7.2 ELECTROCHEMICAL (EC) DETECTION

Another important detection method is based on electrochemical measurements. These include the measurement of voltage (in potentiometry), solution resistance (in conductivity), current (in amperometry), and current-voltage profile (in voltammetry). These electroanalytical techniques are described in subsequent sections.

7.2.1 AMPEROMETRIC DETECTION

The first EC detection, which was carried out on a glass chip, was based on amperometric measurement using integrated working and counter electrodes (see Figure 7.21).[745] This has allowed the facile and stable location of the working electrode at the exit of the separation channel. The reference electrode was in the form of a silver/AgCl wire. This method has been used for analysis of DNA restriction fragments and for sizing of PCR products.[745]

For detection of DNA fragments, Fe(phen)$_3$$^{2+}$ was used as the electrochemically active intercalation reagent. The constant background current from free Fe(phen)$_3$$^{2+}$ decreased in the presence of the DNA-Fe(phen)$_3$$^{2+}$ complexes. Therefore, this is an indirect amperometric detection method. It was found that a distance of 300 µm, instead of 600 µm, between the working electrode and reference

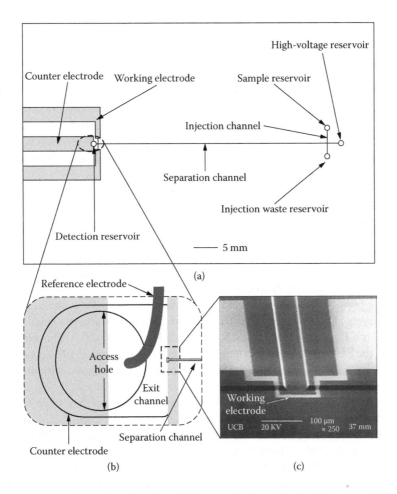

FIGURE 7.21 Capillary electrophoresis chip with integrated electrochemical detection. (a) Full chip view. Injection channel length, 5 mm; separation channel length, 50 mm; channel full width at half depth, 46 μm; channel depth, 8 μm. (b) Expanded view of the integrated electrochemical detector. The working and counter electrodes were radio frequency (rf) plasma sputtered (2,600 Å Pt with 200 Å Ti). (c) Scanning electron micrograph (140×) of the detection region showing the location of the working electrode in the exit channel 30 μm beyond the end of the separation channel. Image has been rotated 90° for viewing clarity.[745] Reprinted with permission from the American Chemical Society.

electrode has produced less electrical interference (in the form of a sloping baseline), allowing the use of a separation voltage up to 1,200 V (240 V/cm).[745]

There are numerous advantages of amperometric detection on-chip. These include: (1) the detector has a minimal dead volume, (2) preparation of electrodes (working) is compatible with the planar micromachining technology, (3) the electrodes can be miniaturized without compromising LOD, and (4) the EC detection has a short response time.[746]

In one report, determination of glucose by amperometric detection was achieved using a gold electrode patterned on glass. This was bonded to a Si chip, whose surface had been oxidized to avoid electrically shorting the electrode. Glucose oxidase, which was covalently immobilized on the inner surface of the Si channel, converted glucose to H_2O_2 to be amperometrically detected.[281]

In another report, unlike CE/EC devices previously reported, not only was the working electrode integrated, but also the reference and counter electrodes were patterned (Pt/Ti) on the glass chip. No external wire electrodes were used. This method has minimized dispersion at the column exit.[747] The electrodes were also situated under a "shelf" (due to the cover plate) so that the detection volume was restricted and the dispersion was further reduced. The stability of the chip was found to be more than 2 months. LOD of dopamine and catechol were reported to be in the 4 to 5 μM range.[747]

In amperometric detection, a reference electrode was usually employed. However, in one report, a platinized Au electrode was used as a pseudoreference electrode in a three-electrode system for amperometric detection. The operation principle follows that of the hydrogen reference electrode.[242]

In amperometric detection, the high separation voltage and the low detection voltage should be decoupled in order to avoid electrical interferences. In one approach, the detector electrode near the buffer waste vial is at ground potential. This is necessary because a finite high voltage applied to the waste vial will cause electrical interferences. Therefore, the gated injection, in which the buffer waste is always at ground, is employed for sample introduction.[748]

Better isolation of the high separation field (similar baseline noise at various separation voltages) was achieved by using a sputtered Au film formed at the exit of the channel, which was perpendicular to the axis of the channel.[749]

A platinized Au electrode was used as a decoupler (the decoupling ground electrode for CE) to reduce the interferences of the high separation electric field on the detection current and of H_2 bubble formation. However, the lifetime of the decoupler was estimated to be ~10 separations.[242]

A Pd film decoupler has also been constructed for amperometric detection of catecholamines. The Pd film has been thermally evaporated onto a plastic chip (without the use of the Cr or Ti adhesion underlayers). Owing to the fast diffusion of H_2 on a Pd surface, gas bubbles will not form. Pd is able to absorb H_2 produced at the cathode up to a Pd/H ratio of 0.6. This reduces one of the interferences to the EC signal, leading to an improvement of LOD to 0.29 μM dopamine.[205,375] With an optimal decoupler size of 500 μm, up to 6 h of operation was achieved with an electric field of 600 V/cm.[375]

In one report, the development of a cellulose acetate decoupler has improved the LOD of dopamine further to 25 nM.[332] In another, a microhole array was constructed in a polyethylene terephthalate (PET) chip and used as an electrical decoupler (see Figure 7.22). The array, which was perpendicular to the separation channel, was 1 mm away from the working electrode. Decoupling was successful for 10 μm holes used in 16 μm channels. However, the decoupling function failed when a deeper channel (60 μm) or smaller holes (5 μm) were used. Moreover, undesirable crosstalk (between the high-voltage separation and low-voltage detection electric fields) occurred. In these failing conditions, 10% of the separation current (10% of 0.3 μA = 30 nA) flowed to the electrochemical cell. With the use of a high-resistance buffer (e.g., 1 mM Tris buffer), the decoupling function was restored. Nevertheless, in this condition, a shift in the detection potential occurred.[191]

It was reported that in-channel amperometric detection was possible without using a decoupler. This has been achieved using an electrically isolated (nongrounded, floating) potentiostat.[750] On the other hand, when the CE power supply was battery operated (12 V battery to power 3 kV HV module), it was electrically isolated from the electrochemical detection system.[765] In another report, a portable HV power supply (using 6 V battery for the HV module and 9 V battery for the electrochemical detection circuit) was constructed for CE-EC detection on a glass chip.[751]

Pulse amperometric detection (PAD) has been used for the detection on a PDMS chip. This method is useful for analysis of underivatized compounds, such as carbohydrates, amino acids, and sulfur-containing antibiotics, which easily caused electrode fouling. In PAD, a high positive potential (1.4 to 1.8 V) is first applied in order to clean the electrode (e.g., Au) surface. This is followed by a negative potential step (−0.5 V) to reactivate the electrode surface. A third moderate potential (+0.5 to +0.7 V) is applied for detection of the target analyte.[752]

To avoid fouling of the Au working electrode, clean by applying a bipolar square wave voltage to the electrode after twenty-five injections; however, more than ten repeated applications will still destroy the electrode.[748] The LOD of catechol (single-electrode detection) was found to be 4 μM,[748] which was three times lower than a previous report,[745] possibly because of the application of the bipolar square wave voltage between injections to clean the electrode.[748]

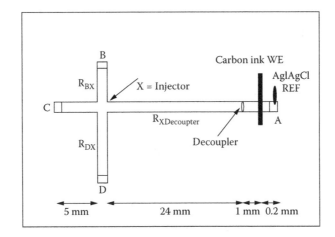

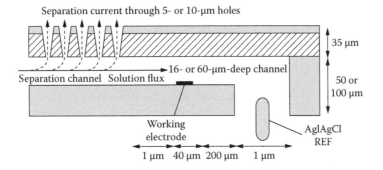

FIGURE 7.22 Top: Schematic representation of the microchip layout used for the CE separation with EC detection. Bottom: Schematic representation of a cross section at the end of the separation channel comprising the microhole array decoupler, the working electrode, and the silver/AgCl reference electrode.[191] Reprinted with permission from Elsevier Science.

Other than metal, carbon was also used to fabricate electrodes for EC detection. Amperometric detection (end column) was achieved at the end side of the chip using the current–time (i-t) curve mode. The screen-printed carbon working electrode was situated at the waste reservoir near the channel outlet side. In addition, a silver/AgCl wire was used as the reference electrode and the platinum wire used as the counter electrode. With the C electrode, a relatively flat baseline with low noise was observed (see Figure 7.23). This occurred despite the use of a high negative detection potential (–0.5 V vs. silver/AgCl), the use of nondeaerated run buffer along the 72 mm long separation channel, and the absence of a decoupling mechanism. However, the use of a higher separation voltage (>2,000 V) will still increase background noise.[625]

In one report, a thick-film C electrode (by screen printing) was constructed. Carbon ink (10 μm thick) was first printed on an alumina plate, and cured thermally. Then the silver ink (28 μm thick) was printed and cured to partially overlap with and hence connect to the C layer. The thick-film C electrode was found to enhance the detection sensitivity, as compared to the thin-film amperometric detector.[753]

A replaceable C disk electrode (two-electrode system) was used for end-column amperometric detection of biogenic amines. The electrode was inserted through a guide tube and was 30 ± 5 μm away from the capillary exit. This configuration allowed easy replacement of the electrode, especially after biofouling. The electrode was situated at a central position to maximize coulombic efficiency with respect to the axis of the capillary.[754] Micro C disk electrodes (30 μm diameter) were also used for amperometric detection of ascorbic acid (LOD = 5 μM) in a PDMS-glass chip.[755]

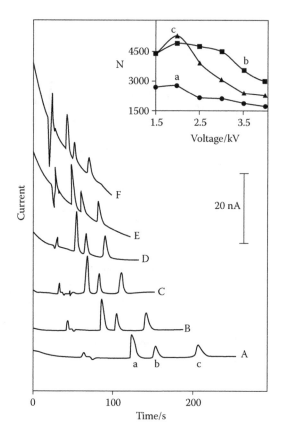

FIGURE 7.23   Influence of the separation voltage upon the amperometric response for a mixture containing 3.0×10^{-5} M paraoxon (a), 3.0×10^{-5} M methyl parathion (b), and 6.0×10^{-5} M fenitrothion (c). Separation performed using (A) +1,500, (B) +2,000, (C) +2,500, (D) +3,000, (E) +3,500, and (F) +4,000 V. Also shown (inset) are the resulting plots of plate number (N) vs. separation voltage. Separation buffer, 20 mM MES (pH 5.0) containing 7.5 mM SDS; injection voltage, +1,500 V; injection time, 3 s; detection potential, −0.5 V (vs. silver/AgCl wire) at the screen-printed carbon electrode.[625] Reprinted with permission from the American Chemical Society.

A glassy C electrode, which was immobilized with tyrosinase, was used for amperometric detection of phenol in a polyimide chip. Phenol was enzymatically converted to catechol, which was then oxidized to quinone during detection. Chlorophenol can also be detected, but this is achieved after a dechlorination step (to phenol) using a Mg/Pd metal catalyst.[229]

Carbon fiber electrodes (~100 nm diameter) were also used to characterize the diffusion between adjacent stream zones at the interface between a microfluidic system (sixteen channels: 50 μm wide, 57 μm deep, separated by a 22 μm wall) and a large volume using 10 mM ferrocyanide and amperometric detection.[756] It was reported that when the carbon fiber electrode was used in a PDMS chip, the in-channel format gave better peak symmetry than the end-channel format.[757]

Conducting carbon polymer ink, which filled a UV-ablated microchannel, was used to construct the integrated microelectrode on a plastic chip. Both chronoamperometry and CV were employed to detect a model compound (ferrocenecarboxylic acid) down to 3 μM, corresponding to 0.4 fmol within a volume of 120 pl.[758] In another report, a carbon-paste electrode was constructed by filling a laser-ablated (PET or polycarbonate [PC]) channel with C ink. The whole structure was then cured at 70°C for 2 h.[189]

Amperometric detection was achieved on two patches of C films (formed by CVD of 3,4,9,10-perylenetetracarboxylic dianhydride) on a glass chip. The microchannels were formed using a 23 μm thick photoresist as a spacer. Glucose oxidase and lactate oxidase were immobilized with HRP on the C films via a coated film of osmium poly(vinylpyriroline) (PVPD) polymer. Simultaneous measurements of glucose and lactate in rat brain cerebrospinal fluid (first perfused with 50 mM veratridine) were achieved. These two films were spatially separated in order to avoid interdiffusion of H_2O_2 formed from the two separate enzymatic reactions. Moreover, the two films were preceded by a third C film immobilized with ascorbate oxidase in order to remove ascorbic acid interference.[759]

Other materials were also employed to construct electrodes for amperometric detection. For instance, a boron-doped diamond (BOD) electrode was used for amperometric detection of nitroaromatic explosives, organophosphate nerve agents, and phenols. The BOD electrode offers enhanced sensitivity, lower noise, negligible adsorption of organic compounds, and low sensitivity to oxygen.[760] In addition, a copper particle-modified carbon composite electrode was used for amperometric detection of glucose in a PDMS chip.[761]

The first dual-electrode amperometric detection was achieved on a PDMS-glass device. An analyte was oxidized at the first electrode, and was then reduced at the second electrode. This method was used to positively identify catechol (100 μM) among a complex mixture containing ascorbic acid.[748]

Another dual electrode was used on a polystyrol chip. A collection efficiency of ~90% was obtained for 1-hydroxyethylferrocene.[210]

Dual LIF and amperometric detection are used to detect a five-component mixture (see Figure 7.24). The labeled amino acids (i.e., NBD-Arg, NBD-Phe, and NBD-Glu) are fluorescently detected. The EOF marker (DA) and internal standard (CAT) are amperometrically detected. The fluorescent peaks are normalized to the internal standard, CAT. As a result, the RSD of migration time was improved from 2.7% to 0.8%.[670]

To avoid deterioration of the Au/Cr electrode, C electrodes have been used. Dual C fiber electrodes were constructed on PDMS. The separation channel (25 μm wide and 50 μm deep) was fabricated on the top plate. Two C fibers (33 μm diameter) were inserted into the PDMS channels (35 μm wide and 35 μm deep) fabricated on the bottom plate. Consecutive injections (up to forty-one) could be performed before the electrode was cleaned with a bipolar square wave voltage. The LOD of catechol was found to be 500 nM. With the use of the C electrode, peptides (e.g., Des-Tyr-Leu-enkephalin), which formed stable Cu(II) and Cu(III) complexes, could be detected.[762] Dual-electrode amperometric detection also allowed the positive detection of two peptides.[763]

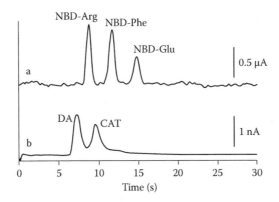

FIGURE 7.24 Dual detection of a five-component sample TES buffer. Electropherograms obtained using (a) LIF and (b) amperometric detection. Peaks are identified as DA, CAT, and NBD-labeled Arg, Phe, and Glu.[670] Reprinted with permission from the American Chemical Society.

7.2.2 VOLTAMMETRIC DETECTION

Voltammetric techniques have also been employed for on-chip EC detection. For instance, anodic stripping voltammetry (ASV) was used to detect Pb in a Si-Pyrex chip[770] or a polyimide chip.[229]

Another voltammetric method, sinusoidal voltammetry (SV), was also employed. This was a frequency-based electrochemical method, which was found to be more sensitive than the usual constant potential (DC) amperometric detection method. SV has been achieved to detect on-chip separated catecholamines. This method is very similar to fast-scan cyclic voltammetry (CV), except that a sine wave of a large amplitude is used as the excitation waveform. Data analysis is performed in the frequency domain in order to better decouple the faradaic signal from the background components.[764]

SV detection was also achieved to detect catecholamines on a PDMS-quartz chip. Pyrolyzed photoresist films (PPFs) were used as planar carbon electrodes. Since the photoresist (AZ4330) must adhere on a substrate for pyrolysis at $1.000°C$, a quartz plate that could withstand high temperature was selected. The LOD of dopamine decreased from 160 nM to 100 nM when the PPF was treated by piranha solution (H_2SO_4/H_2O_2). It was because such a treatment increased the surface oxygen/carbon ratio, and hence the oxidation kinetics of dopamine wave improved.[765]

In addition, the selectivity of SV detection can be enhanced by using appropriate harmonics (see Figure 7.25). When the excitation offset was centered on the half-wave potential of dopamine (134 mV vs. silver/AgCl), dopamine was not detected on the even harmonics (i.e., 6 Hz for second harmonics), but was only detected on the odd harmonics (i.e., 9 Hz for third harmonics)

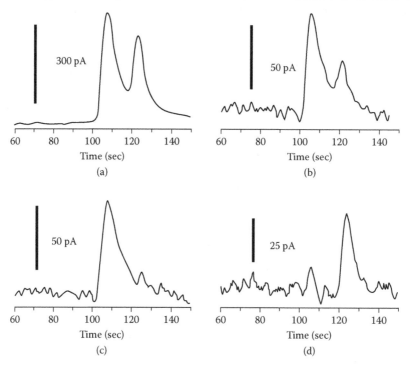

FIGURE 7.25 Manipulation of the applied potential window to enhance selectivity of SV detection. Separation of 5 μM dopamine and 5 μM isoproterenol produced a resolution of 0.9. A 3 Hz sine wave with a 500 mV amplitude with a variable offset was applied for SV detection. 1. Excitation offset centered on the half-wave potential of isoproterenol (180 mV vs. silver/AgCl). (a) First harmonic (3 Hz) time course and (b) fourth harmonic (12 Hz) time course. 2. Excitation offset centered on the half-wave potential of dopamine (134 mV vs. Ag/AgCl). (c) Third harmonic (9 Hz) time course and (d) second harmonic (6 Hz) time course.[765] Reprinted with permission from the American Chemical Society.

(see Figure 7.25c and d). Further selectivity enhancement can be obtained through manipulation of the monitored phase angle determined by the digital lock-in technique. The optimum phase angles for maximum signal occur at 150° and 123° for dopamine and isoproterenol, respectively. Therefore, isoproterenol will not be detected when the phase angle is monitored at 213°, which is +90° out of phase of 123°, and only dopamine is detected (dotted trace), as shown in Figure 7.26.[765]

7.2.3 POTENTIOMETRIC DETECTION

The potentiometric detector or ion-selective electrode (ISE) has been used for on-chip detection. For instance, Ba^{2+} in a flow stream (1 μl/min) has been detected on an ISE chip (see Figure 7.27).[766] The ISE chip comprises two distinct channels (10 μm deep). The sample channel holds the flowing sample solution (Ba^{2+}). The U-channel entraps the polymeric membrane and the barium ionophore (Vogtle): *N,N,N',N'*-tetracyclohexylbis(*o*-phenyleneoxyldiacetamide). The two channels are

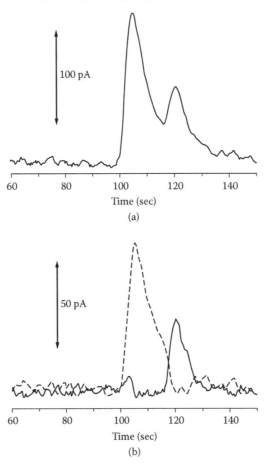

FIGURE 7.26 Utilization of the digital lock-in technique to enhance selectivity of SV detection. Separation of 10 μM dopamine and 10 μM isoproterenol, which produced a resolution of 0.9, was shown at the fifth harmonic (15 Hz). A 3 Hz sine wave from −116 to 384 mV (vs. Ag/AgCl) was applied for SV detection. (a) Phase common signal: Time course monitored at the phase common angle of 137°. (b) Phase nulled signals: Time course monitored at −90° out of phase (60°) with the dopamine optimum phase angle (150°) (solid trace), and time course monitored at +90° out of phase (213°) with the isoproterenol optimum phase angle (123°) (dashed trace).[765] Reprinted with permission from the American Chemical Society.

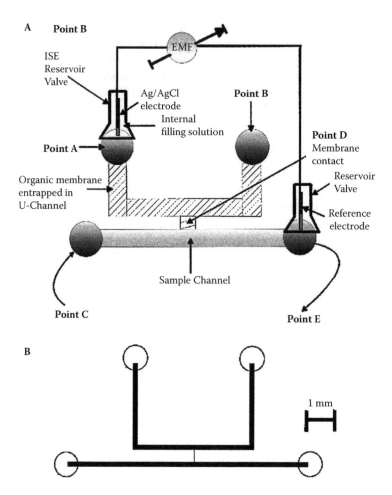

FIGURE 7.27 Micromachined ISE chip. (a) Schematic drawing of a sensor chip design with channels and reservoirs: point A, inlet for U-channel; point B, outlet for U-channel; point C, inlet for sample channel; point D, junction structure, where membrane contacts sample solution; and point E, outlet for sample channel. The diagram illustrates the complete filling of the silanized U-channel with an organic membrane cocktail. (b) Scaled diagram of the 12×6 mm chip.[766] Reprinted with permission from the American Chemical Society.

connected by a junction structure (point D) of 20 μm in width for the sample to come into contact with the membrane. To prevent potential drift, reservoir A has been filled with an internal filling solution (0.1 M $BaCl_2$). This filling solution is usually missing in other types of miniaturized ISE. The response times to changes of Ba^{2+} are rapid (within seconds).[766]

Other than barium, potassium and nitrite ionophores have been incorporated into an optode membrane to detect K^+ and NO_2^- potentiometrically on a plastic disk.[767] In another report, a thin layer of CuS (50 to 200 nm) was used to construct a copper ISE on a Si-glass chip.[768]

Potentiometric measurement using a Pt electrode has been employed in a titration (between $Cr_2O_7^{2-}$ and $Fe(CN)_6^{3-}$) that was carried out on a PDMS-glass chip. An eleven-channel serial dilutor was used to produce the different titrant concentrations, and a chaotic advective mixer was incorporated to facilitate solution mixing.[769]

Finally, a Cl^- ion selective membrane was also used in a microchip for the detection of Cl^-. However, the detection was achieved by absorbance, not potentiometric, measurement.[453]

7.2.4 CONDUCTIVITY DETECTION

Conductivity detection has also been used for on-chip measurements. In one report, the solution (e.g., NaCl) was in contact with the Pt electrodes on a Si-Pyrex chip.[770] Conductivity detection was made possible on a Pt electrode sputtered on a PMMA chip.[631]

An on-column contact conductivity detector was fabricated on a PMMA chip to analyze amino acids, peptides, proteins, and oligonucleotides. To minimize faradaic reaction and to only measure ohmic resistance (or conductivity), a bipolar pulse conductivity detector was developed. LOD of alanine was found to be 8.0 nM, which translates to 3.4 amol (in an injection volume of 425 pl).[208] For comparison, indirect fluorescent detection of amino acids produced an LOD of 1.6 amol.[696] Such a low LOD obtained by conductivity detection results from a number of factors: small spacings of the Pt wire electrodes (20 µm), implementation of the bipolar pulse (5.0 kHz, ±0.5 V), and signal averaging (five thousand data point average).[208]

In a special case, the decrease in resistance of Pd after H_2 absorption was employed for its determination. This decrease, which was due to the decrease in the work function of the Pd layer by H_2, was correlated with the H_2 concentration. For instance, H_2 gas (LOD ~ 0.5%) was detected in a Si microchannel electrodeposited with a Pd layer.[771]

To avoid electrolysis and electrode fouling when the solution was in contact with the measurement electrodes, contactless conductivity detection was proposed. This noncontact method relied on the capacitive coupling of the electrolyte in the channel, and the method has been used to detect inorganic ions that alter the conductivity and capacitance in the electrolyte.[277,638]

A four-electrode capacitively coupled (contactless) detector has been integrated on a Pyrex glass chip for detection of peptides (1 mM) and cations (5 mM K^+, Na^+, Li^+). The Al electrode (500 nm Al/100 nm Ti) was deposited in a 600 nm deep trench and was covered with a thin dielectric layer (30 nm SiC). The other parts of the channel were covered and insulated with Si_3N_4 (160 nm). To avoid gas bubble formation after dielectric breakdown, the electric field for separation was limited to 50 V/cm.[145] This four-electrode configuration allows for sensitive detection at different background conductivities without the need of adjusting the measurement frequency.[328]

In contactless mode, the dielectric thickness should be small. If the insulating layer is thicker, a higher signal frequency must be used to obtain adequate signals. For instance, an excitation frequency of 58 kHz is sufficient for measurement in using the 10 to 15 µm thick glass wall as the dielectric layer. The optimum frequency for conductivity detection was determined from a signal vs. frequency plot shown in Figure 7.28. To reduce on-chip stray capacitance, the detection electrodes are shielded by two in-plane shielding electrodes (grounded).[141]

After frequency optimization, three separated cations were detected, and these are shown in Figure 7.29. The initial elevated signal is attributed to the application of the high voltage used in the separation of the cations.[141]

A contactless conductivity detector was also constructed using 10 µm Al foil strips fixed on a 125 µm thick PMMA chip with epoxy. Here, the PMMA plate was the dielectric layer. The electrodes were arranged in an antiparallel fashion to minimize stray capacitance between them.[772]

In another report, the electrodes were placed in ultrasonically ablated wells so that they were very close (0.2 mm) to the channel. In addition, a faraday shield was placed between the two electrodes to avoid direct coupling between them. A voltage of several hundred volts was used to achieve low detection limits.[773]

Electrodes in a capacitively coupled conductivity detector were made by injection molding carbon-filled polymer into a preformed PS chip. The polymer consisted of three conducting formulations: 8% carbon black–filled PS, 40% C fiber–filled nylon-6,6, and 40% C fiber–filled high-impact PS.[774] In another report, a movable contactless conductivity detector was also developed to allow the distance of the electrode to be adjustable.[775]

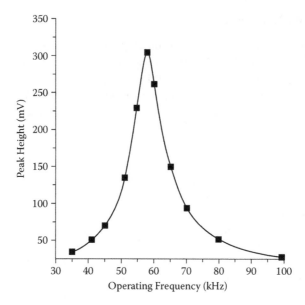

FIGURE 7.28 Determination of the optimum operating frequency in contactless conductivity detection. Sample plugs of 20 mM MES/His were injected at varying excitation frequencies, and the peak height was recorded for each frequency. In this example for a device with 400 μm wide electrodes and a 50 μm wide channel, the optimum operating frequency was found to be 58 kHz. Running buffer, 10 mM MES/His at pH 6.0; separation conditions, 200 V/cm; effective separation length, 3.4 cm.[141] Reprinted with permission from Wiley-VCH Verlag.

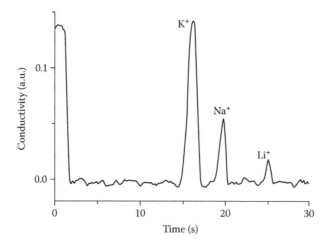

FIGURE 7.29 Microchip CE separation of a sample containing 100 μM of K+, Li+, and Na+ (prepared in the running buffer). The measured plate number for K+ is 43,200 plates/m, with an estimated limit of detection of 18 μM. Running buffer, 10 mM MES/His at pH 6.0; separation conditions, 280 V/cm; effective separation length, 3.4 cm; contactless conductivity detection at 58 kHz (as optimized in Figure 7.28).[141] Reprinted with permission from Wiley-VCH Verlag.

7.3 MASS SPECTROMETRY (MS)

For MS detection, the microfluidic chip has been coupled to various ionization interfaces (ESI, MALDI) and to different mass analyzers (by single quadrupole, triple Q, ion trap, MS/MS, and FTICR). The chip can be single or multichannel, and can be integrated with various sample handling processes such as preconcentration, digestion, and separation. The following two sections describe the two ionization interfaces (ESI and MALDI) that precede the MS analysis.

7.3.1 ELECTROSPRAY IONIZATION (ESI)

The first mode to interface the sample effluent from a microchip to a mass spectrometer (MS) was based on electrospray ionization (ESI). For electrospray generation, a sharp tip is usually used as an emitter. For instance, a sheath flow micro-ion sprayer was used to interface a microchip to a mass spectrometer (see Figure 7.30). CE separation was first carried out, and then the separated components were transferred to the mass spectrometer for MS analysis.[812]

In addition, a pulled capillary tip was inserted and glued to the end of a microchannel to be used as a disposable nanoelectrospray emitter. Membrane proteins from nonpathogenic (Rd) and pathogenic (Eagan) strains of *Haemopilus influenzae* were isolated (in 1D sodium dodecyl sulfate–polyacrylamide gel electrophoresis [SDS-PAGE]) and in-gel digested with trypsin. The digests were subsequently placed on the chip to carry out on-chip separation for cleanup and partial separation and subsequent ESI–quadrupole time of flight (QTOF) detection and peptide mass-fingerprint database search.[809] Various proteins and peptides have also been studied by ESI-TOF.[810]

To reduce the dead volume at the electrospray tip, it was inserted (then glued) into a specially drilled flat-bottom hole, as opposed to a conical-bottom hole.[132] A similar method was used for analysis of various peptide standards and tryptic digests of lectins from *Dolichlos biflorus* and *Pisum sativum*.[812]

A single microchannel has been used for sequential MS analysis of multiple samples. In this case, no sample cross-contamination has been reported. For instance, no sample crossover was observed for 10 μM cytochrome *c* vs. 10 μM ubiquitin,[943] and tryptic digests of β-lac, CA, and BSA.[776] In another report, sequential infusion of tryptic digests of CA (290 nM) and BSA (130 nM) into ESI-MS was achieved using EOF. No cross-contamination resulted when a central flow of buffer confined the other samples in the reservoir and channel by precise voltage control.[801]

Electrospray ionization has been obtained from a flat edge, without using a sharp tip.[195,816,818] For instance, a stable electrospray was generated at the flat edge of the channel exit with a pressure flow (100 to 200 nl/min). However, the flat edge should be coated with a hydrophobic layer (only good for 30 min) to prevent liquid spreading. Detection of recombinant human growth hormone (2 μM), ubiquitin (10 μM), and endorphin (30 μM) has also been demonstrated.[818] Another way to produce

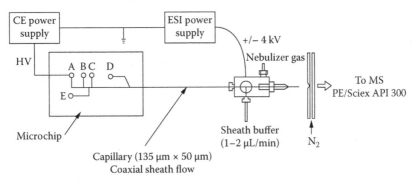

FIGURE 7.30 Schematic of the chip–CE coupled with a sheath flow ESI interface for MS analysis.[812] Reprinted with permission from the American Chemical Society.

a hydrophobic surface was to use a polytetrafluoroethylene (PTFE) membrane (70% porosity) to cover the orifice of a PC emitter tip.[777]

In one report, multiple channels have been constructed on a microchip for interfacing to MS. For instance, a nine-channel glass microchip was interfaced to MS to detect myoglobin. The mass spectrum of myoglobin that was transferred from the microchip appeared to produce comparable results to that transferred using a conventional capillary (see Figure 7.31).[818] Tryptic digests of BSA, horse myoglobin, human haptoglobin, and two-dimensional (2D) gel yeast proteins were sequentially analyzed in a nine-channel chip with a single transfer capillary coupled to an ESI-ITMS instrument.[776]

In another report, a microchip comprising ninety-six channels with ninety-six embedded capillary tips was constructed for high-throughput MS analysis of peptides (see Figure 7.32).[780]

On-line protein digestion has been achieved by trypsin adsorbed in a porous poly(vinylidene fluoride) (PVDF) membrane (0.45 μm pore) coupled in a PDMS chip. In this way, on-chip protein digestion and subsequent MS analysis have been carried out for horse heart cytochrome c.[817]

In another report, trypsin was mixed with the protein (e.g., cytochrome c and hemoglobin) off-chip and tryptic digestion was allowed to continue before MS analysis was conducted.[807] To ensure that the enzyme trypsin was most active, ammonium bicarbonate (pH 8) was employed. This salt was used instead of ammonium acetate (pH 6.5) commonly used for ESI-MS. As shown in Figure 7.33, the use of ammonium bicarbonate resulted in less noise (Figure 7.33b). Sufficient signal was also obtained for protein digestion in a shorter time, if a greater amount of protein was used (Figure 7.33c).[807]

Desalting of samples was also carried out before ESI-MS analysis. To achieve this, a PVDF membrane was sandwiched at the sample reservoir on a chip for dialysis of the sample as it entered the microchannel. This method was used to analyze picomole amounts of propanodol, insulin, and cytochrome c. This process was tolerant to high urea concentration and the presence of a common reductant, DTT.[778] In another report, cleanup separation was carried out by gradient frontal liquid chromatography (LC) performed on a C18 cartridge before MS analysis. The tryptic digest of horse apomyoglobin and yeast proteins was analyzed in this way.[803]

Microchip MS analysis was also employed to study drug metabolism. For instance, metabolism of antidepressant drugs (imipramine, doxepin, amitriptyline) was studied using ESI-MS/MS (see Figure 7.34). Conversion of these drugs was carried out by cytochrome P450 (CYP) enzymes in human liver microsomes. The enzymatic conversion of imipramine (5 to 100 μM) to desipramine as inhibited by tranylcypromine (5 to 15 μM) was also studied. Although P450 reactions are traditionally carried out in an open environment exposed to O_2, it was found that this reaction can also occur in a closed system within a polymer chip.[217]

An 8-mer peptide substrate was used for the study of HIV-1 protease activity in the presence of various inhibitors (e.g., Pepstatin A). With a tripeptide as an internal standard, inhibitors of various concentrations were placed in the ninety-six-channel plate to establish the inhibition constant.[780]

In another report, a microchip was interfaced to MS for the detection of the tetrameric plasma protein (transthyretin), which was involved in the transport of thyroxine. Screening of small molecules, including the natural ligand thyroxine, that could stabilize the tetramer structure was carried out.[779]

Polymeric materials have been used to fabricate the MS chip. These materials, including PMMA,[200] Epofix,[780] and PDMS,[162,800] have been evaluated for possible organic contaminants. When polymeric chips are used, there might be MS signals resulting from the organic contaminants in the polymer substrate, and they are called chemical noises. Some chemical noises are apparent in a PMMA chip, as shown in Figure 7.35A. However, after washing the chip to remove organic contaminants resulting from the chip fabrication the chemical noises are much reduced (see Figure 7.35B). Moreover, good stability (for 4 h) in the MS signal was obtained.[200] For PDMS chips, the chemical background was found to be low for MS analysis.[162]

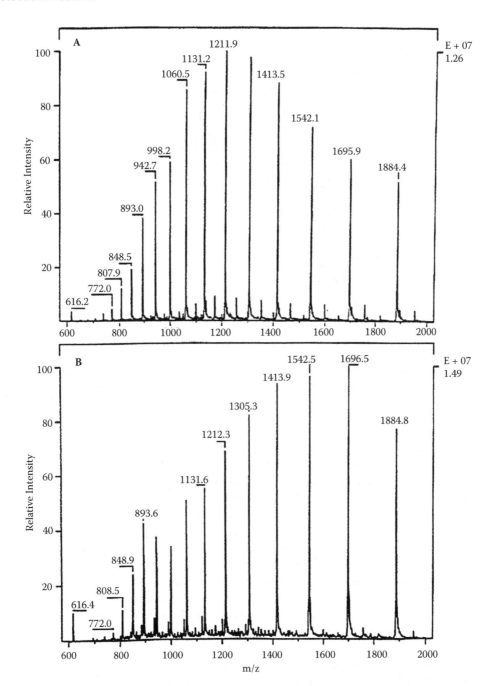

FIGURE 7.31 Comparison of ESI-MS of myoglobin (6 µM in 75% methanol, 0.1% acetic acid) transferred from (A) an HF-etched capillary (50 µm i.d.), 800 nl/min, and (B) a microchip with 60 µm wide and 25 µm deep channels, 200 nl/min.[818] Reprinted with permission from the American Chemical Society

An ESI-MS chip has been constructed on PMMA. The chip consisted of eight open channels and eight sharp ends (see Figure 7.36). The pointed tip and the hydrophobic nature of PMMA allowed the facile formation of an ESI. A flow of 0.5 to 1 µl/min was used to sustain the electrospray. This flow was provided mostly by HDF (using a syringe pump), but the flow could also be generated by EOF or by EOF-induced flow. This chip has been used for the MS analysis of sildenafil in a commercial tablet.[781]

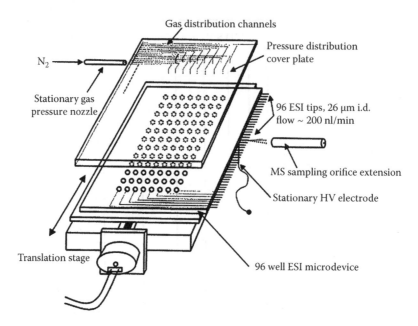

FIGURE 7.32 Exploded view of a ninety-six-channel MS chip design. The plate with individual channels and electrospray tips was positioned on a translation stage in front of the extension of the MS sampling orifice. Individual samples were transferred by sequential pressurization of the sample wells through the pressure distribution cover plate, which was sealed to the ESI device by a silicone rubber gasket. The ESI high voltage was applied via the stationary HV electrode positioned under the ESI device.[780] Reprinted with permission from the American Chemical Society.

An ESI tip was also constructed by laminating two PET films (commonly used in overhead transparency film). The printed toner served as the channel wall structure. The tip border was dipped in a silicone solution, thus forming a hydrophobic layer that prevented the aqueous solution from spreading along the edge during electrospray formation.[238]

Various methods have been reported to fabricate monolithic tips in polymeric materials for MS analysis.[782] The tip fabricated in PC is shown in Figure 7.37.[290] Other materials, such as parylene,[783] PDMS,[784,785] or SU-8,[786] were also used. For instance, an open tip was constructed on a PDMS chip. A lid was used to cover the whole chip except the microchannel tip.[172,787]

An array of triangular-shaped tips was fabricated from a 5 μm polymeric sheet (parylene C) by O_2 plasma etching. The sheet was sandwiched between two plastic plates. An ESI formed at one of the tips was shown in Figure 7.38. The wicking tip protruding from the end of a microchannel aided in forming a triangularly shaped droplet (0.06 nl in volume) and assisted in forming a stable Taylor electrospray cone.[788]

In another report, electrospray emitter tips were mechanically milled from the end of a microchip microfabricated by injection molding (see Figure 7.39). A conducting layer was put on the emitter tip so that the electric potential could be directly applied at the tip. In this manner, a sheathless electrospray could be formed. Since no aiding liquid or gas was used to produce the spray, no dilution of the sample occurred. The conducting layer consisted of either polymer-embedded 2 μm gold particles (dubbed as fairy dust) or 1 to 2 μm graphite particles. This conducting layer produced longer stability, compared to sputtered gold, which suffered from instability due to poor adhesion.[213]

Another sheathless ESI interface was constructed on PET by photoablation. High voltage was supplied close to the outlet of the microchannel, through an embedded thick-film C ink microelectrode. The microspray was generated directly from the flat edge of the substrate (without any tip addition). No solution wetting occurs at the flat edge, thanks to the hydrophobicity of PET. Since the

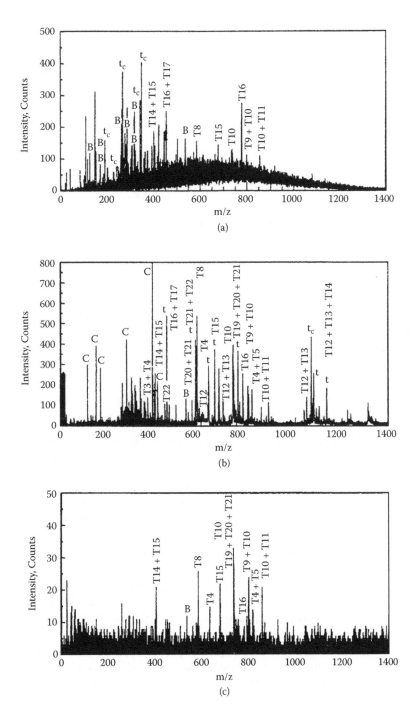

FIGURE 7.33 Microchip ESI mass spectra of tryptic digests of cytochrome *c*. Spectra for cytochrome c (0.8 μM)/trypsin (10 μg/ml) in (a) 10 mM ammonium acetate containing 20% methanol, 25 min digestion, and in (b) 10 mM ammonium bicarbonate containing 20% methanol, 15 min digestion. TOF-MS acquisition: 5,000 Hz, 25,600 summed spectra. (c) Spectrum for cytochrome *c* (0.4 μM)/trypsin in 10 mM ammonium bicarbonate containing 20% methanol, 7 min digestion. TOF-MS acquisition: 5,000 Hz, 1,600 summed spectra, 3.2 spectra/s. Peak notations: B, buffer-related ions; C, system contaminants; t, trypsin autolysis products; tc, trypsin contaminants; T14 etc., trypsinated cytochrome *c* fragments.[807] Reprinted with permission from the American Chemical Society

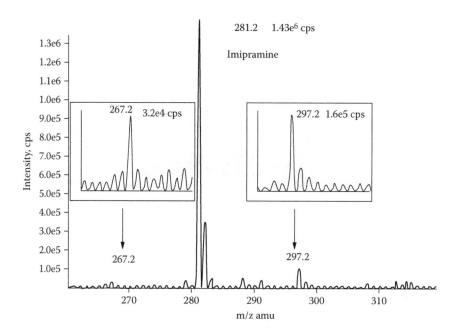

FIGURE 7.34 ESI-MS results from the incubation of imipramine with human liver microsomes in a Zeonor (cyclic olefin copolymer) chip. A monolithic SPE column was integrated for sample cleanup (desalting and removal of enzyme) and preconcentration of the N-demethylated monohydroxylated metabolites. The data were acquired in the m/z range of 250 to 350. The mass spectrum shows the remaining parent drug imipramine ($m/z = 281.2$, 1.43×10^6 cps), the formation of the monohydroxylated isomeric metabolites of imipramine ($m/z = 297.2$, 1.6×10^5 cps), and the formation of the N-demethylated monohydroxylated metabolite, desipramine ($m/z = 267.2$, 3.2×10^4 cps).[217] Reprinted with permission from the American Chemical Society.

microelectrode area was small, this resulted in a low current density during electrospray, and hence bubble formation due to water hydrolysis was minimized.[195]

In one report, the electric potential necessary for ESI generation and EOF pumping was supplied through an electrically permeable membrane (4.6 μm wide) contacting the fluidic channel holding the ESI capillary tip.[121]

In another report, instead of producing a conductive tip for ESI, the entire chip was fabricated from a conductive substrate, glassy C. This allowed HV to be directly applied on the chip, but an ESI tip was still needed to be inserted and glued to the chip. Etching of glassy C was achieved electrochemically in 0.1 M NaOH. Bonding with a glass cover was achieved using an adhesive.[789]

Monolithic nozzles have been fabricated to construct the ESI tips. They were fabricated on Si by DRIE. This approach has showed greater signal stability and intensity.[790]

A Si microneedle has been fabricated, though the application was not in making the ESI tip, but in transdermal drug or vaccine delivery. Figure 7.40 shows how the microneedle penetrates a 10 μm thick Al foil. This needle has the openings in the shaft, rather than in an orifice at the tip.[791]

7.3.2   MATRIX-ASSISTED LASER DESORPTION IONIZATION (MALDI)

Matrix-assisted laser desorption ionization is another ionization mode used for MS analysis. Enzymatically digested peptides have been studied using a ninety-well microchip constructed in a MALDI plate format (see Figure 7.41). Peptide digestion was initiated in the MALDI interface where the peptide hormone, adrenocorticotropin (ACTH), was mixed with the enzyme carboxypeptidase Y. The mixing process was self-activated in the vacuum conditions. Subsequent TOF MS analysis produced kinetic information of the peptide digestion reaction.[820]

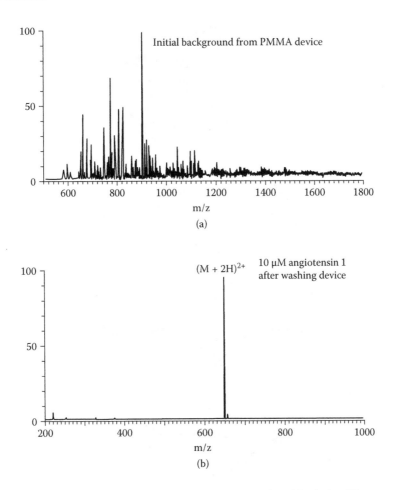

FIGURE 7.35 (a) Initial mass spectrum obtained from PMMA microchip device. The mass spectrum is characterized by a substantial chemical noise background most likely due to residual developer solution present in the microchannels. (b) Electrospray mass spectrum obtained with 10 µM angiotensin I (in 50:50 MeOH/H$_2$O containing 1% acetic acid) after the PMMA microchip has been washed with 50:50 MeOH/H$_2$O. The simple solvent wash completely eliminates the residual chemical contamination arising from the microchip fabrication step.[200] Reprinted with permission from the American Chemical Society.

In another report, an open-access channel (i.e., no cover plate) has been used for CE separation before the MALDI-MS analysis.[792]

One of the requirements in MALDI-MS analysis is the use of a liquid matrix. The electrowetting-on-dielectric (EWOD) method has been used to move and mix droplets containing proteins and peptides with the liquid matrix, all of which were situated at specific locations on an array of electrodes. With this method, insulin (1.75 µM), insulin chain B (2 µM), cytochrome c (1.85 µM), and myoglobin (1.45 µM) have been analyzed.[518]

Microfluidic processing (concentration and desalting) of protein digests (peptides) was carried out on a compact disk (CD) consisting of multiple channels. This CD was compatible with MALDI-MS analysis. Centrifugal force moves the liquid through multiple microstructures via common liquid channels, with the operation of only one such microstructure shown in Figure 7.42. The tryptic digests of the peptides (1 µl) were loaded to the CD, as shown in Figure 7.42A. A common distribution channel was used to add wash solution (200 nl); see Figure 7.42B. To elute the peptides from the column, a volume of 200 nl of eluant (50% CH$_3$CN with 1 mg/ml of the MALDI matrix) was distributed via the common channel (see Figure 7.42B). The MALDI matrix

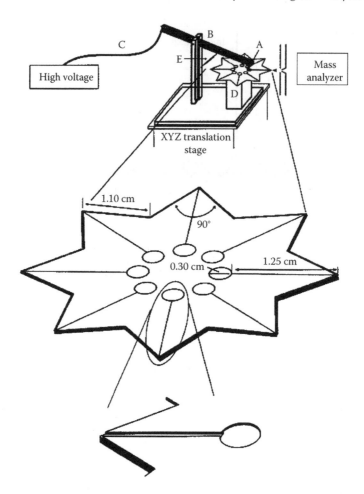

FIGURE 7.36   Schematic of the PMMA chip electrospray arrays used to analyze sequentially eight samples (A, copper electrode; B, acrylic bar; C, electrical cable; D, plastic box; E, spring on an acrylic rod). The insets show the dimensions of the chip and the close-up view of a channel.[781] Reprinted with permission from the American Chemical Society.

is cinnamic acid (CHCA) containing 1% of trifluoroacetic acid (TFA). The eluate was transported to the MALDI target area (400×200 μm) for solvent evaporation and crystallization of the peptide–matrix mixture (see Figure 7.42C). After crystallization, the CD is cut and inserted into the MS system. The total time required to process a CD with ninety-six samples is ~40 min. For comparison, 1 μl of tryptic digest was mixed off-chip with 1 μl of 3 mg/ml CHCA in 70% CH_3CN containing 0.1% TFA. This mixture was applied by the dried droplet technique to form dots of an area of 1 to 1.5 mm^2. The MS spectra obtained from the samples prepared in these two ways are depicted in Figure 7.43 for comparison. It is found that ten additional peptides are detected from the CD-processed sample. Moreover, a higher protein identification score based on seventeen peptides with a higher amino acid sequence coverage (29%) was found on the CD-processed sample, compared to seven peptides with a sequence coverage of 12% in the conventionally processed sample.[793]

The CD-based microfluidic processing method for MALDI-MS analysis was also achieved to analyze phosphorylated peptides. For instance, bovine protein disulfide isomerase (PDI), after phosphorylation, underwent tryptic digestion. The peptide solution (1 μl) was put into the CD for a phosphopeptide enrichment step by immobilized metal affinity chromatography (IMAC). A group

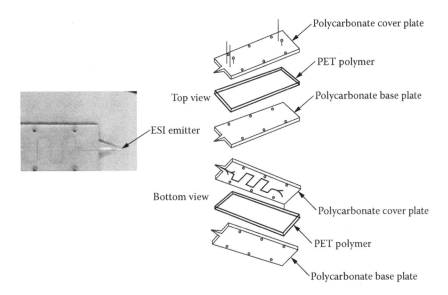

FIGURE 7.37 Left: Photograph of the ESI emitter at the outlet portion of a microfabricated PC chip. Right: Exploded view showing construction of the chip consisting of the PC cover and base plates sealed by a PET polymer layer. The bottom view shows the channel layout, whereas the top view shows the reservoir connection.[290] Reprinted with permission from Wiley-VCH Verlag.

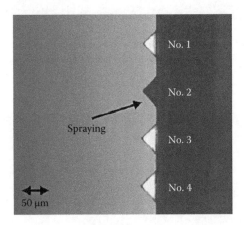

FIGURE 7.38 An ESI array device with triangular tips has been fabricated on parylene (only four tips are shown). The no. 2 tip is spraying buffer solution.[788] Reprinted with permission from the American Chemical Society.

of forty-eight channels in the ninety-six-channel CD was used for this step. Another group of forty-eight channels was used for IMAC with subsequent enzymatic dephosphorylation by alkaline phosphatase. Figure 7.44a shows the tryptic digest of unenriched PDI, whereas Figure 7.44b shows the mass spectrum of the enriched sample, which indicates two phosphopeptides at m/z values of 964 and 2,027. After dephosphorylation, both peptides showed the 80 Da shift to lower masses (see Figure 7.44c). This is a characteristic mass shift of the phosphate group.[794] Another analysis performed in a way similar to that on the ligand-binding domain of the human mineralocorticoid receptor has revealed two novel phosphorylation sites, at Thr 735 and Ser 737.[794]

The research in the area of on-chip MS analysis is very active, and the important differences are summarized in Table 7.1.

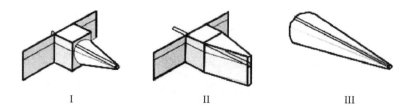

I   II   III

FIGURE 7.39   Schematic representation of three types of ESI emitter tips, which were microfabricated on PC or PMMA. I, machine-milled 3D tip; II, machine-milled 2D tip; III, hand-polished 3D tip.[213] Reprinted with permission from the American Chemical Society.

FIGURE 7.40   Penetration of 10 μm thick aluminum foil by a side-opened monolithic Si microneedle. Note that no damage can be observed on the needle.[791] Reprinted with permission from the Institute of Physics Publishing.

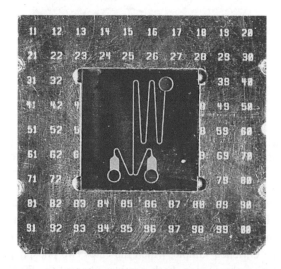

FIGURE 7.41   Picture of the microfabricated fluidic device integrated with a standard MALDI-TOF sample plate. Because of the self-activating character of the microfluidic device, the system can be introduced into the MALDI ionization chamber without any wire or tube for the sample introduction and the flow control.[820] Reprinted with permission from the American Chemical Society.

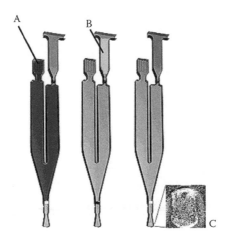

FIGURE 7.42 On-CD processing of samples (ninety-six structures per CD). An individual microstructure element is shown separately for the three consecutive steps of sample application (1 μl), washing/elution, and co-crystallization with a MALDI matrix. First, the dilute and salt-containing crude sample is applied at (A) onto the 10 nl reversed-phase column (white). Second, the washing and elution/matrix solutions are applied via a common distribution channel at (B), with the liquid volume defined to 200 nl after activation of an over-flow channel (not shown). Third, co-crystallization of the concentrated and desalted sample with a MALDI matrix is taking place at (C). This is a 200 × 400 μm target area where the crystalline deposit is accessible to the laser beam of the MALDI instrument as shown by a photograph of the crystals in (C).[793] Reprinted with permission from the American Chemical Society.

7.4 OTHER DETECTION METHODS

7.4.1 THERMAL DETECTION

Noncontact temperature measurement inside microfluidic channels was achieved by using fluorescence quenching of a rhodamine dye. The intensity of the dye fluorescence is temperature sensitive in a range of 5 to 95°C.[795] Another on-chip temperature measurement method was achieved by measuring the ratio of the fluorescent intensities emitted by the monomer-excimer pair. For instance, the ratio of the fluorescent intensities of the monomer (at 415 nm) and the excimer (at 485 nm) of 1-pyrenesulfonic acid sodium salt (1.16 mM) was measured.[275] In addition, thermochromic liquid crystals were loaded into the microchannels to indicate the temperature inside the channels.[357] These methods have been used to characterize the temperature inside the PCR chips during thermal cycling.[275,357]

In another report, a microthermocouple (Ni-Ag) was fabricated on-chip for temperature measurement (see Figure 7.45). The thermocouple metals were fabricated using electrodeless deposition. This thermal measurement method was used to monitor an acid–base neutralization reaction and an enzymatic reaction.[271]

7.4.2 ACOUSTIC WAVE DETECTION

A quartz crystal sensor chip was bonded with a microfluidic glass chip for acoustic wave detection (see Figure 7.46). The sensor was operated in the thickness shear mode (TSM). This has allowed rat heart muscle cell contraction to be studied based on the measurement of the resonant frequency changes.[133]

Gas-phase photoacoustic detection of propane in N_2 was attempted on a Si chip. In this method, a modulated light beam is incident on the sample. If the wavelength of the modulated light couples to an energy transition in a gas, the gas absorbs the modulated light, resulting in periodic gas expansions and contractions, which are manifested as an acoustic wave. This wave can be detected by a

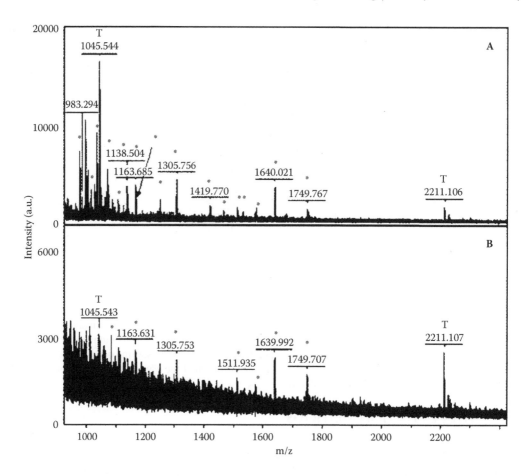

FIGURE 7.43   MALDI mass spectra from analysis of a tryptic peptide extract recovered after in-gel diges-tion of BSA (maximally 12 fmol/µl based on the amount of protein applied to the gel before electrophoresis, staining, excision, digestion, and extraction). (a) 1 µl of the extract processed and analyzed on-CD resulted in seventeen identified peptide peaks (asterisks). (b) 1 µl of the same extract processed and analyzed in a standard manner on a steel target (dried droplet technique), after mixing with 1 µl of 3 mg/ml CHCA in 70% acetonitrile containing 0.1% TFA, resulted in seven identified peptide peaks (asterisks). Autolytic trypsin frag-ments are indicated by T.[793] Reprinted with permission from the American Chemical Society.

microphone. In contrast to conventional absorption spectroscopy, the sensitivity of photoacoustic spectroscopy scales inversely with dimension, and hence this method is favored in the microscale. This is because photoacoustic spectroscopy is a differential technique in which the absorption is measured as the intensity per unit surface area.[796]

7.4.3 Nuclear Magnetic Resonance (NMR)

A planar microcoil-based probe was fabricated on a glass chip for NMR measurement (see Figure 7.47). The ^{1}H NMR spectrum of sucrose in D_2O was shown in Figure 7.48. The noise is mainly due to the thermal fluctuations generated by the series resistance of the coil (e.g., 1 Ω). To reduce the channel wall thickness and to improve the filling factor encompassed by the coil, the glass chip was mechanically thinned down to 65 µm. Two probes of different coil diameters were designed. It was found that the NMR signal for the anomeric proton of sucrose (~5.3 ppm) was strong in the case of the larger probe (2 mm diameter); see Figure 7.48. The greater volume of ~470

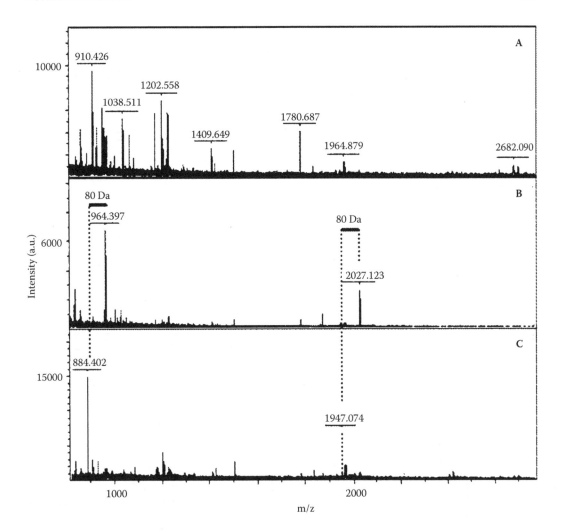

FIGURE 7.44 MALDI mass spectra from on-CD analysis of the phosphopeptide-containing sample. (a) Peptide mass spectrum after concentration/desalting. A database search showed that the sample contained bovine protein disulfide isomerase. (b) Phosphopeptide enrichment by IMAC. Two phosphopeptides at m/z 964 and 2,027 were recognized (c) Phosphopeptide enrichment followed by enzymatic on-column dephosphorylation using alkaline phosphatase. Two phosphopeptides at *m/z* 884 and 1947 were recognized from the mass shifts of 80 Da from (B), which were resulted from dephosphorylation.[794] Reprinted with permission from the American Chemical Society.

nL corresponds to an increased amount of sucrose (160 μg). To improve resolution, a Lorentz-Gauss transformation is applied to the original free induction decay data to yield a spectrum with a higher resolution (see Figure 7.49).[797]

In another report, dual copper microcoils were fabricated on a glass chip sandwiching an etched chamber of 1.4 μl for NMR measurements. A 1D ^{31}P spectrum of phosphoric acid and a ^{13}C spectrum of $^{13}CH_3OH$ were recorded. In addition, a COSY spectrum of ^{13}C-labeled acetic acid was also acquired in ~1 h.[798]

TABLE 7.1
Summary of MS Analysis Using Various Techniques of Chip-MS

Chip Design	Flow	MS Mode	Analytes	Sample Treatment	Separation	Reference
Single-Channel ESI Chip						
Glass chip with a transfer capillary	EOF 100–150 nl/min	ESI-QTOF	Myoglobin tryptic digest (290 fmol/μl)	Trypsin	Nil	799
Glass chip with a transfer capillary	Pressure flow	ESI-LCQ	BSA, cytochrome c, angiotensin peptides (10 μg/ml)	Trypsin	CE	296
Transfer capillary, PMMA chip	Pressure flow 20–200 nl/min	ESI-FTICR	Cytochrome c, 500 nM	Nil	Nil	200
Transfer capillary PDMS chip	EOF plus pressure 2–3 psi	ESI-ITMS	Rat serum albumin separated by 2D gel, angiotensin I (100 nM)	Trypsin	Nil	800
Two samples/one transfer capillary glass chip	EOF 200–300 nl/min	ESI-ITMS	Fibrinopeptide A (33 nM) tryptic digest of CA (290 nM), BSA (130 nM)	Trypsin	Nil	801
Transfer capillary integrated nebulizer, glass chip	Pneumatic nebulization (150 nl/min)	ESI-ITMS	Angiotensin peptides (20 μg/ml) cytochrome c tryptic digest	Trypsin	ITP-CE	802
Transfer capillary, glass chip	EOF	ESI-ITMS	Tryptic digest of horse apomyoglobin (7.4 nM), yeast proteins	Trypsin	Nonchip gradient frontal CEC separation C18 column	803
Coupled microsprayer, glass chip	4 μl/min pressure flow	ESI-triple Q	Carnitine and acylcarnitines in fortified human urine 35–124 μM (10–20 μg/ml)	N/A	CE	804
Glass chip with ESI tip	EOF 30–100 nl/min	ESI-TOF	Gramicidin S (1 μM)	Nil	Nil	121
Glass chip	Pressure flow	ESI	Cytochrome c	Trypsin on beads	CE	805
Inserted/glued capillary tip, glass chip	1 μl/min pressure flow	QTOF	17–62 μM carnitine and acylcarnitines in fortified human urine	N/A	CE	804
Inserted glued tip, glass chip	EOF 100 nl/min, pressure flow 4 μl/min	ESI-triple Q-SIM	In fortified human plasma, carnitine, or acetylcarnitine (1–500 μg/ml) Imipramine or desipramine (5–500 μg/ml)	N/A	CE	806

Interface	Flow	MS	Analyte	Digestion	Separation	Ref.
Inserted glued tip	EOF 20–30 nl/min	ESI-TOF	Tryptic digest of bovine hemoglobin (4 μM), cytochrome c (0.8 μM)	Mixed off-chip/on-chip tryptic digestion	Nil	807
Glass chip			Human hemoglobin (normal and sickle cell, 0.24 μM)			
Inserted glued tip, Zeonor chip	Pressure flow 6 μl/min	ESI-triple Q	Carnitine, acylcarnitines, butyl carnitine	N/A	CE	808
Inserted glued capillary tip, glass chip with an internal standard side channel	EOF 200 nl/min, pressure flow 50 nl/min	ESI-Q TOF	1D gel of tryptic digest of membrane protein of H. influenzae (nonpathogenic Rd and pathogenic Eagan strains)	Trypsin	CE	809
		Triple Q	Mixture of leu-enkephalin, somatostatin, angiotensin II bradykinin, luteinizing hormone releasing hormone (LHRH)			
		Triple Q	Tryptic digest of seed lectin of P. vulgaris L			
Orthogonal inserted glued tip, glass chip	EOF 20–30 nl/min	ESI-TOF	Gramicidin S (0.1–10 μM), cytochrome c (0.1 μM), bradykinin, leu-enkephalin, methionine encephalin	Trypsin	CE	810
Inserted tip, PC chip	Pressure flow 0.3 μl/min	ESI-ITMS	Horse heart cytochrome c (5 μM), 30-mer oligonucleotide (10 μM)	Trypsin	Nil	811
Inserted glued capillary, glass chip	Pressure flow 50 nl/min	ESI-triple Q	Angiotensin I, leu-enkephalin vascoactive intestinal peptide, Glu-fibrinopeptide B	Trypsin	CE	812
Microsprayer, glass chip	Pressure flow 1.5 μl/min		Tryptic digest from D. biflorus, P. sativum lectins, substance P, brakykinin (20 μg/ml) oxytocin, Met-enkephalin, Leu-enkephalin, bombesin, luteinizing hormone releasing hormone (LHRH), Arg^8-vasopressin, bradykinin			
Inserted capillary tip (no glue), glass chip	Pressure flow 100 nl/min	ESI-ITMS (LCQ)	Tryptic digest of cytochrome c myoglobin, β-lactoglobulin A and B, BSA	Trypsin	CE	813

—continued

TABLE 7.1 (continued)
Summary of MS Analysis Using Various Techniques of Chip-MS

Chip Design	Flow	MS Mode	Analytes	Sample Treatment	Separation	Reference
Inserted tip, glass chip	EOF-hydraulic pump 200 nl/min	ESI-TOF	Bovine hemoglobin tryptic digest	Mixed off-chip, on-chip tryptic digestion	Nil	115
Inserted emitter, silver glue coated, PMMA chip		ESI-ITMS	BSA, fibrinopeptide A, osteocalcin fragment 7–19, bradykinin (all 10 pmol/µl)	Trypsin, off-chip	Nil	814
Inserted tip PC chip	200 nl/min	ESI-ITMS	BSA (30 µM), cytochrome c (8 µM), ubiquitin (2.4 µM), $E.\ coli$ lysate	Dialysis	Nil	815
Flat-Edge ESI Sprayer						
Flat-edge emitter, glass chip	EOF induced flow 1.5 nl/s	ESI ITMS	Tetrabutylammonium iodide (10 µM)	N/A	Nil	816
PET, no tip	Sheathless flow 200 nl/min	ESI-ITMS	Reserpine (1 pg/ml to 1 mg/ml) Horse heart myoglobin (5 µM)	Nil	Nil	195
PDMS chip	Pressure flow 0.1–0.3 µl/min	ESI, single Q	Cytochrome c	Trypsin on PVDF membrane	CITP/CZE	817
Hydrophobically coated flat edge, nine-channel glass chip	Pressure flow 100–200 nl/min	ESI-triple Q	Myoglobin (60 nM)	Nil	Nil	818
PDMS chip with no tip	EOF	ESI-triple Q	Psilocin (50 pmol) Buprenorphine (20 pmol)	Nil	Nil	162
Multichannel ESI Chip						
Multichannel, glued capillary tip, glass chip	EOF flow	ESI	Melittin	Trypsin	Nil	819
Nine-channel/one-transfer capillary sheath, glass chip	EOF flow	ESI-ITMS	Tryptic digest of BSA-182 nM	Trypsin	Nil	776
			Horse myoglobin (237 nM) Human haptoglobin (222 nM) 2D gel yeast proteins (40 µg)			
96-channel 96-tip epoxy resin chip	N$_2$ pressure 200 nl/min	ESI-ITMS	8-mer peptide for HIV-1 protease (with inhibitors: e.g., pepstatin A)	N/A	Nil	780

Device	Flow/Conditions	MS	Analyte	Pretreatment	Separation	Ref.
90-channel Si-glass chip	Vacuum 10^{-6} Pa		41-base deoxyoligonucleotide	Digested by snake venom phosphodiesterase	Nil	820
			Adrenocorticotropin (ACTH)	Digested by carboxypeptidase		
Polymeric ESI Tip						
Embedded tip, PDMS chip	Pressure flow 0.4 µl/min	ESI-single Q	0.4 µM phenobarbital	Dialysis	Nil	821
Triangular-shaped tip on parylene C sheet, cyclo-olefin chip	Pressure flow 300 nl/min	ESI-TOF	1 µM desipramine and 1 µM imipramine, chicken cytochrome c	N/A	Nil	788
Monolithic Si nozzle	0–5 psi N_2, 100 nl/min	ESI-TOF	Cytochrome c (10 nM), nine-peptide mix (160 pg/µl each)	Trypsin	Nil	790
Micromachined emitter, parylene chip	2–4 psi gas, 35–77 nl/min	ESI-ITMS	Gramicidin S, cytochrome c	Trypsin	Nil	783
16-channel, monolithic PDMS tip, PDMS	300 nl/min		Adrenocorticotropic hormone fragment 1–17 (ACTH 1–17) Angiotensin I and III (Arg I and Arg III) (all 1 µM)	Trypsin	Nil	784
PC plate with an integrated tip	0.5 psi N_2, EOF 0.05 µl/min Sheath liquid 0.2 µl/min	ESI-ITMS	Myoglobin (0.05 mg/ml)	Trypsin	IEF	290
Polyimide tip	300 nl/min	ESI-ITMS	Mammalian proteins	Trypsin	LC	822
PMMA chip, hand-polished 3D tip	Sheathless 1 µl/min	ESI-TOF	1 µM angiotensin II, $[Arg^8]$-vasopressin, leucine-enkephalin, methionine-enkephalin, luteinizing hormone releasing hormone (LHRH)	Nil	Nil	213
Zeonor chip		ESI-triple Q	Imipramine, doxepin amitriptyline	On-chip human liver microsome metabolism	SPE	217
SU-8 nib	1 µl/min	ESI-LCQ	Gramicidin S (1–10 µM)	Nil	Nil	786
Eight-channel PMMA chip with sharp end	0.5–1 µl/min	ESI-Q	Myoglobin, cytochrome c, insulin, bradykinin (1 µM)	Nil	Nil	781
COC chip with parylene C tip	5 µl/min	ESI-triple Q	Methylphenidate (Ritalin) in urine	Nil	Nil	823
Monolithic Si nozzle	Pressure	ESI-QTOF	O-Glycosylated sialylated amino acids and peptides in urine	Nil	Nil	824

—continued

TABLE 7.1 (continued)
Summary of MS Analysis Using Various Techniques of Chip-MS

Chip Design	Flow	MS Mode	Analytes	Sample Treatment	Separation	Reference
Glassy C chip with an ESI tip	Pressure 0.4 µl/min	ESI-Q	Angiotensin II (26.8 nM)	Nil	Nil	789
PDMS chip, with PDMS tip	3 µl/min	ESI-TOF	Angiotensin I (10 µM), bradykinin (10 µM)	Nil	Nil	785
PTFE membrane over emitter tip exit on a PC chip	EOF 80 nl/min	ESI-QTOF	Myoglobin (1 mg/ml)	Nil	Nil	751
PDMS chip with an open channel tip	Pressure 0.2 µl/min	ESI-TOF	Myoglobin (1 µM)	Nil	Nil	172
		MALDI				
Si chip	N/A	MALDI	Lysozyme (1 µM)	Trypsin on porous Si	Nil	819
Glass chip	N/A	MALDI-FTMS	[Lys']-bradykinin (0.5 mg/ml)	Nil	CE	792

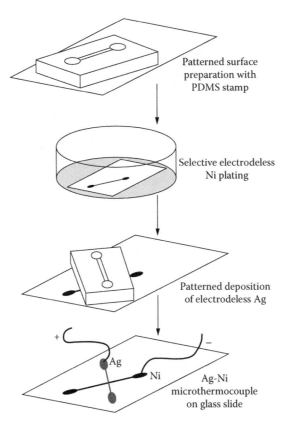

FIGURE 7.45 Schematic showing the procedure used for the microfabrication of a thermocouple. The thermocouple, which consists of intersecting Ni and Ag metal lines, is fabricated by electrodeless metal deposition as masked by the PDMS channels.[271] Reprinted with permission from the American Chemical Society.

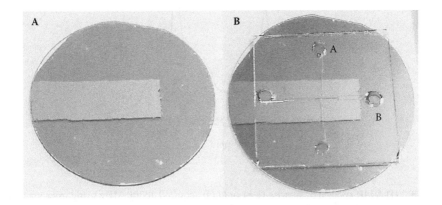

FIGURE 7.46 (a) Image of a TSM sensor consisting of a 4 × 15 mm front-side Au electrode. (b) Image of the TSM sensor bonded with a microfluidic channel plate.[133] Reprinted with permission from Royal Chemical Society.

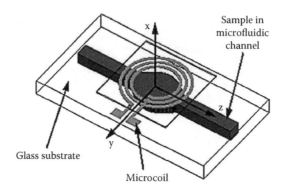

FIGURE 7.47   Schematics of a micromachined NMR probe. The probe consists of a multiturn electroplated planar microcoil integrated on a glass substrate with etched microfluidic channels for sample containment. The coil has typical dimensions of 2 mm or less, with sample containment capability ranging from a few nanoliters to 1 μl, depending on coil size. As a reference, the coil lies in the yz-plane with the static magnetic field B_0 along the z-axis.[797] Reprinted with permission from Elsevier Science.

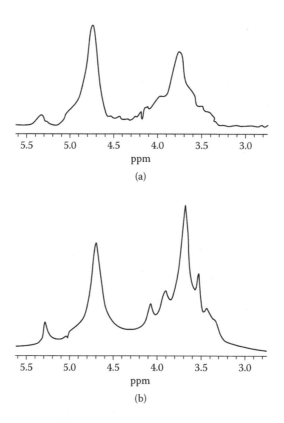

FIGURE 7.48   1H NMR spectra of sucrose (1 M in D_2O) acquired at 300 MHz using micromachined NMR probes. (a) Spectrum acquired using a probe with sample volume ~30 nl (diameter over height ratio 3.3). The corresponding sample mass is 10 g (30 nmol). (b) Spectrum acquired using another probe with sample volume ~470 nl (diameter over height ratio 13.3). The corresponding sample mass is 160 g (470 nmol).[797] Reprinted with permission from Elsevier Science.

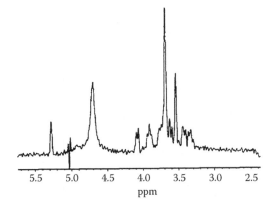

FIGURE 7.49   1H NMR spectrum of sucrose after Lorentz-Gauss resolution enhancement applied to the data shown in Figure 7.48b.[797] Reprinted with permission from Elsevier Science.

7.5   PROBLEM SETS

1. Describe, with a labeled diagram, the principle of confocal fluorescent detection. (4 marks)
2. In order to integrate the optics in the microchip for fluorescent detection (see Figure 7.7), describe how a bandpass emission filter is fabricated. (2 marks)
3. In the ninety-six-channel microchip in Figure 9.22, what are the advantages for the channels to be arranged radially? (2 marks)
4. In one report, DTT was added in the run buffer for the CE separation of fluorescently labeled peptides with fluorescent detection. What is the purpose of adding DTT?[1039] (2 marks)
5. In SCOFT for multipoint fluorescent detection, a Shah function is used (Figure 7.11). What are the two different types of methods in implementing the shah function? (4 marks)
6. Why is decoupling necessary in amperometric detection for CE separations (see Figure 7.22)?[191] (2 marks)
7. What are the advantages of carbon over metal for fabrication of electrodes in electrochemical detection? (2 marks)
8. Why are the electrodes used in contactless conductivity detection arranged in an antiparallel fashion?[1153] (2 marks)
9. How can an increase in the AC frequency in AC conductivity detection prevent the faradaic reaction of electrolysis from occurring? (3 marks)[208]
10. In fluorescent detection, an argon ion laser is usually used for excitation. Suggest one fluorophore that is compatible with this source, and describe why there are two associated wavelengths. (3 marks)
11. What are the light source and detector used in the microchip with integrated fluorescent detection?[692] (2 marks)
12. When AC is used in conductivity detection on-chip, there is no electron transfer or no faradaic current measured. Then what is actually measured? And how is the conductivity value generated from the measurement? (4 marks)
13. Name two analytes that can be detected by the conductivity detector. (2 marks)
14. In amperometric detection, a three-electrode system is used (Figure 7.21). Describe the function of each electrode. (3 marks)
15. In the placement of the working electrode in amperometric detection, the electrode cannot be too close or too far from the channel end. Why? (2 marks)

16. Name two analytes that can be detected by the amperometric detector. (2 marks)

17. In fluorescent detection of an analyte at low concentration, the fluorescent intensity F is given by $F = 2.303\, \Phi_f\, I_0\, \varepsilon\, b\, c$.
    a. What are Φ_f, I_0, and ε ? (2 marks)
    b. What is the assumption for the linear dependence of F to the analyte concentration c? (2 marks)
    c. State one method each to increase detection sensitivity of an analyte based on the various parameters, Φ_f, ε®, and I_0. (2 marks)

18. Why is ITO sometimes used as an electrode for electrochemical detection?[273,727,728,869] (2 marks)

19. Why is desalting of a protein sample necessary in ESI-MS analysis of the protein?[347,981] (2 marks)

20. Why is a pointed tip always used for ESI formation?[290] (2 marks)

21. Why would an increased amount of salt in a DNA sample during CE separation result in a reduction in the detection signal?[982] (2 marks)

22. Describe the two general types of methods used to create the ESI emitter tip for microchip MS analysis. (4 marks)

23. How can the addition of organic solvents in the sample help to achieve a stable electrospray?[290] (2 marks)

24. What are the functions of the sheath liquid and sheath gas used in an ESI chip for interfacing with a mass spectrometer (Figure 7.30)?[290,808] (2 marks)

25. A microfabricated dialysis device was used to clean up the sample (i.e., protein) by removing the matrix interference in an ESI chip before carrying out MS analysis (see Figures 5.10 and 5.11). What substances from the matrix are removed?[811] (2 marks)

26. An electrospray is normally produced using a fine capillary tip. However, ESI was successfully achieved on a flat edge. (i) Describe the critical issue that is overcome for this success. (ii) Suggest two ways to achieve this success.[818] (4 marks)

27. What is the function of the hydrophobic break in the MALDI sample preparation compact disk (Figure 7.42)? (2 marks)

28. In the construction of an ISE in a microchip, only one arm of two joined channels was filled with an organic liquid. State two important steps in the procedure to ensure the filling in one arm but not the other.[766] (2 marks)

29. Briefly describe the operation principle of potentiometric Ba^{2+} detection in a microchip ISE. State two examples of the Ba^{2+} ionophore.[766] (4 marks)

30. In the microchip ISE, the calibration plot of the measured potential vs. log $[Ba^{2+}]$ gives a slope of 36 mV. What is the theoretical value of the Nernstian slope? Explain the difference between the experimental and Nernstian slopes.[766] (2 marks)

31. Describe two advantages and one disadvantage of the contactless and contact conductivity detectors.[1159] (3 marks)

32. In amperometric detection, why should the detector voltage be isolated from the separation voltage? And how?[750] (4 marks)

33. How can a silver/AgCl reference electrode be made from a silver wire?[749] (2 marks)

34. When using a metal electrode for amperometric detection, there is an electrode fouling problem. Describe one procedure and one new material to alleviate the problem.[748] (2 marks)

35. In amperometric detection, there is H_2 evolution at the electrode. How can this problem be prevented by using a Pd decoupler near the grounded cathode?[205] (2 marks)

36. Sometimes a gold electrode is platinized. What is its purpose? And how can this be done?[242] (4 marks)
37. In using PDMS for chip-MS, there are problems of chemical noises. What are the methods used to reduce the noises?[800] (2 marks)
38. What are the singlet and triplet states in relation to the fluorescence phenomenon? (2 marks)
39. What is the definition of the detection limit of an analyte? (2 marks)
40. Explain why two lasers are used in TLM detection.[1021] (2 marks)

8 Applications to Cellular/Particle Analysis

Cellular studies are facilitated in the microfluidic chips because of their small dimensions. In addition, the chip provides excellent optical properties for observation and flexible fluidic control capabilities for reagent delivery.

Cell retention, manipulation, and subsequent cellular analysis can be all achieved on-chip. Cell retention can be achieved by using slit- or weir-type filters, or by cell adhesion. In addition, fluid flow optical trapping and dielectrophoresis (DEP) have been exploited to manipulate and retain cells.

Particles are sometimes used to carry out modeled studies of fluid flow or cell manipulation. In other studies, the particles or beads are actually employed to carry out analysis, e.g., immunoassays.

8.1 RETENTION OF CELLS AND PARTICLES

8.1.1 SLIT-TYPE FILTERS

Slit-type filters have been fabricated to trap micrometer-sized particles and to retain suspension cells (e.g., blood cells).

To trap particles, a filter consisting of a 1 μm deep channel was constructed on a quartz chip. This filter was employed for sample filtration before channel electrochromatography (CEC) separation.[148]

In one report, a filter consisting of slits of 5 μm spacings was fabricated on a Si chip. After sealing with a Pyrex glass plate, this filter was found to be able to retain latex beads of 5.78 μm in diameter within a 500 μm channel (see Figure 8.1).[919] However, in the study of red blood cells (RBCs) using this type of filter, the cells tended to deform and passed through the slits.[835,826]

In another report, a fluid filter was constructed using two silicon membranes (containing numerous 10 μm diameter holes). The two membranes were displaced laterally relative to each other and separated by a distance with silicon dioxide spacers of submicron thickness.[827]

A lateral percolation filter was fabricated on quartz (see Figure 8.2). The filter elements, which were located near the entry port of a channel, had 1.5 μm channel width and 10 μm depth. They were anisotropically etched with an aspect ratio greater than 30:1.[828] In contrast to the usual axial slit filter, in which the filter area is dictated by the channel area, the fluid flow in the lateral percolation filter is at a right angle to the filtering direction. This filter, which is also known as the cross-flow filter,[228] provides a much larger filter area than can be offered by the channel width. Because of the channel depth of 10 μm, even though the channel entrance is blocked by a particle, the liquid is still capable of flowing under the particle. Filtering of 5 μm silica particles, soybean cells (Glycine max var. Kent), human KB cancer cells, and *E. coli* cells has been achieved with the lateral percolation filter.[828]

Retention of oligonucleotide immobilized beads was also achieved on a slit-type filter chamber fabricated on Si. The three different designs are shown in Figure 8.3.[829–831] Although the flow rate decreased when the chamber was filled with beads, clogging of the filter was rare and reversible, especially with the design shown in Figure 8.3c.[831]

Slit filters are usually fabricated on Si or quartz, but they can also be made on glass (but with greater gap spacings of 11 to 20 μm). In one report, a glass slit filter was fabricated essentially to prevent particles from clogging channels downstream.[115]

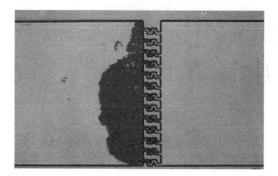

FIGURE 8.1   Filtrations of 5.78 μm diameter latex beads by a slit-type filter. A 500 μm wide and 5.7 μm deep channel was fabricated on a 400 μm thick Si wafer. The slit structure consisting of 5 μm wide slits between 73 μm wide posts was fabricated by deep RIE (DRIE).[919] Reprinted with permission from Elsevier Science.

FIGURE 8.2   Lateral percolation in a microfabricated filter. It is seen that as liquid vertically enters an array of microfabricated cubes attached to an underlying substrate and is drawn laterally to the sides of the array through an interconnecting channel network, particles larger than the 1.5 μm channels separating the cubes will be excluded, similar to axial filters. It should also be noted that as the entrance to a channel is blocked by a particle, liquid is still capable of flowing under the particle, and multiple alternative routes are still available for liquid to enter the bed and migrate through the filter. The significance of this is that the illustrated lateral percolation filter will have a specific filter capacity (particles filtered/cm² at the filter face) equivalent to high-capacity axial filtration systems.[828] Reprinted with permission from the American Chemical Society.

Particle retention has been achieved by a porous polyimide channel (see Figure 8.4). Based on ion-track technology, the photocured polyimide was first irradiated by energetic xenon ions to form ion tracks. By chemical etching using NaOCl, these tracks were enlarged to form nano-sized (500 nm) channel pores. Two different sized beads were detected and differentiated by laser light scattering measurement. It was found that large fluorescent beads (~5 μm) were filtered off, with small fluorescent beads (~300 nm) passing through the porous nanochannels (see Figure 8.5).[228]

The slit-type filter has also been used for cell retention. For instance, a porous filter has been used to retain rabbit blood cells in a whole blood sample. This filter was fabricated inside a glass microchannel using emulsion photopolymerization. The retained cells were then lysed for the assay of G-6-PDH.[832]

In another report, vegetal cells were guided by a curved channel and retained against vertical pillars in a poly(dimethylsiloxane) (PDMS) chip. An average of eight cells were successfully trapped at sixteen sites arranged in a star-like shape.[833]

A muscle cell (freshly isolated mouse ventricular myocyte) was retained at a 5 μm constriction in a PDMS-glass chip. Extracellular potential recordings were measured during the spontaneous contraction of the muscle cell. Experiments were usually performed within 3 h of cell isolation in order to maintain cell viability.[270]

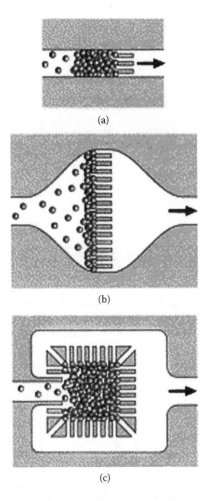

FIGURE 8.3 Different microfilter designs. (a) The filter pillars are placed inside the channel. (b) The channel is widened at the bead-trapping location. (c) The filter pillars define a square reaction chamber where the beads are collected.[831] Reprinted with permission from Elsevier Science.

A heart muscle cell (rabbit cardiomyocyte) was aligned in a narrow PDMS-glass microchannel. The muscle cell was electrically stimulated via a pair of Au electrodes (~60 µm apart) in contact with the cell. Intracellular Ca measurement showed that the cell remained contracted for 60 min within the restricted space. An electric field strength of 20 V/cm translates to 0.12 V, which is lower than the electrolysis threshold of 0.8 V.[834]

In one report, a mouse islet of Langerhans (consisting of two thousand cells) was trapped at a tapered region (from 300 to 60 µm) connecting two microfluidic channels. Stimulation of the islet by 11 mM glucose was conducted on only one side of the islet. It was found that this localized stimulation only caused Ca influx on that side, rather than through the whole islet body.[835]

8.1.2 Weir-Type Filters

Slit-type filters have mostly been fabricated on Si or quartz because of the need for high aspect ratio etching. Therefore, weir-type filters have been fabricated for cell or particle retention. These filters have constrictions along the depth dimension, not along the lateral channel direction as in slit-type filters. For instance, a barrier (or weir) with a gap of 0.1 µm was fabricated on a Si wafer.

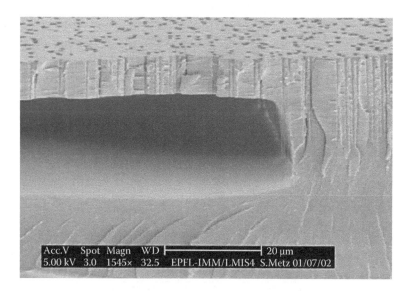

FIGURE 8.4   Cross-sectional scanning electron micrograph (SEM) image of a porous microfluidic channel after being irradiated with ions of low energy. The beam energy determines the length of the tracks allowing perforation of the top layer.[228] Reprinted with permission from the Institute of Physics Publishing.

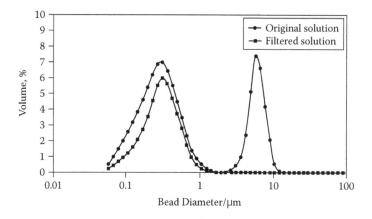

FIGURE 8.5   Size distribution of 300 nm and 5 μm beads in the original solution and in the solution after filtering through the ion-track membrane with 0.5 μm pores.[228] Reprinted with permission from the Institute of Physics Publishing.

After bonding with Pyrex glass, the chip was used as a fluid filter (see Figure 8.6). The barrier was along a V-shaped fluid channel where particle-laden fluid passed from the entry to the exit ports, and the filtered fluid plasma passed over the barrier to a side channel. A vertical wall (i.e., 90°) instead of the V-shaped wall (i.e., 55°) would require a much higher pressure (1 atm) to overcome the surface tension present in a 0.1 μm gap.[836] Using this device, filtration of blood cells in whole blood was achieved (see Figure 8.7).[836] In another report, a similar device was fabricated to retain microspheres.[397]

A weir-type Si microfilter (3.5 μm gap), which was covered by glass, was fabricated to retain white blood cells (WBCs) in whole blood. The cell retention performance was better than with the slit-type filter (7 μm gap). Although the slit-type filter can retain nondeformable latex microbeads, deformable objects such as blood cells squeeze through the gaps.[919]

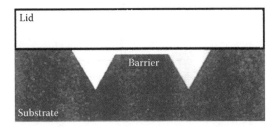

FIGURE 8.6 The cross section of a weir-type filter (not to scale). The channels in the silicon substrate are anisotropically etched using EPW. This gives the characteristic V-shaped grooves. This profile is critical to preventing surface tension lock. The barrier or weir is etched in a different step from the channels and can be anywhere from 0.1 μm to a few micrometers from the lid. The lid is Pyrex glass and is attached to the substrate by anodic bonding.[836] Reprinted with permission from Elsevier Science.

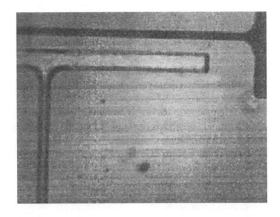

FIGURE 8.7 A reflected-light image of a chip in which whole blood, which appears dark, enters from the top right and flows across the upper channel (100 μm wide) to the left. After filtration of blood cells, the plasma crosses the barrier and fills the region at the bottom channel.[836] Reprinted with permission from Elsevier Science.

Microbial cells were also trapped using a 1 to 2 μm weir gap in a Si-Pyrex chip. This is because these cells (*Cryptosporidium parvum* and *Giardia lamblia*) contain cell walls and are not easily crushed or squeezed through the weir gap.[837] In addition, a 0.5 μm gap was fabricated in a Si-glass chip for retention of yeast cells. The gap was created by the patterned Au dots (0.5 μm high).[838]

V-shaped grooves of 9 μm top opening and 14.4 μm length were fabricated on a Si substrate. This was mechanically pressed to an optically flat Pyrex glass plate for sealing. The V-shaped channels were used to study rheology (flow dynamics) of WBCs. Apparently, the cells, after activation by a chemotactic peptide (formyl-methionyl-leucyl-phenylalanine [FMLP]), showed a greater resistance to channel passage.[839]

A glass microchip in which a chamber was bound by two barriers or weirs was constructed for cell retention (see Figure 8.8). Mouse lymphocytes were introduced into the chamber and retained in the chamber by the weirs, where the main flow of fluid passes over the weirs.[1170]

Two weir-type cell traps were microfabricated in a glass chip to retain yeast cells (see Figure 8.9). Figure 8.9A shows the sequences of how the cells enter the left and right cell traps. Figure 8.9b shows the locations of these cell traps in the large channel. In Figure 8.9a, for the left cell trap, the cells curved upward where the weir was vertically oriented. As for the right cell trap, in which the weir was horizontally situated, the cells entered horizontally; see also the outlines of the weirs in Figure 8.8b.[138]

When the cell trap was microfabricated in PDMS, a small-sized cell trap could be made. For instance, a single mammalian cell (Jurkat T cell, 15 μm) can be trapped in the U-shaped structure (see Figure 8.10). Here, the main channel and the cell dock are 20 μm deep, while the drain channel

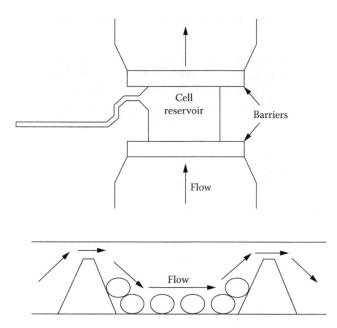

FIGURE 8.8 Top: Schematic diagram of the weir-based device for physical trapping of cells. Bottom: The cross section of the device showing how the cells are retained in the chamber, with the fluid flowing over the two barriers. Reprinted with permission from D.J. Harrison.

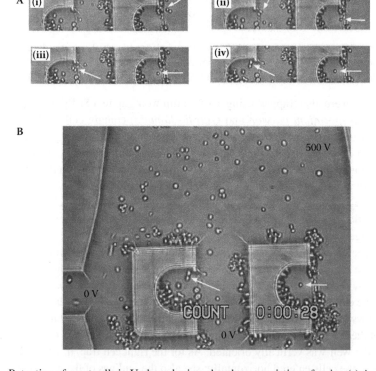

FIGURE 8.9 Retention of yeast cells in U-shaped microchambers consisting of weirs. (a) A series of images indicating the transport, entry, and retention of yeast cells: (i) cell starts to enter, (ii) cells are at the opening, (iii) cells enter completely, (iv) cells are retained. (b) A zoom-out view of (iii) showing the location of the two chambers in a large microchannel.[138] Reprinted with permission from the Royal Society of Chemistry.

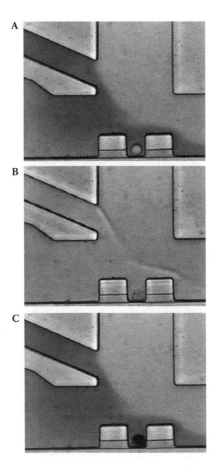

FIGURE 8.10 Chronological sequence of images of a viability assay on a single Jurkat T cell. (a) Live cell perfused with trypan blue dye. Since the cell is alive, it is not stained. (b) Cell perfused with methanol, which causes cell membrane permeabilization, followed by cell death. (c) Permeabilized cell perfused with dye; it is rapidly (<5 s) stained. The entire assay is performed in less than 2 min.[368] Reprinted with permission from the American Chemical Society.

(or weir) has 5 μm clearance. In the case of small U937 cells (10 μm), sometimes two cells were trapped together. The location of the cell dock was intentionally set at the point of flow stagnation. Therefore, it was possible to stabilize the flow stream so that a cell that lands in the dock became trapped. Using this method, ionomycin-mediated $[Ca^{2+}]_i$ flux (ringing or sustained elevation) in a trapped Jurkat cell has been measured.[368] For U937 cell experiments, the PDMS channels were typically incubated with bovine serum albumin (BSA) to prevent cell adhesion, and sulfinpyrazone (0.1 mM) was used to treat the cells to prevent Fluo-4 dye leakage. It is found that $[Ca^{2+}]_i$ flux occurs when the U937 cell is first treated with human IgG (primary Ab) and then with goat anti-human IgG (secondary Ab). This $[Ca^{2+}]_i$ flux is an $F_{c\gamma}R$-mediated response, which can be altered by the expression level of these receptors (for the constant region of IgG) on U937 cells. The expression level can be varied when the cells are differentiated with interferon gamma (IFN-γ) or dibutyryl cyclic AMP during cell culture.[368]

HL-60 cells have been docked along a parallel weir fabricated in a quartz chip for cell immobilization. In contrast to a usual perpendicular (or axial) dam structure, the cells docked along a parallel dam suffered less shear stress because the main fluid flow was not reduced (see Figure 8.11).[840] This design provided a sideways flow to trap cells while allowing the main flow to go on, and so this method did not result in cell squeezing or crushing as observed in an earlier study.[841] According to

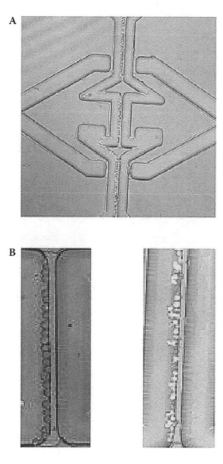

FIGURE 8.11 Docking and alignment on a microchip. (a) Fluorescence image of the microchannels show-ing the alignment of mammalian HL-60 cells along the dam structures. (b) Left: Photograph of a section of the dam with docked cells. A single layer of HL-60 cells was immobilized along the dam with good integrity. Right: Live/dead cell staining of a line of HL-60 cells docked at the dam structure for 20 min. Most of the cells were stained by acridine orange (yellow), indicating that the cells were viable after docking. Some cells were stained by ethidium bromide (red), indicating that these cells were dead.[840] Reprinted with permission from the American Chemical Society.

a mathematical model, an at least fortyfold increase in liquid pressure occurred across the particles when immobilized at a perpendicular dam, leading to a large stress. The threshold ATP concentra-tion for Ca uptake (as measured by intracellular Fluo-3) was estimated to be within a range of 0.16 ± 0.02 μM. The same batch of docked cells (~80) was used to measure the response to the ATP stimulus at four consecutive concentrations (0, 2, 5, and 10 μM). Although the ATP-stimulated Ca uptake reaction was reversible, desensitization of the calcium ion channel occurred after multiple ATP stimulations.[840] Other dam structures that combine the advantages of perpendicular and paral-lel dams have been proposed.[165]

Mouse macrophage-like cells (J774.1) were retained by a weir (10 μm clear) in a Pyrex chip. Upon stimulation by lipopolysaccharide (LPS), nitric oxide (NO) was released. NO was detected by thermal lens microscopy (TLM) using the following procedure. Dissolved O_2 first oxidized NO to form NO_2^- and then NO_3^- for detection was subsequently reduced to NO_2^- (by nitrate reductase), which reacted with Griess reagents to produce a colored product to be detected by TLM.[1077]

A single neuron-like PC12 cell was trapped in an etched glass (30 μm deep) pocket sealed against a PDMS channel layer (20 μm). Quantal release of dopamine (in transient exocytosis) from the cell as stimulated by nicotine was amperometrically detected with a carbon fiber electrode. The cells flow into the channels caused by the liquid pressure, which was provided by a liquid height at the sample reservoir (e.g., 0.5 to 2 mm). To facilitate transport of cells in the microchannels, the cell density should not be higher than 10^4/ml. Serious cell adhesion occurred if the transport speed was low (as provided by liquid height below 0.5 mm), whereas a speed of 0.55 m/s produced reasonable cell transport by a liquid height difference of 2 mm.[134]

A single human lymphoblast (MOLT-3) was retained by a weir (5 µm) in a PDMS-glass chip. This cell channel was separated from another channel containing sodium dodecyl sulfate (SDS) lysing solution by a gas bubble (introduced by a thermopneumatic actuator). Removal of the gas bubble initiated the liquid mixing, and cell lysis occurred.[842]

A two-dimensional (2D) microwell array was used to trap CHO cells in a chamber containing a weir. Flow control was achieved by pneumatic valving on a PDMS chip. Once the cells were introduced into the chamber, its opening was sealed by valve closure to retain the cells. To show a cell reaction, the cells were lysed by deionized water.[843]

In a cell manipulation experiment, a constriction together with two narrow channels (~50 µm) at the corner of a large channel was fabricated on a PDMS chip (see Figure 8.12). This microfluidic method was devised to remove cumulus cells from individual bovine cumulus–oocyte complexes before *in vitro* fertilization. In natural fertilization, sperms release hyaluronidase that attack and

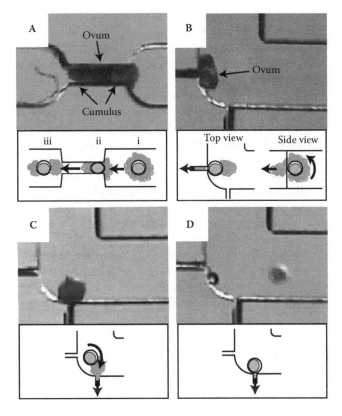

FIGURE 8.12 Pictures at the four main stages in the cumulus removal process in *in vitro* fertilization. Insets show the schematics of the pictures. (a) Picture of the cumulus–oocyte complex (CC) being reoriented as it passes through a constricted region. Insert shows a cross section of reorientation of the cumulus cells around the CC (i) before, (ii) during, and (iii) after the CC passes through the constricted region. (b) The CC is at the left removal port. Inset shows top and side views of the cumulus being sucked into the removal port. (c) The cumulus rotates as it moves from the left port to the bottom port. (d) Remaining cumulus is removed at the bottom port, leaving a clean oocyte.[1171] Reprinted with permission from Springer Science and Business Media.

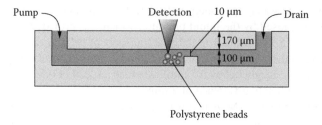

FIGURE 8.13   Cross section of the fused silica glass microchip for immunosorbent assay on the trapped polystyrene beads (4.5 μm diameter).[846] Reprinted with permission from the American Chemical Society.

remove the cumulus mass, and allowed the penetration of a sperm into the oocyte (or ovum). The microfluidic operation sequence is illustrated as follows. As shown in Figure 8.12a, the complex (~400 μm diameter) was reoriented. In Figure 8.12b and c, the small cells (~10 μm) making up the cumulus easily entered the removal port (~50 μm), whereas the oocyte (~100 μm) was retained (see Figure 8.12d).[1171]

In neurophysiology experiments, neural (brain) tissues (600 μm thick) were adhered to an array of micropillars (instead of a nylon mesh) in a microchamber. The micropillars (400 μm high SU-8) uphold the neural tissue and enhance the exposure to oxygen and nutrients, i.e., four times faster perfusion exchange rate than the conventional perfusion chamber. Epileptiform activities (spontaneous single and multiple electrical spike bursts) in the CA3 region of a rat hippocampal brain slice were observed. These activities, which were stimulated by zero-Mg^{2+} in an oxygenated artificial cerebrospinal fluid (ACSF), lasted for more than 5 h.[236]

Microbeads are shown to be retained by the weirs (dams) around the chambers fabricated in a Si chip,[844] a glass chip,[639] or a PDMS chip.[845,987] For instance, polystyrene beads have been retained in a microchannel consisting of a weir (or shallow channel) of 10 μm clearance (see Figure 8.13). The trapped beads were used for immunosorbent assay of human secretory immunoglobulin A (s-IgA) (~200 μg/ml in human saliva).[846] Detection of colloidal gold, which has been conjugated to anti-s-IgA (goat antiserum), was achieved by thermal lens microscopy (TLM). Antigen-antibody reaction was completed in 10 min, which was comparable to liquid-phase (homogeneous) immunoassay at room temperature. The total assay time was shortened from 24 h to less than 1 h.[846] A similar strategy has also been used to detect the carcinoembryonic antigen (CEA) (see Section 10.1.2 in Chapter 10).[1021]

In another report, a weir-type filter constructed on a PDMS-glass chip, shown in Figure 8.14, was used to destack microbeads (or microbarcodes) for easy observation. The suspension containing stacked microbeads (20 μm thick) was loaded in the center reservoir (see Figure 8.14a, top). Upon spinning, the beads moved outward due to centrifugal force and were retained in a single layer without stacking on top of each other (see Figure 8.14b, bottom).[1172]

8.1.3   CELL ADHESION

For adherent cells, the retention of cells for measurements is straightforward. For instance, a microphysiometer was constructed to detect the change in cellular acidity (due to lactate and CO_2 formation) of mammalian cells. The acidity was measured using the light-addressable potentiometer sensor (LAPS) fabricated on a Si chip (see Figure 8.15).[848]

Various types of adherent cells were grown on a coverslip, which was then laid on top of the LAPS chip. Measurements were made for acidification of (1) normal human epidermal keratinocytes stimulated by epidermal growth factor (EGP) or organic chemicals, and (2) human uterine sarcoma cells as a response to doxorubicin and vincristine (chemotherapeutic drugs). In addition, the inhibition (by ribavirin) of the viral infection of murine fibroblastic L cells by vesicular stomatitis virus (VSV)

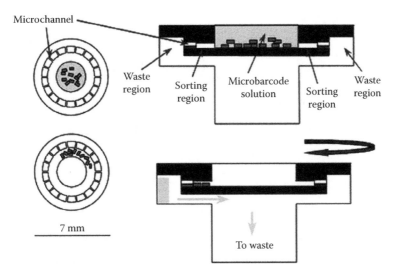

FIGURE 8.14 A microbarcode/microbead sorting plate using a weir-type filter at the sorting region. Top: Loading of microbeads that were stacked together, in the center of the sorting plate. Bottom: After spinning, the microbeads were destacked and collected at the perimeter.[1172] Reprinted with permission from the Royal Society of Chemistry.

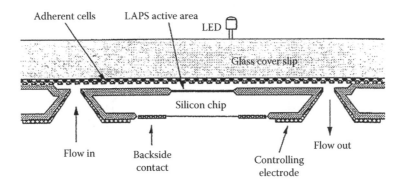

FIGURE 8.15 Cross section of the light-addressable potentiometer sensor (LAPS) chip integrated with a flow channel for study of adherent cells.[848] Reprinted with permission from the American Association for Clinical Chemistry, Inc.

was investigated by following the acidification rate. A limitation of these studies is the requirement for a low-buffered medium (low bicarbonate content) to achieve maximum sensitivity.[847]

In another report, adherent CHO-K1 cells, which expressed the m1-muscarinic acetylcholine receptors, were studied as a response to stimulation by the agonist, carbachol. The cells were first grown on a coverslip, then sealed to a LAPS chip with eight fluidic channels for measurement (see Figure 8.16).[848,849]

Another adherent cell, rat basophilic leukemia cell (RBL-2H3), has been studied in a PDMS reaction chamber.[250,850] The cell is a tumor analog of rat mucosal mast cells. The RBL-2H3 cells were used by an antigen for study of histamine release from the cellular acidic histamine granules. To measure histamine, quinacrine (5 μM), a basic aminoacridine fluorescent dye, was first loaded into the histamine granules by mass action. When stimulated by an antigen (dinitrophenylated BSA), excoytosis of histamine occurred, and quinacrine was released together with histamine, generating a fluorescent signal. For the cells to respond to the antigen, they were first sensitized with monoclonal mouse IgE antibody against the above antigen (0.25 μg/ml).[250]

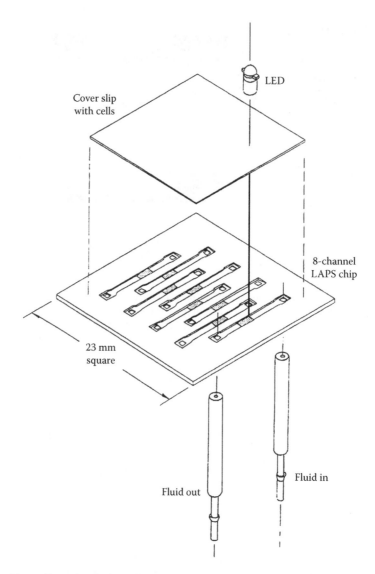

FIGURE 8.16   Three-dimensional view of the assembly of an eight-channel LAPS chip with coverslip (with adherent cells), LEDs, and fluid connections.[848] Reprinted with permission from the American Association for Clinical Chemistry Inc..

In another report, the release of cytochrome c from mitochondria to cytosol in neuroblastoma-glioma hybrid cells was studied on a quartz chip. The released cytochrome c was measured by scanning thermal lens microscopy.[851]

The channel of a PDMS chip was featured with human anti-α_5-integrin (HAI) to study cell adhesion. The effect of a cell adhesion inhibitor, echistain (1 to 33 µg/ml), which was a drug candidate, on the adhesion of bovine aortic endothelial cells (BAECs) was studied. HAI was immobilized in a PDMS chip by the following procedure. First, the PDMS chip surface was functionalized with pentafluorophenol groups (from PPX-PPF). These groups were substituted by the amino-terminated biotin ligands. Then streptavidin was used as a linker to conjugate both the surface-immobilized biotin and the biotinylated HAI.[262]

A cell adhesion study of platelets was also conducted in PDMS channels. These channels have first been coated with fibrinogen, using precursor polyelectrolyte layers (polyethylenimine [PEI], poly(styrene sulfate) [PSS], polydimethyldiallylammonium [PDDA]) for platelets to adhere. The platelets were doubly fluorescently labeled with FITC-tagged antibodies (anti-glycoproteins: GPIIb and GPIIIa) and with acridine orange-tagged anti-CD41. It was found that the platelet adhesion depended on the speed or shear rate of the liquid flow.[852]

In one report, a single human erythrocyte was adhered in a glass chip for cell lysis, and subsequent CE analysis of the released glutathione (GSH) was carried out. This analysis was achieved after hydrodynamic cell loading, electrokinetic cell docking, and electrical cell lysis (in 40 ms). A total of fourteen cells were docked, lysed, and analyzed for GSH consecutively. A value of 0.70 ± 0.3 mM of GSH was determined in the human erythrocyte. Detection of GSH by fluorescence was achieved after labeling. This was achieved by incubating the cells with the hydrophobic NDA dye before cell loading in the chip.[853] In another report, transiently EGFP-transfected Jurkat cells in microchannels were treated with TNF-a for a signaling protein study.[138]

To pattern adherent cells on surfaces, a novel method, microcontact printing, was developed by means of PDMS stamping. A glass slide was first patterned with a contoured surface using a Au-coated polyurethane film with ridges and grooves. Then two kinds of self-assembled monolayers (SAMs) were created on the contoured surface. For instance, the ridges or plateaus were patterned with a methyl-terminated SAM (hexadecanethiol), and the grooves were patterned with triethyleneglycol-terminated SAM. Therefore, the hydrophilic ridges allowed fibronectin to adsorb, whereas the hydrophilic grooves resisted the protein. Thereafter, bovine capillary endothelial (BCE) cells, which attached only to the fibronectin surfaces, were patterned along the ridges for cell apoptosis studies.[372,854] A similar approach was used to pattern NIH-3T3 murine embryonic fibroblasts for live/dead cell assays.[855]

In one report, cell growth of chicken embryo spinal cord neurons was studied. These neuron cells have been deposited in a micromachined Si chip coated with a synthetic additive protein (polylysine). It was found that groups of neurites grew toward the channel (50 μm wide) connecting between pits where the neurons were deposited.[856]

In another report, adherent cells were patterned within PDMS channels that were coated with proteins, such as collagen (0.2 mg/ml) or fibronectin (10 mg/ml). The channels were sealed against various substrates, such as Cr, Au, PDMS, polystyrene (PS), PMMA, or poly(methylpentene). It was found that fibroblasts (3T3-J2) and rat hepatocytes adhere to all these surfaces. However, PDMS did not adhere to other surfaces, such as UV-cured polyurethane (PU) film (unless hydroxylated by O_2 plasma) agarose.[908]

Adherent cells have also been patterned using cell-adhesive and non-cell-adhesive microdomains created on a glass coverslip. This coverslip was then bonded with a PDMS channel layer for cell patterning.[857] For instance, endothelia or astrocyte-neuron co-cultures were plated only on the cell-adhesive poly-l-lysine microdomains, but not on the nonadhesive agarose (1%) microdomains. Subsequently, calcium wave measurements were made to study calcium signaling of cells in two modes: (1) within confluent cell microdomains and (2) across neighboring, but spatially disconnected, microdomains.[857]

On the other hand, multiple laminar flow was used to pattern cells and their environments in PDMS chips.[858] For instance, two cell types (chicken erythrocyte and *Escherichia coli*) have been shown to deposit next to each other on fibronectin-treated surfaces (see Figure 8.17). Moreover, cell detachment occurred only to the cells (BCE) that were passed with a patterned stream containing trypsin/EDTA.[858]

By streaming different solutions inside a PDMS chip, selective labeling of subpopulations of mitochondria in the left and right regions of the same live BCE cell was achieved. This was conducted by flowing a solution of Mitotracker Green FM over the right pole of the BCE cell, and another solution of Mitotracker Red CM-H_2XRos over the cell left pole. After 2.5 h, the two labeled subpopulations could be seen moving and intermixing inside the cell.[859] Similarly, selective disruption of actin filaments in selected cytoskeletal regions of the BCE cell was achieved by treatment of latrunculin A.[859]

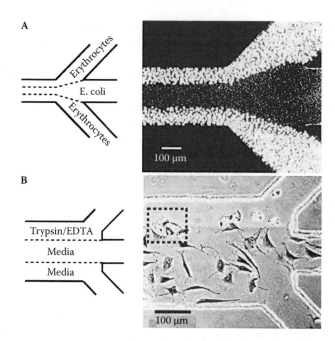

FIGURE 8.17 Examples of types of patterns that can be formed by laminar flow patterning in PDMS channels sealed to a polystyrene Petri dish. (a) Fluorescent images of two different cell types patterned next to each other. A suspension of chicken erythrocytes was placed at the top and bottom inlets, and PBS at the middle inlet, and was allowed to flow by gravitational force for 5 min followed by a 3 min PBS wash; this flow formed the pattern of bigger chicken cells (outer lanes). Next, a suspension of *E. coli* (RB 128) was placed in the middle inlet and PBS in inlets 1 and 3 and was allowed to flow by gravitational forces for 10 min followed by a 3 min PBS wash; this flow created the pattern of smaller cells (middle lane). Both cell types adhered to the Petri dish by nonspecific adsorption. Cells were visualized with Syto 9 (15 µM in PBS). (b) Phase-contrast image of patterned detachment of BCE cells by treatment with trypsin/EDTA. Cells were allowed to adhere and spread in a fibronectin-treated capillary network for 6 h and any nonadherent cells were removed by washing. Trypsin/EDTA and media were allowed to flow from the designated inlet for 12 min.[858] Reprinted with permission from the National Academy of Sciences.

Chick embryo heart muscle cells were patterned and grown on a fibronectin (FN) surface patterned by PDMS stamping. The phosphate buffered saline (PBS) solution (containing Ca^{2+} and K^+) was used to stimulate spontaneous muscle contraction.[198] Laminar flows provide a reaction path (buffer plus 1-octanol) and a control patch (buffer only) for study of communication between excitable cells (cardiomyocytes) through gap junctions (see Figure 8.18).[198]

Further, drug transport through epithelial cells has been studied. For instance, Madin-Darby canine kidney (MDCK) cells have been grown into tight monolayers on a permeable polycarbonate membrane. This membrane (0.4 µm pores, 1 to 4 mm² area) was glued in between two micromachined silicon or glass wafers.[860] The glue (about 1 µm thick) was applied using a paraffin foil and a rolling procedure to ensure that there was no blocking of the membrane nanopores by excess glue, but there was sufficient glue for bonding.[860]

8.1.4 POLYMER ENTRAPMENT

Another method to retain cells is to trap the cells inside a polymer matrix. For instance, mammalian cells were entrapped in PEG-based hydrogel fabricated in glass-Si chips (see Figure 8.19). The cells included murine 3T3 fibroblasts and SV-40-transformed murine cells (i.e., hepatocytes,

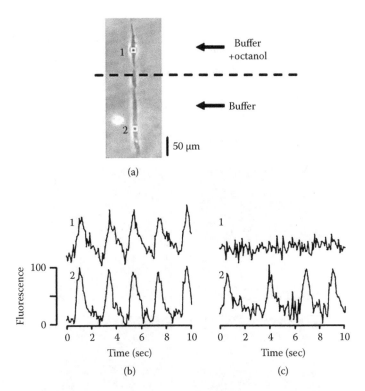

FIGURE 8.18 (a) Phase-contrast micrograph of cardiac myocytes patterned in the acrylic channel. Time courses of fluorescence of cytosolic Ca^{2+} measured at points 1 and 2 denoted in the micrograph: (b) when all myocytes were placed in laminar flows of the buffer, (c) 2 min after localized delivery of 100 μM octanol only to the upper part of the image (point 1).[198] Reprinted with permission from the Royal Society of Chemistry.

macrophages). A cell adhesion peptide (Gly-Arg-Gly-Asp-Ser) was grafted into the hydrogel matrix in order to promote spreading and growth of fibroblasts.[861]

Escherichia coli cells (BL21) were entrapped in hydrogel micropatches. The cells were found to remain viable in the patches. Furthermore, diffusion of small molecules through the polymer (with 1 to 10 nm pores) to the cells allows cell reactions to be studied. For example, the fluorescent dye BCECF-AM diffused to the cells and the dye was converted to BCECF by intracellular enzymes present only in live cells.[1053]

A layer-by-layer microfluidics technology was used to construct a 3D microscale hierarchical tissue-like structure. For instance, three layers of tissues, fibroblasts (human lung), myocytes (smooth muscle cells), and endothelial cells (human umbilical vein) were cultured on top of each other using consecutive microchannel cell matrix delivery. Cell viability was confirmed by fluorescent staining.[862]

Microwells defined in agarose were used to culture nerve cells. The dendrite growth of the nerve cells was studied in narrow tunnel-shaped channels in agarose. The tunnels were fabricated by photothermal etching the agarose.[863]

8.1.5 THREE-DIMENSIONAL FLOW CONTROL

A 3D liquid flow has been exploited to retain single biological cells for long-term real-time study in a glass chip. With the use of a special microstructure, a zero-speed point (ZSP) can be established as shown in Figure 8.20. The liquid flow in the 2D plane of the channel dimension is shown

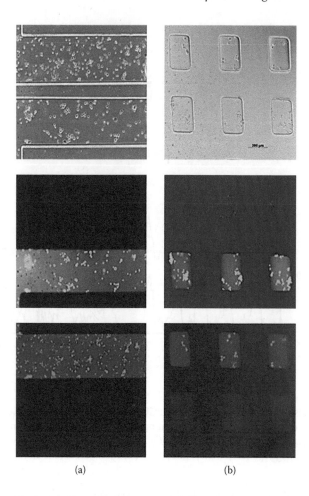

(a)   (b)

FIGURE 8.19   Array of hydrogel microstructures encapsulating two kinds of cells: (a) microchannels filled with fibroblasts and hepatocytes; (b) hydrogel microstructures containing both cells. Top: Bright-field image. Middle: Green fluorescence due to fibroblasts. Bottom: Blue fluorescence due to hepatocytes.[861] Reprinted with permission from the American Chemical Society.

in Figure 8.20A–F. The liquid flow along the depth dimension is shown in Figure 8.20h–k. The force balance that is critical for cell retention is shown in Figure 8.20g. Manipulation of the ZSP by the 3D liquid flow control has been employed for scanning the cell around a detection window. By scanning the cell (yeast cell) around a detection window, both the cellular fluorescent signal and background can be detected as a series of peaks, as shown in Figure 8.21a. In this way, a weak cellular signal can be easily discerned (see Figure 8.21b). Moreover, different regions of a cell (mother and daughter) can be distinguished when the detection window is smaller than the cell size (see Figure 8.21c). The detection of background is particularly important in multiple-stimuli studies, as shown in Figure 8.22. For instance, it can be confirmed that fluorescent change is due to cellular change, but not fluorescent background change. The different background signals from different solutions also indicated the success in reagent changes, and changes in excitation light conditions (see Figure 8.22b). After background subtraction, data interpretations become easier; see Figure 8.22c for the corrected peaks and Figure 8.22d for the peak envelopes.[680]

With cell retention accomplished by 3D flow control, calcium mobilization in a single yeast cell has been studied over a long period of time, and in the manner of multiple stimuli and multiple responses. Loading of the Ca^{2+}-sensitive dye Fluo-4 in the cell was accomplished either (1) after cell

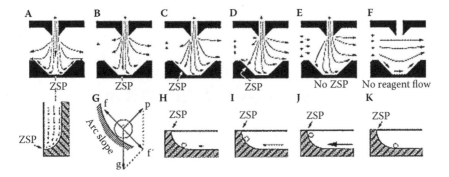

FIGURE 8.20 Three-dimensional flow control for cell retention. (a) The two-dimensional channel flow field was created by the flow from the middle channel. There was a zero-speed point (ZSP) where the flow speed decreased to near zero. When there was no flow between left and right, the ZSP was in the middle. The third-dimensional flow field was along the arc slope as shown in the cross section diagram in the inset. (b–e) As the fluid potential from the left increased, the shape of the flow field changed and the ZSP moved to the left. (f) The flow field as driven by the fluid potential from the left was shown when there was no reagent flow from the middle channel. (g) The third-dimensional flow field caused the cell to be balanced on the arc slope. The balance was between the upward force exerted by the liquid flow (f), downward gravitational force (g), and reaction force from the sloping wall (P). (h–j) The position of the cell changed as the reagent flow from the middle channel became increasingly strong. (k) The position of the cell was shown when there was no flow.[680] Reprinted with permission from the American Chemical Society.

wall removal or (2) without cell wall removal, but with a high dye concentration. It was discovered that great Ca changes in response to glucose or pH change only occurred in deprived cells that were either dormant cells directly obtained from solid agar culture plates or cultured cells after a no-glucose treatment. Metabolism of a model substrate, fluorescein diacetate (FDA), which produces fluorescein, has also been studied on a single yeast cell. Again, a high metabolic rate was associated with deprived cells. In addition, efflux of the fluorescent metabolite was observed because the presence of the liquid flow helped remove the excreted fluorescein. Using a refined kinetic model, the whole process was modeled (see Figure 8.23). The detailed study of these biochemical processes (metabolism and excretion) was rendered possible, thanks to the success in the retention of the single cell in a continuous-flow system.[864]

Bidirectional liquid flow in a Pyrex chip was created by the opposing electroosmotic flow (EOF) and pressure-driven flow (PF).[479,865] This can lead to the generation of a controlled recirculating flow or vortex, as shown in Figure 8.24. Small polymer beads can be trapped to give the controlled rotating flow patterns, depending on whether the channel is a diverging channel or converging channel. The section between the first and second diverging channel is called the trapping channel. This channel was employed for the study of the binding of the trapped streptavidin-coated beads (0.4 to 4 μm) with a flowing stream of fluorescent-labeled biotin.[479]

In another report, microparticles and biological cells were trapped in microvortices produced in the microchambers constructed in a PDMS chip. Figure 8.25 shows different designs of the microchambers, in which only Figure 8.25c and d demonstrates stable microvortices. Rotational motion of either particles (10 μm) or cells (B lymphocytes) was studied in these microvortices.[1173]

8.1.6 OPTICAL TRAPPING OF CELLS

Optical traps have also been used to retain cells.[176,866–868,1174] For instance, manipulation of polystyrene beads by optical gradient force (attractive) and scattering force (repulsive) was achieved in a PDMS chip (see Figure 8.26). A polystyrene bead was first retained by an optical trap. Then the bead was moved and released so that it flowed to a desired channel downstream.[1174]

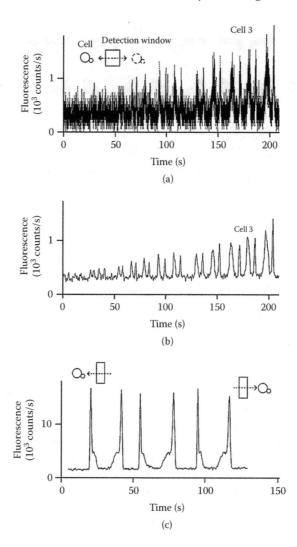

FIGURE 8.21 Cell scanning and noise filtering. (a) A fluorescent yeast cell going through the detection window back and forth generated peaks over the background. (b) The peak signal became clearer after filtering the noise (>2.5 Hz). (c) The use of a narrower detection window on another yeast cell could distinguish the mother yeast cell from its daughter yeast cell by showing the peak and the shoulder, respectively.[680] Reprinted with permission from the American Chemical Society.

In one report, the optical gradient field ($\lambda = 808$ nm) was used to interact with cells and beads and to measure the resultant time of flight. This method (known as optophoresis) has successfully differentiated between cancerous and noncancerous cells (breast cancerous melanoma).[867]

Single *E. coli* cells were isolated and retained using an optical trap (by the use of the Nd:YAG laser). Four daughter cells (divided from a single *E. coli* cell) were isolated and transported into each of their microchambers in a glass chip (see Figure 8.27). Cell growth for twelve generations was conducted in these microchambers. The cell length, interdivision time (cell cycle), and growth speed were measured with this method, called single-cell-based differential cell assay. It was found in one case that the cell length of a daughter cell became extraordinarily long (14 μm), suggesting a possible change in phenotype. However, the cell recovered subsequently to provide a daughter cell of the usual cell length (6 μm). This alteration in phenotype during cell division would have been

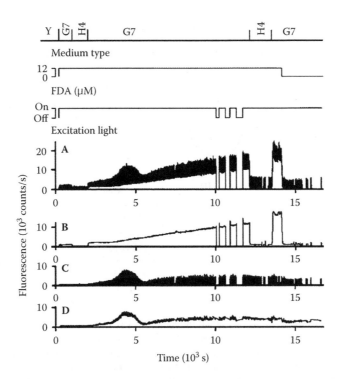

FIGURE 8.22 Background correction applied to an experiment with a yeast cell. (a) Raw data showing peaks due to cell fluorescence plus background. (b) Background baseline extracted from (a). (c) Cell fluorescence peaks after background subtraction. (d) Peak envelope of all fluorescence peaks. Two reagent scales showed the medium types and the fluorescein diacetate (FDA) concentrations. In addition, one excitation light scale showed when the excitation light was shut off or turned on.[680] Reprinted with permission from the American Chemical Society.

missed in batch cell experiments, but in the single-cell experiment, such an observation may reveal new information.[176]

Optical trap or tweezers were used to select single suspension cells (acute myeloid leukemia [AML]) in a PET film-sealed PMMA chip. The use of the thin PET film (35 μm) was necessary in order to allow the use of a high-magnification microscopic objective needed for optical trapping. This cell selection usually took about 5 min. Subsequent cell lysis and CE separation of the cell contents were carried out.[197]

In one report, the optical tweezers were used to hold a bend in order to stretch a DNA molecule. First, a DNA molecule (λDNA) was biotin congugated so that it could be attached to a streptavidin-coated polystyrene bead. The bead was held by optical tweezers inside a PDMS chip. The other end of the DNA was labeled with digoxigenin to be immobilized on the channel coated with anti-digoxigenin antibody. The DNA was stretched by moving the chip, with the bead being held by the optical tweezers.[868]

8.1.7 DIELECTROPHORESIS (DEP)

Dielectrophoresis refers to the force acting on induced polarizations or charge dipoles (e.g., in a particle or cell) in a nonuniform electric field of low voltage (1 to 3 V). The nonuniform electric field or alternating current (AC; 1 to 20 MHz) is produced by embedded electrodes. Usually, in highly conductive aqueous media, the cells will be less polarizable than the media at all frequencies. In this case, negative DEP force was developed in which the particles moved toward the field minima. This

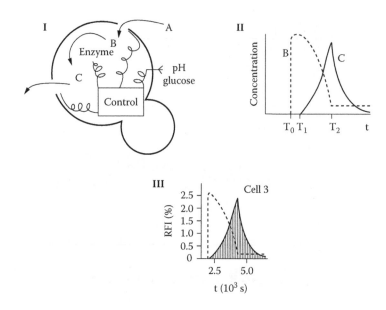

FIGURE 8.23 Experiments and mathematical modeling for FDA metabolism in a single cell. (I) There are three cellular processes: influx, hydrolysis, and efflux. The yeast cell exerts control over the influx of FDA (S), hydrolysis of FDA (N), to form fluorescein (C) and efflux of fluorescein in response to the stimuli of pH and glucose. (II) Intracellular concentrations of FDA (B) and fluorescein (C). Only C was experimentally measured. T_0, T_1, and T_2 represent the time(s) when buffer change, peak increase, and peak decrease occurred, respectively. (III) Curve fitting: In the graph, stripes represent the measured amount of fluorescein; dotted and solid lines represent the modeled amount of intracellular FDA and fluorescein, respectively, which were calibrated with a fluorescent bead of known intensity. (RFI: relative fluorescent intensity in which 1% represents the fluorescence resulting from full hydrolysis products from 6×10^{-19} mol of FDA.)[864] Reprinted with permission from the American Chemical Society.

force has been employed to trap and sort latex beads (of diameter 3.4 to 15 μm) and live mammalian cells (mouse L929) (see Figure 8.28).[869,870] Due to the high permittivity of water, polymer particles and living cells also show negative DEP at high field (0.5 MV/m) and high frequencies (1 MHz). The DEP force, which is about 0.2 to 15 pN for latex spheres (diameter 3.4 to 15 μm), is sufficient to hold particles at a flow rate up to 10 mm/s.[869]

Some heat can be generated in this operation, but it can be reduced in the microscale. In addition, the use of a high-frequency (>100 kHz) AC field (1) eliminates any electrophoretic movement of the cell due to its charged membrane, (2) eliminates the electrochemical reactions (bubble formation or electrode corrosion), and (3) minimizes the alternating voltage imposed upon the resting (static) transmembrane voltage (at 3 V and 20 MHz, the alternating voltage was 12 mV, which was twenty times less than the value at direct current [DC]).[871]

DEP has been employed to trap or manipulate cells such as HL-60,[871] Jurkat cells,[872] mouse fibroblast (3T3 cells),[873] and rabbit heart cells.[234]

In one report, DEP was used for the separation of biological warfare bacterial simulants in a laminated device (five layers including polyimide). For instance, *Bacillus cereus*, *E. coli*, and *Listeria monocytogenes* were separated from human blood cells. All separations were run with an AC voltage (10 Vpp at 10 kHz). Bacteria are collected at the electric field maxima by positive DEP forces, and blood cells (RBCs and WBCs) are collected at field minima by negative DEP forces. By applying a suitable fluid flow, the blood cells are swept away, but the bacterial cells are retained.[874] In another report, spores of *Bacillus globigii* were trapped on a DEP chip.[486]

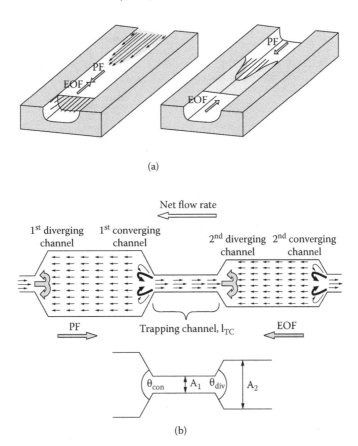

(a)

(b)

FIGURE 8.24 (a) Left: Schematic diagram of the velocity profiles of EOF and PF in a microchannel of a Pyrex chip. Right: These two flows can be combined to give rise to a bidirectional flow. (b) Top schematic view of the microfluidic vortex device with streamlines shown. In the large channels, the flow streamlines follow the EOF. In the small channels, the flow streamlines follow the PF. EOF and PF conditions are such that net flow rate is close to zero. A pair of symmetrical vortices is generated at each end of the trapping channel, and flow recirculates between the diverging and converging channel elements through this channel. θ_{con}, opening angle of the converging channel element; θ_{div}, opening angle of the diverging channel element; A_1, cross-sectional area of the narrow trapping channel; A_2, cross-sectional area of the wide channel segment after the diverging channel element.[479] Reprinted with permission from the Royal Society of Chemistry.

To achieve DEP over a large region, a large-area traveling wave DEP was developed. This was conducted on a glass chip consisting of one thousand electrodes. In the DEP separation of rabbit blood cells, it was observed that rabbit erythrocytes traveled faster than leukocytes.[266]

So far, DEP work has been conducted using planar quadrupoles. However, they are inadequate to trap the cells against a greater fluid flow velocity (e.g., 12 µl/min), which results in a greater drag force (~50 pN). So a series of four 50 µm high cylindrical Au electrodes were fabricated to trap the cells (see Figure 8.29). These electrodes could confine particles over one hundred times more strongly than planar quadrupole electrodes. Luminescent dynamics of a surrogate assay was conducted in the trapped HL-60 cells.[871] A similar DEP method was also used to study the uptake and cleavage of calcein-AM in a trapped Jurkat cell.[872]

The nonuniform field needed for DEP has so far been provided directly by numerous embedded electrodes. In one report, a single DC electric field was provided across a channel, but it was provided to the channel consisting of an array of insulating posts. Such an arrangement resulted in a spatially nonuniform electric field around the posts needed for DEP. Figure 8.30 shows how the fluorescent beads undergo streaming DEP. This mode occurred when the DEP force could not

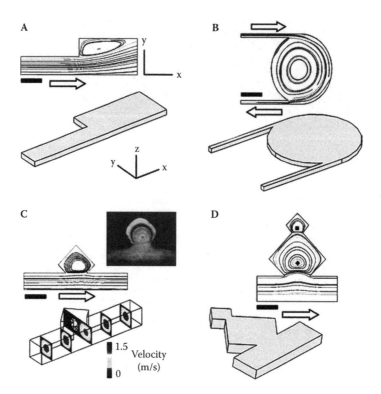

FIGURE 8.25 Generation of microvortices in microfluidic systems to study rotation of particles. Recirculation flows occur under a wide range of conditions and channel geometries, including (a) sharp corners, (b) circular cavities that we call "rotors," (c) side chambers, and (d) interconnect chambers. For rotation of microparticles, vortices with stable centers are required. This can only be achieved in the geometries shown in (c) and (d), but not in (a) and (b). In (c), the upper inset is a fluorescence micrograph that shows the flow profile inside the diamond-shaped cavity as traced by 1 μm in diameter fluorescent beads. Although the straight channel flow velocity is high, the recirculation flow velocity around the bead is low. (d) By placing two microchambers adjacent to each other, counterrotating flows that we call fluidic gears can be created (counterclockwise in lower diamond and clockwise in the upper diamond). (→) Center of rotation of lower diamond. (n) Center of rotation of upper diamond. The *Re* at the opening of each microchamber was (a) 75, (b) 285, (c) 30, and (d) 40. Scale bars: (a, c) 30 μm; (b, d) 60 μm.[1173] Reprinted with permission from the American Chemical Society.

overcome the force caused by the EK flow, but it could overcome diffusion. For comparison, there is no DEP trapping when EK flow >> DEP or diffusion, and there is DEP trapping when DEP >> EK flow and diffusion.[321]

DEP trapping of nanoparticles (100 nm) is more difficult than that of larger microparticles. The greater electric field gradient needed for successful trapping of smaller-sized nanoparticles can only be achieved by smaller-sized electrodes. Unfortunately, Joule heating and hydrodynamic effect become more of concerns when a higher field is used. To increase the DEP force, a higher proportion of voltage drop should be achieved in the bulk solution, rather than in the electric double layer. This situation can be achieved at a higher frequency.[875]

Other than cells and particles, DEP has also been used to trap DNA molecules. In one report, DEP trapping of *Escherichia coli* chromosomal DNA (chromatin) that followed cell lysis was carried out on a PDMS chip.[687] In another report, DEP (1 MHz) was used to stretch λDNA molecules (48 kbp).[323] There are other reports on DEP work.[866]

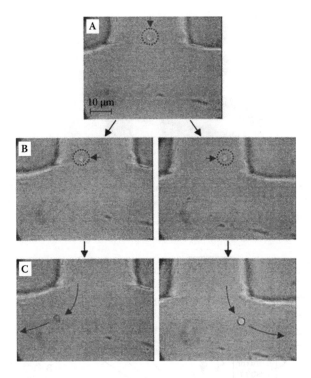

FIGURE 8.26 Images demonstrating the optical gradient force used for flow switching. (a) Microspheres flow into the downstream channel from a reservoir at the top. The microspheres are first captured by the optical trap. (The dotted circle indicates the position of the optical trap.) (b) The microsphere is directed manually to either left or right. (c) After releasing the optical trap, the microsphere is allowed to follow the fluid stream toward either left or right. The laser source was an 850 nm vertical cavity surface-emitting laser (VCSEL). To create sufficient output power and trapping strength, the laser spatial mode should be changed to the Laguerre mode (or donut-shaped mode).[1174] Reprinted with permission from Springer Science and Business Media.

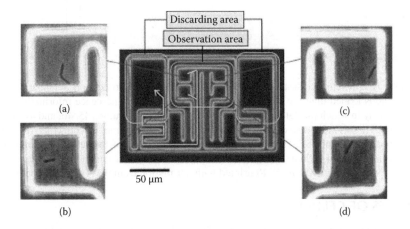

FIGURE 8.27 Observation of growth of *E. coli* cells in a microchamber fabricated on a glass chip consisting of SU-8 structures (5 μm high). Center: Micrograph of the microchamber. Insets a–d: Micrographs of four *E. coli* cells growing in individual chambers, as taken at 347 min after cultivation started.[176] Reprinted with permission from Elsevier Science.

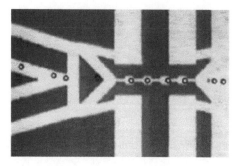

FIGURE 8.28   Latex particles (15 μm) have been confined from a particle jet from the left by DEP. The high field was provided by 10 V (1 MHz) across a 20 μm electrode gap, or 0.5 MV/m.[869] Reprinted with permission from the American Chemical Society.

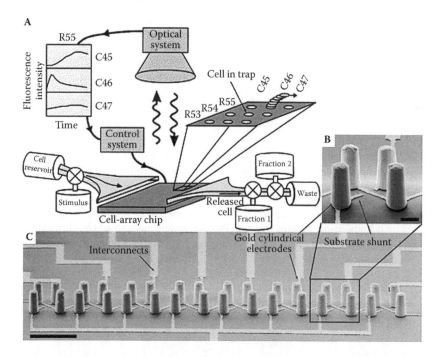

FIGURE 8.29   A DEP chip consisting of nonplanar cylindrical electrodes. (a) Schematic representation showing single cells loaded onto the cell-array chip. The dynamic luminescence information was obtained from the entire array in which the data of three cells in row 55 and in columns 45, 46, and 47 are shown. The information is used to sort individual cells by the control system into the fraction collectors. (b) Pseudocolored scanning electron micrograph (SEM) showing a single trap consisting of four electroplated gold electrodes arranged trapezoidally along with the substrate interconnects. Scale bar: 20 μm. (c) SEM of a completed 1 × 8 array of DEP traps. Scale bar: 100 μm.[871] Reprinted with permission from the American Chemical Society.

8.2   STUDIES OF CELLS IN A FLOW

Sometimes, cellular analysis was performed with cells in a flow stream, which is also termed flow cytometry, or FACS in the field of cell biology. For instance, human blood cell (WBC, RBC) rheology was studied in channels fabricated on the Si-Pyrex substrates. The channels were either uncoated or coated with albumin.[825]

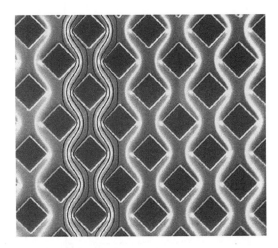

FIGURE 8.30 Fluorescence image of DEP achieved on a microchip. Streaming dielectrophoresis of 200 nm fluorescent latex spheres (white) is shown. The glass microchannel is 7 μm deep and consists of an array of 36 μm square posts (the black diamonds) with 63 μm spacing between centers. The DC electric field (80 V/mm) was applied over the two ends of the microchannel. The resulting electroosmotic flows are from top to bottom. The black lines indicate the streamlines predicted using simulations. The bright spots at the left and right tips of the posts are regions of trapping dielectrophoresis.[321] Reprinted with permission from the American Chemical Society.

Mouse sperm cells were allowed to swim through a meander channel in a glass chip (Pyrex or soda lime glass) to fertilize an egg for *in vitro* fertilization (IVF).[876] Later, sperm studies were carried out in PMMA chips.[877]

In another report, a PDMS chip was constructed with a main flow channel for sorting of motile human sperm (~60 μm in length). A side channel allowed motile sperm (moving at 20 μm/s at 25°C) to divert from the main liquid flow and to be collected at the side channel. In contrast, nonmotile sperm diffused slowly ($D = 1.5 \times 10^{-13}$ m^2/s); i.e., they diffused for 10 μm in 690 s.[878]

Yeast cells (*Saccharomyces cerevisiae*) were mobilized in a microfluidic glass chip using electrokinetic (EK) flow (see Figure 8.31).[879] In addition, erythrocytes were lysed using a detergent (e.g., 3 mM SDS), then light scattering and video imaging were used to monitor the lysing reaction.[879,880] However, the use of EK flow to manipulate lymphocytes in microchannels may inactivate the cells. This problem, which may be caused by the pH change occurring at the cell reservoir, can be alleviated by introducing a salt bridge between the cell reservoir and the electrode.[350]

The flow of leukocytes was studied in square capillaries fabricated on a Si chip, and sealed with a PDMS or Pyrex cover plate. This capillary size (cross section of 4 μm^2) is similar to the diameter of a human blood capillary, but is less than both the average diameter of a leukocyte cell (10 μm) and its nucleus (6 μm). Figure 8.32 shows the difference in the flow behavior of two leukocytes (possibly neutrophils).[1175] Deformation-induced release of ATP from erythrocytes in PDMS channels was studied. The released ATP was detected by chemiluminescence using the luciferin/luciferase system.[169]

Rheology of human erythrocytes has been studied in a Si-Pyrex chip (3.2 μm deep, 3.5 μm wide, and 10–μm long). The chip was fabricated with a series of eight microchannels using a pressure of 3 mm of water (see Figure 8.33). The cells were revealed by optical absorption (at 400 nm) due to intracellular hemoglobulin. The mean erthyrocyte flow velocities of normal cells at 37°C were measured and plotted against their cell length (see Figure 8.34). The artifact-impaired transits were caused by partial obstruction or platelet adhesion. Therefore, the impaired transits showed lower velocity than, and were clearly distinguished from, clear transits. The plot of velocities of abnormal erythrocytes, which were obtained from iron deficiency anemia sufferers, is shown in Figure 8.35. Both plots clearly showed the difference in the flow behaviors between the normal and abnormal cells.[304]

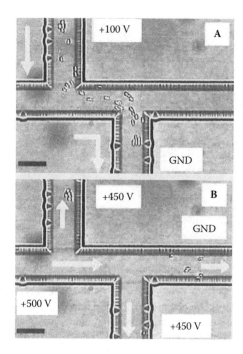

FIGURE 8.31   Photomicrographs showing yeast cell transport in a glass chip. White arrows show the direction of buffer flow. Scale bar: 40 μm. (a) Cell loading vertically downwards. (b) Cell selection to the right.[879] Reprinted with permission from the American Chemical Society.

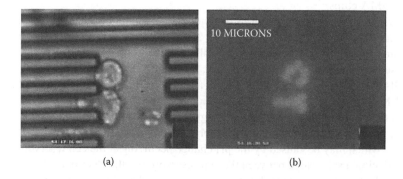

FIGURE 8.32   (a) A bright field epi-fluorescence image of two polymorphonuclear white cells (probably neutrophils). (b) The same cells shown in (a) imaged via the Hoechst stain. Note the differences in conformation between the two nuclei. The cell on the top is stuck at the entrance to a channel, and the cell at the bottom is being deformed into a channel. The channels are coated with polyurethane to reduce cell adhesion and to enhance white cell penetration.[1175] Reprinted with permission from Springer Science and Business Media.

Rolling of cells (neutrophils) initiated by cell adhesion was studied in a Y-channel. Inhibition by various anti-cell-adhesion molecules (anti-E-selectin, anti-P-selectin, sialyl Lewis X, Furoidan) was studied. Comparison can easily be made for cells with rolling and without rolling (control) in the two arms of the Y-channel.[881]

Jurkat cells have been lysed in a flow stream in a glass microchip for cell content analysis. After cell lysis, the two preloaded fluorescent dyes and their metabolites were released from the cells and separated by CE (see Figure 8.36). To prevent cell adhesion, the glass channel surface was modified by adsorbing Pluronic F-127 to the channels. In addition, to avoid blockage of adhered cell debris

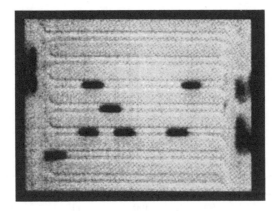

FIGURE 8.33 Erythrocytes flowing through microchannels with background subtraction applied to the images. Single, double, and triple transits are observed in some of the microchannels.[304] Reprinted with permission from IEEE.

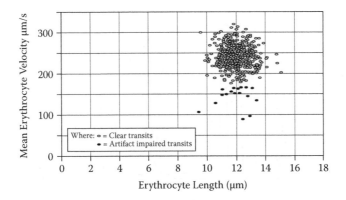

FIGURE 8.34 Scatterplot of mean velocity vs. cell length of normal erythrocytes.[304] Reprinted with permission from IEEE.

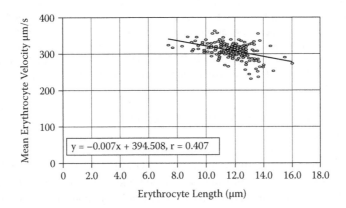

FIGURE 8.35 Scatterplot of mean velocity vs. cell length of abnormal erthrocytes.[304] Reprinted with permission from IEEE.

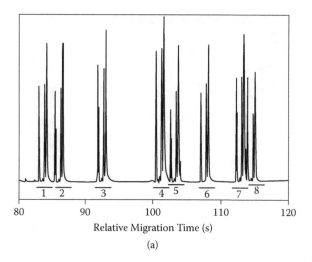

(a)

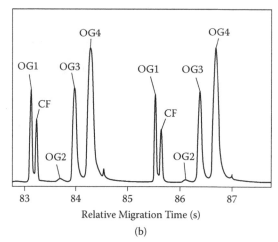

(b)

FIGURE 8.36 (a) CE separation of fluorescent dyes released from individual cells. The separation is a 40 s segment of a 2 min run. The peak envelopes resulting from eight cells are shown and are denoted by the thick black horizontal lines and labeled 1 to 8. (b) Expanded view showing the peak envelopes from two cells (cells 1 and 2). The dyes to which the peaks are attributed are indicated by the peak tags (OG, Oregon green; CF, carboxyfluorescein). Cell lysis and subsequent dye separation were achieved by applying an AC field (Vpp of 450 V/cm and 50% duty cycle) with a DC offset (675 V/cm). The DC offset was used to maintain separation, and the AC field was used to avoid undesirable gas bubble formation.[1176] Reprinted with permission from the American Chemical Society.

and to improve migration time stability, an emulsification agent, such as Pluronic P84, was added to the separation buffer.[1176]

Flow cytometry of *E. coli* has been demonstrated on a glass chip. Electrokinetic focusing was used to confine the cells. Cell counting rates of 30–85 Hz were achieved. The membrane-permeable nucleic acid stain (Syto 15, λ_{em} = 546 nm) and membrane-impermeable nucleic acid stain (propidium iodide [PI], λ_{em} = 617 nm) were used to detect viable and nonviable cells, respectively.[882,883] The fluorescein-labeled (λ_{em} = 520 nm) polyclonal antibody to *E. coli* was also used. Light scattering (forward) was used to provide information about the cell size (of length 0.7 to 1.5 µm) of *E. coli*. Besides using a PDMA surface coating, BSA (1 mg/ml) was used to coat the channel to prevent cell adhesion and cell clumping.[882]

Another flow cytometry study involved the measurement of capacitance changes due to varied amounts of polarizable cellular DNA.[884]

On the other hand, electrical impedance measurement of bacterial suspensions (*Listeria innocua*) was carried out on a Si-glass chip. This method was used to perform a cell viability test. This is because the metabolic products produced from viable cells modify the ionic strength of a low-conductivity medium, significantly altering its electrical characteristics. Later work also involved the detection of the presence of small numbers of bacterial cells in a Si chip: 100 *L. innocua* cells, 200 *L. monocytogenes* cells, and 40 *E. coli* cells.[93,885] In another report, impedance spectroscopy has been used for analysis of erythrocytes in a glass-polyimide chip.[886]

A flow cytometer with integrated waveguides and lens was fabricated on a PDMS chip. Forward scattering (0.5 to 5°), large-angle scattering (15 to 150°), and extinction were simultaneously measured. These measurements provided the information about the size and surface roughness of the particles to be studied.[887] Another flow cytometer was made on an SU-8-glass chip, and it has been incorporated with the light-emitting diode (LED) source (InGaN), microprism, holographic diffraction grating, avalanche photodiode detector, and fiber optics (transmission and collection) for the study of beads.[888]

Flow cytometry of fluorescent (0.972 µm diameter) and nonfluorescent (1.94 µm diameter) latex beads was achieved on a glass chip using dual-channel laser light scattering and fluorescent measurements.[889] A maximum sample throughput of thirty-four particles was obtained using pinched injection, which was akin to the use of electrokinetic focusing (cf. hydrodynamic focusing used in conventional flow cytometry).[889]

An integrated microfabricated cell sorter has been constructed using a control layer and a fluidic layer fabricated on PDMS.[890] The control layer consists of valves that will be pneumatically controlled by pressurized N_2. This device is superior to an electrokinetic sorter, which suffers from buffer incompatibilities and frequent voltage adjustments because of ion depletion or pressure imbalance due to evaporation. *E. coli* cells expressing EGFP were sorted and selected out from a cell mixture that also contained *E. coli* cells expressing the *p*-nitrobenzyl (PNB) esterase. About 480,000 cells have been sorted at a rate of 44 cells/s in 3 h. The recovery yield was 40% and the enrichment ratio was about eighty-three-fold.[890]

A PDMS-glass chip was used to hydrodynamically focus fluorescent beads to flow as a narrow stream (see Figure 8.37). The beads can also be electrokinetically sorted by applying a deflecting voltage (20 to 30 V/cm) via the two side channels. The narrowly focusing stream matched the size of the confocal volume located at a downstream position. The chip was employed to show the binding reaction between R-phycoerythrin (a fluorescent protein extracted from red algae) and a short peptide sequence (expressed on *E. coli* cells, BMH 71-18).[161]

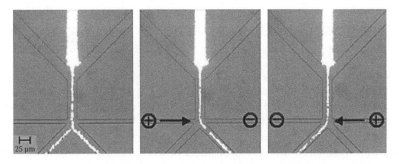

FIGURE 8.37 Hydrodynamic focusing and sorting of a highly concentrated suspension of fluorescent beads. After hydrodynamic focusing, the beads entered both output channels with equal probability (left image). By activation of a perpendicular electroosmotic flow, the suspension could be deflected into one desired output channel, depending on the polarity of the electrodes (middle and right images).[161] Reprinted with permission from the American Chemical Society.

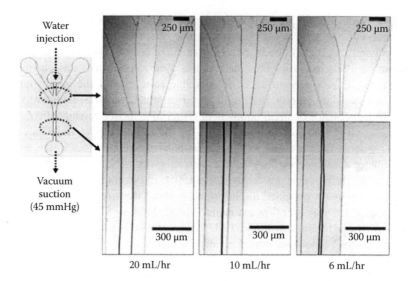

FIGURE 8.38 Top: Stable aerodynamic focusing in the sample focusing chamber. Bottom: Stably maintained air–water two-phase stratified flow in the observation channel. Left inset shows the location of the sample focusing chamber and observation channel on the microchip. The left, middle, and right images depict the flow behaviors at the flow rates of 20, 10, and 6 ml/h, respectively. It is observed that as the flow injection rate of the syringe pump is lowered, the width of the water column decreases.[382] Reprinted with permission from Springer Science and Business Media.

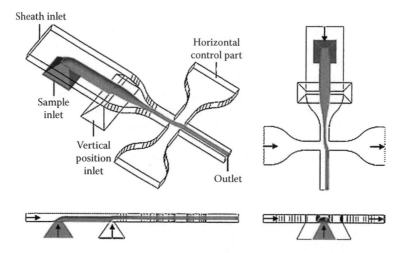

FIGURE 8.39 A flow cell constructed for noncoaxial sheath flow. The additional vertical position inlet of this flow cell allows positioning of the sample flow in the center of the channel. The horizontal control ports are used to position the sample flow with a horizontal shift. The perspective view, top view, and two cross section views are shown.[891] Reprinted with permission from the Royal Society of Chemistry.

A two-phase air–liquid flow was developed on a PDMS chip (see also Chapter 3, Section 3.1.2). It was found that the focused two-phase flow was stable in hydrophobic channels (down to 6 ml/h); see Figure 8.38. Below this flow rate, the sample column no longer maintained its integrity and broke up. This method was used to provide aerodynamic focusing of myoblast cells (C_2C_{12}) for the flow cytometry study. The cells, which were labeled with Syto 9, were focused and counted in the chip at a rate of 100 cells/s.[382]

Another configuration for hydrodynamic focusing is to create a nonaxial sheath flow, as shown in Figure 8.39. The sheath inlet resembles the one used in usual microchip focusing experiments. The nonaxial sheath flows are supplied from the sheath inlet and the vertical positioning inlet. Additional controls of the position and width of the sample flow are created from the two horizontal control ports.[891] A similar approach, termed the smoke chimney principle, was reported. This is a pressure-driven microcell sorter that was constructed on a Si-glass chip. It has been shown that fluorescent latex beads were enriched one-hundred-fold from a mixture containing chicken RBCs. The sample throughout was 12,000 cells/s, which was comparable to the conventional FACS sorter. The high throughput resulted from hydrodynamic focusing on four sides (i.e., both top/bottom and left/right). This top/bottom focusing allowed the use of a channel depth of 50 to 200 µm and a flow rate up to 0.11 µL/s.[838]

Other methods have been developed for fractionation of cells or particles. For instance, in an electrical field-flow fractionation chip, particles with higher charge density pack closer to the oppositely charged wall than do particles with lower charge density. Therefore, less tightly packed low-charge particles will have higher average velocities.[892]

Isoelectric focusing (IEF), which is usually employed for protein separation, has been used to separate mitochondria and nuclei from lysate of NR6wt cells (murine fibroblast) on a Pyrex chip.[893]

Flow study has been achieved in a thermal field-flow fractionation chip. In this method, the temperature gradient is at a right angle to the fluid flow. Higher MW particles are compacted more tightly against the cold channel wall, compared to lower MW particles. Therefore, the less compacted low MW particles will have higher average flow velocities. Fractionation of 154 and 394 nm polystyrene particles in toluene was achieved in a 50 µm high microchannel over a temperature difference of 30°C.[894]

A different diffusion rate of particles along a laminar flow boundary in a Si-Pyrex chip has been employed for fractionation. With this method, fluorescein and polystyrene beads (0.5 µm diameter) were separated from each other.[90]

A magnetic cell separator was constructed on a Si wafer to separate cells that were labeled with paramagnetic beads (FeO nanocrystals, 50 nm) from unlabeled ones. The magnetic force was generated from thin magnetized wires (10 µm wide, 0.2 µm thick) formed by depositing a cobalt-chrome-tantalum alloy in pre-etched 0.2 µm deep trenches. These wires were parallel and were oriented at 45° to the hydrodynamic flow direction of cells.[278]

In another report, arrays of nickel posts (~7 µm high, ~15 µm diameter) were fabricated in a PDMS-Si chip. Once magnetized by a magnetic field, these posts can efficiently trap paramagnetic beads.[895]

Magnetic beads (50 nm) were used in a PC chip to fractionate *E. coli* cells (K12) from sheep blood cells. A biotinylated rabbit polyclonal anti-*E. coli* antibody was first attached to the magnetic beads coated with streptavidin. Then *E. coli* became attached to the beads and was retained by the magnetic field. On the other hand, the other nonretained blood cells were washed away. The cell capture efficiency was evaluated after subsequent PCR of *E. coli* DNA.[227]

8.3 OTHER CELL OPERATIONS

8.3.1 CELL CULTURE

Cell cultivation has been conducted within aqueous fluid droplets embedded in a nonimmiscible carrier liquid. This method that was conducted in a Si-glass microchannel was used to generate monoclonal antibodies.[312] In addition, human connective tissue progenitor (CTP) cells and human bone marrow-derived cells were cultured within PDMS channels.[164] Cell culture for fibroblasts has been possible even under an AC field.[896]

In one report, kidney cells of *Xenopus laevis* (A6) were cultured in a ninety-six-well Si microphysiometer sensor array.[897] Mouse fibroblasts (3T3-J2) were cultured in stacked PU microstructures that mimic the bone architecture. PU is used because it is biocompatible.[160]

Polymer scaffolds that have been constructed by replica mold a poly(lactcic/glycolic acid) (PLGA) sheet to culture endothelial cells. The device was first perfused with a solution (poly-l-lysine, collagen, gelatin, or fibronectin) used to aid cell adhesion.[898] In another report, bovine aortic endothelial cells (BAECs) were cultured on a Pyrex-Si chip precoated with collagen (type I). Cell elongation and alignment with or without liquid flow were studied. The significance of this study is that round cells are prone to atherosclerosis, while elongated cells are not. Whole cell patch-clamp measurements were performed on single (subconfluent) cells in an open-channel configuration (with no lid).[308]

Another PDMS chip was constructed with 3D microstructure to mimic *in vivo* conditions for cell culture. The chip was used to culture hepatocarcinoma (liver cancer) cells (HepG2). The inner PDMS surface was precoated with type I collagen (0.03%). The cell activity was monitored by measuring albumin production in the culture medium.[361] The supply of O_2 was calculated to be adequate (3.5×10^{-6} mol/device per day), thanks to the high O_2 permeability of PDMS (4.1×10^{-5} cm^2/s).[163,361]

Human hepatoma (HepG2) and murine fibroblast (L929) cells were cultured on deep UV-irradiated PS surfaces. This surface UV treatment appears to be more convenient than other treatments (silanization on glass or thiolation on gold). Photochemical modification of polymer surfaces yielded active peroxide groups, which allow graft coupling of acrylamide. The peroxide group has a half-life of 2 to 3 days, which is quantified by the iodide method. Although this UV method is also applicable to other polymers, such as PMMA or PC, these substrates exhibit high cell adhesion even in the native polymer. Accordingly, thorough cleaning was required prior to irradiation in order to achieve high contrast of cell patterning between irradiated and nonirradiated regions.[899]

Cultures of primary cardiac myocytes (chick embryo) were formed on fibronectin-patterned acrylic surfaces.[198] A microtextured PDMS chip with 20 μm wide pegs was used to promote cell culture. After coating the PDMS chip with a thin layer of laminin, neonatal rat cardiac myocytes were cultured on it. The cultured cells are typically 50 μm in length and 10 to 15 μm in diameter. The PDMS chip was cast on a mold with parylene structures patterned on a Si wafer. Using the same mold, a PLGA chip could also be made for culture of rat cardiac fibroblasts.[900]

A mouse neuronal cell line (PCC7-Mz1) was cultured on laminin deposited onto PS substrates using PDMS microcontact printing.[901] Growth of rat hippocampal neuron was found to be more favorable to closely spaced PS pillar structures.[85]

A PDMS chip was used as a culture vessel for ovary cells of fall array worm (Sf 9). These cells are typically 20 to 30 μm in diameter and have a doubling time of ~24 h.[163]

Cell culture has also been conducted for bacterial and fungal cells. For instance, growth of *L. innocua* and GFP-recombinant *E. coli* has been carried out in a PDMS-Si hybrid chip.[249]

Filamentous fungus (*Colletotrichum graminicola*) was found to grow on closely spaced PS pillar structures.[85] Culture of yeast cells was carried out in a Si-glass chip after sorting.[838] Single yeast cell culture (for four generations in 10 h) has been rendered possible thanks to 3D liquid flow control for cell retention in a glass chip. In addition, on-chip cell wall removal by an enzyme (zymolase) has been achieved.[680]

There are several advantages of carrying out cell culture in the microscale; e.g., (1) medium stirring is not necessary, (2) aeration by air bubbling is not needed if a gas-permeable polymer is used to construct the chip, and (3) the ratio of cell volume to extracellular fluid volume is close to 1 (i.e., closer to the *in vivo* environment).[902]

The use of a 2.5:1 ratio (instead of 10:1) of PDMS prepolymer to prepare the PDMS chip has reduced liquid loss by absorption into PDMS by 2.5 times. This has delayed the time for complete liquid absorption from 2.7 h to 6.3 h. Presaturating the PDMS with the liquid has also prolonged the liquid loss by absorption up to 22 h. This leads to less liquid loss problems in performing cell culture in PDMS chips.[249]

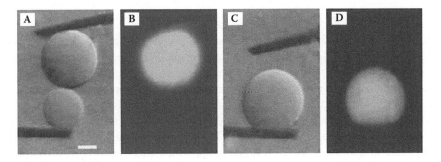

FIGURE 8.40 Sequential pairwise fusion of two different liposomes. (a, b) Images of the liposomes before fusion. (c, d) Images after the fusion between a "plain" liposome (no membrane dye in the membrane) and a "dyed" liposome, with the membrane fluorescent dye DiO. It is observed that the membrane dye distributes evenly over the entire membrane surface in the fused liposome. (a, c) Bright-field images. Scale bar: 10 μm. (b, d) Fluorescence images are enhanced digitally.[903] Reprinted with permission from the American Chemical Society.

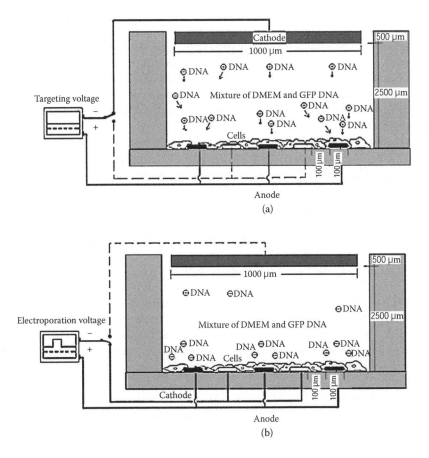

FIGURE 8.41 Illustrations of experimental setup and mechanism of the electroporation chip: (a) DNA attraction and targeting and (b) electroporation processes.[370] Reprinted with permission from the Royal Society of Chemistry.

8.3.2   Electroporation

Another cell operation is to use a pulsed electric field to conduct electroporation or electrofusion. For instance, electrofusion of liposomes (5 μm diameter) and RBCs has been conducted on a microchip (see Figure 8.40).[903]

An electroporation chip was constructed by sealing a PDMS layer (3 mm) onto a glass chip patterned with interdigital electrodes (see Figure 8.41). The chip was used for both electrical preconcentration of plasmid DNA and subsequent electroporation. The PEGFP-NI plasmid (4 μg) coded for GFP was transfected into various cells (e.g., 293T, HepG2, Chang liver, and MC3T3E-1 cells). These cells were first cultured for 24 h on the chip, which was coated with 0.01% poly-l-lysine. In the electrophoresis (targeting) mode (Figure 8.41a), 0.6 V was applied across the top cathode and bottom anode. Then in the electroporation mode (Figure 8.41b), 3 V was applied across the interdigital electrodes. An increase in transfection efficiency was found only after conducting the preconcentration step.[370]

Electroporation has also been achieved on a porous membrane. For instance, holes (2 to 10 μm) were fabricated on a 1 μm thick transparent Si_3N_4 membrane using reactive ion etching (RIE). The membrane was sandwiched between two 500 nm thick poly-Si layers serving as electrodes. When a human prostate adenocarcinoma (ND-1) cell of size 20 μm was trapped at the hole (5 μm), the cell plugged the hole to form an insulated seal. For a smaller-sized rat hepatocyte, a hole diameter of 4 μm was used. Electroporation of the trapped cell was performed by applying a potential across the two electrodes. The current (0 to 500 μA) was measured and plotted against the electrical pulse amplitude (0 to 40 V). From the plot, the onset of electroporation was indicated by the rapid rise in current. It was found that even for the same batch of cells, the onset voltage can differ by as much as 20%. This explains why optimizing batch electroporation experiments has been difficult.[904]

In another report, electroporation of ND-1 cells for transfection of EGFP gene was conducted via an aperture (2 to 6 μm) fabricated on a Si-SU-8-glass chip. Electroporation was performed by applying an electric pulse (10 V, 100 ms).[905] A review on microchip cell electroporation has been published.[906]

8.3.3   Chip Cleaning and Sterilization

To reuse the glass microchip after cell studies, the chip is cleaned using various cleansing solutions. For instance, clogged channels by cell debris can be cleaned by using 1% SDS or 1 N NaOH.[368] In another report, in order to help restore the treated surface after cell studies, the channel was cleaned with decanol. However, a better approach is to heat the glass chip at 500°C for 8 h to oxidize the surface (and to decompose cell debris), and then recoat the glass chip. Alternatively, the chip is placed on a ceramic base into a furnace for thermal treatment: temperature is raised slowly (within 1 h) to 300°C and held for 1 h, and then allowed to cool to room temperature. This process burns the organic materials inside the channel. Subsequently, 65% HNO_3 is aspirated into the channel to remove the burnt organic material and then washed thoroughly with chloroform.[907]

For study of cells within the microchannels, sterilization is necessary. This was performed by treatment of (1) 70% ethanol,[160,164,249,250,361,908] (2) bleach,[893] or (3) ethylene oxide.[361] Moreover, various methods such as UV irradiation[160,299,908,909] or autoclaving[898,910,911] have also been used for sterilization.

8.3.4   Prevention of Cell Adhesion

To reduce cell adhesion during cell studies in microchips, various coating methods have developed. For instance, a polyacrylamide film was photopatterned on acrylate-coated glass, using 3-(trimethyoxysilyl)propylacrylate. This coating reduced the fraction of adhesion as follows: CATH.a cells (from 1.05 to 0.25), U937 cells (from 0.99 to 0.11), and Jurkat T cells (from 0.05 to 0.04).[623]

In one report, Pluronics surfactant (triblock copolymer of PEO-PPO-PEO) was used to prevent cell adhesion in a chip. This is because the center PPO block is hydrophobic and thus shields any hydrophobic surface on the chip. In addition, the end PEO groups are neutral but very hydrophilic, and will not interact with proteins and cells.[278]

In another report PEG, which is a water-soluble, nontoxic, and nonimmunogenic polymer, has been evaluated for its use as a nonbiofouling coating for Si-based chips.[912]

8.4   PROBLEM SETS

1. When DEP is used to retain cells, a nonuniform electric field is needed. What are the two strategies to create a nonuniform electric field in a microchannel?[321] (4 marks)
2. What are the sterilization methods used in chips for cellular work? (2 marks)
3. In cell culture, O_2 is needed. The amount of O_2 that could be supplied into a PDMS device per day (F_{max}) can be estimated by the equation

$$F_{max} \approx D_{PDMS} \frac{\Delta c}{\Delta z}$$

where D_{PDMS} is the permeability of O_2 in PDMS (4.1×10^{-5} cm^2/s) and Δc is the difference in O_2 across the device thickness (2.0×10^{-7} mol/cm^3). Calculate F_{max} for a PDMS device with a 200 μm thick cover slab.[361] (2 marks)

9 Applications to Nucleic Acids Analysis

DNA amplification (mostly by polymerase chain reaction [PCR]) and other subsequent DNA analysis (including hybridization, sequencing, and genotyping) have been facilitated by the use of the microfluidic chip. These applications are described in detail in subsequent sections.

9.1 NUCLEIC ACIDS EXTRACTION AND PURIFICATION

DNA extraction has been achieved using silica beads (5 μm), which were packed in a glass microchannel and held by a sol-gel.[913] This method provided more reproducible extraction than a previous method in which extraction was performed without physically fixing the beads in microchambers.[639]

DNA was also extracted from lysed *Escherichia coli* (*E. coli*) cells using immobilized beads in a chip. The extraction efficiency was found to be high, especially when the number of *E. coli* cells was small.[914]

A DNA purification chip was constructed on poly(dimethylsiloxane) (PDMS). The chip consisted of mixing valves and a rotary mixer with fluidic and valve actuation channels. The chip had twenty-six access holes, one waste hole, and fifty-four valves to simultaneously carry out three purification procedures of *E. coli* genomic DNA. After removing DNA from the chip, the gene that encoded the prelipin protein peptidase-dependent protein D (ppdD) was PCR amplified externally.[915]

A mRNA purification from cells was conducted on a PDMS chip with pneumatic valve activation channels and fluidic channels. A single NIH 3T3 cell was first isolated and lysed. The liberated mRNA was captured by oligo-dT-coated magnetic beads. Using this method, highly abundant β-actin mRNA was isolated from a single cell, and moderately abundant zinc finger gene mRNA was isolated from two to ten cells. After purification, the mRNA was amplified by off-chip reverse transcription (RT)–PCR.[915]

Other nucleic acid extraction methods, which may have been integrated in a microsystem, will be described in subsequent sections.

9.2 NUCLEIC ACIDS AMPLIFICATION

9.2.1 DNA AMPLIFICATION

9.2.1.1 Polymerase Chain Reaction (PCR)

The polymerase chain reaction is the prevalent method for DNA amplification. Much effort has been made to integrate PCR chambers on microchips to carry out amplifications of DNA molecules prior to their analysis. For instance, PCR was first achieved on a Si-based reaction chamber (25 or 50 μl) integrated with a polysilicon thin-film (2,500 Å thick) heater for the amplification of the GAG gene sequence (142 bp) of HIV (cloned in bacteriophage M13).[997]

In another report, PCR of bacteriophage λDNA (500 bp) was performed in a 10 μl chamber fabricated in a Si-Pyrex chip. Finally, the temperature stayed at 72°C for 5 min. Subsequently, off-chip agarose gel electrophoresis was performed. Nevertheless, it was found that PCR on-chip was not as efficient as conventional PCR.[916]

On-chip PCR benefits from the high surface-to-volume ratio (SVR) of the PCR chamber. For instance, SVRs of Si chambers are 10 mm^2/μl[916] and 17.5 mm^2/μl,[917] which are greater than 1.5 mm^2/μl in conventional plastic reaction tubes and 8 mm^2/μl in glass capillary reactive tubes.[917,918]

In one report, anti-Taq DNA polymerase antibody was employed to avoid loss of PCR efficiency. The antibody inhibited the Taq polymerase before PCR reagents attained a high temperature, and this procedure is thus called hot-start PCR. In this procedure, the loss of Taq polymerase due to nonspecific binding was reduced. This on-chip hot-start PCR resulted in a more consistent and higher yield than that obtained in the PCR chip without hot start, and even that in the conventional PCR tube (with hot start).[917]

PCR has not only been achieved from extracted DNA, but also directly from DNA in the cells. For instance, the human cystic fibrosis transmembrane conductance regulator (CFTR) (~100 bp) was amplified from isolated human lymphocytes. The results were comparable with those of PCR from human genomic DNA extracted from the cells. The results indicated that tedious extraction of the DNA template might not be necessary, and PCR can be conducted directly on lysed cells.[917] In a dual-function Si-glass microchip, the isolation of white blood cells (WBCs) from whole blood (3.5 μl) using weir-type (3.5 μm gap) filters was followed by PCR of the exon 6 region of the dystrophin gene (202 bp).[919] After removal of red blood cells (RBCs) (hemoglobin in RBCs is a PCR inhibitor), the DNA from the retained WBCs was released by the initial high-temperature denaturation step (94°C).[919]

Faster thermal cycling for PCR can be achieved by reducing the thermal mass on a Si-glass chip.[378,920] For instance, this was accomplished by creating grooves (1 mm wide, 280 μm deep) in the Si chip where the thin-film heater was located. As a result, high rates of heating (36°C/s) and cooling (22°C/s) were obtained.[378]

In one report, a thin-walled PCR chamber was constructed on a PET chip. The thin membrane (200 μm) between the chamber and the Al heater resulted in only a short 2 s thermal delay. Fast heating (34 to 50°C/s) and cooling (23 to 31°C/s) rates were obtained.[921] In another report, a thin Si membrane (~50 μm) was formed between the Pt thin-film heater and the Si-based PCR chamber. This led to a heating rate of 15°C/s and a cooling rate of 10°C/s.[922] Furthermore, a thin-walled PCR chamber was constructed on a Si-PMMA chip. The Si membrane between the chamber and heater is 0.8 μm SiO$_2$/0.3 μm Si$_3$N$_4$/0.8 μm SiO$_2$. This combination produces a stress-reduced membrane because SiO$_2$ is heat compressive and Si$_3$N$_4$ is heat tensile. This chamber design allows for fast rates of heating (80°C/s) and cooling (60°C/s).[923]

A noncontact heating method called infrared-mediated temperature control has been employed for PCR. Because the chip material (polyimide) does not absorb infrared (IR), only the solution absorbs IR. Therefore, the low thermal mass of the solution allows for fast thermal cycling, and fifteen cycles have been achieved in 240 s! Amplification of λ phage DNA (500 bp) was first conducted at 94°C for 10 s, followed by fifteen cycles of 94°C (2 s), 68°C (2 s), and 72°C (2 s), and then finally stopped at 72°C for 10 s.[192]

Integration of the PCR chamber and subsequent on-chip CGE analysis has been reported. First, the target DNA, which is a β-globin target cloned in M13 (268 bp) or genomic salmonella DNA (159 bp), was amplified in a microfabricated PCR reactor. The reactor (~20 μl) was heated by a polysilicon heater (3,000 Å Si doped with boron). Then the PCR products were directly injected into a glass CE chip for CGE separations. The heating and cooling rates are 10°C/s and 2.5°C/s, respectively, compared to typical rates of 1°C/s in conventional thermal cyclers.[924]

In another report, PCR and subsequent CGE separation were integrated on a glass microchip (see Figure 9.1). PCR of λ phage DNA was conducted in a sample reservoir with thermal cycling by a Peltier heater/cooler. Subsequent CGE separation was conducted immediately after PCR because the PCR reservoir led to the CE channel. In addition, an on-chip DNA preconcentration device was included. This reduced the analysis time to 20 min (by decreasing the number of thermal cycles required to ten cycles), and the starting DNA copy number to fifteen (0.3 pM).[925]

The use of a thin-film heater permitted a PCR cycle time as fast as 30 s in a 280 nl PCR chamber. Stochastic DNA amplification has been demonstrated using one to three templates. The temperature

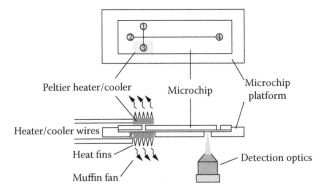

FIGURE 9.1   Dual Peltier assembly for rapid on-chip thermal cycling of PCR followed by on-chip CE analysis.[925] Reprinted with permission from the American Chemical Society.

program started at 95°C for 60 s, followed by thirty cycles of 94°C (5 s), 64°C, touch-down (in 2°C increments) to 50°C (15 s), and 72°C (10 s). Ten representative electropherograms (out of sixty) demonstrate the stochastic amplification from zero to three starting template copies (M13, 136 bp); see Figure 9.2. Note the presence of the control peak (S65C, 231 bp) in all cases. Before separation, the PCR mixture was prevented from flowing into the CE channel using the passive barrier formed by the hydroxyethylcellulose (HEC) sieving medium.[926] A Si-based PCR chamber was also interfaced to a glass chip using a PDMS gasket. PCR and subsequent on-chip CE separation were conducted for the DNA-based determination of bird sex.[927] In another report, PCR and CGE were integrated on a PDMS-glass chip. The PCR chamber is an open PDMS reservoir covered by silicone oil during PCR. The connection channel between the PCR chamber and separation channel was narrowed down in width, forming a stop valve between them until CGE separation started.[253] Furthermore, IR-mediated PCR amplification of the β-globin gene was performed on a glass chip sample reservoir, followed by on-chip CGE within the glass microchannels.[928]

For complex samples containing several different DNAs, multiplex PCR has been carried out. For instance, on-chip multiplex PCR was achieved on four DNAs representing regions in the bacteriophage λDNA (199 and 500 bp), *E. coli* genomic DNA (346 bp), and *E. coli* plasmid DNA (410 bp). After PCR, the fluorescent intercalating dye (TO-PRO) was added to the PCR reservoir, and CGE separation was performed downstream.[929]

Lysis of *E. coli*, multiplex PCR amplification, and electrophoretic sizing were executed sequentially on a glass chip.[929,930] Multiplex PCR was achieved for five regions of the genomic and plasmid DNA of *E. coli*, and four regions in the λDNA. Subsequent CGE separation was performed.[929] From the separation results, the 3% PDMA sieving medium appeared to be stable, even though it was present during thermal cycling. Moreover, the components in the PCR mixture did not appear to adversely affect the performance of the CGE analysis of the PCR products.[930]

Multiplex PCR of the muscular dystrophin gene was performed on-chip. The method of degenerate oligonucleotide-primed PCR (DOP-PCR) was first used to increase the number of DNA templates before multiplex PCR. Therefore, DOP-PCR can provide the template DNA from the whole human genome, but this procedure is feasible only when the amplicon size is less than 250 bp.[931]

Research on the integration of PCR and DNA analysis has been very active. For comparison, the PCR chip designs and PCR mixtures are tabulated in Tables 9.1 and 9.2, respectively.

9.2.1.2   Surface Passivation of PCR Chambers

To avoid inhibition of the enzyme (polymerase) and adsorption of DNA (template and product), surface passivation on the PCR was needed. It was found that native silicon is an inhibitor of PCR and an oxidized silicon (SiO_2) surface was required to provide the best passivation.[943]

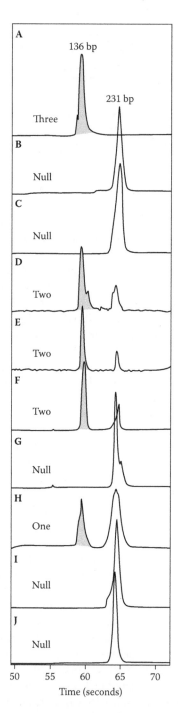

FIGURE 9.2   PCR amplifications/analyses of single-molecule templates with internal controls. Ten representative electropherograms from a total of sixty are shown. In each electropherogram, the 231 bp control peak is seen, indicating PCR amplifications have occurred. In some electropherograms (b, c, g, i, j), only the control peak is seen, indicating unsuccessful amplification of the 136 bp template. In one case (h), a 136 bp peak is seen, but it is small in proportion to the control peak, indicating a one-template PCR. In some cases (d, e, f), the 136 bp peak is larger, indicating a two-template PCR. In another case (a), this 136 bp peak dominates over the control peak, indicating a three-template PCR. These ten amplifications were conducted in this order over a single day.[926] Reprinted with permission from the American Chemical Society.

TABLE 9.1
Design of PCR Chambers

PCR Chamber Material	Chamber Volume	Surface Treatment	Heating Rate	Cooling Rate	Heating Methods	Ref.
Si-Pyrex	5 μl	Thermal oxide 4,000 Å	5°C/s	5°C/s	Peltier heater	943
Si	20 μl	Thin-walled polypropylene liner	10°C/s	2.5°C/s	Polysilicon (3,000 Å) heater doped with boron	924
Si-glass	50 μl	Silanization	13°C/s	35°C/s	Polysilicon (2,500 Å) heater	997
Si-Pyrex	10 μl	Thermal oxide 1,000 Å			Peltier heater/cooler	917
Glass	10–20 μl	BSA (2.5 mg/ml) treatment	2°C/s	3–4°C/s	Peltier heater/cooler	925
Glass	280 nl	Coated	10°C/s	10°C/s	1 cm² resistive heater N₂ cooling	926
PC	38 μl	—	7.9°C/s	~4.6°C/s	Peltier heater	937
PC	20 μl	UV			Commercial thermal cycler	938
Plastic	—	—	2.4°C/s	2°C/s	Resistive heater plus Peltier heater/cooler	998
Polyimide	1.7 μl	—	10°C/s	10°C/s	Infrared heating	192
PDMS	30–50 μl	BSA 150 μg/ml	—	—	Peltier heater	253
PDMS	1.5 μl	—	~2.4°C/s	~2.0°C/s	Resistive wire heater, Peltier cooler	964
Si-glass	2 μl	PECVD SiO₂ (200 nm), BSA	~60–90°C/s	74°C/s	Resistive Pt heater	920
PMMA/Si	200 nl	—	80°C/s	60°C/s	Thin-film Pt heater	923, 999
Glass	200 nl	—	20°C/s	>10°C/s	Resistive heater, forced-air cooling	282
PDMS	1.9 μl	Parylene (4.5 μm)	2°C/s	1.2°C/s	Cartridge heater, DC blower fan	1000
Si-Pyrex	3 μl	Silanized	2.9°C/s	2.4°C/s	Resistive Pt heater	94
Si-Pyrex	3 μl	BSA 1.0 mg/ml	11°C/s	11°C/s	Resistive Al heater	927
PC	29 nl	—	~12°C/s	~2°C/s	Resistive C heater	291
PC	—	—	—	—	Peltier heater	462
—	—	—	7°C/s	6°C/s	Peltier heater/cooler	916
Si-glass	—	—	36°C/s	22°C/s	Resistive Pt heater, cooling fan	378
—	—	—	7–8°C/s	5–6°C/s	Peltier heater, cooling fan	1001
Si	—	—	15°C/s	10°C/s	Thin-film Pt heater	922
PET	—	—	34–50°C/s	23–31°C/s	Thin-film Al heater	921

TABLE 9.2
Composition of PCR Mixtures

Primer Concentration	Nucleotide Conc.	Polymerase	Amount of DNA Template	Additives for Surface Modifications[a]	Volume of Mixture	Ref.
1 μM	200 μM	Taq 2.5 U	10^8 copies	Zwitterion buffer (tricine), 1.4 μM PVP, and nonionic surfactant (0.01% w/v Tween-20) as dynamic coating	10 μl, flow through	946
1 μM	200 μM	Taq 25 U/ml	*E. coli* cells	0.001% gelatin, 250 μg/ml BSA	10–25 μl	930
0.1–1 μM	200 μM	Taq 25 U/ml	10 ng/ml phage DNA	0.001% gelatin, 250 μg/ml BSA	6 μl	925
0.2 μM 0.25 μM	200 μM	Taq 2.5 U/50 μl	200 copies (M13/pUC, 136 bp)/50 μl 1,000 copies (S65C, 231bp)/ 50 μl	1.5 μM BSA	280 nl	926
1.0 μM	200 μM	Taq 25 U/ml	10 ng/ml	0.001% gelatin, 250 μg/ml BSA	20 μl	938
Forward primer, 12 nM Reverse primer, 1.2 μM	125 μM	Taq 25 U/ml	50 pg/μl	0.001% gelatin, 250 μg/ml BSA	38 μl	937
0.5 μM	2 mM	(Taq start antibody 2.2 μg) Taq 10 U	White blood cells	Possibly coated by proteins from white blood cells	50 μl	919
1.0 μM	200 μM	Taq 25 U/ml	10 ng/ml λ DNA, *E. coli* cells	0.001% gelatin, 250 μg/ml BSA	12 μl	929
1.0 μM	200 μM	Taq 25 U/ml	*E. coli* cells (346 bp PCR product)	0.001% gelatin, 250 μg/ml BSA	1.5 μl	998
80 μM	2.5 mM	(Taq start antibody 132 ng) Taq 1.2 U	9.6 ng human genomic DNA	0.1% Triton X-100, 0.01% gelatin	12 μl	931
0.2 μM	200 μM	Taq 10 U	0.1 ng λ phage DNA (500 bp)	0.001% gelatin, 0.75% (w/v) PEG	50 μl	192
Forward primer, 4 nM Reverse primer, 0.08–0.4 μM	0.2 mM	1.25 U/50 μl Taq	0.015–50 ng/50 μl TCM herbs: *P. padatisecta, D. innoxia*	0.5 μg/μl BSA		94
1 μM	0.2 mM	0.025 U/μl Taq	Bird genomic DNA	0.1% Triton X-100	Flow through	927
1 μM	0.2 mM	25 U/ml Taq	—	0.25% Tween-20, 25% glycerol	Flow through	275
0.5 μM	0.2 mM	0.5U/μl Taq	PSA gene	0.2 g/L BSA, 1 ml/L Triton X-100, 1 ml/L Tween-20, 2.8 μmol/L PVP	Flow through	1002
1 μM	0.2 mM	0.1 U/μl Taq	Various DNA templates 0.2 mM	Tween-20 (1–40% w/v), BSA (0.1 to 20 μg per 50 μl)		1003

—*continued*

TABLE 9.2 (continued)
Composition of PCR Mixtures

Primer Concentration	Nucleotide Conc.	Polymerase	Amount of DNA Template	Additives for Surface Modifications[a]	Volume of Mixture	Ref.
Forward primer, 0.1 µM Reverse primer, 10 µM	10 mM	5 U/µl Taq	1 ng/µl TCM herb: *F. thunbergii*	20 µg/µl BSA		97
2.5 µM	0.2 mM	0.5 U/µl Taq	4.5×10^3 copies of bacterial DNA	0.1 mg/ml BSA		291
0.2 µM	0.2 mM	2.5 U/100 µl Taq	1 ng/100 µl λDNA	100 µg/ml BSA		253
0.25 µM	5 mM	0.01 U/µl Taq	10 ng/20 µl	Nil		920
Forward primer, 0.05 µM Reverse primer, 5 µM	0.4 mM	5 U/20 µl Taq	*E. coli* K12-specific gene fragment	Gelatin 0.001%, BSA 0.1%		462
0.5 µM	200 µM	0.5 U/µl Taq	Human β-globin gene (0.23 kbp), PSA DNA (0.23 kbp)	0.1% Triton X-100, 0.1% Tween-20, 0.2 g/L BSA, 28 µM PVP		907
0.25 µM	200 µM	2.5 U/50 µl Taq	5 ng human chromosome, X: 157 bp, Y: 200 bp	1.5 µM BSA		282
0.5 pM	50 µM	0.5 U/µl Taq	4.95 ng/µl (490 bp) angiotensin-converting enzyme gene	Nil		1000
0.1 µM	1.6 mM	1.5 U/10 µl Taq	0.2 ng human genomic DNA	0.75% w/v PVP 40 or PEG 8000		936
0.2 µM	100 µM	1.5 U hot-start Taq	*E. coli* genes	100 µg/ml BSA		269
0.5 µM	0.4 mM	5 U/40 µl Taq	*E. coli* K12 gene, 10^6 cells	0.01%, gelatin, 0.75% BSA (250 µg/ml), PEG 8000		1001
400 µM	200 µM	Taq 1.25 U	10^8 copy		50 µl	997
	200 µM	Taq 1.0 U	0.2 ng		10 µl	916
0.3 µM	200 µM	Taq 2.5 U	1 ng			918
0.5 µM	400 µM	Taq 2.5 U	500 ng (4×10^5 copies)/ Salmonella DNA		50 µl	924
0.6 µM	200 µM	Taq 0.6 U	1.2 ng *C. jejuni* DNA		12 µl	917
0.6 µM	200 µM	Taq 0.4 U	125 ng human genomic DNA			
0.6 µM	200 µM	Taq 0.6 U	1.2 ng *C. jejuni* DNA			
0.2 µM	200 µM	(Taq start antibody 132 ng) Taq 0.4 U	Human lymphocytes (1,500 and 3,000 cells)			

[a] To prevent adsorption of PCR polymerase or nucleic acids, surface treatment is necessary.

In a later report, inhibition of PCR in Si-glass PCR chips was found to be mainly caused by the adsorption of the Taq polymerase, rather than by the adsorption of DNA.[299] It was also found by x-ray photoelectron spectroscopy (XPS) analysis that the primary PCR inhibitor in a glass PCR chamber was Cr, which was involved in the microfabrication process. The highest Cr concentration that could be tolerated was found to be 0.1 M.[932]

Hemoglobin and heparin present in blood samples are also PCR inhibitors. In one report, these inhibitors were removed from yeast cells (used as the model for blood cells) after dielectrophoresis (DEP) cell retention. The yeast cells were retained by DEP, whereas bovine hemoglobin (1 mg/ml) and heparin (13 μg/ml) were washed away.[933]

A polymeric coating, poly(vinylpyrrolidone) (PVP), has been developed and used in glass microchips for robust CGE analysis of PCR products containing a high salt content.[934] The coating was applied after silanization (using trimethylsilane [TMS]) of the channel wall. Among other coatings tested, PVP has produced good separation performance even after 635 analyses.[934]

Four coatings, such as polyethylene glycol or polyethylene oxide (PEG; 8.5% w/v), PVP (2.5% w/v), HEC, and epoxy poly(dimethylacrylamide) (EPDMA), were evaluated for PCR on glass chips. It was found that the amounts of PCR products were 13, 78, 0, and 72%, respectively, when the amount was 100% using a conventional PP tube. These values were also compared to 50 to 120% when only silanization was used for passivation. While the first three are dynamic coatings, the last one is an adsorbed coating.[935]

In another report, it was found that the addition of PEG 8000 or PVP 40 in the PCR mixture produced a significant surface passivation effect in either native or SiO_2-precoated Si-glass chips. Using this passivation method, PCR of the nitric oxide synthase gene (human endothelial) in human lymphocytes was demonstrated.[936]

9.2.1.3 Integrated DNA Analysis Microsystems

In addition to integrating PCR and CE separation, a complete nucleic acid analysis system requires the inclusion of many more operations. For instance, PCR amplification and subsequent hybridization have been achieved in a disposable polycarbonate (PC) chip (see Figure 9.3).[462,937] Analysis was focused on the *E. coli* DNA sequence (221 bp) and the *Enterococcus faecalis* E gene (195 bp). The higher transition temperature of PC (145°C) than that of other plastic materials allows thermal cycling conducted at high temperature (e.g., 95°C).[938] The chip has a well-positioned window (see PCR in Figure 9.3) to provide liquid access. Through the window (1.7 cm²), surface treatment and oligonucleotide probe attachment were conducted on the hybridization channel (HC). Afterwards, the window was enclosed with a second PC piece using adhesive tape and epoxy.[937,939] The DNA sample was loaded through SL (see Figure 9.3) into PCR for DNA amplification. Subsequently, the liquid was transferred to HC for hybridization. To bypass a postamplification step, i.e., denaturing of dsDNA, asymmetrical PCR was employed. This approach utilized strand biasing to preferentially amplify the target strand of choice. Hybridization was performed on a microarray spotted onto HC downstream of the PCR chamber. To evaluate the probe attachment, the probes were Cy3 labeled; to evaluate the hybridization chemistry, the targets were Cy5 labeled. While the normal approach is not to label the probe, this strategy allows the visualization of probes even though no hybridization has occurred.[937]

A Si-based device was fabricated to integrate DNA amplification and hybridization analysis. Oligonucleotide probes were first spotted into an array of multiple silanized PCR microreactors. DNA samples were introduced into these microreactors for asymmetric PCR. Right after PCR, hybridization occurred in the same locations by changing the solution conditions. This device was applied for the genotyping of Chinese medicinal plants, which was based on the variations of different plant species in their noncoding regions of the 5S rRNA gene.[94] Multiple PCR chambers (i.e., 1,064 microreactors) have also been constructed on a PDMS chip.[940]

A fully integrated eight-channel genomic analysis microsystem has been constructed on a glass wafer (see Figure 9.4). The microsystem includes microfabricated heaters, resistance temperature detectors (RTDs), PCR chambers, and a CE separation channel. Both the heater and RTD are

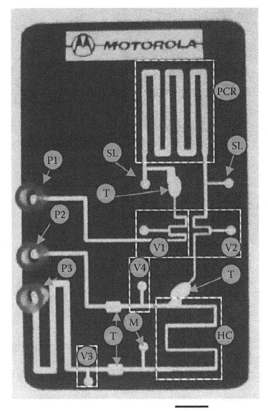

10 mm

FIGURE 9.3 Monolithic integrated polycarbonate device for DNA assay that involves PCR and DNA hybridization. The device consists of the following components: serpentine PCR channel (PCR), 7 µl hybridization channel (HC), Pluronics valves (V1 to V4), Pluronic traps (T), hydrophobic air-permeable membrane (M), PCR reagent-loading holes (SL), sample driving syringe pump (P1), waste-withdrawing syringe pump (P2), and wash syringe pump (P3).[937] Reprinted with permission from the American Chemical Society.

fabricated from Pt/Ti layers. Fast heating and cooling rates of 20°C/s were achieved for PCR of human genomic DNA for sex determination. In the CGE analysis shown in Figure 9.5, female sex was identified by the presence of the 157 bp PCR product derived from the X chromosome, whereas male sex was identified by the presence of the 157 and 200 bp PCR products derived from both X and Y chromosomes.[282]

A PCR chamber (200 nl) constructed on a glass chip was integrated with a Pt heater. In addition, PDMS valves were integrated to enclose the PCR mixture during PCR. Triplex PCR amplification of target genes that encode 16S ribosomal RNA, the fliC flagellar antigen, and SltI shigatoxin was carried out for the detection of pathogenic *E. coli*. Subsequent CE separations of the three *E. coli* strains are shown in Figure 9.6. In Figure 9.6a, only benign *E. coli* strain K12 produced the 280 bp 16S product. In Figure 9.6b, the enteropathogenic *E. coli* O55:H7 produced both the 16S product and the 625 bp fliC, product. In Figure 9.6c, the enterohemorrhagic shigatoxin-producing strain O157:H7 produced all the 16S, fliC, and 348 bp SltI products. This method can identify pathogenic *E. coli* down to a level of two to three bacterial cells.[269]

Using multilayer soft lithography, PCRs were achieved in 3 nl reactors at the vertex of an $N \times N$ microfluidic matrix ($N = 20$) in a PDMS-glass chip. Using pneumatic control in the PDMS control layer, twenty primers and twenty concentrations of DNA template were introduced into the fluidic

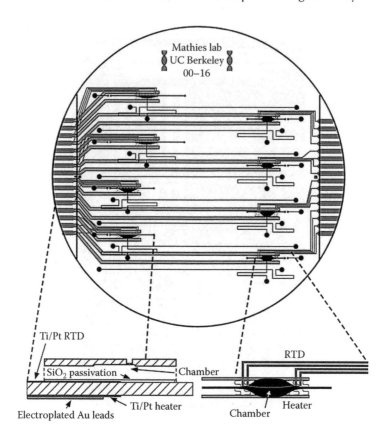

FIGURE 9.4　(See color insert following page 144.) Overview of the PCR-CE mask design including microfabricated heaters (pink) and green-colored resistance temperature detectors (RTDs). Bottom right: An expanded top view of the chamber area, showing the chamber, the four-lead RTD, and the heater. The leads for the heater should be electroplated with gold (see also Chapter 2, Section 2.6). The four-lead RTD provides more accurate temperature measurement than the two-lead RTD, since the four-lead device separates the measurement due to Joule heating in the device from that due to the actual temperature change within the PCR chamber. Bottom left: An expanded view of a partial cross section of the device, showing the two glass layers and their relative alignment.[282] Reprinted with permission from the Royal Society of Chemistry.

channels. A single 2 μl aliquot of polymerase is distributed over all four hundred independent reactions, thus dramatically reducing sample and reagent consumption. Here, $N^2 = 400$ distinct reactions were carried out with only $2N + 1 = 41$ liquid pipetting steps. This was compared to the $3N^2 = 1,200$ steps required with conventional fluid handling. A 294 bp segment of the human β-actin cDNA was amplified inside a flat-bed thermocycler. By titrating the template DNA concentrations, the detection limit was ~60 template copies per reactor.[941]

Integration of PCR and CE was achieved on a COC chip for the detection of bacteria (e.g., *E. coli* and *Salmonella typhimurium*). PCR was conducted in a 29 nl chamber, which was temporarily sealed off by gel valves during PCR. The gel valve can withstand a hydrostatic pressure of 100 psi, which is higher than the water vapor pressure of 12 psi achieved in PCR up to 95°C. After PCR and the electric field was applied, the DNA migrated through the gel and commenced CGE separation. The COC substrate was selected over acrylic because COC has a higher glass transition temperature, and over PC because COC has a lower fluorescent background.[291]

Micro-PCR is normally achieved in PC, Si, and glass but not PDMS because of its porosity. However, with a parylene (or poly-para-xylene) coating, problems of bubble formation, sample evaporation, and even protein adsorption are solved. In addition, no BSA or PEG additive is needed in

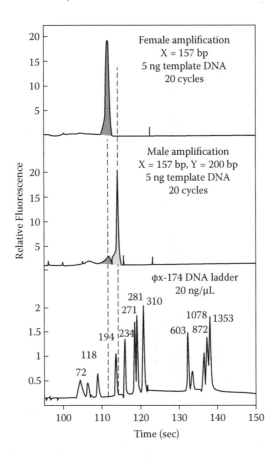

FIGURE 9.5 Sex determination from human genomic DNA using the PCR-CE microdevice. Top: Amplification from female DNA. Only the 157 bp peak representing amplification from the X chromosome is seen. Middle: Amplification from male DNA; both the 157 bp X chromosome and the 200 bp Y chromosome amplicons are observed. Bottom: DNA sizing ladder for amplicon size reference, run separately on the same device.[282] Reprinted with permission from the Royal Society of Chemistry.

the PCR mixture for dynamic coating because the parylene coating (4.5 μm thick) is hydrophobic. This can be seen in the contact angle measurements in Figure 9.7. On the other hand, since the thermal conductivity of PDMS is lower than that of Si or glass, the heating rate in a PDMS device is lower (see Figure 9.8). However, with both top and bottom heating, a heating rate of 2°C/s and cooling rate of 1.2°C/s can be achieved.[1000]

An integrated DNA analysis system has been constructed to perform various tasks, such as measuring of nanoliter-sized reagents and DNA samples, solution mixing, DNA amplification, DNA digestion, CGE separation, and fluorescent detection. Both the heater (boron doped) and temperature sensor (diode photodetector (with the TiO_2/SiO_2 interference filter) are also integrated. Hydrophobic regions are present for fluid control. The only external electronic component not integrated with the chip is a blue light-emitting diode.[942]

9.2.1.4   Real-Time PCR

Real-time PCR for DNA amplification and detection was facilitated using the microfluidic chip.

Real-time monitoring of PCR amplification was achieved by sequential CGE after 15, 20, 25, and 30 s (see Figure 9.9).[924] In addition, PCR of β-actin DNA (294 bp) has been integrated with real-time detection at 518 nm using a fluorescent reporter probe.[943]

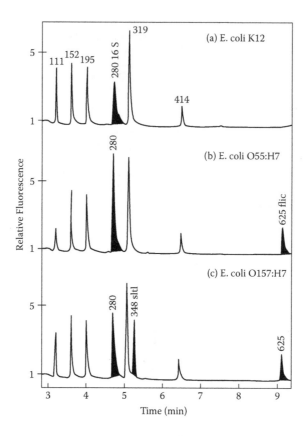

FIGURE 9.6 The portable PCR-CE microsystem for pathogenic organism analysis conducted directly from intact E. coli cells. (a) Analysis conducted from *E. coli* K12 cells, showing only the presence of the co-injected DNA ladder and the 280 bp 16S amplicon specific for *E. coli*. (b) Analysis conducted from *E. coli* O55:H7 cells, showing the ladder, the 280 bp amplicon, and the 625 bp *fliC* amplicon characteristic of cells presenting the H7 surface antigen. (c) Analysis conducted from *E. coli* O157:H7 cells, showing the DNA ladder, the 280 bp amplicon, the 625 bp *fliC* amplicon, and the 348 bp *sltI* amplicon, characteristic of *E. coli* both possessing an H7 antigen and expressing shigatoxin. Each analysis was conducted from a starting concentration of forty cells in the reactor in a time of 30 min.[269] Reprinted with permission from the American Chemical Society.

The use of an intercalating dye, SYBR Green I, for real-time PCR can only be performed in an all-glass device, rather than one of PDMS-glass, because the dye and DNA appeared to migrate into the PDMS (Sylard 184) polymer.[447]

Real-time PCR was also performed by on-chip thermal cycling for the detection of Hantavirus, HIV, orthopoxviruses (266 to 281 bp), *Borrelia burgdorferi*, human β-actin (294 bp), and the human complement C6 gene (73 bp).[944] In another report, real-time PCR was performed for DNA extracted from *Bacillus subtilis* in a plastic microfluidic cassette. The spore samples (10^4/ml) were sampled, filtered, and sonicated in the presence of 6 μm glass beads for spore disruption to release the DNA.[945]

9.2.1.5 Flow-Through PCR

Alternative methods have been reported to achieve temperature variations needed for PCR.

For instance, flow-through PCR was performed by moving the PCR mixture in a glass chip through three separate heating zones for melting, annealing, and extension (see Figure 9.10).[946,947] A twenty-cycle PCR amplification of the gyrase gene (176 bp) of *Neisseria gonorrhoeae* was performed. This was carried out at flow rates of 5.8 to 15.9 nl/s, corresponding to cycling times of 18.8 to 1.5 min for twenty cycles.[946,947]

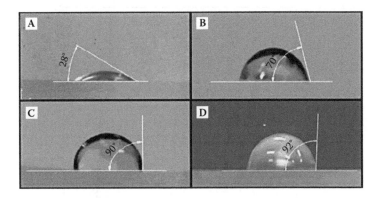

FIGURE 9.7 The contact angle of PDMS after various surface treatments: (a) 3 h after O$_2$ plasma treatment (28°C), (b) 48 h after O2 plasma treatment (70°C), (c) 144 h after O$_2$ plasma treatment (90°C), and (d) parylene-coated PDMS (92°C).[1000] Reprinted with permission from the Institute of Physics Publishing.

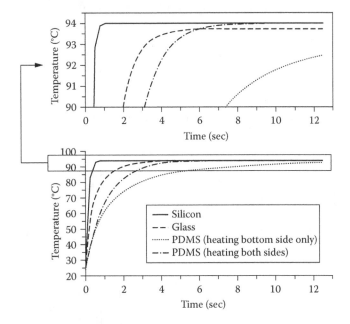

FIGURE 9.8 Temperature profile of the PCR chamber (center) fabricated on different substrates: silicon (solid line), glass (broken line), PDMS with bottom heating (dotted line), and PDMS with both top and bottom heating (chain line). Top inset shows the temperature profile in the expanded scale (90 to 94°C).[1000] Reprinted with permission from the Institute of Physics Publishing.

Similar methods of flow-through PCR chips were also reported.[212,907,948] Temperature analysis of continuous-flow PCR has been carried out on both a glass-Si and a glass-glass chip.[949]

In another report, heating in a flow-through quartz PCR chip was achieved by the integrated transparent ITO heater,[275] rather than by opaque copper heating blocks.[946,947] In this report, only a two-stage PCR was performed to amplify a 450 bp product.[275] Moreover, a flow-through PCR chip was fabricated on a Si wafer so that the thin-film Pt heaters and sensors could be integrated. The Si wafer was bonded with a glass plate with etched channels.[1003]

A sophisticated integrated flow-through PCR chip was fabricated (see Figure 9.11). Three reaction chambers, connection channels, thin-film resistive heaters, temperature sensors, and optical detectors are fabricated on a Si wafer. A thin pump membrane (200 μm), three lead zirconate titanate

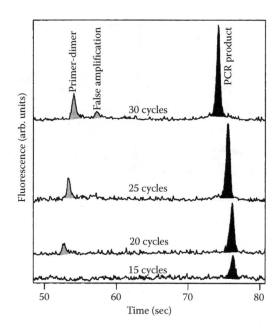

FIGURE 9.9 Real-time PCR of β-globin using an integrated PCR-CE microdevice. Consecutive chip CE separations of the same sample were performed in the integrated chip after 15, 20, 25, and 30 cycles. Thermal cycle was 96°C for 30 s and 60°C for 30 s. The PCR product peak that is shaded with dark gray increased with the number of cycles. The false amplification and primer–dimer peaks are shaded with light gray.[924] Reprinted with permission from the American Chemical Society.

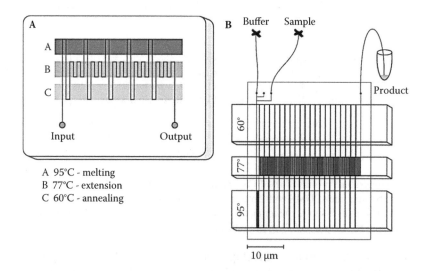

FIGURE 9.10 Schematic diagram of the flow-through PCR chip. By means of thermostated copper blocks, three well-defined temperature zones are kept at 95, 60, and 77°C for melting (dissociation), annealing, and extension of DNA. The glass chip incorporates twenty identical cycles. However, the first cycle includes a threefold increase in time for DNA melting by using a longer serpentine channel length.[946] Reprinted with permission from the American Association for the Advancement of Science.

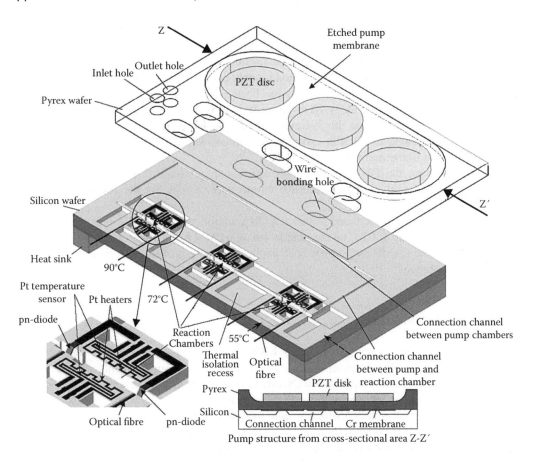

FIGURE 9.11 Schematic of the pumping PCR chip in which the solution mixture is pumped back and forth over three temperature zones (90, 55, and 72°C). For simplification the upper Pyrex glass wafer and the lower silicon wafer are illustrated apart, although in the actual device both wafers are anodically bonded together. The glass wafer consists of the etched region to house the three PZT disks for piezoelectric actuation. The Si wafer consists of three heating regions that are thermally isolated by the recess structures. Left lower inset shows an expanded view of the reaction chamber consisting of heaters, temperature sensors, optical fiber, and pn diode. Right lower inset shows the cross section of the micropump.[950] Reprinted with permission from the Institute of Physics Publishing.

(PZT) transducers, and access holes are constructed on a Pyrex plate. A liquid droplet (1 μl) can be pumped back and forth between the chambers by the microperistalic pumping action actuated by PZT. This arrangement can be used for carrying out spatial thermal cycling used for PCR.[950]

In another report, a PDMS rotary device (12 nl) was used for PCR both spatially (moving over three temperature zones) and temporally (at three temperatures with rotary mixing). PCRs of a β-actin gene (123 bp) and λ phage DNA (199 bp) were demonstrated. Real-time PCR was also performed as detected by an intercalating dye (Sybr Green I).[357]

9.2.1.6 Other DNA Amplification Techniques

There are other techniques that can be used to amplify DNA. For instance, the ligase chain reaction (LCR) was achieved on a Si-glass chip for detecting known point mutation in the *Lac*I gene sequence.[951] The yield from the chip was found to be 44% of that obtained from using the conventional reaction tube. This lower yield was possibly caused by nonspecific binding of both the template DNA and ligase enzyme to the chip surface.[951] The LCR method was also used to detect K-ras mutations.[212]

An isothermal cycling probe technology (CPT) reaction has been achieved on-chip to perform DNA amplification but without thermal cycling.[606,952] This technique works by producing an increased amount of the cleaved chimeric probe for fluorescent detection. However, the DNA template does not increase in amount. This method is applied for the detection of the 391 bp DNA used for diagnosis of the Newcastle disease. This has been achieved by measuring the amount of the labeled 12-mer cleaved probe (dA_{10}-rA_2) resulting from the cleavage of the chimeric 24-mer probe (dA_{10}-rA_4-dA_{10}) after binding to the 391-mer target.[952] In another report, detection of the DNA sequence derived from the methicillin-resistant *Staphylococcus aureus* was also achieved, using the CPT reaction at an isothermal temperature of 50°C.[606]

In another report, the DNA amplification was based on strand displacement amplification (SDA). This method worked at an isothermal temperature of 50°C for the 106 bp product from *Mycobacterium tuberculosis*.[942]

Another method is loop-mediated isothermal DNA amplification (LAMP). This was carried out at a constant temperature of 65°C on the sample well of a PMMA chip. Subsequent CGE separation was carried out. The LAMP method has been applied for analysis of the PSA gene with a 23 fg/μl template concentration.[953,954]

9.2.2 RNA AMPLIFICATION

Amplification of nucleic acids also includes RNA amplifications. For instance, the heat-shock mRNA (103-mer) in viable *Cryptosporidium parvum* in water treatment plants was detected by nucleic acid sequence-based amplification (NASBA). After amplification, a sandwich hybridization assay was subsequently conducted within a PDMS channel.[955]

In another report, real-time NASBA was used to amplify RNA in a Si-Pyrex chip containing 50 nl reaction chambers. This is an isothermal (41°C) technique. A fluorescent molecular beacon probe was used to detect the RNA derived from the human papillomavirus (HPV16).[337]

An integrated PC chip was developed to extract and concentrate nucleic acids (RNA) from milliliter-volume aqueous serum samples (see Figure 9.12). Microliter-volume amplification, liquid metering, mixing, and DNA hybridization are carried out for the mutation detection in the 1.6 kb region of the HIV genome. Serum samples containing as little as five hundred copies of RNA can be detected. DNA hybridization (20 min at 37°C) was performed on a GeneChip array integrated in the chip.[246]

Purification of mRNA and subsequent single-strand cDNA synthesis by reverse transcription (RT) were achieved all on the same chip.[956,957] Reverse transcription of an mRNA and subsequent PCR of the cDNA formed were also carried out in a continuous flow. This was achieved in a Pyrex glass chip (see Figure 9.13). Usually RT and PCR are performed separately because the reagents used in the two steps interfere with each other. For instance, the Mg^{2+} concentration in the RT mixture is four times higher than the optimum for PCR. Moreover, the reverse transcriptase used in RT inhibits PCR, and so the enzyme must be thermally inactivated (95°C) prior to PCR. To reduce the Mg concentration, the flow rates of RT and PCR are in a ratio of 1:9, and thus the RT mixture constitutes only about 10% of the total PCR volume. So, the heating block (d in Figure 9.13) for RT was set at 42°C for 30 min, then at 95°C to inactivate the RT enzyme. Thereafter, PCR was conducted when the mixture was flowing via the three heating blocks, a, b, and c. With this RT-PCR method, four 0.7 μl RT mixtures (coded for the PSA cDNA, 0.23 kbp) were run simultaneously while being segmented by intervening air bubbles (0.4 μl).[907] Subsequent development of the RT-PCR chip involved real-time off-chip LIF detection of the PCR products.[1002]

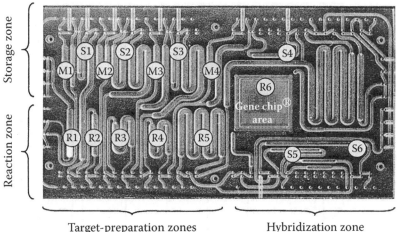

Target-preparation zones   Hybridization zone

FIGURE 9.12 An integrated device that automatically performs a multistep HIV genotyping assay. Three independently controlled temperature zones are indicated as storage zone, reaction zone, and hybridization zone on this PC device with dimensions of (8 × 40 × 70 mm). At the beginning of the process reagent mixtures are loaded into storage chambers S1 to S6. The device automatically carries out various operations in respective chambers, such as R1, RNA purification from a serum lysate (extraction); R2, RT-PCR; R3, nested PCR; R4, DNase fragmentation and dephosphorylation; R5, terminal transferase labeling; R6, dilution and hybridization, washing, phycoerythrin staining, and washing. Intermediate product samples from reaction chambers R1 to R4 are stored in chambers M1 to M4 for later analysis.[246] Reprinted with permission from Oxford University Press.

9.3 DNA HYBRIDIZATION

9.3.1 MICROCHANNEL DNA HYBRIDIZATION

Solid-phase DNA hybridization has been carried out between the immobilized probes and their complementary oligonucleotides. Immobilization of oligonucleotides on glass microchannels (20 μm deep, 1,000 μm wide) was achieved after continuously flowing reagents through the microchannels. A good reproducibility in the surface density resulted. Hybridizaton time was shortened from hours to minutes. To measure the temperature inside the microchannel, molecular beacons were used. As the temperature increased, the molecular beacons denatured (or melted) to release the fluorophore from its quencher, leading to an increase in the fluorescent intensity. With this temperature measurement, melting curves for the hybridized oligonucleotides in the microchannels were obtained.[83]

Four spots of DNA probes were applied in a PMMA microchannel (500 μm wide, 50 μm deep). This is intended for detection of low-abundant point mutation in the K-ras gene, which is diagnostic for colorectal cancer. Aminated 24-mer DNA probes were immobilized via glutaraldehyde linkage on the amine groups on the PMMA surface. Hybridization time was reduced from 3 h in a conventional array to less than 1 min in the microchannel at 50°C. Detection was achieved by using a near-IR dye (IRD 800) conjugated to the target sequence extracted from three types of cancer cells (HT29, wild-type K-ras; C12V, mutant K-ras).[958]

It was also found that the hybridization kinetics were faster in a moving sample than in a stationary sample.[939] In another report, active acoustic mixing was used to achieve a fivefold faster DNA hybridization rate. Hybridization was detected electrochemically (by AC voltammetry) based on the ferrocene redox chemistry.[62]

DNA hybridization has been carried out on beads. For instance, dynamic DNA hybridizations were performed by pumping fluorescently labeled DNA probes through microchannels that contained target-bearing paramagnetic polystyrene beads. Successive hybridizations have been performed (at 37°C) by new probe DNA, after the thermal denaturation (87°C) of the duplex formed by

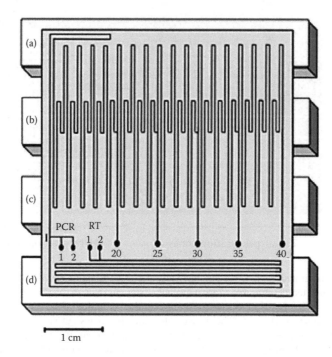

FIGURE 9.13 Schematic diagram of the microfabricated device for continuous-flow PCR and RT-PCR. There are two inlets for PCR (PCR1 and 2), two inlets for RT (RT1 and 2), and five outlets for product collection at 20, 25, 30, 35, and 40 cycles. The chip is placed over four heating blocks. Blocks a, b, and c are at appropriate temperatures for denaturation, extension, and annealing in PCR, respectively. Block d is for RT. The intersection between the RT and the PCR channels is denoted I.[907] Reprinted with permission from the American Chemical Society.

the old probe DNA. Four DNA targets (25- to 50-mer) were immobilized on paramagnetic beads. These immobilizations were achieved either by interaction between the streptavidin-coated beads and biotinylated DNA targets (25- to 30-mer) or by base pairing between $(dT)_{25}$ oligonucleotide beads and poly-A-tailed target DNA. Simultaneous interrogations of four DNA targets (in duplicates) consecutively by five probes were achieved in eight microchannels within a microfluidic chip.[959]

In another report, hybridization of the DNA target to the biotinylated DNA probes immobilized on streptavidin-coated microbeads was performed. The beads (15.5 μm diameter) were retained in serial microchambers as defined by weirs and hydrogel (poly(ethyleneglycol) diacrylate) plugs (see Figure 9.14).[960] Subsequently, fluorescently labeled targets were electrophoretically transported to the chamber consecutively through the hydrogel plugs. Detection was achieved by fluorescent imaging of the fluorescein-labeled DNA targets that were hybridized to the probe immobilized on the beads (see Figure 9.15). Hybridization was 90% complete within 1 min. After hybridization, the targets in a particular chamber could be released for subsequent analysis after denaturation using a pressure-driven flow of 0.1 N NaOH into that chamber. Therefore, multiple analysis could be performed using the same bead set. However, there was degradation of performance because extended exposure of the microchambers to high-pH solution leads to (1) shrinkage of the hydrogel plugs and hence leakage of the targets, and (2) deterioration of streptavidin-biotin linkage and hence loss of the probes.[960]

In one report, DNA probes were directly immobilized into hydrogel plugs formed in a PC chip for hybridization. Acrylamide-modified 20-mer oligomers were incorporated in the hydrogel during its photopolymerization. The hydrogel was porous to a fluorescently tagged DNA target for hybridization under an electrophoretic flow.[961] Furthermore, DNA hybridization was conducted

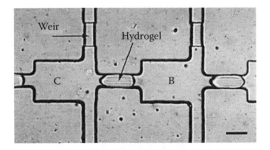

FIGURE 9.14 Optical image of the xPEG-DA hydrogel microstructures photopolymerized within the microchannels for sealing off microchambers (B, C), which also contained weirs to retain the beads. Scale bar: 100 μm.[960] Reprinted with permission from the American Chemical Society.

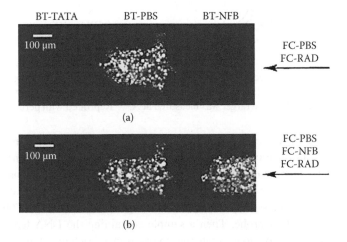

FIGURE 9.15 Fluorescence images of microbeads after hybridization with 100 ng/L of a mixture of (a) FC-PBS and FC-RAD targets and (b) FC-PBS, FC-NFB, and FC-RAD targets. Microbeads conjugated with BT-TATA (left), BT-PBS (middle), and BT-NFB (right) are packed in each microchamber shown in Figure 9.14. FC, fluorescein-labeled target; BT, biotinylated probe; PBS, NFB, RAD, and TATA are four different DNA sequences.[960] Reprinted with permission from the American Chemical Society.

with DNA-immobilized beads, which were embedded in a PDMS channel. The beads were partially embedded by pouring PDMS prepolymer on a molding master with the bead suspension spotted on the ridges of the master.[1024]

A DNA hybridization test that did not require labeling the target was reported. The test was based on the displacement of a fluorescently tagged indicator oligonucleotide, which was immobilized in a hydrogel plug, by the unlabeled target DNA. This is because the binding strength of the immobilized DNA probe (20-mer) to the unlabeled longer target DNA (20-mer) is stronger than that to the shorter indicator oligonucleotide (10-mer). The electrophoretic movement of the target DNA through the hydrogel plug resulted in a short sensing time (~2 min).[962]

In another report, detection of hybridization of an unlabeled target was achieved by exploiting laminar flows in microchannels. Two streams of FITC-labeled oligonucleotide probe and its unlabeled complementary target were created in an acrylic chip, as shown in Figure 9.16. Hybridization of the complementary target and probe occurred at the solution interface. However, at the many curving parts in a serpentine channel, the internal force of the fluid produced secondary flow, which disrupted the interface and moved the heavier duplex to the outer side of the curve (see Figure 9.16). This disruption occurring at one curve will be compensated at the next curve in the serpentine channel. Therefore, such a disruption or higher concentration of the duplex in one flow stream vs.

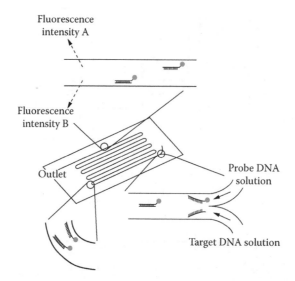

FIGURE 9.16 Principle and procedure of the sequence-specific DNA detection in the microfluidic system under laminar flow conditions. The labeled DNA was indicated by a circular dot. Fluorescence of the duplex was measured at B, and that of labeled target was measured at A.[1177] Reprinted with permission from the Royal Society of Chemistry.

the other (i.e., not at the interface) will prevail only after an odd number of turns. This is the basis of detection of DNA hybridization for the labeled duplex. Fluorescent detection of 0.50 pM target DNA was achieved and single-base mismatch for 10- to 70-mer DNA was demonstrated.[1177]

DNA hybridization was also conducted after PCR. Subsequent detection was done amperometrically in a chamber in a Si-glass chip. First, the working electrode (Au or ITO) in the chamber was immobilized with a DNA probe. Then a sample containing the DNA templates was added. Asymmetric PCR was carried out to generate the single-stranded rich target amplicon. If the amplicon is complementary to the immobilized probe, hybridization occurs. After washing off unhybridized PCR products, labeling (Au or Hoechst 33258) was conducted for detection. When the metal complex intercalator (Hoechst 33258) was used, differential pulse voltammetry was used to detect the presence of double-stranded DNA, indicating successful hybridization. In another method, a biotinylated primer was used in PCR to create the biotinylated amplicon. Hybridization was indicated by the presence of streptavidin-coated gold nanoparticles (10 nm). They were detected after silver deposition on the gold was conducted. Then linear sweep voltammetry was used to detect the silver dissolution current.[97]

9.3.2 MICROARRAY DNA HYBRIDIZATION

DNA hybridization in microchannels has also been conducted in the low-density microarray format. For instance, DNA hybridization was carried out in thirty-two chambers each spotted with sixteen DNA probes for the detection of bacteria (e.g., *Campylobacter jejuni*, *E. coli*) on a PMMA/PDMS-glass chip. PDMS was used as a bonding layer to seal the PMMA plate onto the glass microarray plate.[963]

In another report, DNA probes were spotted on a glass plate, which was then aligned and covered by the PDMS channel plate for hybridization to occur.[964]

A DNA microarray (512 spots) was spotted in the microchannel in a PC chip. After spotting, the channel was closed by lamination with a polyolefin foil for hybridization test. The volume of the hybridization channel (~2 cm long, 250 μm wide, and 100 μm deep) was 10 μl.[965]

On-chip hybridization to a high-density gene chip has been described in Section 9.2.2.

9.4 OTHER NUCLEIC ACID APPLICATIONS

9.4.1 DNA Sequencing

Because of the need for fast and high-throughput DNA sequencing in the human genome project, research in this area using the microfluidic chip has been very active. For instance, DNA sequencing of M13mp18 has been achieved on a glass chip using a denaturing sieving medium (9% T, 0% C polyacrylamide). Using the one-color detection, separation to 433 bases was achieved in 10 min. Using the four-color detection, separation to 150 bases was achieved in 540 s. In contrast, the sequencing rates by conventional slab gel and capillary electrophoresis are slower, 50–60 bases/h and 250–500 bases/h, respectively. However, twelve times more primers and templates are needed for the sequencing reactions performed in the CE chip than in conventional CE.[554]

Various experimental conditions have been optimized for DNA sequencing on a glass chip. These conditions include separation matrix (denaturing 3 to 4% linear polyacrylamide [LPA]), separation temperature (35 to 40°C), channel length (7.0 cm), channel depth (50 μm), and injector parameters (100 or 250 μm double-T injector, 60 s loading time). These optimal conditions facilitated the one-color detection of separation of five hundred bases of M13mp18 ssDNA in 9.2 min, and four-color detection of five-hundred-base separation in 20 min.[548]

Further optimization of DNA sequencing of M13mp18 DNA (up to 550 bases) on a microfabricated glass chip was achieved on separation buffer composition (no borate), sieving polymer concentration (>2% LPA), device temperature (≤50°C), electric field strength (lower E for longer bases), and channel length (>6 cm).[966] In another report, the dependence of various parameters, such as selectivity, diffusion, injector size, device length, and channel folding on the resolution of the LPA sieving medium for ssDNA separation, was investigated on a glass chip. Separations of 400 bases in under 14 min at 200 V/cm, and of 350 bases in under 7 min at 400 V/cm were achieved.[967]

When a longer CE channel (40 cm) on a glass chip was used, the average DNA read length (four-color detection) was successfully increased to eight hundred bases in 80 min (98% accuracy) for M13mp18 DNA (or the DNA with a 2 kbp human chromosome 17 insert). For comparison, a commercial capillary-based DNA sequencer took 10 h to complete the same eight-hundred-base analysis.[968] Moreover, DNA sequencing up to 700 bp (with 98% accuracy) has been achieved in 40 min using an 18 cm plastic microchannel.[969]

In one report, DNA sequencings up to 275 bases using one-color detection and up to 320 bases using four-color detection have been achieved on microchannels (4.5 cm long) constructed on a COC chip.[245] Moreover, DNA sequencing of clinical samples of human p53 cDNA was carried out on a fused silica chip using four-color detection. It took 20 min to complete the analysis, compared to 3.5 h using a commercial sequencer.[970] DNA sequencing was also carried out on a glass chip, with single-base resolution greater than six hundred bases achieved.[140]

A hybrid device was constructed for DNA sequencing. Other than a microchannel, a capillary tube was interfaced to a glass chip for separation.[114]

Whether removal of DNA templates was essential for high-quality separations in sequencing applications was investigated.[265] In another report, purification of DNA sequencing products to improve separation performance was attempted. For instance, a low-viscosity gel capture matrix, containing an acrylamide-copolymerized oligonucleotide, was loaded into a 60 nl capture chamber to purify DNA sequencing products prior to separation of sequencing fragments on a glass chip (see Figure 9.17).[971] The capture oligonucleotide, which was a twenty-base complementary sequence, bound to the sequencing products, while chloride, excess primer, and excess DNA template were unretained and were thus washed off. The process (at 50°C) took only 120 s to complete. The sequencing products were then released at 67°C and subsequently introduced into a 15.9 cm CE microchannel. CGE separation completed in 32 min to produce the sequencing information up to 560 bases. For comparison, conventional sample purification or cleanup has been performed by ethanol precipitation of DNA.[971]

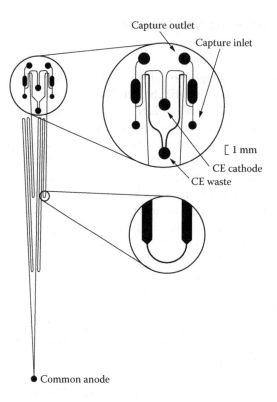

FIGURE 9.17 Design of the DNA sequencing channel integrated with a capture chamber. The injector has been modified to include the capture chamber consisting of capture inlet and capture outlet wells for oligonucleotide purifications. The CE waste and CE cathode provide voltage access points for electrokinetic injection and CE analysis. Four tapered turns allow an elongated serpentine separation channel geometry (effective length 15.9 cm) that minimizes turn-induced dispersion of the sequencing bands. The high-voltage anode well is located at the bottom.[971] Reprinted with permission from the American Chemical Society.

High-throughput sequencing has been achieved using microchips that consist of multiple channels. For instance, a sixteen-channel glass chip was used for automated parallel DNA sequencing. The use of a four-color confocal fluorescence scanner allows sequencing of 450 bases in 15 min in all sixteen channels. As shown in Figure 9.18, the channels radiate from the bottom. To ensure identical channel length, the bottom part was extended as shown in the inset of Figure 9.18. The top sample reservoirs are arranged in the same spacings as an eight-tip automatic pipette. Such channel arrangement is needed to ensure long enough channel length on the glass wafer for sufficient read length in DNA sequencing. Various experimental conditions have also been tested on such a chip. For instance, in order to ensure unbiased EK injection using the on-chip twin-T injector, the effect of injection time on separation was studied, as shown in Figure 9.19. Appreciable separation occurs using an injection time of at least 20 s, but fragments beyond 267 bases were not injected and not seen. At 30 s injection, fragments up to 550 bases could be separated, but fragments higher than 550 bases did not reach steady-state injection. This steady-state injection for all fragments was seen at 50 s injection, which remained steady even at 70 s injection. These experiments were performed at low field strength and at room temperature to magnify the injection time effects, and so the separation time was long. Actual separation was conducted at elevated temperature (i.e., 50°C) to not only accelerate separation, but also to reduce compressions and improve resolution because higher temperature increased denaturation of DNA.[265]

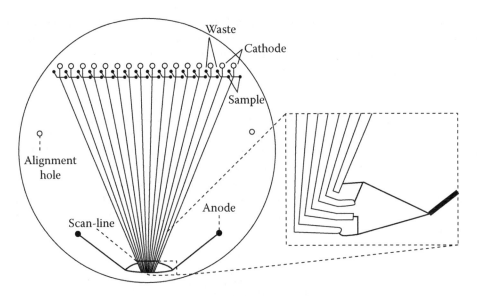

FIGURE 9.18 Mask design of the sixteen-channel CE chip for parallel DNA sequencing. The wafer used has a diameter of 10 cm and a thickness of 1.1 mm. Sixteen identical 250 μm twin-T injectors are used for all channels. All lines on the mask have a width of 10 μm. A final channel width of ~110 μm is obtained with a depth of 50 μm. All reservoirs are formed by access holes drilled on the etched wafer. The diameter of the access holes is ca. 1.4 mm for sample and waste, and 2.1 mm for cathode and anode, which respectively correspond to ca. 1.7 μl and ca. 3.8 μl volume reservoirs. The alignment holes have a diameter of 2.1 mm. The scan line represents the scanning location of the detector. Inset: Detail of the compensation of different channel lengths for the group of eight channels on the right.[265] Reprinted with permission from the National Academy of Sciences.

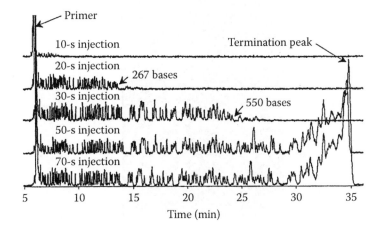

FIGURE 9.19 Effect of injection times on steady injection states of different-sized DNA fragments separated on the chip shown in Figure 9.18. The arrows point to the last peaks that reached steady injection states. The separations were performed at ambient temperature by using 4% LPA and an electric field strength of ca. 0.180 V/cm for injection and separation. The data shown were obtained from a channel with an effective separation distance of 7.46 cm.[265] Reprinted with permission from the National Academy of Sciences.

In a later report, a ninety-six-lane CE glass chip was used to carry out DNA sequencing. Tapered turns were employed to increase the channel length, and hence the DNA read length.[972] More recently, a 384-lane plate for DNA sequencing has also been reported.[973]

9.4.2   Genetic Analysis

The microfluidic chip uniquely facilitates high-throughput genetic analysis. For instance, twelve different samples were analyzed in parallel by capillary array electrophoresis (CAE); see Figure 9.20. The two-color method was used for the multiplex fluorescent detection of the HLA-H DNA and the sizing ladder, pBR322 Msp1. While the sizing ladder was prelabeled with butyl TOTIN (emitting in both green and red), the DNA sample was on-column labeled with thiazole orange (emitting in green).[974] A reciprocating confocal fluorescence scanner was used to detect twelve channels at a rate of 0.3 s/scan.[974] The use of the thirteen-channel[975] or sixteen-channel chips[976] for DNA sequencing was also reported.

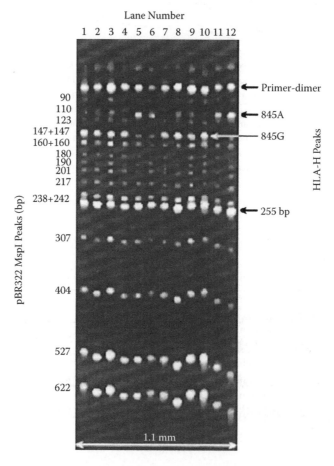

FIGURE 9.20   Separation of twelve HLA-H genotyping samples. The 845G type shows a single band at 140 bp; the 845A type shows a single band at 111 bp; the heterozygote type, 845A/G type, exhibits both the 140 bp and 111 bp bands The samples contain the DNA ladder, pBR322 MspI (2 ng/µl). The migration times of the lanes have been adjusted to align the HLA-H peaks, and the fluorescence signals have been normalized for improved viewing clarity. The standard ladder peaks are labeled on the left side of the image and the HLA-H peaks are labeled on the right side. Genotyping results: Lanes 1, 4, 7, and 10 are 845G/G; lanes 2, 3, 8, and 9 are 845G/A; lanes 5, 6, 11, and 12 are 845A/A.[974] Reprinted with permission from the American Chemical Society.

Thereafter, a 10 cm glass wafer with forty-eight separation channels (but with ninety-six samples) was used to perform DNA genetic analysis in less than 8 min, which translated to <5 s/sample (see Figure 9.21). The variants of the HFE gene, which were correlated with hereditary hemochromatosis (HHC; iron storage disorder), were detected using DNA that was isolated from peripheral blood leukocytes.[557] An elastomer sample loading array and electrode array were used for sample loading to the forty-eight-lane chip.[108]

Following this, ninety-six-channel CAE was achieved on a microchip by adopting a radial design in the channel arrangement. The channels shared a common anode reservoir at the center of a 10 cm glass microplate (see Figure 9.22). Grouping the sample, cathode, and waste reservoirs reduced the number of reservoirs per plate to 193. A four-color rotary confocal fluorescence scanner was used for detection (see Figure 7.3 in Chapter 7). CGE separation of the pBR322 restriction digest samples was achieved in 120 s.[977] CGE separation of the variants of the HFE gene was also achieved.[978] Loading of ninety-six samples to this chip from a ninety-six-well microplate was achieved using a capillary array loader.[979]

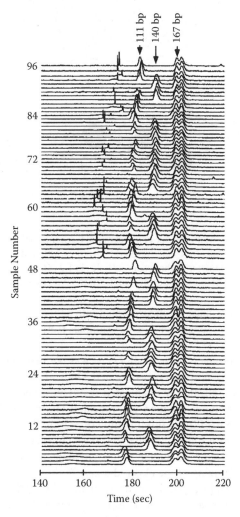

FIGURE 9.21 Electropherograms generated from ninety-six concurrent separations obtained in a glass chip with forty-eight channels. The various samples show the genotyping of the HFE gene. The 845G type shows a single peak at 140 bp; the 845A type shows a single peak at 111 bp; the heterozygote type exhibits both the 140 and 111 bp peaks.[557] Reprinted with permission from the National Academy of Sciences.

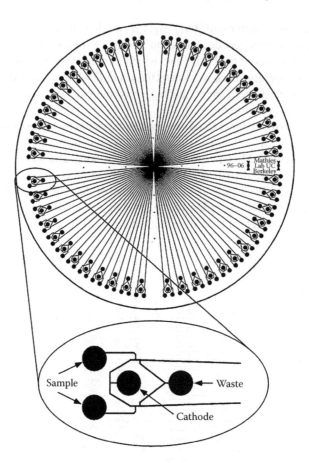

FIGURE 9.22 Mask pattern for the ninety-six-channel radial CAE microplate (10 cm in diameter). Separation channels with 200 μm double-T injectors were masked to 10 μm width and then etched to form 110 μm wide by ~50 μm deep channels. The diameter of the reservoir holes is 1.2 mm. The distance from the injector to the detection point is 33 mm.[977] Reprinted with permission from the American Chemical Society.

Furthermore, 384 capillary lanes are incorporated into a 20 cm diameter glass wafer to perform simultaneous genotyping. CGE separation of DNA samples from 384 individuals was performed on a 8 cm long channel in only 325 s.[980]

CGE separation of PCR products of the FMRI gene was performed on a PMMA chip. This gene was to genotype the fragile X syndrome consisting of (CGG)n alleles. To achieve a fourfold increase in fluorescent detection, both covalently conjugated dye (Cy 5) and intercalating dye (TOPRO-3) were used. Based on the copy number in the CGG-repeat, (CGG)n, the FMRI is classified as normal ($n < 54$), premutation ($n = 55$ to 200), or full mutation ($n > 200$).[981]

DNA analyses on a twelve-sample one-channel glass chip have been performed for (1) CGE separation of the PCR products of the human monocyte antigen (CD14) gene, adenovirus 2 DNA, and human p53 gene, and (2) sizing of the plasmid digest.[982]

Single nucleotide polymorphism (SNP) analysis of the p53 cDNA from clinical samples was analyzed by CGE separation on a fused silica chip. No sample cleanup was necessary because of the more robust EK injection conducted on-chip. The analysis was completed in 102 s, which was fifty times faster than slab gel electrophoresis, and ten times faster than conventional CE, all with similar accuracy.[970]

A multiplex genetic analysis of short tandem repeats (STRs) was carried out in a glass chip.[543,983,984] As required by the FBI's Combined DNA Index System (CODIS) for forensic identification, all thir-teeen loci of STR are needed.[543]

Point mutation detection by restriction fragment length polymorphism (RFLP) mapping was conducted in a glass chip.[982] The RFLP method was also used on a glass chip to detect homozygous mutant genotypes. These are commonly encountered with the gene implicated in hereditary diseases such as hemochromatosis. The RFLP method involves a restriction enzy-matic digestion followed by capillary gel electrophoretic sizing. This method complements the heteroduplex analysis, which is usually used to detect mutation in heterozygous, but not homozygous, mutants.[534]

Digestion of pBR322 (125 ng/μl) by HinfI (4 U/μl) and subsequent CE analysis were integrated in a chip at room temperature (20°C). Fluorescent detection of the restriction fragments was achieved using an intercalating dye TOTO-1 (1 μM).[316]

In one report, plasmid DNA (supercoiled Bluescript SK) was digested by *Taq*I restriction enzyme in a Si-glass device with two channels. The flow streams of DNA and enzymes were first introduced in individual channels, and then moved to mix and stopped to react (at 65°C for 10 min).[392]

In another report, restriction digest fragments (*Kpn*I) of a 62 bp DNA sequence were separated on a five-channel microchip. A sampling capillary allowed for continuous monitoring of the restric-tion digestion of the DNA sample.[320]

9.4.3 Separation of Large DNA Molecules

Single-molecule DNA fragments (2 to 200 kbp) were sized on a PDMS chip based on the differences in the molecule size (length), but not in the electrical mobility.[985] DNA fragments were obtained from λ phage DNA, which was digested with *Hin*dIII or ligated with the T4 ligase. DNA molecules were allowed to pass through a narrow channel of 5 μm wide (see Figure 9.23), so only one mol-ecule could pass through the channel at a time.[1178] Fragment lengths were measured by the amount of intercalating dye, YOYO-1 (one molecule per 4 bp), using quantitative fluorescent measurement. It was found that 28 fg of DNA (about three thousand molecules) was detected in a volume of 375 fl in 10 min. This method is a hundred times faster than pulsed-field gel electrophoresis for DNA separations.[985]

Single λDNA molecules (41 pM) were detected as bursts on-chip using the intercalating dye YOYO-1.[159] Single-molecule detection of λDNA (310 fM) and other small DNA molecules was also performed on PC or PMMA chips, using TOPRO-5.[986]

With micromachining, a physical sieving matrix was designed for the analysis of long DNA molecules. An array of microfabricated microposts facilitated the study of the electrical mobility of long DNA molecules under an electrical field in a well-controlled sieving matrix (unlike the polymeric gel matrix).[987] An array of 0.15 μm high posts (of 1.0 μm diameter and 2.0 μm spacing) was microfabricated on a Si wafer, which was later bonded to Pyrex glass. The effective pore size of 1.0 μm corresponds roughly to a 0.05% agarose gel, which is never realized because it is physically unstable. The DNA used in the study was a highly purified 100 kbp DNA (~30 μm contour length) from *Micrococcus luteus*.[987] For information, the pore diameter of a 1% agarose gel is ~500 nm, which is applicable for separation of DNA of a size range of 250 to 12,000 bp.[351] Tribranched DNA molecules (from bacteriophage λDNA) were also studied in a microfabricated micropost array.[988] A similar array was also fabricated on a Si chip to sort DNA based on diffusion.[989]

A microfabricated entropic trap array was constructed for separating long DNA molecules. The channel comprised narrow constrictions (of width 75 to 100 nm) and wider regions (of width 1.5 to 3 μm) that allowed size-dependent trapping and separation of two DNA molecules (37.9 and 164 kbp). Surprisingly, it was found that a longer DNA molecule had a higher probability to escape the constriction, and hence had a greater mobility. The separation was completed in 15 min, which was

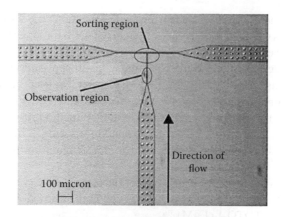

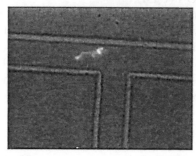

FIGURE 9.23 Optical micrograph of a T-channel device used for DNA separation. The large channels have lateral dimensions of 100 μm, which narrow down to 5 μm at the T-junction. The depth of the channels is 3 μm. In early prototypes, we found that because of the large aspect ratio (100 μm in width by 3 μm in depth), some of the elastomer channels would bow and pinch off by sealing directly to the glass. This problem was remedied in later versions by adding support pillars to the mask that would prop up the large channels and prevent bowing. Bottom inset: Magnified view of the T-junction in which the fluorescent image of a DNA molecule is apparent. The channels are 5 μm wide at this point. Note the high fidelity of the elastomer replica.[1178] Reprinted with permission from S. Quake.

much faster than the 12 to 24 h achieved in pulse-field slab gel electrophoresis.[990] A modeling study to rationalize the entropic trapping mechanism was reported.[991]

A Brownian ratchet array (with micrometer-sized obstacles) was constructed for separation of DNA molecules. These ratchets permit Brownian motion in only one direction. As shown in Figure 9.24, while the main flow is vertically down (see inset), the ratchets are tilted at an angle to the flow direction. Any Brownian motion to the left will easily be deflected back to the main flow. However, when the Brownian motion is to the right, the molecule will be deflected to slide along another ratchet and displaced to the gap to the right, delaying the migration time. Since smaller DNA (λDNA, 48.5 kb) has a greater diffusion coefficient, it has a higher probability of being ratcheted away and migrates more slowly. But the success of separation relies on sufficient time for the molecules to diffuse over the distance required for ratcheting. This can be achieved by either reducing the fluid flow or increasing the tilting angle, θ_{tilt}, or both, as shown in Figure 9.25. However, the process is slow, and it takes ~70 min to achieve a separation resolution of ~3.8.[686]

Different nanostructured regions were designed in a Si chip for separation of small and large DNA molecules (i.e., 2, 5, 10 kbp). The separation was based on either the sieve-type separation or size-exclusion-type separation. The sieve-type separation was achieved using a regular nanostructured matrix (see Figure 9.26). Long DNA molecules, which meandered through the regular array of pillars, migrated late (see Figure 9.26b). On the other hand, the size-exclusion-type separation

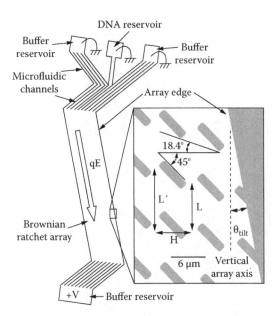

FIGURE 9.24 The device consisting of a Brownian ratchet array for DNA separation. The flow angle, θ_{tilt}, is 10.8° with respect to the vertical array axis. H, L, and L' were 6, 8, and 10 μm, respectively. The obstacles were ~5.6 μm long, ~1.4 μm wide, and ~3.2 μm tall.[686] Reprinted with permission from the American Chemical Society.

was achieved via the wide and narrow gaps (see Figure 9.27). Small DNA molecules enter the narrow gaps and slow down, so they migrate later (see Figure 9.27c).[992]

Nanopillars were fabricated on a quartz chip to carry out separation of large DNA fragments (see Figure 9.28). For instance, λDNA (48.5 kbp) and T4 DNA (166 kbp) were separated without the use of a pulsed-field technique. The nanopillars consist of 500 nm diameter posts of 500 nm spacings. Figure 9.29 illustrates the separation of the two DNA molecules over the first 380 μm long nanopillar region. In Figure 9.29A, fluorescent images show that the separation becomes apparent at 8 s, with the 166 kbp DNA migrating later and at lower fluorescent intensity. This is also depicted in Figure 9.29B, in which the x-axis is the distance from the injection point. The first detection point is in the first nanopillar-free region, which is at the end of the first nanopillar region (see Figure 9.28). This is designed to avoid loss of fluorescent intensity due to light scattering. At this detection point, the two DNA molecules are well resolved, as shown in Figure 9.29C. Such a DC separation for the two DNA molecules cannot be achieved in agarose because it loses its resolving power for DNA molecules longer than ~40 kbp. Using the nanopillar structures with wide spacings, even the 48.5 kbp λDNA molecule exists essentially as a compact conformation and migrates faster, as shown in Figure 9.30A. As for the bigger T4 DNA molecule, it occasionally loops around the nanoposts and forms a U-shape conformation (see Figure 9.30B). However, if an agarose gel with smaller pore size is used, both molecules will appear as U shapes in conformation and migrate with similar velocity. On the other hand, the nanopillar structure cannot be used to separate shorter DNA molecules (i.e., 200, 1,000 bp), unless much smaller nanopillars (~100 nm) with smaller spacings are constructed.[351]

DNA separation was also achieved in nanostructured regions that were constructed by magnetic beads (700 nm). These beads were self-assembled with 5.7 μm spacings in the presence of a magnetic field. Then separation of DNA molecules (15, 33.5, and 48.5 kbp) was conducted in such an environment.[993,994] In another report, separation of λ and T4 DNA was conducted in a sieving matrix formed by smaller-sized magnetic beads (570 nm) that were self-assembled in a PDMS chip.[995]

A medium that contained core-shell-type nanospheres (30 nm) was employed to separate DNA fragments (up to 15 kbp) within 100 s. The nanospheres contain a hydrophobic poly(lactic acid)

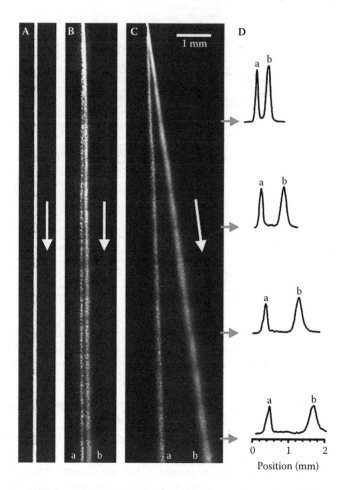

FIGURE 9.25 Fluorescent micrograph of the separation of 48.5 and 164 kb DNA in Brownian ratchet arrays. The molecules were injected at a single point at the top of the array and run at different flow conditions. The array axis is vertical in all micrographs. The arrow shows the direction of the ionic flow θ_{tilt}. (a) Zero flow tilt angle ($\theta_{tilt} = 0°$) and high speed (~24 µm/s). (b) Zero flow tilt angle ($\theta_{tilt} = 0°$) and low speed (~1.5 µm/s). (c) Tilted flow ($\theta_{tilt} = 7.2°$) at ~1.5 µm/s. (d) Electropherograms of C measured 3, 6, 9, and 12 mm from the injection point. Band assignment for DNA: (a) 164 kb, (b) 48.5 kb.[686] Reprinted with permission from the American Chemical Society.

(PLA) core and a hydrophilic PEG shell. In addition, a double pressurization technique was employed for stacking DNA fragments before separation. As shown in Figure 9.31, such a stacking effect only occurs when using the nanosphere medium (left), but not the conventional methylcellulose medium (right). The DNA sample was introduced into the vertical channel by initial pressure (P_{1st}) application. Just before the electrophoretic separation step, a second pressure (P_{2nd}) was applied, causing the sample at the intersection to advance further as a dispersed band in the absence of an electric field. Subsequently, CGE separation was conducted without using pressure. The electrofocusing, which is observed only in the nanosphere medium, seems to be a stacking effect of the packed nanospheres (close-packed structure occurs at 1% concentration). Figure 9.32 shows the separations of various DNA molecules using the nanosphere and conventional media.[996]

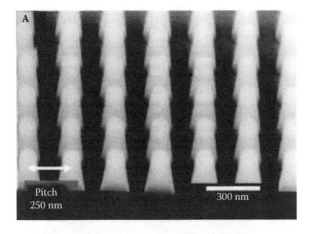

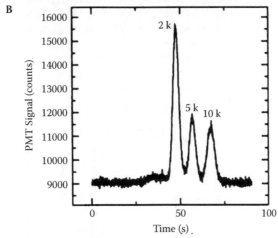

FIGURE 9.26 Nanostructured region constructed of regular pillar array for sieve-type separation. (a) Scanning electron micrograph of regular pillar array with 250 nm pitch. (b) Electropherogram of DNA molecules with sizes of 2, 5, and 10 kbp, applied voltage of 30 V, and separation length of 0.5 mm.[992] Reprinted with permission from the American Institute of Physics.

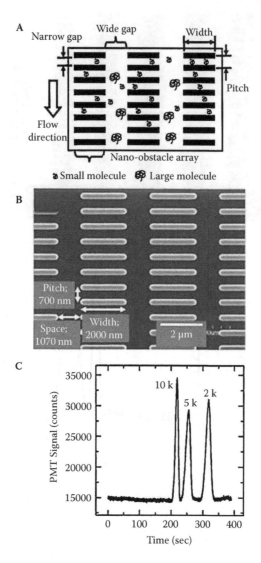

FIGURE 9.27 Nanostructured region for size-exclusion-type separation. (a) Principle of separation in the nanostructure region consisting of narrow and wide gaps. Larger molecules move only in the wide gaps, while smaller ones move in both the narrow and wide gaps. (b) Scanning electron micrograph of nano-obstacle arrays with 700 nm pitch and 1,070 nm wide gaps. (c) Electropherogram of three DNA molecules with sizes of 2, 5, and 10 kbp, applied voltage of 40 V, and separation length of 4 mm.[992] Reprinted with permission from American Institute of Physics.

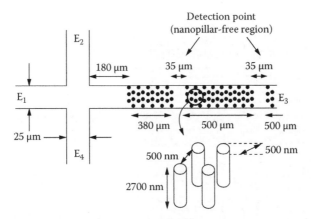

FIGURE 9.28 Schematic representation of a microchannel equipped with nanopillars for separation of large DNA molecules. Each nanopillar (a black dot) diameter, spacing, and height used here were 500, 500, and 2,700 nm, respectively. Short nanopillar-free regions were placed at regular intervals for detection without the problem of fluorescent light scattering.[351] Reprinted with permission from the American Chemical Society.

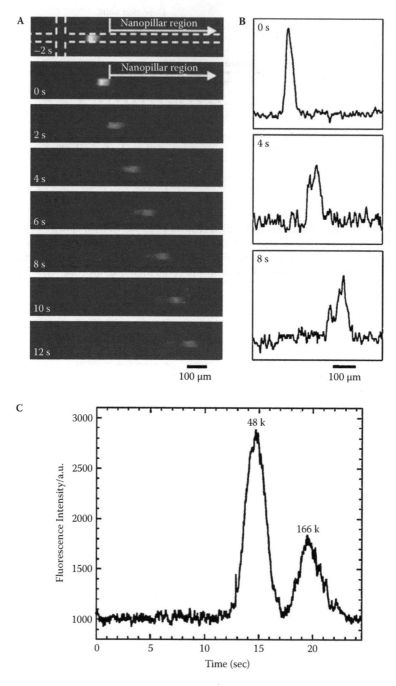

FIGURE 9.29 (a) Consecutive fluorescence images of λDNA (right) and T4 DNA (left) separation at 5 kV after injection into a nanopillar region. White dotted lines show an external view of the microchannel. At 0 s, sample plug began to inject into the nanopillar region. (b) Fluorescence intensity profiles along a migration direction. Each profile corresponds to the left fluorescence images. (c) Electropherogram of λDNA (48 kbp) and T4 DNA (166 kbp) separation at the first detection point (380 μm below from the nanopillar entrance). Conditions: In sample loading phase, potential at E_2 well was 300 V, and E_1, E_3, and E_4 wells were grounded. See Figure 9.28 for the locations of E_{1-4}. In separation phase, potentials at E_2, E_3, and E_4 wells were 500, 5,000, and 500 V, respectively, and the E_1 well was grounded.[351] Reprinted with permission from the American Chemical Society.

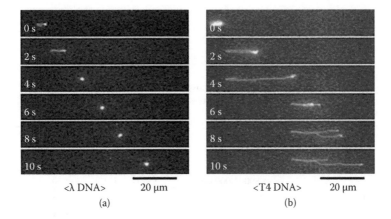

FIGURE 9.30 Fluorescence images of (a) a λDNA and (b) a T4 DNA migrating in a nanopillar region at 7 V/cm.[351] Reprinted with permission from the American Chemical Society.

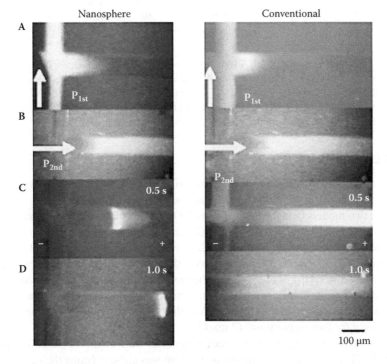

FIGURE 9.31 Fluorescent images of DNA separations using a nanosphere-containing medium and a double pressurization technique. Left: Separations in 1% nanosphere solution. Right: Separations in conventional 0.5% methylcellulose solution. Fluorescence images taken during application of the double pressurization technique. We observed the sample plug formation (20 ng/μl of 200 bp DNA with 0.8 μM of the fluorescent dye YOYO-1 in 4% 2-mercaptoethanol in TE buffer). Arrows show the pressure application. The horizontal channel is a separation channel. (i) P_{1st}: 2.5 kPa, 1 s; after filling all channels with solution the sample was injected by P_{1st} for 1 s using a syringe from the bottom well of the loading channel, with the top and right wells open. (ii) P_{2nd}: 2.5 kPa, 1 s; the sample band introduced into the cross section was advanced by P_{2nd} for 1 s from the left well and in the right direction with the top and bottom wells open. In the case of the nanosphere system, this operation places the sample between the layers of the nanospheres (left). (iii) Electrophoretic separation: 160 V/cm, 0.5 s. (iv) Electrophoretic separation: 160 V/cm, 1 s.[996] Reprinted with permission from Nature Publishing Group.

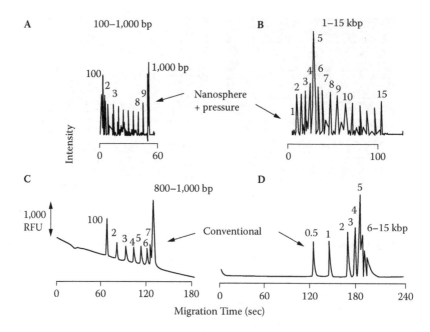

FIGURE 9.32   Electropherograms of DNA separations conducted using (a) nanosphere solutions (1% or 10 mg/ml at pH 9) vs. (b) conventional polymers. (a) A solution of a 100 bp dsDNA ladder containing ten fragments (100 to 1,000 bp, 1 µg/ml); P_{1st}, 2.5 kPa; P_{2nd}, 2.5 kPa. (c) A solution of a 1 kbp dsDNA ladder containing 15 fragments (1 to 15 kbp, 1 µg/ml); P_{1st}, 10 kPa; P_{2nd}, 10 kPa. (b) The 100 bp dsDNA ladder. (d) A solution of a 1 kbp dsDNA ladder.   Reprinted with permission from Nature Publishing Group

9.5   PROBLEM SETS

1. There is a limit of DNA size in CGE separation of DNA. Describe two conventional methods and one microchip method to overcome such a limit. (4 marks)
2. Name the two models (and the assumptions for the molecular shape) for explaining separation of DNA of different lengths in a gel. (2 marks)
3. In Figure 9.22 in Chapter 9, Section 9.4.2, showing the ninety-six-channel chip for high-throughput CGE analysis of oligonucleotides, the sample injection area is expanded in the inset. Describe, in terms of the voltage schemes, how the samples are introduced and how the subsequent separations are carried out. Note the voltage polarity. (2 marks)
4. Why cannot oligonucleotides of different lengths be resolved in free-solution CE, and a gel must be used for their separation? (2 marks)
5. In high-throughput CGE, ninety-six radial channels were created on a 200 mm diameter glass disk (Figure 9.22). State why the number 96 was chosen. Name the next higher numbers used. (2 marks)

10 Applications to Protein Analysis

Research on microchip protein analysis has been very active for cellular protein functional assay, clinical diagnostics, and proteomics studies. Once again, the microfluidic technology plays an important role in protein assays. Immunoassay, protein separation, and enzymatic assay will be described in detail in subsequent sections.

10.1 IMMUNOASSAY

Immunoassay involves the use of the highly specific antibody–antigen interaction for assay of the antigen or the antibody. This method is generally classified into the homogeneous and heterogeneous formats.

10.1.1 Homogeneous Immunoassay

Homogeneous immunoassay is conducted entirely in solutions. In this format, separation of the labeled antibody (Ab) and the antibody–antigen (Ab-Ag) complex is required, and this is usually achieved using capillary electrophoresis (CE) separations. For instance, separation of Cy5-bovine serum albumin (BSA) from unreacted Cy5 and the complex (formed from Cy5-BSA and anti-BSA) was achieved in a flow-through sampling chip.[567]

In addition, affinity CE separation was carried out to resolve monoclonal anti-BSA and BSA in dilute mouse ascites fluid. Then the affinity constant of this antigen–antibody interaction was measured.[1004]

Separation of antigen and antibody–antigen complex was also achieved by dialysis with a polymeric microfluidic chip containing a poly(vinylidene fluoride) (PVDF) dialysis membrane.[821]

In competitive homogeneous immunoassay, separation and quantitation of free and bound labeled antigen (cortisol) were carried out in a fused silica chip. Since the antibody–antigen complex was not detected, an internal standard (fluorescein) was added to aid quantitation. In addition, since most of the total cortisol was bound in the serum, a releasing agent, 8-anilino-1-naphthalenesulfonic acid (ANS), should be added.[1006] In other reports, competitive immunoassay for BSA was demonstrated after performing a CE separation on-chip.[105,1005]

Competitive immunoassay was also conducted for serum T4 (3,5,3',5'-tetraiodo-l-thyroxine) on a fused silica chip. Again a T4-releasing agent (TAPS) was used to release T4 from serum.[152]

A robust assay for theophylline was established within the clinical range of 10 to 20 µg/ml.[330] Since the Ab-Ag complex can be mobilized, no additional internal standard was needed,[330] as in an immunoassay study for cortisol.[1006]

Simultaneous immunoassays for ovalbumin and anti-estradiol were performed on a six-channel microfluidic device within 60 s (see Figure 10.1). The limit of detection of anti-estradiol was determined to be 4.3 to 6.4 nM.[1007] Later work reduced the LOD of the antibody to 310 pM, which equaled 2,100 molecules. This compared with a theoretical detection limit of 125 to 525 pM.[1008] In another report, chicken egg ovalbumin (600 nM) was determined by immunoassay on a Pyrex glass chip using Cy5-labeled anti-ovalbumin antibody (200 nM).[678]

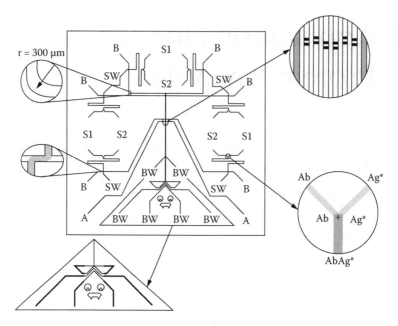

FIGURE 10.1 Overall layout of six units of reaction cell/separation channel in the glass SPIDY device used for immunoassay. Each reaction cell has reservoirs for sample (S1), antibody (S2), sample waste (SW), and run buffer (B). The two alignment channels are given as (A). Right bottom inset illustrates a mixer in the reaction cell in which the antibody (Ab) and antigen (Ag) are mixed. Left middle inset illustrates the double-T injector design for loading sample from the reactor into the separation column. Left upper inset shows the smooth turn at the separation channel. Left bottom inset depicts the pattern of channels at the buffer waste (BW) reservoirs used for separations. The thicker lines identify channel segments that are 300 μm wide, while the thinner lines represent 50 μm wide segments. Right upper inset shows the detection zone, across which a laser beam is swept. Separation of two components in the six separation channels is depicted. Two outermost channels are optical alignment channels (A) that are filled with fluorescent dye.[1007] Reprinted with permission from the American Chemical Society.

In one report, a glass chip consisting of a precolumn reactor, a CE separation channel, and a postcolumn reactor was used for an electrochemical enzyme immunoassay (see Figure 10.2). While the precolumn reactor was used for the binding reaction of mouse IgG by the ALP-labeled antibody (goat anti-mouse IgG), the postcolumn reactor was used for the enzymatic conversion of the ALP substrate, 4-aminophenyl phosphate (APP). Amperometric detection of APP was achieved using a carbon ink electrode. To prevent protein adsorption during CE separation, a surfactant, Tween-20, was added to the run buffer, resulting in an approximate threefold shorter analysis time. Since the free antibody was well separated from the complex, the amounts of both species could be used for quantitation. A remarkably low LOD for mouse IgG was achieved, 2.5×10^{-16} g/ml (or 1.7×10^{-18} M).[1009]

Human IgM was determined using a homogeneous immunoassay by mouse anti-human IgG on a poly(methylmethacrylate) (PMMA) chip. CE separation of IgM, IgG, and their complex was achieved and they were detected using a contactless conductivity detector (see Figure 10.3). An LOD value of 34 ng/ml IgM was obtained on the chip, which was different from the 0.15 ng/ml obtained in a capillary tube.[1010]

Immunoassay of goat IgG can also be achieved on a PMMA chip.[1011] The PMMA chip should be placed on a piece of electrically grounded aluminum foil to reduce the electrostatic charges acquired by the plastic surface; otherwise, the fluid flow by voltage control cannot be performed. PMMA was selected because it was reported to be the least hydrophobic of the commonly available plastic materials.[1011]

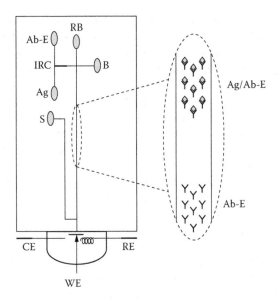

FIGURE 10.2 Schematic of the immunoassay chip. Notations for chip layout: RB, run buffer; Ab-E, enzyme-labeled antibody; Ag, antigen; S, substrate; IRC, immunoreaction chamber; B, unused reservoir. Notations for detector: RE, reference electrode; CE, counterelectrode; WE, working electrode. The inset shows a schematic diagram of the separation of Ab-E from the Ag/Ab-E complex.[1009] Reprinted with permission from the American Chemical Society.

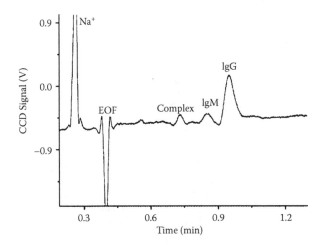

FIGURE 10.3 Electropherogram of a homogeneous immunoassay obtained with the glass chip. Conditions: 10 µg/ml IgG and 10 µg/ml IgM in 20 mM TAPS/AMPD buffer containing 0.01% of Tween-20; injection, 1 kV for 5 s; actuator voltage, 400 V_{p-p}, frequency, 50 kHz; separation potential, 4 kV, 8 cm length.[1010] Reprinted with permission from the American Chemical Society.

Other than protein, histamine in human blood samples was determined by the immunoassay method. It was detected by its binding with anti-histamine IgG, which was coupled to ferrocene (Fc-IgG). Separation of histamine and its complex with antibody was based on pI differences, and it was achieved in a PMMA chip consisting of a multichannel matrix column coated with a cation-exchange resin (Nafion). Histamine was detected electrochemically; a decrease in the current due to Fc-IgG occurred after its binding to histamine.[1012]

An immunoassay for 2,4,6-trinitrotoluene (TNT) and its analogs was carried out on a glass chip. A competitive assay was adopted in which TNT or its analogs competed with fluorescein-labeled TNB (TNB-Fl) for the anti-TNT antibody. The ratio of fluorescent intensity of the free TNB-Fl to that of the complex was plotted against the TNT concentration, and the LOD was determined to be 1 ng/ml.[285] Dissociation kinetics of the antigen-antibody complex was also studied, thanks to the use of multiple detection points (1 to 6) along a serpentine channel (see Figure 10.4). An excess of anti-TNT antibody was added to completely bind with TNB-Fl, and fluorescein was used as an internal standard. The dissociation of the complex produced a decrease in the fluorescent intensity due to the complex (see Figure 10.5). It was observed that along the channel from points 1 to 6, not only did the amount of the complex decrease (relative to fluorescein), but the separation between these two species also increased. Although TNB-Fl was also released upon complex dissociation, the continuous nature of the release did not result in a detectable sharp peak of TNB-Fl.[285]

Homogeneous immunoassay for TNT and its analogs was also achieved in a multichannel format using a ninety-six-lane capillary array electrophoresis (CAE) chip.[1013]

A human serum albumin (HSA) assay was performed on a T-sensor (on a Si-Pyrex chip) in a flow stream (see Figure 10.6).[442,443] The left reference solution (1 μM HSA) was used as a control. The right sample solution (HSA) was used at various concentrations (0, 2, 4, 6, 8 μM). The center stream contained the indicator solution (Albumin Blue, AB 580: λ_{ex} = 580 nm, λ_{em} = 606 nm). Since AB 580 produced an enhanced fluorescence upon binding to HSA because the indicator has

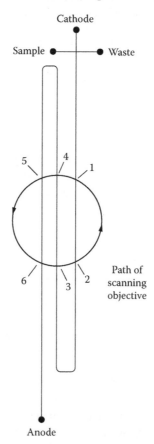

FIGURE 10.4   Layout of microfabricated capillaries employed for immunoassays. Serpentine capillary with 22 cm separation length. The path of the rotary scanning objective for fluorescent detection is presented with the six detection points marked 1 to 6.[285] Reprinted with permission from the American Chemical Society.

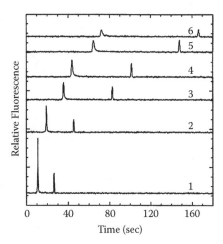

FIGURE 10.5 Electropherograms monitoring the dissociation of the Ab–TNB-Fl complex monitored at six consecutive detection points along the separation CE channel (see Figure 10.4 for locations) by using the rotary fluorescent scanner.[285] Reprinted with permission from the American Chemical Society.

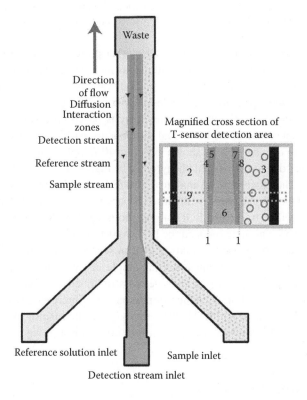

FIGURE 10.6 Schematic diagram of flow and diffusion within the T-sensor at a 1:1:1 flow ratio. A reference solution enters the device from the left, a detection solution from the middle, and a particle-laden sample stream from the right. The inset shows (1) original flow boundaries, (2) reference stream, (3) particle-laden sample stream, (4) diffusion of detector substance into reference stream, (5) diffusion of reference substance into detector stream, (6) detection stream, (7) diffusion of sample analyte into detection stream, (8) diffusion of detector substance into sample stream, and (9) detector window.[443] Reprinted with permission from B. Weigl.

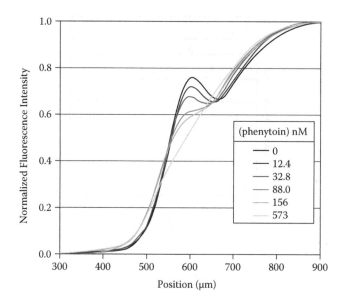

FIGURE 10.7 Competitive diffusion immunoassay analysis for phenytoin using a T-sensor as shown in Figure 10.6. Plots of fluorescence intensity vs. channel width dimension, as the concentration of unlabeled phenytoin increased from 0 to 573 nM. Reprinted with permission from B. Weigl.

a high affinity for serum albumin, but a low one for other proteins,[442] the three streams interacted with each other based on diffusion, and so simultaneous measurements of the reference and sample solutions were feasible.[443,444,1014] This T-sensor method was also used for a competitive diffusion immunoassay of an antiepileptic drug molecule, phenytoin. The assay was achieved with a blood sample without the removal of blood cells. This assay is based on a slower diffusion rate of the antigen–antibody complex than of the individual components, causing the accumulation of the complex along the center of the channel. As shown in Figure 10.7, high fluorescent intensity on the right side of the channel was due to the labeled phenytoin. The high level of the labeled antibody–antigen complex was present when there was no analyte (i.e., 0 nM unlabeled phenytoin); see the blue trace. A higher level of unlabeled phenytoin will cause less accumulation of labeled antigen complex at the interface; see the yellow trace. The LOD of phenytoin was 8.1 nM when the antibody used was 10%. A reduction to 7.5% antibody reduced the LOD to 0.43 nM.[1015]

Immunoassay has also been used to analyze insulin secreted from mouse islet of Langerhans. An islet was placed in a reservoir that led to a sample channel in a chip. Insulin, which was secreted from the pancreatic β cells in the islet as stimulated by glucose (3 to 20 mM), was analyzed using a competitive homogeneous immunoassay. Fluorescein isothiocyanate (FITC)-insulin (150 nM) and anti-insulin (75 nM) were mixed with the insulin secreted from the β cell. Tween-20 was added to the run buffer to prevent protein adsorption and to enhance CE separation reproducibility.[139]

10.1.2 Heterogeneous Immunoassay

One form of heterogeneous immunoassay is called enzyme-linked immunosorbent immunoassay (ELISA). In one instance, electrochemical immunoassay was performed for anti-ferritin (antibody) in a poly(dimethylsiloxane) (PDMS)/PMMA chip. First, 3,3-dithiopropionic acid bis-*N*-hydroxy-sulfosuccinimide ester (DTSSP) was self-assembled on the gold electrode deposited on the PMMA plate. Then horse spleen ferritin (antigen) was attached to the DTSSP layer. A 100 μg/ml solution of anti-horse ferritin (rabbit serum) was added. Then a secondary anti-rabbit antibody (horseradish peroxidase [HRP] linked) was introduced. A substrate (4-CN) was finally added that was converted to a precipitate product. The precipitate caused a reduction in the electrode surface, and hence a reduction in the electrochemical current (due to the reaction of ferrocene methanol in the electrolyte) was recorded.[178]

Heterogeneous immunoassay has also been conducted with the antibody immobilized on beads. For instance, mouse IgG (50 to 100 ng/ml) was detected by ELISA in a glass chip. First, mouse IgG (antigen) was captured by magnetic beads coated with sheep anti-mouse antibody (1.02×10^7 beads/ml). Then the secondary antibody, which was rat anti-mouse conjugated with alkaline phosphatase (0.7 μg/ml), was delivered. Thereafter the substrate, PAPP, was added. It was enzymatically converted to *p*-aminophenol (PAP), which was electrochemically detected by the on-chip interdigital microelectrodes.[1016]

Heterogeneous immunoassay has also been conducted without the use of an enzyme label. For instance, electrochemical immunoassay of mouse IgG (antigen) was carried out in glass chip. The chip contained magnetic beads coated with the sheep anti-mouse antibody. After flowing in the secondary antibody (rat anti-mouse conjugated with PAPP), electrochemical oxidative detection of PAP was achieved (i.e., PAP was oxidized to *p*-quinoneimine).[1016]

Fluorescent immunoassay was also conducted. For instance, physisorption of goat anti-mouse IgG (biotin conjugated) was first made to the PDMS surface for the immunoassay of IgG. Then neutravidin was introduced to bind to the biotin on the surface. Immobilization of biotinylated goat anti-human IgG on the neutravidin-coated channel was first achieved by confining the reagent flow at a T-intersection on the PDMS chip. The antigen (Cy5-human IgG) was applied via a perpendicular channel. Fluorescence measurement allowed the antibody–antigen binding (signal), background, and nonspecific binding (NSB) to be accomplished simultaneously (see Figure 10.8).[170]

In other reports, heterogeneous immunoassay was performed in a glass microchip with protein A (0.1 mg/ml) first immobilized on the microchannel wall.[1017–1019] After unlabeled rabbit IgG (the analyte) was introduced using the electrokinetic flow it was incubated in the channel to bind with protein A. After washing, Cy5-labeled IgG was admitted, which competed with the unlabeled IgG

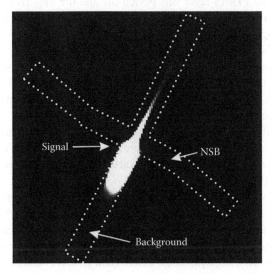

FIGURE 10.8　Fluorescent image of labeled antigen bound to immobilized antibody observed at the intersection of a microchip. The antibody, which is biotin-conjugated goat anti-human IgG, is first flowed from the bottom channel and confined by the left/right channels. Then the antigen, which is Cy5-human IgG, is introduced from the left channel and confined by the top/bottom channels. Information on background levels (background) and nonspecific binding (NSB) can be extracted from the same experiment. Quantitation is performed by averaging the fluorescence intensity in a small area in the region of interest.[170] Reprinted with permission from the American Chemical Society.

for the protein A sites. After a second washing, the amount of Cy5-IgG was eluted (using a chaotropic buffer containing glycine at pH 2.0) and detected fluorescently. The whole assay procedure (incubation, washing, and elution) was completed within 5 min using the electrokinetic flow. A concentration down to 50 nM for rabbit IgG has been detected.[1017]

ELISA was also carried out in a polyethylene terephthalate (PET) chip to detect D-dimer (DDi) using an HRP-conjugated antibody for detection. HRP catalyzed the oxidation of hydroquinone by H_2O_2 to benzoquinone, which was detected by amperometric detection. It was found that the enzymatic reaction was seventy-eight-fold more efficient in the microchannel than in a standard mircrowell because of a much higher surface-to-volume ratio in the microchannel. In this connection, more BSA (higher concentration and longer time) was needed in treatment for blocking of nonspecific binding.[1020] In other reports, DDi was similarly assayed using an ALP-conjugated antibody.[189,190]

In another report, an assay of a cardiac marker (human C-reactive protein [CRP]) was achieved on a Si-PDMS chip based on a solid-phase sandwich immunoassay.[459]

Polystyrene beads coated with a capture (mouse) antibody against the carcinoembryonic antigen (CEA) were introduced in a microchannel consisting of a weir structure (see Figure 8.13 in Chapter 8).[1021] Thereafter, a serum sample containing CEA (from human colon adenocarcinoma) was introduced. Then the first detection antibody (rabbit anti-human) was allowed to bind with the bound CEA. Finally, a second detection antibody (goat anti-rabbit) conjugated to colloidal gold was used. Detection was achieved by thermal lens microscopy (TLM). The assay time was shortened from 45 h (conventional ELISA) to 35 min. A detection limit of ~0.03 ng/ml CEA was achieved, which was adequate for normal serodiagnosis of colon cancer (5 ng/ml).[1021] A similar method was used to assay a protein, interferon (IFN).[1022,1023]

Heterogeneous immunoassay of anti-IgG was achieved in PDMS channels where the antigen IgG was coated on beads that were partially embedded in PDMS (see also Chapter 9, Section 9.3.1 for the use of embedded beads for DNA hybridization).[1024]

PDMS microfluidic networks (µFNs) were used to pattern biomolecules (mouse IgG, chicken IgG) in low volumes (µl) and on small areas (mm²). These channels are 3 µm wide and 1.5 µm deep.[1025] Subsequently, the PDMS layer was peeled off and the underivatized areas were blocked by BSA. Then the patterned surface was exposed to three fluorescently labeled antibodies: tetramethylrhodamine-conjugated anti-chicken IgG (red), fluorescein-labeled anti-mouse IgG (green), and R-phycoerythrin-labeled anti-goat IgG (orange). Only the binding for chicken IgG and mouse IgG was observed through the red and green channels, and the nonbinding for goat IgG (orange) was not observed.[1025] In another report, µFNs were also formed from PDMS chips to create patterns of immobilized chicken IgG. Heterogeneous immunoassay with anti-species IgG (fluorescently labeled) was then conducted.[247]

After patterning antibodies as an array using the PDMS microchannel, 2,4,6-trinitrotoluene (TNT) was detected using the heterogeneous immunoassay format.[1026] Different modes (direct, competitive, displacement, sandwich) produced various limits of detection (5 to 20 ng/ml) and dynamic ranges (20 to 200 ng/ml).[1026]

In one report, an array of antigens patterned on a substrate was constructed for detection of IgG antibodies from various species. This method, termed miniaturized mosaic immunoassay, is rendered possible by using the PDMS channel for patterning materials.[1027] In another report, various antibodies were patterned via the PDMS channel on a substrate for detection of three antigens: F1 antigen (from *Yersinia pestis*), staphylococcal enterotoxin B, and D-dimer (DDi, a marker of sepsis and thrombotic disorder).[1028]

In a heterogeneous immunoassay format, the antigen dinitrophenyl (DNP) was presented as a DNP-conjugated lipid bilayer, and it was coated on a twelve-channel glass-PDMS chip (see Figure 10.9). Then twelve concentrations (0.15 to 13.2 µM) of anti-DNP IgG antibodies (labeled with Alexa fluor 594) were introduced to the twelve channels simultaneously.[1029] Total internal reflection fluorescence microscopy (TIRFM) was used for detection. This method enabled an entire binding curve to be obtained in a single experiment.[1029] In another report, PDMS channels were

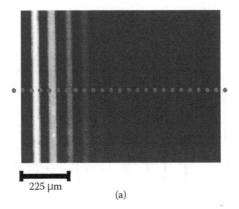

225 µm

(a)

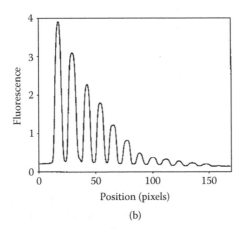

(b)

FIGURE 10.9 (a) Bulk-phase epifluorescence image of Alexa 594 dye-labeled anti-DNP inside twelve bilayer-coated microchannels. Starting from the left-hand side, the antibody concentrations in these channels are 13.2, 8.80, 5.87, 3.91, 2.61, 1.74, 1.16, 0.77, 0.52, 0.34, 0.23, and 0.15 mM. A line scan of fluorescence intensity across the twelve microchannels is plotted in (b).[1029] Reprinted with permission from the American Chemical Society.

used to pattern mouse IgG (in every other channel) in a twelve-channel chip that conformed to the standard liquid introduction format using a twelve-channel pipettor.[184]

ELISA of sheep IgM (antigen) was conducted in a PDMS chip. The capture antibody (rabbit anti-sheep IgM-HRP) was added. Then the fluorogenic substrate (3-(p-hydroxyphenyl)-propionic acid [HPPA]) was added for fluorescent detection. The conventional blocking reagents (0.5% w/v BSA, 0.5% w/v caesin, 0.5% v/v Tween-20), which normally worked for the polystyrene ELISA plate, did not work with the hydrophobic PDMS surface. To solve this problem, 4% PEG and 7% normal rabbit serum were included in the above solution for effective blocking.[173]

Serial dilution was achieved in a PDMS microchannel network for immunoassay, as shown in Figure 10.10. Immunoassay of IgG antibodies present in HIV+ human serum was conducted. Using a 1:1 dilution ratio and ten sequential dilutions (only three are shown in Figure 10.10), a dynamic range of 2^{10} or ~1,000 of the serum concentration was obtained. The HIV antigens (gp41, gp120) were first adsorbed on a PC membrane, which was then sealed by the PDMS chip for immunoassay. The amounts of anti-gp 120 and anti-gp 41 in the serum were found to be 2.0 ± 0.7 and 1.0 ± 0.5 mg/ml, respectively.[248]

Anti-atrazine antibodies were coupled to a Si microchip comprising forty-two coated porous flow channels. The porous Si layer was achieved by anodizing the channel in 40% HF/96% ethanol

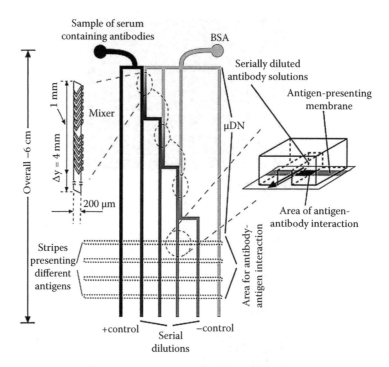

FIGURE 10.10 Schematic representation of the microfluidic network (μFN). At the top of the figure, a serial dilution of analytes (antigen-containing serum) using the serum blank (BSA) is generated. Dilution is assisted by a chaotic advection mixer. Left inset shows that each mixer includes four cycles of herringbone patterns and is 4 mm long. The bottom part of the figure suggests how the serially diluted antibody interacts with spatially segregated antigens, immobilized on a polycarbonate membrane. The membrane is called the antigen-presenting membrane (see right inset).[248] Reprinted with permission from the American Chemical Society.

(1:1). Detection of atrazine was achieved in a competitive format. The chemiluminescence signal decreased when a known amount of horseradish peroxidase-labeled atrazine was added. This decrease was used to determine the amount of unlabeled atrazine in the sample.[1030]

A multiple-channel device consisting of twenty-four channels was fabricated on a compact disk (CD); see Figure 10.11. Multiple ELISA tests were performed on it for rat IgG from a hybridoma cell culture. Centrifugal force was generated when the disk was spun, and this force was employed for liquid pumping. In addition, capillary burst valving (see Chapter 3, Section 3.2.4) was used for consecutive liquid delivery using increasingly high rotation speeds. The first antibody (goat anti-rat, 2.5 μg/ml) was introduced from an outer reservoir (3) into the measurement site (2) using a low disk rotation speed (see Figure 10.11b). Site 2 had been surface treated to produce a higher protein affinity. Since the centrifugal force was weaker at a smaller radius, by using a higher rotation speed, a washing was delivered from reservoir 4 to site 2. Then a blocking reagent (BSA) was delivered from reservoir 5 (10 mg/ml). After a second wash from reservoir 6, the antigen (2 to 2.5 μg/ml of rat IgG) was delivered from reservoir 7. After the third wash from reservoir 8, the second antibody (HRP-labeled goat anti-rat) from reservoir 9 and a final wash from reservoir 10 were conducted. Then the substrate HPPA (3 mg/ml) from reservoir 11 was added. The LOD of rat IgG is found to be 5 mg/L (or 31 nM), which is adequate for IgG analysis (1 to 100 mg/L) in hybridoma cell cultures. The whole procedure took only 200 s to complete, compared to the 120 min needed for the same enzymatic reaction to complete in a microwell plate. This is because the surface area in the microchannel is ~36 times larger than that in the microplate well.[226]

A resistive pulse method of particle sizing was used to detect antibody–antigen binding events at a pore fabricated on a PDMS chip. The pore was typically 7 to 9 μm long and 1 μm in diameter.

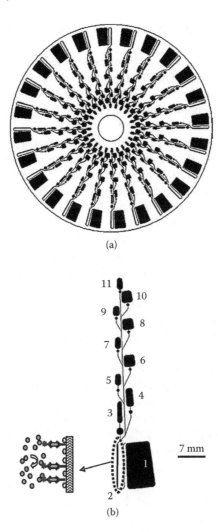

(a)

(b)

FIGURE 10.11 Schematics of (a) a 12.7 cm diameter PMMA CD disk used for ELISA with twenty-four sets of assays, and (b) a single assay (1, waste; 2, detection; 3, first antibody; 4, 6, 8, 10, washing; 5, blocking protein; 7, antigen/sample; 9, second antibody; and 11, substrate).[226] Reprinted with permission from the American Chemical Society.

Mouse monoclonal anti-streptavidin antibody (0.75 to 10 µg/ml) was supposed to bind to the surface of latex colloidal particles coated with streptavidin (the antigen). This binding, which caused an increase of 1 to 9 nm of the particle diameter, was measured by the resistive pulse method.[1031]

10.2 PROTEIN SEPARATION

In protein separations, adsorption of protein molecules to the channel wall is always a problem. For instance, adsorption of green fluorescent protein (GFP) to the walls of plastic polycarbonate microchannels is problematic.[597] To avoid protein adsorption, the microchannels normally need to be treated with coatings, such as (acryloylaminopropyl)trimethylammonium chloride (BCQ)[806,809,812] methacrylocyloxyethylphosphorylcholine (MPC),[1032] or pyrolyzed PDMS.[608] In protein separation for homogeneous immunoassay, a zwitterionic buffer (tricine) combined with a neutral surfactant (Tween-20) was used to produce a dynamic coating.[330] In addition, a three-layer coating was applied to a PDMS channel to reduce analyte and reagent adsorption while maintaining a modest cathodic electroosmotic flow (EOF).[170]

Separations of various proteins, such as bovine carbonic anhydrase II, insulin, and lysozyme, have been achieved on a PDMS chip. The PDMS surface was oxidized to become hydrophilic. Since lysozyme is positively charged, the oxidized PDMS channel was coated with a cationic polymer, polybrene, to prevent wall adsorption of protein.[1033] Other than proteins, fluorescently labeled peptides have been separated on a PDMS chip.[159,924]

A protein sizing assay was performed according to separation of protein–SDS complexes. In this case, a universal noncovalent fluorescent labeling method was used for detection. This involved the loading of a fluorescent dye to the SDS–protein complex.[1034,1035] However, the presence of excess SDS resulted in the problem of high fluorescent background because of the binding of the dye to SDS micelles in the solution. This problem was alleviated by a postcolumn dilution step. Eleven samples were applied to reservoirs A1–A3, B1–B3, C1–C3, and D1–D2 (see Figure 10.12). On-chip dilution helped break up the SDS micelles, thereby allowing more dye molecules to bind to the protein. This is a faster process (0.3 s) than the conventional destaining process (1 h) based on diffusion. Good separation was found when the dilution ratio (DR) was high, i.e., 10.5 and 14 (see Figure 10.13). This microchip method produced a much greater speed in separation and destaining. With this method, a sizing range of 9 to 200 kDa can be achieved, with good accuracy (5%) and sensitivity (30 nM for carbonic anhydrase). Since dilution from D4 (see Figure 10.12) was needed to remove the SDS–dye complexes in the solution after protein separation, the eleven samples could only be analyzed sequentially.[1034]

Protein separation has commonly been achieved by isoelectric focusing (IEF). In various reports, IEF separations have been achieved with BSA,[1036] bovine hemoglobin,[1036] cytochrome c,[189] β-lactoglobulin,[189] protein markers,[975,1037] and bacteria.[1038]

In addition, IEF of naturally fluorescent proteins was achieved on a glass chip. Subsequent mobilization was carried out hydrodynamically.[684] In one report, IEF of standard proteins was carried out before mass spectrometry (MS) analysis.[290]

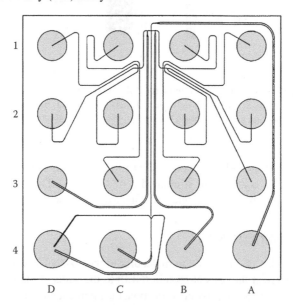

FIGURE 10.12 Chip design for protein separation. The wells are shown in light gray. Well D4 is the SDS dilution well and is connected to both sides of the dilution intersection. Wells A4 and C4 are the separation buffer and waste wells, and B4 and D3 are used as load wells. All other wells contain samples.[1034] Reprinted with permission from the American Chemical Society.

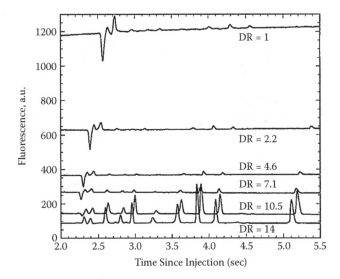

FIGURE 10.13 Data showing a series of separations of a protein ladder at various dilution ratios (DR) of 1 to 14. The electric field strength ranged from 219 to 221 V/cm. The total protein concentration is 0.4 mg/ml for eight fragments.[1034] Reprinted with permission from the American Chemical Society.

One-step IEF of several Cy5-labeled peptides was achieved on a glass chip without simultaneous focusing and mobilization (by EOF).[626] In another report, enhanced green fluorescent protein (EGFP)- and FITC-labeled peptides were separated by IEF in a PC chip. When a dynamic approach in the electrokinetic analyte introduction was employed, an enhancement in peak intensity was obtained, compared to the conventional IEF in which EOF is suppressed.[1039]

Other than producing the pH gradient using ampholyte solutions, IEF of BSA was achieved on-chip using a pH gradient naturally developed under the electrokinetic flow conditions. The pH gradient was produced by transverse electrodes that were constructed by sandwiching two gold electrodes by Mylar, held together by a pressure-sensitive adhesive.[545]

Transverse IEF was also conducted in a pressure-driven flow for BSA and soybean lectin separation on-chip.[1040] Here, Pd electrodes were used (in preference to Au) because of the nongassing character of Pd. In addition, the protein sample was sandwiched between two buffer streams and was prevented from direct contact with the channel wall (and hence the electrode), a process akin to hydrodynamic focusing.[1040]

Imaging of IEF separation has been carried out. For instance, IEF of EGFP was achieved in a PMMA chip sealed with PDMS. Observation of fluorescence was achieved through the optically transparent PDMS layer.[177] In another report, IEF of pI markers was demonstrated on a quartz chip using whole column UV (280 nm) absorption imaging.[1041]

10.3 ENZYMATIC ASSAYS

10.3.1 Assay of the Enzymes

Another area of protein assays involves the use of enzymes. For instance, multiple enzymatic assays were achieved on a forty-eight-channel disk to study the effect of an inhibitor on the enzyme (see Figure 10.14).[1042] In particular, alkaline phosphatase (ALP) and p-nitrophenol phosphate (5 mM) were mixed with fifteen concentrations (0.01 to 75 mM) of theophylline (inhibitor). The tests were carried out in triplicate, which summed up to forty-five assays. These, together with three calibrations (for the product p-nitrophenol), were simultaneously accomplished in a single PDMS/PMMA disk. A complete isotherm for the inhibition of ALP (0.1 mg/ml) by theophylline was obtained in

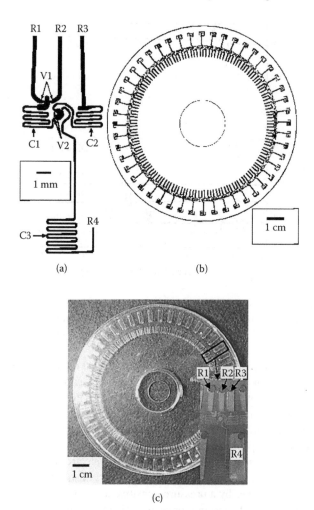

(a)   (b)

(c)

FIGURE 10.14   (a) Design of the microfluidic structure used to perform an assay composed of mixing an enzyme with an inhibitor, followed by mixing with a substrate, and detection. Enzyme, inhibitor, and substrate were loaded in reservoirs that were connected to channels labeled R1, R2, and R3, respectively. After being released by capillary burst valves (V1), enzyme (R1) and inhibitor (R2) were mixed in a meandering 100 μm wide channel (C1). After being released by burst valve (V2) the enzyme-inhibitor mixture was mixed with the substrate (R3) in a meandering channel (C3). Then the mixture was emptied into a cuvette (not shown) connecting to R4. (b) The complete chip layout showing forty-eight of the structures shown in (a). (c) Photograph of the disk, which is a PDMS replica sealed to a PMMA layer. The inset shows the magnified detail of one of the fluidic structures of the disk.[1042] Reprinted with permission from the American Chemical Society.

one single experiment, giving rise to $K_i = 9.7 \pm 0.9$ mM (see Figure 10.15). In addition, fifteen (in triplicate) substrate concentrations ($K_m/2$ to $8\,K_m$) were used to generate the necessary data to establish the Michaelis–Menten kinetic parameters (K_m and V_m) in a single experiment.[1042]

Michaelis–Menten kinetics of ALP was established using a continuous-flow method in a glass chip (shown previously in Figure 2.37 in Chapter 2). A continuous-flow titration between the enzyme and a fluorogenic substrate was performed to obtain substrate conversion after a fixed short residence time as controlled by a fixed fast flow. This was the fixed-time method, as compared with the usual initial-slope method. Cleavage of a fluorogenic peptide substrate (DiFMUP) by ALP produced the fluorescent signal. To account for the background, the substrate was titrated for ten different concentrations from 0 to 70% of the total flow (with zero enzyme concentration). After

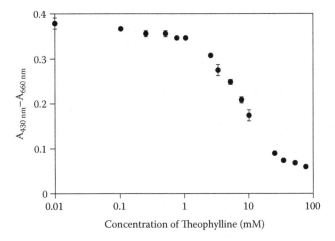

FIGURE 10.15 Enzyme inhibition assay with substrates of multiple concentrations running simultaneously on a disk shown in Figure 10.14. In each case, 3 µl of 0.1 mg/ml alkaline phosphatase (enzyme) was mixed with 3 µl of theophylline (inhibitor) at various concentrations. This mixture was then combined with 6 µl of 5 mM PNPP (substrate). The enzymatic products (*p*-nitrophenol) were monitored by measuring the absorbance at 430 nm. A total of fifteen concentrations of theophylline (0.01 to 75 mM) were run in triplicate, giving rise to forty-five assays. The results ($A_{430\ nm} - A_{660\ nm}$) are plotted as the absorbance at 660 nm (background) subtracted from the absorbance at 430 nm (signal) as a function of the concentration of theophylline. $A_{430\ nm} - A_{660\ nm}$ is proportional to the amount of product, and hence the activity of the enzyme. The error bars are standard deviations of the average over three measurements at each concentration. A fit of these data to the Langmuir equation yielded an inhibition constant, K_i, of (9.7 ± 0.9) mM.[1042] Reprinted with permission from the American Chemical Society.

obtaining ten substrate-only signals, the flow from the enzyme was set to 30% and the same ten-point titration was repeated to obtain the substrate conversion at ten substrate concentrations. After the continuous-flow experiment, the raw data were processed to generate the Lineweaver–Burk plot to obtain a K_m of 1.6 mM.[295]

In another report, Michaelis–Menten parameters of ALP have been obtained using a one-shot Lineweaver–Burk (reciprocal) plot. This can be achieved by simultaneously measuring the conversion of twelve independent concentrations of the substrate (4-methylumbelliferyl phosphate) created on-chip. The enzyme was streptavidin-conjugated ALP that was linked to biotinylated phospholipid bilayers coated inside PDMS microchannels. The blue fluorescence of the enzymatic product, 7-hydroxy-4-methyl coumarin, was measured.[1043] The surface-bound enzyme was found to have a lower (sixfold) turnover rate than the free enzyme in solutions. After diffusion mixing between two streams (substrate and buffer) for a fixed distance (2 cm), the main channel was partitioned into a series of smaller channels that emanated from the main flow stream (previously shown in Figure 3.24 in Chapter 3). This method created a concentration gradient that resulted in a dilution series that ranged over a factor of 33 (i.e., 0.103 to 3.41 mM) of the substrate.[1043] The one-shot experiment was done in one order of magnitude less time to establish Michaelis–Menten kinetics (needs twelve concentrations) and with one-third less error in the kinetic raw data.[1043]

In one report, ALP was patterned at desired locations in a PDMS-glass channel via the streptavidin–biotin interaction. ALP was streptavidin linked. Instead of a simple coating, biotin was formed covalently on the channel surface using a three-step process. First, bovine fibrinogen was adsorbed in the channel, and a biotin 4-fluorescein (B4F) solution was flowed in. Second, upon irradiation by an Ar⁺/Kr⁺ laser through a photomask, the fluorophore B4F was bleached to create highly reactive free-radical species. Third, this reactive species allowed the attachment of B4F to the free amino groups on the adsorbed fibrinogen via a singlet-oxygen-dependent mechanism.[1044]

In another report, several acetylcholinesterase (AChE) inhibitors, including tacrine, edrophonium, tetramethyl- and tetraethyl-ammonium chloride, carbofuran, and eserine, were assayed on a chip.[1045] AChE converted the substrate, acetylthiocholine, to thiocholine. This product reacted with coumarinylphenylmaleimide (CPM) to form thiocholine-CPM (a thioether) for LIF detection. Since the acetonitrile solvent used to dissolve CPM inhibited AChE activity, the CPM solution was added after the enzymatic reaction.[1045] Another enzyme inhibition assay using a peptide substrate was performed on a PMMA chip.[1046]

Activities of liver transaminases (GOT and GPT) were determined by monitoring the product, l-glutamate. With GOT, l-aspartate and 2-oxoglutarate were enzymatically converted to pyruvate and l-glutamate. An elevated GOT level in serum indicates severe damage to the heart (e.g., myocardial infarction) and to the liver. With GPT, l-alanine and 2-oxoglutarate were enzymatically converted to oxaloacetate and l-glutamate. An elevated level of GPT indicates viral hepatitis. Enzymatic activities of both enzymes in the range 6 to 192 U/L were detected. In both cases, l-glutamate was detected amperometrically using a thin-film glutamate biosensor constructed on a glass chip.[1047]

In one report, the activity of the HRP enzyme was studied by chemiluminescence (CL) detection using the xanthine/xanthine oxidase (XOD)/luminol system. The enzymes were immobilized on glass beads trapped in a reaction chamber by weirs. It was found that when HRP was immobilized, a greater CL signal was generated than with free-solution HRP. However, when XOD (not HRP) was immobilized, enzymatic reaction products were not detectable. This was probably because the immobilized XOD enzyme had a low specific activity.[721]

Enzyme kinetics were evaluated in a PDMS-glass chip using a continuous-flow system. A biotinylated enzyme (HRP or β-galactosidase) was coupled to streptavidin-coated beads via the amide coupling of an aminocaproyl spacer. These beads (15.5 μm) were retained by a weir in the chip. The channel wall was passivated by 1 mg/ml BSA. The apparent enzyme kinetic parameters were evaluated using the Lilly–Hornby model, as developed for the packed-bed enzymatic reactor systems. It was found that the apparent Michaelis constant (K_m) approached the true K_m value of the free enzyme at zero-flow rate of a homogeneous reaction.[845]

The activity of β-galactosidase (β-Gal) was studied on a quartz chip using a static micromixer to mix the enzyme and substrate on the millisecond timescale. Inhibition by phenylethyl-β-d-thiogalactoside was also studied.[1048] In another report, the enzyme β-Gal was assayed on a chip in which β-Gal would convert a substrate, resorufin-β-d-galactopyranoside (RBG), to resorufin to be detected fluorescently.[1049] By varying the substrate concentrations and monitoring the amount of resorufin by LIF, Michaelis-Menten constants could be determined. In addition, the inhibition constants of phenylethyl-β-d-thiogalactoside, lactose, and p-hydroxymercuribenzoic acid to the enzyme β-Gal were determined.[1049]

A microchip-based assay for the enzymatic activity of protein kinase A (PKA) was conducted.[1050] A mixture of PKA (enzyme) and the fluorescently labeled substrate (kemptide [LeuArgArgAlaSerLeuGly]) was successively injected on a glass chip. The time-course changes in the enzymatic reaction were studied by performing on-chip electrophoretic separation. It was found that the amount of product increased at the expense of the substrate (see Figure 10.16).[1050] The effect of a protein kinase A inhibitor (H89: N-[2-((p-bromocinnamyl)amino)ethyl]-5-isoquinoline sulfonamide hydrogen chloride) was then studied on-chip.[1050]

The study of the activity of an intracellular enzyme (β-galactosidase) extracted from cells was also reported. In this report, lysis of *E. coli*, enzyme extraction, and its detection (using RBG as the substrate) were all achieved on-chip using pressure-driven flow (see Figure 10.17).[1051,1052] The first two steps were achieved in an H-filter using a stream of cells and a stream of lytic agent (a proprietary mild detergent). Since the cells were large, they remained in the left half of the channel. Upon contact with the lytic agent, the intracellular enzyme was released, and it diffused further to the right channel. The H-filter was coupled to a T-sensor to complete the third step.[1051]

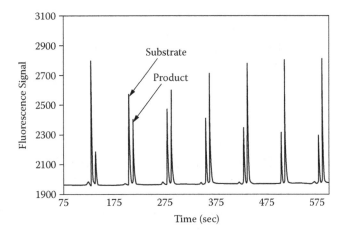

FIGURE 10.16   Phosphorylation of kemptide by PKA in a reagent well on a glass chip. Successive electropherograms of unphosphorylated (substrate) and phosphorylated fluorescein-labeled kemptide (product) over a time course of ~600 s. Conditions: 13.3 mM Fl-kemptide, 24.5 nM PKA in 100 mM HEPES, pH 7.5, 1 M NDSB-195, 5 mM MgCl$_2$, 100 mM ATP, 50 mM cAMP, 0.1% Triton X-100, 10 mM DTT. NDSB-195 (dimethylethylammonium propanesulfonate) is a zwitterionic compound to prevent enzyme adsorption; ATP is to provide the phosphate group to be transferred to the serine residue of kemptide during phosphorylation; cAMP is to enable the function of PKA, as it is a cAMP-dependent kinase; DTT is to protect the enzyme from electrochemical damage in CE.[1050] Reprinted with permission from Elsevier Science.

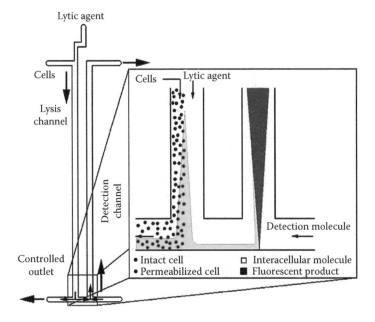

FIGURE 10.17   Schematic of microfluidic device consisting of an H-filter for cell lysis/fractionation coupled to a T-sensor for detection of an intracellular enzyme (β-galactosidase). Pump rates are controlled at all inlets and one outlet. In the H-filter (left), lytic agent diffuses into the cell suspension and lyses the cells. Intracellular components (enzyme) then diffuse away from the cell stream and some are brought around the corner into the detection channel. In the T-sensor (right), the presence of β-Gal is detected fluorescently.[1051] Reprinted with permission from the American Chemical Society.

There are several advantages in studying enzymatic reactions in a cell-based system as opposed to in a solution-based system: (1) no enzyme extraction and purification are required, (2) the enzymes in cells may have a longer lifetime than soluble enzymes, and (3) individual cells provide a confined picoliter volume for measurement.[1053]

10.3.2 ASSAY OF ANALYTES AFTER ENZYMATIC REACTIONS

Enzymes were also used to determine the analytes after their enzymatic conversion to some detectable species. For instance, to detect glucose and ethanol, dual precolumn enzymatic reactions, followed by CE separation and amperometric detection, were achieved on-chip.[1054,1055] A sample containing glucose and ethanol was placed in the sample reservoir, and an enzyme mixture containing glucose oxidase (GOx), alcohol dehydrogenase (ADH), and NAD^+ was placed in the reagent reservoir. Glucose was converted to hydrogen peroxide by GOx, whereas ethanol converted NAD^+ to NADH by ADH. After separation of H_2O_2 and NADH (see Figure 10.18), they were detected amperometrically at the anode.[1054]

In one report, a single analyte, glucose, was determined using two enzymes, GOx and GDH (glucose dehydrogenase). They converted glucose into two products for subsequent detection.[1055] In addition, glucose was determined by CL detection in PDMS channels where glucose oxidase was partially embedded.[1024]

Poly(ethyleneglycol) (PEG) hydrogel-based micropatches were first molded within two small PDMS channels (see Figure 10.19). Glucose oxidase and HRP were physically entrapped during the

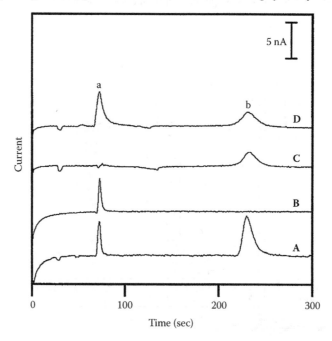

FIGURE 10.18 Electropherograms of hydrogen peroxide (a) and NADH (b) for detection of glucose and ethanol, respectively. (A) a control mixture containing 5×10^{-5} M hydrogen peroxide and 5.0×10^{-4} M NADH; (B) a sample of 4.0×10^{-4} M glucose plus 10 units/ml GOx; (C) 2.0×10^{-2} M ethanol plus 10 units/ml ADH and 1×10^{-2} M NAD^+; and (D) a sample of 4.0×10^{-4} M glucose and 2.0×10^{-2} M ethanol plus 10 units/ml GOx, 10 units/ml ADH, and 1×10^{-2} M NAD^+. Injection potential, +2,500 V; injection time, 3 s; separation potential, +2,000 V. Electrophoresis buffer, 20 mM phosphate buffer (pH 7.8). Amperometric detection was conducted with a gold-plated screen-printed electrode held at +1.0 V.[1054] Reprinted with permission from the American Chemical Society.

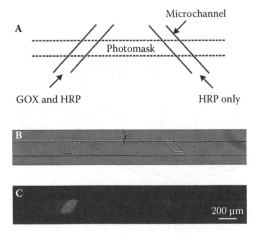

FIGURE 10.19 Enzymatic reaction in PEG-based hydrogel micropatches for detection of glucose. (a) Schematic diagram illustrating how two micropatches containing different enzymes can be immobilized within two 17 μm high and 135 μm wide channels (solid line) using a photomask (dotted line). (b) Optical micrograph of the two hydrogel micropatches subsequently enclosed within a single larger horizontal channel (32 μm high and 200 μm wide). The hydrogel on the left contains GOx and HRP, and the one on the right contains only HRP. (c) Fluorescence micrograph of the two micropatches after a flow (1 L/min) of glucose, oxygen, and Amplex Red. Strong fluorescence is apparent only in the left micropatch containing both enzymes.[1056] Reprinted with permission from the American Chemical Society.

polymerization of micropatches for the detection of glucose.[1056] Then the PDMS molding channel plate was removed and the micropatches were enclosed within a larger PDMS channel. First, glucose was enzymatically converted to gluconolactone and H_2O_2. Then in the presence of HRP, H_2O_2 reacted with a dye (Amplex Red) to form resorufin for fluorescent detection. The enzymes GOx and HRP were entrapped in the hydrogel, whereas glucose (the analyte), O_2, and Amplex Red were able to diffuse into the hydrogel within a reasonable timescale. As a control, a hydrogel micropatch containing only HRP, but no GOx, was also formed inside the channel.[1056] Another glucose detection based on the same two-step enzymatic reaction to form resorufin was achieved in a PDMS channel under a flow.[1057]

In one report, a two-step enzymatic detection of glucose was achieved using two immobilized enzymes. Avidin-conjugated glucose oxidase was immobilized in A1 and B1 in one region, and streptavidin-conjugated HRP was immobilized in A2 and B2 in another region (see Figure 10.20). After admitting an O_2-saturated glucose-containing sample into A1 (only buffer into B1 as a control channel), H_2O_2 was formed only in A1. After the enzymatic reaction was completed, A1 and A2 (also B1 and B2) were connected with a reversibly attachable U-shaped plastic tube (700 μm o.d.). The H_2O_2 formed (only in A1) subsequently flowed to A2 to convert Amplex Red (in the presence of HRP) to resorufin for fluorescent detection.[1043]

In a report for the determination of insulin, a glass microchip was integrated with a precolumn enzymatic reacting channel, a CE separation channel, and a postcolumn labeling region, as shown in Figure 10.21. Localized thermal control of the reacting channel was achieved using a resistive heating element. Tryptic digestion of bovine insulin (B-chain) was first carried out at 37°C under various conditions (continuous flow or stop flow). CE separation of the reactant and products is shown in Figure 10.22. It was found that the two digested products were not separated, but they comigrated earlier than the reactant. Detection was achieved after postcolumn derivatization with naphthalene-2,3-dicarboxaldehyde (NDA). It was found that stop flow for 15 min resulted in complete digestion.[1058] In another experiment, reduction of the disulfide bond of insulin by DTT to two peptide chains (A-chain and B-chain) was carried out at various temperatures. It was found that

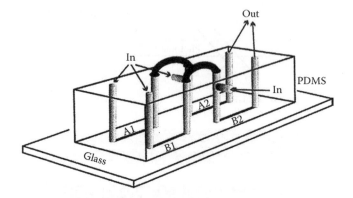

FIGURE 10.20 Schematic representation of the microfluidic device used for the two-step enzymatic detection of glucose.[1043] Reprinted with permission from the American Chemical Society.

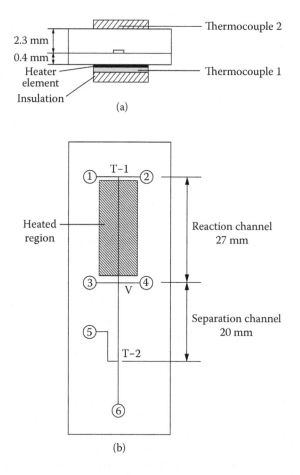

FIGURE 10.21 (a) Cross-sectional view of the microchip, heating element, and the thermocouples. (b) Schematic of the microchip used for on-chip reactions, separations, and postcolumn labeling. The fluid reservoirs are: (1) substrate, (2) enzyme or DTT, (3) buffer, (4) sample waste, (5) NDA, and (6) waste.[1058] Reprinted with permission from Elsevier Science.

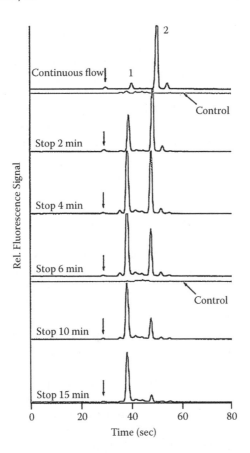

FIGURE 10.22 Electropherograms of the products following on-chip tryptic digestion of bovine insulin B-chain at 37°C. The flow in the reaction channel was stopped for different times to study the effect on the reaction. Control runs (see the flat curves) without insulin B-chain were performed in the continuous-flow mode and in a flow with a stop time of 6 min. The arrows indicate the migration time of benzylamine, which was added as an EOF marker. All electropherograms are plotted on the same scale with an offset for clarity.[1058] Reprinted with permission from Elsevier Science.

as the reaction temperature (35 to 55°C) increased, the reactant (insulin) peak decreased, and no insulin peak was detected at 55°C.[1058]

Enzymatic conversion of NBD-labeled ArgOEt to NBD-Arg was monitored in a PMMA microchannel. The enzyme, trypsin, was immobilized via the reactive ester groups on a phospholipid polymer adsorbed on the PMMA surface.[1059]

It was found that cytochrome c, β-caesin, and BSA can be digested by trypsin-immobilized beads in a microchamber on a glass chip. The protein (1 mg/ml) was mobilized by EOF. Although without external heating, Joule heating may assist in the enzymatic digestion reaction of the proteins.[116]

Enzymatic microreactors (7.5 nl) have been fabricated in the microfluidic chip to prepare the tryptic digest of equine (horse) myoglobin (14.2 pmol/μl) for subsequent off-line MALDI-TOF analysis. The enzyme, trypsin, is immobilized on porous polymer monoliths that are made *in situ*. This is achieved by reacting 2-vinyl-4,4-dimethylazlactone, ethylene dimethacrylate, and acrylamide (or 2-hydroxyethyl methacrylate). The azalactone functionality on the polymer reacts readily with the amine and thiol groups of the enzyme (trypsin) to form stable covalent bonds so that the enzyme can be immobilized.[1060] In addition, trypsin and chymotrypsin were immobilized on an enzyme reactor fabricated on Si to digest proteins.[1061]

Trypsin-encapsulated sol-gel (alkoxysilane-based) was fabricated *in situ* onto the sample reservoir of a PMMA chip. This was employed for enzymatic conversion of NBD-labeled ArgOEt and bradykinin, followed by CE separation of the products. The enzymatic activity of the encapsulated trypsin as given by the Km value was found to be nineteen times higher than that of the free trypsin. The stability of trypsin was 1 week at 4°C. This enhanced enzyme stability was possibly caused by the prevention of enzyme autolysis by the sol-gel matrix.[1062]

Other than proteins, the immobilized enzyme was also used to digest DNA. For instance, the restriction endonuclease enzyme *Hae*III was immobilized on an amine-modified PMMA surface to digest DNA.[1063]

10.4   PROBLEM SETS

1. In immunoassay, what are antigen and antibody? Can an antibody act like an antigen? (3 marks)
2. Compare and contrast direct and competitive immunoassays. (4 marks)
3. In determining cortisol in serum, most of the cortisol is bound to corticosteroid-binding globulin and other serum proteins. How then is cortisol released for detection?[1006] (2 marks)
4. What are the two buffer additives used in the on-chip theophylline immunoassay? Explain the functions of these two additives.[330] (4 marks)
5. In the on-chip theophylline homogeneous immunoassay, the antibody–antigen complex can be separated from the antigen. Why can such a separation be successful, but not in the case of the cortisol assay?[330,1006] (2 marks)
6. Describe what are employed in the three streams in a T-sensor for immunoassay. (3 marks)
7. For an ELISA enzymatic reaction occurring for 1 min, calculate the product concentration in the following two cases (given that the surface enzyme concentration is 1 fmol/cm^2 and a turnover rate is 1,000 substrate molecules per second):
   a. A standard microwell ($S = 1$ cm^2, $V = 100$ µl)
   b. A microchannel ($S = 0.031$ cm^2, $V = 40$ nl)
   Hence, show that the enzyme reaction is ~78 times more efficient in the microchannel than in a standard microwell plate.[1020] (4 marks)
8. Describe two detection methods used in ELISA. (2 marks)
9. Give two reasons why colloidal gold is suitable as a labeling material in immunoassay using laser-induced photothermal spectroscopy.[1021] (2 marks)
10. In one report, there are three blocking agents used for channel wall, colloidal gold, and polystyrene bead to reduce nonspecific binding. What are they?[1021] (3 marks)
11. State two reasons why the microchannel bead-based immunoassay shortened the reaction time to ~1% of the conventional microplate-based immunoassay method.[1021] (2 marks)
12. In the calibration of an antigen (e.g., theophylline) in fluorescent homogeneous immunoassay, both labeled antigen and unlabeled antibody are added to the samples containing different amounts of the antigen (unlabeled) to create a calibration plot. Why cannot only a labeled antibody be simply added to the samples containing different amounts of the antigen (unlabeled) to perform the calibration?[330] (2 marks)
13. In the on-chip heterogeneous immunoassay for human carcinoembryonic antigen (CEA) in serum, three antibodies (mouse anti-human CEA, rabbit anti-human CEA, and anti-rabbit IgG-colloidal gold) were used. (a) What is the purpose of colloidal gold? (b) Why isn't the colloidal gold directly attached to rabbit anti-human CEA to save a step?[1021] (3 marks)

References

1. Terry, S.C., Jerman, J.H., Angell, J.B. 1979. A gas chromatographic air analyzer fabricated on a Silicon wafer. *IEEE Trans. Electron Devices* 26:1880–86.
2. Manz, A., Miyahara, Y., Miura, J., Watanabe, Y., Miyagi, H., Sato, K. 1990. Design of an open-tubular column liquid chromatograph using silicon chip technology. *Sensors Actuators B* 1:249–55.
3. Manz, A., Harrison, D.J., Verpoorte, E.M.J., Fettinger, J.C., Paulus, A.L.H., Widmer, H.M. 1992. Planar chips technology for miniaturization and integration of separation techniques into monitoring system. *J. Chromatogr.* 593:253–58.
4. Hodge, C.N., Bousse, L., Knapp, M.R. 2001. Microfluidic analysis, screening, and synthesis. In *High-throughput synthesis principles and practice*, 303–30. New York: Marcel Dekker.
5. Reyes, D.R., Iossifidis, D., Auroux, P., Manz, A. 2002. Micro total analysis systems. 1. Introduction, theory, and technology. *Anal. Chem.* 74:2623–36.
6. Auroux, P.-A., Iossifidis, D., Reyes, D.R., Manz, A. 2002. Micro total analysis systems. 2. Analytical standard operations and applications. *Anal. Chem.* 74:2637–52.
7. Becker, H., Gärtner, C. 2000. Polymer microfabrication methods for microfluidic analytical applications. *Electrophoresis* 21:12–26.
8. Haswell, S.J. 1997. Development and operating characteristics of micro flow injection analysis systems based on electroosmotic flow. *Analyst* 122:1R–10R.
9. Effenhauser, C.S., Bruin, G.J.M., Paulus, A. 1997. Integrated chip-based capillary electrophoresis. *Electrophoresis* 18:2203–13.
10. Polson, N.A., Haves, M.A. 2001. Microfluidics controlling fluids in small places. *Anal. Chem.* 73:312A–319A.
11. Bergveld, P. 2000. Bedside clinical chemistry: From catheter tip sensor chips towards micro total analysis systems. *Biomed. Microdevices* 2:185–95.
12. van der Lindern, W.E. 1987. Miniaturisation in flow injection analysis practical limitations from a theoretical point of view. *Trends. Anal. Chem.* 6:37–40.
13. Service, R.F. 1995. The incredible shrinking laboratory. *Science* 268:26–27.
14. Burbaum, J. 1998. Engines of discovery. *Chemistry in Britain*, June, 38–41.
15. Service, R.F. 1996. Can chip devices keep shrinking? *Science* 274:1834–36.
16. Jakeway, S.C., de Mello, A.J., Russell, E.L. 2000. Miniaturized total analysis systems for biological analysis. *Fres. J. Anal. Chem.* 366, 525–39.
17. Freemantle, M. 1998. Downsizing chemistry. *Chem. Eng. News* 77:27–36.
18. van de Berg, A., Lammerink, T.S.J. 1998. Micro total analysis system: Microfluidic aspect, integration concept and applications. In *Microsystem technology in chemistry and life science*, 21–50. Heidelberg: Springer-Verlag.
19. Manz, A., Fettinger, J.C., Verpoorte, E., Lüdi, H., Widmer, H.M., Harrison, D.J. 1991. Micromachining of monocrystalline silicon and glass for chemical analysis systems. *Trends. Anal. Chem.* 10:144–49.
20. Manz, A., Graber, N., Widmer, H.M. 1990. Miniaturized total chemical analysis system: A novel concept for chemical sensing. *Sensors Actuators* 1B:244–48.
21. Manz, A., Harrison, D.J., Verpoorte, E., Widmer, H.M. 1993. Planar chips technology for miniaturization of separation systems: A developing perspective in chemical monitoring. In *Advances in chromatography*, 2–66. Vol. 33. New York: Marcel Dekker.
22. Manz, A., Verpoorte, E., Raymond, D.E., Effenhauser, C.S., Burggraf, N., Widmer, H.M. 1994. µTAS-miniaturized total chemical analysis system. In *Micro Total Analysis Systems '94*, University of Twente, Netherlands, November 21–22, pp. 5–27.
23. Shoji, S. 1998. Fluids for sensor systems. In *Microsystem technology in chemistry and life science*, 163–88. Heidelberg: Springer-Verlag.
24. Kopp, M.U., Crabtree, H.J., Manz, A. 1997. Developments in technology and applications of microsystems. *Curr. Opin. Chem. Biol.* 1(3).
25. Erickson, D., Li, D.Q. 2004. Integrated microfluidic devices. *Anal. Chim. Acta* 507:11–26.
26. Campaña, A.M.G., Baeyens, W.R.G., Aboul-Enein, H.Y., Zhang, X. 1998. Miniaturization of capillary electrophoresis system using micromachining techniques. *J. Microcolumn Separations* 10:339–55.

27. Kutter, J.P. 2000. Current developments in electrophoretic and chromatographic separation methods on microfabricated devices. *Trends Anal. Chem.* 19:352–63.

28. Effenhauser, C.S. 1998. Integrated chip-based microcolumn separation systems. In *Microsystem technology in chemistry and life science*, 51–82. Heidelberg: Springer.

29. Jacobson, S.C., Ramsey, J.M. 1997. Microfabricated devices for performing capillary electrophoresis. In *Handbook of capillary electrophoresis*, 827–39. 2nd ed. Boca Raton, FL: CRC Press.

30. Manz, A., Verpoorte, E., Effenhauser, C.S., Burggraf, N., Raymond, D.E., Harrison, D.J., Widmer, H.M. 1993. Miniaturization of separation techniques using planar chip technology. *J. High Res. Chromatogr.* 16:433–36.

31. Ramsey, J.M., Jacobson, S.C., Knapp, M.R. 1995. Microfabricated chemical measurement systems. *Nat. Med.* 1:1093–96.

32. Jacobson, S.C., Ramsey, J.M. 1998. Microfabricated chemical separation devices. In *HPLC: Theory, techniques and applications,* 613–33. New York: Wiley.

33. Manz, A., Effenhauser, C.S., Burggraf, N., Harrison, D.J., Seiler, K., Fluri, K. 1994. Electroosmotic pumping and electrophoretic separations for miniaturized chemical analysis systems. *J. Micromech. Microeng.* 4:257–65.

34. Regnier, F.E., He, B., Lin, S., Busse, J. 1999. Chromatography and electrophoresis on chips: Critical elements of future integrated, microfluidic analytical systems for life science. *Trends Biotechnol.* 17:101–106.

35. Kricka, L.J., Wilding, P. 1996. Micromachining: A new direction for clinical analyzers. *Pure Appl. Chem.* 68:1831–36.

36. Landers, J.P. 2003. Molecular diagnostics on electrophoretic microchips. *Anal. Chem.* 75:2919.

37. Tüds, A.J., Besselink, G.A.J., Schasfoort, R.B.M. 2001. Trends in miniaturized total analysis systems for point-of-care testing in clinical chemistry. *Labchip* 1:83–95.

38. Liu, Y., Garcia, C.D., Henry, C.S. 2003. Recent progress in the development of μTAS for clinical analysis. *Analyst* 128:1002–8.

39. Schulte, T.H., Bardell, R.L., Weigl, B.H. 2002. Microfluidic technologies in clinical diagnostics. *Clin. Chim. Acta* 321:1–10.

40. Sundberg, S.A. 2000. High-throughput and ultra-high-throughput screening: Solution- and cell-based approaches. *Curr. Opin. Biotechnol.* 11:47–53.

41. Fuhr, G., Shirley, S.G. 1998. Biological application of microstructures. In *Microsystem technology in chemistry and life science*, 83–116. Heidelberg: Springer-Verlag.

42. Beebe, D.J., Mensing, G.A., Walker, G.M. 2002. Physics and applications of microfluidics in biology. *Annu. Rev. Biomed. Eng.* 4:261–86.

43. Andersson, H., van den Berg, A. 2004. Microtechnologies and nanotechnologies for single-cell analysis. *Curr. Opin. Biotechnol.* 15:44–49.

44. Dolník, V., Liu, S., Jovanovich, S. 2000. Capillary electrophoresis on microchip. *Electrophoresis* 21:41–54.

45. Sanders, G.H.W., Manz, A. 2000. Chip-based microsystems for genomic and proteomic analysis. *Trends Anal. Chem.* 19:364–78.

46. Khandurina, J., Guttman, A. 2002. Bioanalysis in microfluidic devices. *J. Chromatogr.* 943A:159–83.

47. Colyer, C.L., Tang, T., Chiem, N., Harrison, D.J. 1997. Clinical potential of microchip capillary electrophoresis systems. *Electrophoresis* 18:1733–41.

48. Wang, J. 2002. On-chip enzymatic assays. *Electrophoresis* 23:713–18.

49. Bousse, L., Cohen, C., Nikiforov, T., Chow, A., Kopf-Sill, A.R., Dubrow, R., Parce, J.W. 2000. Electrokinetically controlled microfluidic analysis systems. *Annu. Rev. Biophys. Biomol. Struct.* 29:155–81.

50. Lion, N., Reymond, F., Girault, H., Rossier, J.S. 2004. Why the move to microfluidics for protein analysis? *Curr. Opin. Biotechnol.* 15:31–37.

51. O'Donnell-Maloney, M.J., Little, D.P. 1996. Microfabrication and array technologies for DNA sequencing and diagnostics. *Genet. Anal. Biomol. Eng.* 13:151–57.

52. Figeys, D., Pinto, D. 2000. Lab-on-a-chip: A revolution in biological and medical sciences. *Anal. Chem.* 72:330A–35A.

53. Bruin, G.J.M. 2000. Recent development in electrokinetically driven analysis on microfabricated devices. *Electrophoresis* 21:3931–51.

54. Mathies, R.A., Simpson, P.C., Woolley, A.T. 1998. DNA analysis with capillary array electrophoresis microplates. In *Micro Total Analysis Systems '98*, Banff, Canada, October 13–16, pp. 1–6.

55. Medintz, I.L., Paegel, B.M., Blazej, R.G., Emrich, C.A., Berti, L., Scherer, J.R., Mathies, R.A. 2001. High-performance genetic analysis using microfabricated capillary array electrophoresis microplates. *Electrophoresis* 22:3845–56.

56. Abramowitz, S. 1998. DNA analysis in microfabricated formats. *Biomed. Microdevices* 1:107–12.

57. Burke, D.T., Burns, M.A., Mastrangelo, C. 1997. Microfabrication technologies for integrated nucleic acid analysis. *Genome Res.* 7:189–97.

58. Paegel, B.M., Blazej, R.G., Mathies, R.A. 2003. Microfluidic devices for DNA sequencing: Sample preparation and electrophoretic analysis. *Curr. Opin. Biotechnol.* 14:42–50.

59. Chou, C.F., Austin, R.H., Bakajin, O., Tegenfeldt, J.O., Castelino, J.A., Chan, S.S., Cox, E.C., Craighead, H., Darnton, N., Duke, T., Han, J., Turner, S. 2000. Sorting biomolecules with microdevices. *Electrophoresis* 21:81–90.

60. Henry, C.M. 1999. The incredible shrinking mass spectrometer: Miniaturization is on track to take MS into space and the doctor's office. *Anal. Chem.* 71:264A–68A.

61. Wang, J. 2002. Electrochemical detection for microscale analytical systems: A review. *Talanta* 56:223–31.

62. Schwarz, M.A., Hauser, P.C. 2001. Recent developments in detection methods for microfabricated analytical devices. *Labchip* 1:1–6.

63. de Mello, A.J., Chip, M.S. 2001. Coupling the large with the small. *Labchip* 1:7N.

64. Oleschuk, R.D., Harrison, D.J. 2000. Analytical microdevices for mass spectrometry. *Trends Anal. Chem.* 19:379–88.

65. Lacher, N.A., Garrison, K.E., Martin, R.C., Lunte, S.M. 2001. Microchip capillary electrophoresis/electrochemistry. *Electrophoresis* 22:2526–36.

66. Henry, C. 1997. Micro meets macro: Interfacing microchips and mass spectrometers. *Anal. Chem. News Features* 69:359A–61A.

67. Vandaveer, W.R., IV, Pasas, S.A., Martin, R.S., Lunte, S.M. 2002. Recent developments in amperometric detection for microchip capillary electrophoresis. *Electrophoresis* 23:3667–677.

68. Verpoorte, E. 2003. Chip vision-optics for microchips. *Labchip* 3:42N–52N.

69. Baldwin, R.P. 1999. Electrochemical determination of carbohydrates: Enzyme electrodes and amperometric detection in liquid chromatography and capillary electrophoresis. *J. Pharm. Biomed. Anal.* 19:69–81.

70. Wang, J. 2004. Microchip devices for detecting terrorist weapons. *Anal. Chim. Acta* 507:3–10.

71. Alarie, J.P., Jacobson, S.C., Culbertson, C.T., Ramsey, J.M. 2000. Effects of the electric field distribution on microchip valving performance. *Electrophoresis* 21:100–6.

72. Shoji, S., Esashi, M. 1994. Bonding and assembling methods for realising a μTAS. In *Micro Total Analysis Systems '94*, University of Twente, Netherlands, November 21–22, pp. 165–79.

73. Qin, D., Xia, Y., Rogers, J.A., Jackman, R.J., Zhao, X., Whitesides, G.M. 1998. Microfabrication, microstructures and microsystems. In *Microsystem technology in chemistry and life science*, 1–20. Heidelberg: Springer-Verlag.

74. McDonald, J.C., Duffy, D.C., Anderson, J.R., Chiu, D.T., Wu, H., Schueller, O.J.A., Whitesides, G.M. 2000. Fabrication of microfluidic systems in poly(dimethylsiloxane). *Electrophoresis* 21:27–40.

75. McDonald, J.C., Whitesides, G.M. 2002. Poly(dimethylsiloxane) as a material for fabricating microfluidic devices. *Acc. Chem. Res.* 35:491–49.

76. Xia, Y.N., Whitesides, G.M. 1998. Soft lithography. *Angew. Chem. Int. Ed. Engl.* 37:550–75.

77. Soper, S.A., Ford, S.M., Qi, S.Z., McCarley, R.L., Kelly, K., Murphy, M.C. 2000. Polymeric microelectromechanical systems. *Anal. Chem.* 72:643A–51A.

78. Xia, Y.N., Whitesides, G.M. 1998. Soft Lithography. *Annu. Rev. Mater. Sci.* 28:153–84.

79. Kenis, P.J.A., Ismagilov, R.F., Takayama, S., Whitesides, G.M., Li, S., White, H.S. 2000. Fabrication inside microchannels using fluid flow. *Acc. Chem. Res.* 33:841–47.

80. Elwenspoek, M., Lammerink, T.S.J., Miyake, R., Fluitman, J.H.J. 1994. Towards integrated microliquid handling systems. *J. Micromech. Microeng.* 4:227–45.

81. Cheng, J., Kricka, L.J., Sheldon, E.L., Wilding, P. 1998. Sample preparation in microstructure device. In *Microsystem technology in chemistry and life science*, 215–32. Heidelberg: Springer-Verlag.

82. Gravesen, P., Branebjerg, J., Jensen, O.S. 1993. Microfluidics—A review. *J. Micromech. Microeng.* 3:168–82.

83. Dodge, A., Turcatti, G., Lawrence, I., de Rooji, N.F., Verpoorte, E. 2004. A microfluidic platform using molecular beacon-based temperature calibration for thermal dehybridization of surface-bound DNA. *Anal. Chem.* 76:1778–87.

84. Nilsson, A., Petersson, F., Jönsson, H., Laurell, T. 2004. Acoustic control of suspended particles in micro fluidic chips. *Labchip* 4:131–35.

85. Russo, A.P., Apoga, D., Dowell, N., Shain, W., Turner, A.M.P., Craighead, H.G., Hoch, H.C., Turner, J.N. 2002. Microfabricated plastic devices from silicon using soft intermediates. *Biomed. Microdevices* 4:277–83.

86. Zhan, W., Alvarez, J., Crooks, R.M. 2002. Electrochemical sensing in microfluidic systems using electrogenerated chemiluminescence as a photonic reporter of redox reactions. *J. Am. Chem. Soc.* 124:13265–70.

87. Nieuwenhuis, J.H., Lee, S.S., Bastemeijer, J., Vellekoop, M.J. 2001. Particle-shape sensing-elements for integrated flow cytometer. In *Proceedings of the 5th Micro Total Analysis Systems Symposium,* Monterey, CA, October 21–25, pp. 357–58.

88. Mogensen, K.B., Petersen, N., Hübner, J., Kutter, J. 2001. In-plane UV absorbance detection in silicon-based electrophoresis devices using monolithically integrated optical waveguides. In *Proceedings of the 5th Micro Total Analysis Systems Symposium*, Monterey, CA, October 21–25, pp. 280–82.

89. Raymond, D.E., Manz, A., Widmer, H.M. 1994. Continuous sample pretreatment using a free-flow electrophoresis device integrated onto a silicon chip. *Anal. Chem.* 66:2858–65.

90. Brody, J.P., Yager, P. 1997. Diffusion-based extraction in a microfabricated device. *Sensors Actuators* 58A:13–18.

91. Wei, J., Xie, H., Nai, M.L., Wong, C.K., Lee, L.C. 2003. Low temperature wafer anodic bonding. *J. Micromech. Microeng.* 13:217–22.

92. Roulet, J.-C., Volkel, R., Herzig, H.P., Verpoorte, E., de Rooij, N.F., Dandliker, R. 2001. Fabrication of multilayer systems combining microfluidic and microoptical elements for fluorescence detection. *J. Microelectromech. Syst.* 10:482–91.

93. Gómez, R., Bashir, R., Sarikaya, A., Ladisch, M.R., Sturgis, J., Robinson, J.P., Geng, T., Bhunia, A.K., Apple, H.L., Wereley, S. 2001. Microfluidic biochip for impedance spectroscopy of biological species. *Biomed. Microdevices* 3:201–9.

94. Trau, D., Lee, T.M.H., Lao, A.I.K., Lenigk, R., Hsing, I.-M., Ip, N.Y., Carles, M.C., Sucher, N.J. 2002. Genotyping on a complementary metal oxide semiconductor silicon polymerase chain reaction chip with integrated DNA microarray. *Anal. Chem.* 74:3168–73.

95. Ugaz, V.M., Brahmasandra, S.N., Burke, D.T., Burns, M.A. 2002. Cross-linked polyacrylamide gel electrophoresis of single-stranded DNA for microfabricated genomic analysis systems. *Electrophoresis* 23:1450–59.

96. Lin, R.S., Burke, D.T., Burns, M.A. 2003. Selective extraction of size-fractioned DNA samples in microfabricated electrophoresis devices. *J. Chromatogr.* 1010A:255–68.

97. Lee, T.M.H., Carles, M.C., Hsing, I.M. 2003. Microfabricated PCR-electrochemical device for simultaneous DNA amplification and detection. *Labchip* 3:100–5.

98. Pal, R., Yang, M., Johnson, B.N., Burke, D.T., Burns, M.A. 2004. Phase change microvalve for integrated devices. *Anal. Chem.* 76:3740–48.

99. Niklaus, F., Andersson, H., Enoksson, P., Stemme, G. 2001. Low temperature full wafer adhesive bonding of structured wafers. *Sensors Actuators* 92A:235–41.

100. Hsueh, Y., Collins, S.D., Smith, R.L. 1998. DNA quantification with an electrochemiluminescence microcell. *Sensors Actuators* 49B:1–4.

101. Sobek, D., Young, A.M., Gray, M.L., Senturia, S.D. 1993. A microfabricated flow chamber for optical measurements in fluids. *Proc. IEEE, 2*, 219–224.

102. Fan, Z.H., Harrison, D.J. 1994. Micromachining of capillary electrophoresis injectors and separators on glass chips and evaluation of flow at capillary intersections. *Anal. Chem.* 66:177–84.

103. Kim, H., Ueno, K., Chiba, M., Kogi, O., Noboru, K. 2000. Spatially-resolved fluorescence spectroscopic study on liquid/liquid extraction processes in polymer microchannels. *Anal. Sci.* 16:871.

104. Diepold, T., Obermeier, E. 1996. Smoothing of ultrasonically drilled holes in borosilicate glass by wet chemical etching. *J. Micromech. Microeng.* 6:29–32.

105. Lin, C.H., Lee, G.B., Lin, Y.H., Chang, G.L. 2001. A fast prototyping process for fabrication of microfluidic systems on soda-lime glass. *J. Micromech. Microeng.* 11:726–32.

106. Jacobson, S.C., Hergenröder, R., Moore, A.W., Ramsey, J.M. 1994. Precolumn reactions with electrophoretic analysis integrated on a microchip. *Anal. Chem.* 66:4127–32.

107. Moore, A.W., Jr., Jacobson, S.C., Ramsey, J.M. 1995. Microchip separation of neutral species via micellar electrokinetic capillary chromatography. *Anal. Chem.* 67:4184–89.

108. Simpson, P.C., Woolley, A.T., Mathies, R.A. 1998. Microfabrication technology for the production of capillary array electrophoresis chips. *Biomed. Microdevices* 1:7–25.

109. Fu, L.-M., Yang, R.-J., Lee, G.-B. 2003. Electrokinetic focusing injection methods on microfluidic devices. *Anal. Chem.* 75:1905–10.

110. Holden, M.A., Kumar, S., Castellana, E.T., Beskok, A., Cremer, P.S. 2003. Generating fixed concentration arrays in a microfluidic device. *Sensors Actuators* 92B:199–207.
111. Stjernström, M., Roeraade, J. 1998. Method for fabrication of microfluidic systems in glass. *J. Micromech. Microeng.* 8:33–38.
112. Jacobson, S.C., Hergenröder, R., Koutny, L.B., Ramsey, J.M. 1994. Open channel electrochromatography on a microchip. *Anal. Chem.* 66:2369–73.
113. Liu, S., Zhang, J., Ren, H., Zheng, J., Liu, H. 2001. A microfabricated hybrid device for DNA sequencing. In *Proceedings of the 5th Micro Total Analysis Systems Symposium*, Monterey, CA, October 21–25, pp. 99–100.
114. Liu, S.R. 2003. A microfabricated hybrid device for DNA sequencing. *Electrophoresis* 24:3755–61.
115. Lazar, I.M., Karger, B.L. 2002. Multiple open-channel electroosmotic pumping system for microfluidic sample handling. *Anal. Chem.* 74:6259–68.
116. Jin, L.J., Ferrance, J., Sanders, J.C., Landers, J.P. 2003. A microchip-based proteolytic digestion system driven by electroosmotic pumping. *Labchip* 3:11–18.
117. Attiya, S., Jemere, A.B., Tang, T., Fitzpatrick, G., Seiler, K., Chiem, N., Harrison, D.J. 2001. Design of an interface to allow microfluidic electrophoresis chips to drink from the fire hose of the external environment. *Electrophoresis* 22:318–27.
118. Dutta, D., Leighton, D.T., Jr. 2003. Dispersion in large aspect ratio microchannels for open-channel liquid chromatography. *Anal. Chem.* 75:57–70.
119. Dutta, D., Leighton, D.T., Jr. 2001. Dispersion reduction in pressure-driven flow through microetched channels. *Anal. Chem.* 73:504–13.
120. Schwarz, M.A., Hauser, P.C. 2003. Chiral on-chip separations of neurotransmitters. *Anal. Chem.* 75:4691–95.
121. Lazar, I.M., Ramsey, R.S., Jacobson, S.C., Foote, R.S., Ramsey, J.M. 2000. Novel microfabricated device for electrokinetically induced pressure flow and electrospray ionization mass spectrometry. *J. Chromatogr.* 892A:195–201.
122. Corman, T., Enoksson, P., Stemme, G. 1998. Deep wet etching of borosilicate glass using an anodically bonded silicon substrate as mask. *J. Micromech. Microeng.* 8:84–87.
123. Bien, D.C.S., Rainey, P.V., Mitchell, S.J.N., Gamble, H.S. 2003. Characterization of masking materials for deep glass micromachining. *J. Micromech. Microeng.* 13:S34–40.
124. Baechi, D., Buser, R., Dual, J. 2001. High-density microvalve arrays for sample processing in PCR chips. *Biomed. Microdevices* 3:183–90.
125. Rodriguez, I., Spicar-Mihalic, P., Kuyper, C.L., Fiorini, G.S., Chiu, D.T. 2003. Rapid prototyping of glass microchannels. *Anal. Chim. Acta* 496:205–15.
126. Wensink, H., Elwenspoek, M.C. 2001. New developments in bulk micromachining by powder blasting. In *Proceedings of the 5th Micro Total Analysis Systems Symposium*, Monterey, CA, October 21–25, pp. 393–94.
127. Brivio, M., Tas, N.R., Fokkens, R.H., Sanders, R.G.P., Verboom, W., Reindhoudt, D.N., van de Berg, A. 2001. Chemical microreactors in combination with mass spectrometry. In *Proceedings of the 5th Micro Total Analysis Systems Symposium*, Monterey, CA, October 21–25, pp. 329–30.
128. Schasfoort, R., van Duijn, G., Schlautmann, S., Frank, H., Billiet, H., van Dedem, G., van den Berg, A. 2000. Miniaturized capillary electrophoresis system with integrated conductivity detector. In *Proceedings of the 4th Micro Total Analysis Systems Symposium*, Enschede, Netherlands, May 14–18, pp. 391–94.
129. Schlautmann, S., Besselink, G.A.J., Radhakrishna Prabhu, G., Schasfoort, R.B.M. 2003. Fabrication of a microfluidic chip by UV bonding at room temperature for integration of temperature-sensitive layers. *J. Micromech. Microeng.* 13:S81–84.
130. Belloy, E., Pawlowski, A.-G., Sayah, A., Gijs, M.A.M. 2002. Microfabrication of high-aspect ratio and complex monolithic structures in glass. *J. Microelectromech. Syst.* 11:521–27.
131. Ueno, Y., Horiuchi, T., Morimoto, T., Niwa, O. 2001. Microfluidic device for airborne BTEX detection. *Anal. Chem.* 73:4688–93.
132. Bings, N.H., Wang, C., Skinner, C.D., Colyer, C.L., Thibault, P., Harrison, D.J. 1999. Microfluidic devices connected to fused-silica capillaries with minimal dead volume. *Anal. Chem.* 71:3292–96.
133. Li, P.C.H., Wang, W.J., Parameswaran, M. 2003. An acoustic wave sensor incorporated with a microfluidic chip for analyzing muscle cell contraction. *Analyst* 128:225–31.
134. Huang, W.-H., Cheng, W., Zhang, Z., Pang, D.-W., Wang, Z.-L., Cheng, J.-K., Cui, D.-F. 2004. Transport, location, and quantal release monitoring of single cells on a microfluidic device. *Anal. Chem.* 76:483–88.

135. Lee, E.S., Howard, D., Liang, E.Z., Collins, S.D., Smith, R.L. 2004. Removable tubing interconnects for glass-based micro-fluidic systems made using ECDM. *J. Micromech. Microeng.* 14:535–41.

136. Jacobson, S.C., Hergenröder, R., Koutny, L.B., Warmack, R.J., Ramsey, J.M. 1994. Effects of injection schemes and column geometry on the performance of microchip electrophoresis devices. *Anal. Chem.* 66:1107–13.

137. Jacobson, S.C., Hergenröder, R., Koutny, L.B., Ramsey, J.M. 1994. High-speed separation on a microchip. *Anal. Chem.* 66:1114–18.

138. Li, P.C.H., de Camprieu, L., Cai, J., Sangar, M. 2004. Transport, retention and fluorescent measurement of single biological cells studied in microfluidic chips. *Labchip* 4:174–80.

139. Roper, M.G., Shackman, J.G., Dahlgren, G.M., Kennedy, R.T. 2003. Microfluidic chip for continuous monitoring of hormone secretion from live cells using an electrophoresis-based immunoassay. *Anal. Chem.* 75:4711–17.

140. Backhouse, C., Caamano, M., Oaks, F., Nordman, E., Carrillo, A., Johnson, B., Bay, S. 2000. DNA sequencing in a monolithic microchannel device. *Electrophoresis* 21:150–56.

141. Lichtenberg, J., de Rooij, N.F., Verpoorte, E. 2002. A microchip electrophoresis system with integrated in-plane electrodes for contactless conductivity detection. *Electrophoresis* 23:3769–80.

142. Yan, K.Y., Smith, R.L., Collins, S.D. 2000. Fluidic microchannel arrays for the electrophoretic separation and detection of bioanalytes using electrochemiluminescence. *Biomed. Microdevices* 2:221–29.

143. Ito, T., Sobue, K., Ohya, S. 2002. Water glass bonding for micro-total analysis system. *Sensors Actuators* 81B:187–95.

144. Berthold, A., Nicola, L., Sarro, P.M., Vellekoop, M.J. 1999. Microfluidic device for airborne BTEX detection. *Transducers* 99:1324–27.

145. Guijt, R.M., Baltussen, E., van der Steen, G., Frank, H., Billiet, H., Schalkhammer, T., Laugere, F., Vellekoop, M., Berthold, A., Sarro, L., van Dedem, G.W.K. 2001. Capillary electrophoresis with on-chip four-electrode capacitively coupled conductivity detection for application in bioanalysis. *Electrophoresis* 22:2537–41.

146. Namasivayam, V., Larson, R.G., Burke, D.T., Burns, M.A. 2003. Transpiration-based micropump for delivering continuous ultra-low flow rates. *J. Micromech. Microeng.* 13:261–71.

147. Divakar, R., Butler, D., Papautsky, I. 2001. Room temperature low-cost UV-cured adhesive bonding for microfluidic biochips. In *Proceedings of the 5th Micro Total Analysis Systems Symposium*, Monterey, CA, October 21–25, pp. 385–86.

148. Broyles, B.S., Jacobson, S.C., Ramsey, J.M. 2003. Sample filtration, concentration, and separation integrated on microfluidic devices. *Anal. Chem.* 75:2761–67.

149. Lee, G.B., Chen, S.H., Huang, G.R., Sung, W.C., Lin, Y.H. 2001. Microfabricated plastic chips by hot embossing methods and their applications for DNA separation and detection. *Sensors Actuators* 75B:142–48.

150. Becker, H., Lowack, K., Manz, A. 1998. Planar quartz chips with submicron channels for two-dimensional capillary electrophoresis applications. *J. Micromech. Microeng.* 8:24–28.

151. Swinney, K., Markov, D., Bornhop, D.J. 2000. Chip-scale universal detection based on backscatter interferometry. *Anal. Chem.* 72:2690–95.

152. Schmalzing, D., Koutny, L.B., Taylor, T.A., Nashabeh, W., Fuchs, M. 1997. Immunoassay for thyroxine (T4) in serum using capillary electrophoresis and micromachined devices. *J. Chromotogr.* 697B:175–80.

153. Fluri, K., Fitzpatrick, G., Chiem, N., Harrison, D.J. 1996. Integrated capillary electrophoresis devices with an efficient postcolumn reactor in planar quartz and glass chips. *Anal. Chem.* 68:4285–90.

154. Ericson, C., Holm, J., Ericson, T., Hjerten, S. 2000. Electroosmosis- and pressure-driven chromatography in chips using continuous beds. *Anal. Chem.* 72:81–87.

155. Kim, E., Xia, Y., Whitesides, G.M. 1995. Polymer microstructures formed by moulding in capillaries. *Nature* 376:581–84.

156. Xia, Y., Kim, E., Whitesides, G.M. 1996. Micromolding of polymers in capillaries: Applications in microfabrication. *Chem. Mater.* 8:1558–67.

157. Jeon, N.L., Choi, I.S., Xu, B., Whitesides, G.M. 1999. Large-area patterning by vacuum-assisted micromolding. *Adv. Mater.* 11:946–50.

158. Kim, E., Xia, Y., Whitesides, G.M. 1996. Micromolding in capillaries: Applications in materials science. *J. Am. Chem. Soc.* 118:5722–31.

159. Effenhauser, C.S., Bruin, G.J.M., Paulus, A., Ehrat, M. 1997. Integrated capillary electrophoresis on flexible silicone microdevices: Analysis of DNA restriction fragments and detection of single DNA molecules on microchips. *Anal. Chem.* 69:3451–57.

160. Folch, A., Mezzour, S., Düring, M., Hurtado, O., Toner, M., Müller, R. 2000. Stacks of microfabricated structures as scaffolds for cell culture and tissue engineering. *Biomed. Microdevices* 2:207–14.

161. Dittrich, P.S., Schwille, P. 2003. An integrated microfluidic system for reaction, high-sensitivity detection, and sorting of fluorescent cells and particles. *Anal. Chem.* 75:5767–74.

162. Huikko, K., Östman, P., Grigoras, K., Tuomikoski, S., Tiainen, V.M., Soininen, A., Puolanne, K., Manz, A., Franssila, S., Kostiainen, R., Kotiaho, T. 2003. Poly(dimethylsiloxane) electrospray devices fabricated with diamond-like carbon-poly(dimethylsiloxane) coated SU-8 masters. *Labchip* 3:67–72.

163. Walker, G.M., Ozers, M.S., Beebe, D.J. 2002. Insect cell culture in microfluidic channels. *Biomed. Microdevices* 4:161–66.

164. Mata, A., Boehm, C., Fleischman, A.J., Muschler, G., Roy, S. 2002. Analysis of connective tissue progenitor cell behavior on polydimethylsiloxane smooth and channel micro-textures. *Biomed. Microdevices* 4:267–75.

165. Yang, J., Li, C.W., Yang, M.S. 2004. Hydrodynamic simulation of cell docking in microfluidic channels with different dam structures. *Labchip* 4:53–59.

166. Zhao, D.S., Roy, B., McCormick, M.T., Kuhr, W.G., Brazill, S.A. 2003. Rapid fabrication of a poly(dimethylsiloxane) microfluidic capillary gel electrophoresis system utilizing high precision machining. *Labchip* 3:93–99.

167. Unger, M.A., Chou, H., Thorsen, T., Scherer, A., Quake, S.R. 2000. Monolithic microfabricated valves and pumps by multilayer soft lithography. *Science* 288:113–16.

168. Liu, B.-F., Ozaki, M., Utsumi, Y., Hattori, T., Terabe, S. 2003. Chemiluminescence detection for a microchip capillary electrophoresis system fabricated in poly(dimethylsiloxane). *Anal. Chem.* 75:36–41.

169. Price, A.K., Fischer, D.J., Martin, R.S., Spence, D.M. 2004. Deformation-induced release of ATP from erythrocytes in a poly(dimethylsiloxane)-based microchip with channels that mimic resistance vessels. *Anal. Chem.* 76:4849–55.

170. Linder, V., Verpoorte, E., Thormann, W., de Rooij, N.F., Sigrist, H. 2001. Surface biopassivation of replicated poly(dimethylsiloxane) microfluidic channels and application to heterogeneous immunoreaction with on-chip fluorescence detection. *Anal. Chem.* 73:4181–89.

171. Wang, B., Chen, L., Abdulali-Kanji, Z., Horton, J.H., Oleschuk, R.D. 2003. Aging effects on oxidized and amine-modified poly(dimethylsiloxane) surfaces studied with chemical force titrations: Effects on electroosmotic flow rate in microfluidic channels. *Langmuir* 19:9792–98.

172. Svedberg, M., Veszelei, M., Axelsson, J., Vangbo, M., Nikolajeff, F. 2004. Poly(dimethylsiloxane) microchip: Microchannel with integrated open electrospray tip. *Labchip* 4:322–27.

173. Eteshola, E., Leckband, D. 2001. Development and characterization of an ELISA assay in PDMS microfluidic channels. *Sensors Actuators* 72B:129–33.

174. Xiao, D., Le, T.V., Wirth, M.J. 2004. Surface modification of the channels of poly(dimethylsiloxane) microfluidic chips with polyacrylamide for fast electrophoretic separations of proteins. *Anal. Chem.* 76:2055–61.

175. Jo, B.-H., Van Lerberghe, L.M., Motsegood, K.M., Beebe, D.J. 2000. Three-dimensional micro-channel fabrication in polydimethylsiloxane (PDMS) elastomer. *J. Microelectromech. Syst.* 9:76–81.

176. Wakamoto, Y.C., Umehara, S., Matsumura, K., Inoue, I., Yasuda, K. 2003. Development of non-destructive, non-contact single-cell based differential cell assay using on-chip microcultivation and optical tweezers. *Sensors Actuators* 96B:693–700.

177. Xu, J., Locascio, L., Gaitan, M., Lee, C.S. 2000. Room-temperature imprinting method for plastic microchannel fabrication. *Anal. Chem.* 72:1930–33.

178. Ko, J.S., Yoon, H.C., Yang, H., Pyo, H.B., Chung, K.H., Kim, S.J., Kim, Y.T. 2003. A polymer-based microfluidic device for immunosensing biochips. *Labchip* 3:106–13.

179. Lee, S., Kim, Y. 2001. Metering and mixing of nanoliter liquid in the microchannel networks driven by fluorocarbon surfaces and pneumatic control. In *Proceedings of the 5th Micro Total Analysis Systems Symposium, Monterey,* CA, October 21–25, pp. 205–6.

180. Anderson, J.R., Chiu, D.T., Jackman, R.J., Cherniavskaya, O., McDonald, J.C., Wu, H., Whitesides, S.H., Whitesides, G.M. 2000. Fabrication of topologically complex three-dimensional microfluidic systems in PDMS by rapid prototyping. *Anal. Chem.* 72:3158–64.

181. Hofmann, O., Niedermann, P., Manz, A. 2001. Modular approach to fabrication of three-dimensional microchannel systems in PDMS—Application to sheath flow microchips. *Labchip* 1:108–14.

182. Chen, X., Wu, H., Mao, C., Whitesides, G.M. 2002. A prototype two-dimensional capillary electrophoresis system fabricated in poly(dimethylsiloxane). *Anal. Chem.* 74:1772–78.

183. Sudarsan, A.P., Ugaz, V.M. 2004. Printed circuit technology for fabrication of plastic-based microfluidic devices. *Anal. Chem.* 76:3229–35.

184. McDonald, J.C., Chabinyc, M.L., Metallo, S.J., Anderson, J.R., Stroock, A.D., Whitesides, G.M. 2002. Prototyping of microfluidic devices in poly(dimethylsiloxane) using solid-object printing. *Anal. Chem.* 74:1537–45.

185. Deng, T., Wu, H., Brittain, S.T., Whitesides, G.M. 2000. Prototyping of masks, masters, and stamps/molds for soft lithography using an office printer and photographic reduction. *Anal. Chem.* 72:3176–80.

186. McCormick, R.M., Nelson, R.J., Alonso-Amigo, M.G., Benvegnu, D.J., Hooper, H.H. 1997. Microchannel electrophoretic separation of DNA in injection-molded plastic substrates. *Anal. Chem.* 69:2626–30.

187. Lee, L.J., Madou, M.J., Koelling, K.W., Daunert, S., Lai, S.Y., Koh, C.G., Juang, Y.J., Lu, Y.M., Yu, L.Y. 2001. Design and fabrication of CD-like microfluidic platforms for diagnostics: Polymer-based microfabrication. *Biomed. Microdevices* 3:339–51.

188. Sjoelander, S., Urbaniczky, C. 1991. Integrated fluid handling system for biomolecular interaction analysis. *Anal. Chem.* 63:2338–45.

189. Rossier, J.S., Schwarz, A., Reymond, F., Ferrigno, R., Bianchi, F., Girault, H.H. 1999. Microchannel networks for electrophoretic separations. *Electrophoresis* 20:727–31.

190. Schwarz, A., Rossier, J.S., Bianchi, F., Reymond, F., Ferrigno, R., Girault, H.H. 1998. Micro-TAS on polymer substrates micromachined by laser photoablation. In *Micro Total Analysis Systems '98*, Banff, Canada, October 13–16, pp. 241–44.

191. Rossier, J.S., Ferrigno, R., Girault, H.H. 2000. Electrophoresis with electrochemical detection in a polymer microdevice. *J. Electroanal. Chem.* 492:15–22.

192. Giordano, B.C., Ferrance, J., Swedberg, S., Huhmer, A.F.R., Landers, J.P. 2001. Polymerase chain reaction in polymeric microchips: DNA amplification in less than 240 seconds. *Anal. Biochem.* 291:124–32.

193. Johnson, T.J., Ross, D., Locascio, L.E. 2002. Rapid microfluidic mixing. *Anal. Chem.* 74:45–51.

194. Roberts, M.A., Rossier, J.S., Bercier, P., Girault, H. 1997. UV laser machined polymer substrates for the development of microdiagnostic systems. *Anal. Chem.* 69:2035–42.

195. Rohner, T.C., Rossier, J.S., Girault, H.H. 2001. Polymer microspray with an integrated thick-film microelectrode. *Anal. Chem.* 73:5353–57.

196. Kancharla, V.V., Chen, S.C. 2002. Fabrication of biodegradable polymeric micro-devices using laser micromachining. *Biomed. Microdevices* 4:105–9.

197. Munce, N.R., Li, J., Herman, P.R., Lilge, L. 2004. Microfabricated system for parallel single-cell capillary electrophoresis. *Anal. Chem.* 76:4983–89.

198. Kaji, H., Nishizawa, M., Matsue, T. 2003. Localized chemical stimulation to micropatterned cells using multiple laminar fluid flows. *Labchip* 3:208–11.

199. Ford, S.M., Kar, B., McWhorter, S., Davies, J., Soper, S.A., Klopf, M., Calderon, G., Saile, V. 1998. Microcapillary electrophoresis device fabricated using polymeric substrates and x-ray lithography. *J. Microcolumn Separations* 10:413–22.

200. Meng, Z., Qi, S., Soper, S.A., Limbach, P.A. 2001. Interfacing a polymer-based micromachined device to a nanoelectrospray ionization Fourier transform ion cyclotron resonance mass spectrometer. *Anal. Chem.* 73:1286–91.

201. Goranovic, G., Klank, H., Westergaard, C., Geschke, O., Telleman, P., Kutter, J.P. 2001. Characterization of flow in laser-machined polymeric microchannels. In *Proceedings of the 5th Micro Total Analysis Systems Symposium*, Monterey, CA, October 21–25, pp. 623–24.

202. Klank, H., Kutter, J.P., Geschke, O. 2002. CO_2-laser micromachining and back-end processing for rapid production of PMMA-based microfluidic systems. *Labchip* 2:242–46.

203. Malmstadt, N., Yager, P., Hoffman, A.S., Stayton, P.S. 2003. A smart microfluidic affinity chromatography matrix composed of poly(n-isopropylacrylamide)-coated beads. *Anal. Chem.* 75:2943–49.

204. Wang, S.-C., Perso, C.E., Morris, M.D. 2000. Effects of alkaline hydrolysis and dynamic coating on the electroosmotic flow in polymeric microfabricated channels. *Anal. Chem.* 72:1704–6.

205. Chen, D., Hsu, F., Zhan, D., Chen, C. 2001. Palladium film decoupler for amperometric detection in electrophoresis chips. *Anal. Chem.* 73:758–62.

206. Locascio, L.E., Perso, C.E., Lee, C.S. 1999. Measurement of electroosmotic flow in plastic imprinted microfluid devices and the effect of protein adsorption on flow rate. *J. Chromatogr.* 857A:275–84.

207. Chen, Y.H., Chen, S.H. 2000. Analysis of DNA fragments by microchip electrophoresis fabricated on poly(methyl methacrylate) substrates using a wire-imprinting method. *Electrophoresis* 21:165–70.

208. Galloway, M., Stryjewski, W., Henry, A., Ford, S.M., Llopis, S., McCarley, R.L., Soper, S.A. 2002. Contact conductivity detection in poly(methyl methacylate)-based microfluidic devices for analysis of mono- and polyanionic molecules. *Anal. Chem.* 74:2407–15.

209. Esch, M.B., Kapur, S., Irizarry, G., Genova, V. 2003. Influence of master fabrication techniques on the characteristics of embossed microfluidic channels. *Labchip* 3:121–27.

210. Ueno, K., Kim, H.-B., Kitamura, N. 2003. Characteristic electrochemical responses of polymer microchannel-microelectrode chips. *Anal. Chem.* 75:2086–91.

211. Becker, H., Dietz, W., Dannberg, P. 1998. Microfluidic manifolds by polymer hot embossing for μ-TAS applications. In *Micro Total Analysis Systems '98,* Banff, Canada, October 13–16, pp. 253–56.

212. Soper, S.A., Murphy, M.C., McCarley, R.L., Nikitopoulos, D., Liu, X., Vaidya, B., Barrow, J., Bejat, Y., Ford, S.M., Goettert, J. 2001. Fabrication of modular microsystems for analyzing K-ras mutations using LDR. In *Proceedings of the 5th Micro Total Analysis Systems Symposium,* Monterey, CA, October 21–25, pp. 459–61.

213. Svedberg, M., Pettersson, A., Nilsson, S., Bergquist, J., Nyholm, L., Nikolajeff, F., Markides, K. 2003. Sheathless electrospray from polymer microchips. *Anal. Chem.* 75:3934–40.

214. Griebel, A., Rund, S., Schönfeld, F., Dörner, W., Konrad, R., Hardt, S. 2004. Integrated polymer chip for two-dimensional capillary gel electrophoresis. *Labchip* 4:18–23.

215. Kelly, R.T., Woolley, A.T. 2003. Thermal bonding of polymeric capillary electrophoresis microdevices in water. *Anal. Chem.* 75:1941–45.

216. Barker, S.L.R., Ross, D., Tarlov, M.J., Gaitan, M., Locascio, L.E. 2000. Control of flow direction in microfluidic devices with polyelectrolyte multilayers. *Anal. Chem.* 72:5925–29.

217. Benetton, S., Kameoka, J., Tan, A., Wachs, T., Craighead, H., Henion, J.D. 2003. Chip-based P450 drug metabolism coupled to electrospray ionization-mass spectrometry detection. *Anal. Chem.* 75:6430–36.

218. Beebe, D.J., Mensing, G., Moorthy, J., Khoury, C.M., Pearce, T.M. 2001. Alternative approaches to microfluidic systems design, construction and operation. In *Proceedings of the 5th Micro Total Analysis Systems Symposium,* Monterey, CA, October 21–25, pp. 453–55.

219. Khoury, C., Mensing, G.A., Beebe, D.J. 2002. Ultra rapid prototyping of microfluidic systems using liquid phase photopolymerization. *Labchip* 2:50–55.

220. Harrison, C., Cabral, J.T., Stafford, C.M., Karim, A., Amis, E.J. 2004. A rapid prototyping technique for the fabrication of solvent-resistant structures. *J. Micromech. Microeng.* 14:153–58.

221. Beebe, D.J., Moore, J.S., Bauer, J.M., Yu, Q., Liu, R.H., Devadoss, C., Jo, B. 2000. Functional hydrogel structures for autonomous flow control inside microfluidic channels. *Nature* 404:588–90.

222. Eddington, D.T., Liu, R.H., Moore, J.S., Beebe, D.J. 2001. An organic self-regulating microfluidic system. *Labchip* 1:96–99.

223. Muck, A., Jr., Wang, J., Jacobs, M., Chen, G., Chatrathi, M.P., Jurka, V., Vyborny, Z., Spillman, S.D., Sridharan, G., Schoning, M.J. 2004. Fabrication of poly(methyl methacrylate) microfluidic chips by atmospheric molding. *Anal. Chem.* 76:2290–97.

224. Lai, S., Cao, X., Lee, L.J. 2004. A packaging technique for polymer microfluidic platforms. *Anal. Chem.* 76:1175–83.

225. Lee, L.P., Berger, S.A., Pruitt, L., Liepmann, D. 1998. Key elements of a transparent Teflon microfluidic system. In *Micro Total Analysis Systems '98,* Banff, Canada, October 13–16, pp. 245–48.

226. Lai, S., Wang, S., Luo, J., Lee, L.J., Yang, S.-T., Madou, M.J. 2004. Design of a compact disk-like microfluidic platform for enzyme-linked immunosorbent assay. *Anal. Chem.* 76:1832–37.

227. Grodzinski, P., Yang, J., Liu, R.H., Ward, M.D. 2003. A modular microfluidic system for cell pre-concentration and genetic sample preparation. *Biomed. Microdevices* 5:303–10.

228. Metz, S., Trautmann, C., Bertsch, A., Renaud, Ph. 2004. Polyimide microfluidic devices with integrated nanoporous filtration areas manufactured by micromachining and ion track technology. *J. Micromech. Microeng.* 14:324–31.

229. Lin, Y.H., Timchalk, C.A., Matson, D.W., Wu, H., Thrall, K.D. 2001. Integrated microfluidics/electrochemical sensor system for monitoring of environmental exposures to lead and chlorophenols. *Biomed. Microdevices* 3:331–38.

230. Henry, C.S., Vandaveer, W.R., IV, Mubarak, I., Gray, S.R., Fritsch, I. 2001. Self-contained microelectrochemical detectors for analysis in small volumes of static and flowing fluids. In *Proceedings of the 5th Micro Total Analysis Systems Symposium*, Monterey, CA, October 21–25, pp. 321–22.

231. Webster, J.R., Burns, M.A., Burke, D.T., Mastrangelo, C.H. 1998. An inexpensive plastic technology for microfabricated capillary electrophoresis chips. In *Micro Total Analysis Systems '98,* Banff, Canada, October 13–16, pp. 249–52.

232. Renaud, P., Lintel, H.V., Heuschkel, M., Guérin, L. 1998. Photo-polymer microchannel technologies and applications. In *Micro Total Analysis Systems '98,* Banff, Canada, October 13–16, pp. 17–22.

233. Madou, M.J., Lu, Y., Lai, S., Lee, J., Daunert, S. 2000. A centrifugal microfluidic platform—A comparison. In *Proceedings of the 4th Micro Total Analysis Systems Symposium,* Enschede, Netherlands, May 14–18, pp. 565–70.

234. Cui, Li., Holmes, D., Morgan, H. 2001. The dielectrophoretic levitation and separation of latex beads in microchips. *Electrophoresis* 22:3893–901.

235. Jackman, R.J., Floyd, T.M., Schmidt, M.A., Fensen, K.F. 2000. Development of methods for on-line chemical detection with liquid-phase microchemical reactors using conventional and unconventional techniques. In *Proceedings of the 4th Micro Total Analysis Systems Symposium,* Enschede, Netherlands, May 14–18, pp. 155–58.

236. Passeraub, Ph.A., Almeida, A.C., Thakor, N.V. 2003. Design, microfabrication and analysis of a microfluidic chamber for the perfusion of brain tissue slices. *Biomed. Microdevices* 5:147–55.

237. Lin, C.H., Lee, G.B., Chang, B.W., Chang, G.L. 2002. A new fabrication process for ultra-thick microfluidic microstructures utilizing SU-8 photoresist. *J. Micromech. Microeng.* 12:590–97.

238. Lago, C.L., da Silva, T.H.D., Neves, C.A., Alves Brito-Neto, A.J.G., Fracassi da Silva, J.A. 2003. A dry process for production of microfluidic devices based on the lamination of laser-printed polyester films. *Anal. Chem.* 75:3853–58.

239. Vervoort, N., Clicq, D., Baron, G.V., Desmet, G. 2003. Experimental Van Deemter plots of shear-driven liquid chromatographic separations in disposable microchannels. *J. Chromatogr.* 987A:39–48.

240. Mizukami, Y., Rajniak, D., Rajniak, A., Nishimura, M. 2002. A novel microchip for capillary electrophoresis with acrylic microchannel fabricated on photosensor array. *Sensors Actuators* 81B:202–9.

241. Hisamoto, H., Nakashima, Y., Kitamura, C., Funano, S.I., Yasuoka, M., Morishima, K., Kikutani, Y., Kitamori, T., Terabe, S. 2004. Capillary-assembled microchip for universal integration of various chemical functions onto a single microfluidic device. *Anal. Chem.* 76:3222–28.

242. Wu, C.-C., Wu, R.-G., Huang, J.-G., Lin, Y.-C., Chang, H.-C. 2003. Three-electrode electrochemical detector and platinum film decoupler integrated with a capillary electrophoresis microchip for amperometric detection. *Anal. Chem.* 75:947–52.

243. Chi, J., Kim, S., Trichur, R., Cho, H.J., Puntambekar, A., Cole, R.L., Simkins, J., Murugesan, S., Kim, K., Lee, J., Beaucage, G., Nevin, J.H., Ahn, C.H. 2001. A plastic micro injection molding technique using replaceable mold-disks for disposable microfluidic system and biochips. In *Proceedings of the 5th Micro Total Analysis Systems Symposium,* Monterey, CA, October 21–25, pp. 411–12.

244. Grover, W.H., Skelley, A.M., Liu, C.N., Lagally, E.T., Mathies, R.A. 2003. Monolithic membrane valves and diaphragm pumps for practical large-scale integration into glass microfluidic devices. *Sensors Actuators* 89B:315–23.

245. Shi, Y.N., Anderson, R.C. 2003. High-resolution single-stranded DNA analysis on 4.5 cm plastic electrophoretic microchannels. *Electrophoresis* 24:3371–77.

246. Anderson, R.C., Su, X., Bogdan, G.J., Fenton, J. 2000. A miniature integrated device for automated multistep genetic assays. *Nucleic Acids Res.* 28:e60.

247. Delamarche, E., Bernard, A., Schmid, H., Bietsch, A., Michel, B., Biebuyck, H. 1998. Microfluidic networks for chemical patterning of substrates: Design and application to bioassays. *J. Am. Chem. Soc.* 120:500–8.

248. Jiang, X., Ng, J.M.K., Stroock, A.D., Dertinger, S.K.W., Whitesides, G.M. 2003. A miniaturized, parallel, serially diluted immunoassay for analyzing multiple antigens. *J. Am. Chem. Soc.* 125:5294–95.

249. Chang, W.J., Akin, D., Sedlak, M., Ladisch, M.R., Bashir, R. 2003. Poly(dimethylsiloxane) (PDMS) and silicon hybrid biochip for bacterial culture. *Biomed. Microdevices* 5:281–90.

250. Matsubara, Y., Murakami, Y., Kobayashi, M., Morita, Y., Tamiya, E. 2004. Application of on-chip cell cultures for the detection of allergic response. *Biosens. Bioelectronics* 19:741–47.

251. Lee, J.N., Park, C., Whitesides, G.M. 2003. Solvent compatibility of poly(dimethylsiloxane)-based microfluidic devices. *Anal. Chem.* 75:6544–54.

252. Adams, M.L., Enzelberger, M., Quake, S., Scherer, A. 2003. Microfluidic integration on detector arrays for absorption and fluorescence micro-spectrometers. *Sensors Actuators* 104A:25–31.

253. Hong, J.W., Fujii, T., Seki, M., Yamamoto, T., Endo, I. 2001. Integration of gene amplification and capillary gel electrophoresis on a polydimethylsiloxane-glass hybrid microchip. *Electrophoresis* 22:328–33.

254. Hu, S., Ren, X., Bachman, M., Sims, C.E., Li, G.P., Allbritton, N.L. 2004. Surface-directed, graft polymerization within microfluidic channels. *Anal. Chem.* 76:1865–70.

255. Vaidya, B., Soper, S.A., McCarley, R.L. 2002. Surface modification and characterization of microfabricated poly(carbonate) devices: Manipulation of electroosmotic flow. *Analyst* 127:1289–92.

256. Ren, X.Q., Bachman, M., Sims, C., Li, G.P., Allbritton, N. 2001. Electroosmotic properties of microfluidic channels composed of poly(dimethylsiloxane). *J. Chromotogr.* 762B:117–25.

257. Tohnson, T.J., Ross, D., Gaitan, M., Locascio, L.E. 2001. Laser modification on channels to reduce band broadening or to increase mixing. In *Proceedings of the 5th Micro Total Analysis Systems Symposium*, Monterey, CA, October 21–25, pp. 603–4.

258. Bianchi, F., Wagner, F., Hoffmann, P., Girault, H.H. 2001. Electroosmotic flow in composite microchannels and implications in microcapillary electrophoresis systems. *Anal. Chem.* 73:829–36.

259. Ross, D., Johnson, T.J., Locascio, L.E. 2001. Imaging of electroosmotic flow in plastic microchannels. *Anal. Chem.* 73:2509–15.

260. Yamada, M., Seki, M. 2004. Nanoliter-sized liquid dispenser array for multiple biochemical analysis in microfluidic devices. *Anal. Chem.* 76:895–99.

261. Jo, B., Moorthy, J., Beebe, D. 2000. Polymer microfluidic valves, membranes and coatings. In *Proceedings of the 4th Micro Total Analysis Systems Symposium,* Enschede, Netherlands, May 14–18, pp. 335–38.

262. Lahann, J., Balcells, M., Lu, H., Rodon, T., Jensen, K.F., Langer, R. 2003. Reactive polymer coatings: A first step toward surface engineering of microfluidic devices. *Anal. Chem.* 75:2117–22.

263. Lee, G.B., Hwei, B.H., Huang, G.R. 2001. Micromachined pre-focused M×N flow switches for continuous multi-sample injection. *J. Micromech. Microeng.* 11:654–61.

264. Takamura, Y., Onoda, H., Inokuchi, H., Adachi, S., Oki, A., Horiike, Y. 2003. Low-voltage electroosmosis pump for stand-alone microfluidics devices. *Electrophoresis* 24:185–92.

265. Liu, S.R., Ren, H.J., Gao, Q.F., Roach, D.J., Loder, R.T., Armstrong, T.M., Jr., Mao, Q.L., Blaga, I., Barker, D.L., Jovanovich, S.B. 2000. Automated parallel DNA sequencing on multiple channel microchips. *Proc. Natl. Acad. Sci. USA* 97:5369–74.

266. Koch, M., Chatelain, D., Evans, A.G.R., Brunnschweiler, B. 1998. Two simple micromixers based on silicon. *J. Micromech. Microeng.* 8:123–26.

267. Stark, R.W., Stalder, M.S., Stemmer, A. 2003. Microfluidic etching driven by capillary forces for rapid prototyping of gold structures. *Microelectron. Eng.* 67–68:229–36.

268. Manica, D.P., Ewing, A.G. 2002. Prototyping disposable electrophoresis microchips with electrochemical detection using rapid marker masking and laminar flow etching. *Electrophoresis* 23:3735–43.

269. Lagally, E.T., Scherer, J.R., Blazej, R.G., Toriello, N.M., Diep, B.A., Ramchandani, M., Sensabaugh, G.F., Riley, L.W., Mathies, R.A. 2004. Integrated portable genetic analysis microsystem for pathogen/infectious disease detection. *Anal. Chem.* 76:3162–70.

270. Werdich, A.A., Lima, E.A., Ivanov, B., Ges, I., Anderson, M.E., Wikswo, J.P., Baudenbacher, F.J. 2004. A microfluidic device to confine a single cardiac myocyte in a sub-nanoliter volume on planar microelectrodes for extracellular potential recordings. *Labchip* 4:357–62.

271. Allen, P.B., Rodriguez, I., Kuyper, C.L., Lorenz, R.M., Spicar-Mihalic, P., Kuo, J.S., Chiu, D.T. 2003. Selective electroless and electrolytic deposition of metal for applications in microfluidics: Fabrication of a microthermocouple. *Anal. Chem.* 75:1578–83.

272. Hilmi, A., Luong, J.H.T. 2000. Electrochemical detectors prepared by electroless deposition for microfabricated electrophoresis chips. *Anal. Chem.* 72:4677–82.

273. Qiu, H., Yan, J., Sun, X., Liu, J., Cao, W., Yang, X., Wang, E. 2003. Microchip capillary electrophoresis with an integrated indium tin oxide electrode-based electrochemiluminescence detector. *Anal. Chem.* 75:5435–40.

274. Zhan, W., Alvarez, J., Crooks, R.M. 2003. A two-channel microfluidic sensor that uses anodic electrogenerated chemiluminescence as a photonic reporter of cathodic redox reactions. *Anal. Chem.* 75:313–18.

275. Sun, K., Yamaguchi, A., Ishida, Y., Matsuo, S., Misawa, H. 2002. A heater-integrated transparent microchannel chip for continuous-flow PCR. *Sensors Actuators* 84B:283–89.

276. Jeong, Y.W., Kim, S.Y., Chung, S., Paik, S.J., Han, Y.S., Chang, J.K., Cho, D.D., Chuang, D.S., Chun, K. 2001. Methodology for junction dilution compensation pattern and embedded electrode in CE separator. In *Proceedings of the 5th Micro Total Analysis Systems Symposium*, Monterey, CA, October 21–25, pp. 159–160.

277. Lichtenberg, J., Verpoorte, E., De Rooij, N.F. 2001. Operating parameters for an in-plane, contactless conductivity detector for microchip-based separation methods. In *Proceedings of the 5th Micro Total Analysis Systems Symposium*, Monterey, CA, October 21–25, pp. 323–24.

278. Berger, M., Castelino, J., Huang, R., Shah, M., Austin, R.H. 2001. Design of a microfabricated magnetic cell separator. *Electrophoresis* 22:3883–92.

279. Schlautmann, S., Wensink, H., Schasfoort, R., Elwenspoek, M., van den Berg, A. 2001. Powder-blasting technology as an alternative tool for microfabrication of capillary electrophoresis chips with integrated conductivity sensors. *J. Micromech. Microeng.* 11:386–89.

280. Guenat, O.T., Ghiglione, D., Morf, W.E., de Rooij, N.F. 2001. Partial electroosmotic pumping in complex capillary systems. Part 2. Fabrication and application of a micro total analysis system (TAS) suited for continuous volumetric nanotitrations. *Sensors Actuators* 72B:273–82.

281. Murakami, Y., Takeuchi, T., Yokoyama, K., Tomiya, E., Karube, I. 1993. Integration of enzyme-immobilized column with electrochemical flow cell using micromachining techniques for a glucose detection system. *Anal. Chem.* 65:2731–35.

282. Lagally, E.T., Emrich, C.A., Mathies, R.A. 2001. Fully integrated PCR-capillary electrophoresis microsystem for DNA analysis. *Labchip* 1:102–7.

283. Greenwood, P.A., Merrin, C., McCreedy, T., Greenway, G.M. 2002. Chemiluminescence μTAS for the determination of atropine and pethidine. *Talanta* 56:539–45.

284. Nelstrop, L.J., Greenwood, P.A., Greenway, G.M. 2001. An investigation of electroosmotic flow and pressure pumped luminol chemiluminescence detection for cobalt analysis in a miniaturised total analytical system. *Labchip* 1:138–42.

285. Bromberg, A., Mathies, R.A. 2003. Homogeneous immunoassay for detection of TNT and its analogues on a microfabricated capillary electrophoresis chip. *Anal. Chem.* 75:1188–95.

286. Kutter, J.P., Jacobson, S.C., Ramsey, J.M. 1997. Integrated microchip device with electrokinetically controlled solvent mixing for isocratic and gradient elution in micellar electrokinetic chromatography. *Anal. Chem.* 69:5165–71.

287. Fang, Q., Xu, G.-M., Fang, Z.-L. 2001. High throughput continuous sample introduction interfacing for microfluidic chip-based capillary electrophoresis systems. In *Proceedings of the 5th Micro Total Analysis Systems Symposium*, Monterey, CA, October 21–25, pp. 373–74.

288. Vanysek, P., Boone, T., Dang, T., Geiger, H., Klapperich, C.M., Lee, H., Nicewarner, D., Kurnik, R., Singh, S., Xiao, V., Zhao, M. 2001. Multiwell microfluidic plates for evaporation-controlled sub-microliter assays: Design and results. *Electrochem. Soc. Proc.* 376:383.

289. Wallenborg, S.R., Bailey, C.G., Paul, P.H. 2000. On-chip separation of explosive compounds—Divided reservoirs to improve reproducibility and minimize buffer depletion. In *Proceedings of the 4th Micro Total Analysis Systems Symposium,* Enschede, Netherlands, May 14–18, pp. 355–58.

290. Wen, J., Lin, Y.H., Xiang, F., Matson, D.W., Udseth, H.R., Smith, R.D. 2000. Microfabricated isoelectric focusing device for direct electrospray ionization-mass spectrometry. *Electrophoresis* 21:191–97.

291. Koh, C.G., Tan, W., Zhao, M.Q., Ricco, A.J., Fan, Z.H. 2003. Integrating polymerase chain reaction, valving, and electrophoresis in a plastic device for bacterial detection. *Anal. Chem.* 75:4591–98.

292. Sanders, J.C., Breadmore, M.C., Mitchell, P.S., Landers, J.P. 2002. A simple PDMS-based electro-fluidic interface for microchip electrophoretic separations. *Analyst* 127:1558–63.

293. Erickson, D., Sinton, D., Li, D.Q. 2004. A miniaturized high-voltage integrated power supply for portable microfluidic applications. *Labchip* 4:87–90.

294. Pattekar, A.V., Kothare, M.V. 2003. Novel microfluidic interconnectors for high temperature and pressure applications. *J. Micromech. Microeng.* 13:337–45.

295. Chien, R.L., Parce, W.J. 2001. Multiport flow-control system for lab-on-a-chip microfluidic devices. *Fres. J. Anal. Chem.* 371:106–11.

296. Zhang, B., Foret, F., Karger, B.L. 2001. High-throughput microfabricated CE/ESI-MS: Automated sampling from a microwell plate. *Anal. Chem.* 73:2675–81.

297. Rohlíček, V., Deyl, Z. 2002. Versatile tool for the manipulation of electrophoresis chips. *J. Chromotogr.* 770B:19–23.

298. Yang, Z., Maeda, R. 2003. Socket with built-in valves for the interconnection of microfluidic chips to macro constituents. *J. Chromatogr.* 1013A:29–33.

299. Erill, I., Campoy, S., Erill, N., Barbé, J., Aguiló, J. 2003. Biochemical analysis and optimization of inhibition and adsorption phenomena in glass-silicon PCR-chips. *Sensors Actuators* 96B:685–92.

300. Harrison, D.J., Glavina, P.G., Manz, A. 1993. Towards miniaturized electrophoresis and chemical analysis systems on silicon: An alternative to chemical sensors. *Sensors Actuators* 10B:107–16.

301. Hatch, A., Kamholz, A.E., Holman, G., Yager, P., Bohringer, K.F. 2001. A ferrofluidic magnetic micropump. *J. Microelectromech. Syst.* 10:215–21.

302. Ocvirk, G., Munroe, M., Tang, T., Oleschuk, R., Westra, K., Harrison, D.J. 2000. Electrokinetic control of fluid flow in native poly(dimethylsiloxane) capillary electrophoresis device. *Electrophoresis* 21:107–15.

303. Lee, H.H., Kuo, Y. 2001. A new micro-fluidic device for protein separation fabricated on a silicon substrate. *Electrochem. Soc. Proc.* 18:395–98.

304. Tracey, M.C., Greenaway, F.S., Das, A., Kaye, P.H., Barnes, A.J. 1995. A silicon micromachined device for use in blood cell deformability studies. *IEEE Trans. Biomed. Eng.* 42:751–61.

305. Harrison, D.J., Manz, A., Glavina, P.G. 1991. Electroosmotic pumping within a chemical sensor system integrated on silicon. *Transducers* 91:792–95.

306. Jeong, Y.W., Kim, B.H., Lee, J.Y., Park, S.S., Chun, M.S., Chun, K., Kim, B.G., Chung, D.S. 2000. A cyclic capillary electrophoresis separator on silicon substrate with synchronized-switching. In *Proceedings of the 4th Micro Total Analysis Systems Symposium,* Enschede, Netherlands, May 14–18, pp. 375–78.

307. Blankenstein, G., Larsen, U.D. 1998. Modular concept of a laboratory on a chip for chemical and biochemical analysis. *Biosens. Bioelectronics* 13:427–38.

308. Gray, B.L., Lieu, D.K., Collins, S.D., Smith, R.L., Barakat, A.I. 2002. Microchannel platform for the study of endothelial cell shape and function. *Biomed. Microdevices* 4:9–16.

309. Böhm, S., Timmer, B., Olthuis, W., Bergveld, P. 2000. A closed-loop controlled electrochemically actuated micro-dosing system. *J. Micromech. Microeng.* 10:498–504.

310. Jorgensen, A.M., Mogensen, K.B., Kutter, J.P., Geschke, O. 2003. A biochemical microdevice with an integrated chemiluminescence detector. *Sensors Actuators* 90B:15–21.

311. Jeong, Y.W., Kim, S.Y., Chun, K.J., Chang, J.K., Chung, D.S. 2001. Methodology for miniaturized CE and insulation on a silicon substrate. *Labchip* 1:143–47.

312. Martin, K., Henkel, T., Baier, V., Grodrian, A., Schön, T., Roth, M., Köhler, J.M., Metze, J. 2003. Generation of larger numbers of separated microbial populations by cultivation in segmented-flow microdevices. *Labchip* 3:202–7.

313. Weiller, B.H., Ceriotti, L., Shibata, T., Rein, D., Roberts, M.A., Lichtenberg, J., German, J.B., de Rooij, N.F., Verpoorte, E. 2002. Analysis of lipoproteins by capillary zone electrophoresis in microfluidic devices: Assay development and surface roughness measurements. *Anal. Chem.* 74:1702–11.

314. Solignac, D., Gijs, M.A.M. 2003. Pressure pulse injection: A powerful alternative to electrokinetic sample loading in electrophoresis microchips. *Anal. Chem.* 75:1652–57.

315. Woolley, A.T., Mathies, R.A. 1994. Ultra-high-speed DNA fragment separations using microfabricated capillary array electrophoresis chips. *Proc. Natl. Acad. Sci. USA* 91:11348–52.

316. Jacobson, S.C., Ramsey, J.M. 1996. Integrated microdevice for DNA restriction fragment analysis. *Anal. Chem.* 68:720–23.

317. Jacobson, S.C., Koutny, L.B., Hergenröder, R., Moore, A.W., Ramsey, J.M. 1994. Microchip capillary electrophoresis with an integrated postcolumn reactor. *Anal. Chem.* 66:3472–76.

318. Yan, J., Du, Y., Liu, J., Cao, W., Sun, X., Zhou, W., Yang, X., Wang, E. 2003. Fabrication of integrated microelectrodes for electrochemical detection on electrophoresis microchip by electroless deposition and micromolding in capillary technique. *Anal. Chem.* 75:5406–12.

319. Hibara, A., Tokeshi, M., Uchiyama, K., Hisamoto, H., Kitamori, T. 2001. Integrated multilayer flow system on a microchip. *Anal. Sci.* 17:89.

320. Roddy, E.S., Price, M., Ewing, A.G. 2003. Continuous monitoring of a restriction enzyme digest of DNA on a microchip with automated capillary sample introduction. *Anal. Chem.* 75:3704–11.

321. Cummings, E.B., Singh, A.K. 2003. Dielectrophoresis in microchips containing arrays of insulating posts: Theoretical and experimental results. *Anal. Chem.* 75:4724–31.

322. Handique, K., Burke, D.T., Mastrangelo, C.H., Burns, M.A. 2000. Nanoliter liquid metering in microchannels using hydrophobic patterns. *Anal. Chem.* 72:4100–9.

323. Namasivayam, V., Larson, R.G., Burke, D.T., Burns, M.A. 2002. Electrostretching DNA molecules using polymer-enhanced media within microfabricated devices. *Anal. Chem.* 74:3378–85.

324. Harrison, D.J., Fluri, K., Seiler, K., Fan, Z., Effenhauser, C.S., Manz, A. 1993. Micromachining a miniaturized capillary electrophoresis-based chemical analysis system on a chip. *Science* 261:895–97.

325. Razunguzwa, T.T., Timperman, A.T. 2004. Fabrication and characterization of a fritless microfabricated electroosmotic pump with reduced pH dependence. *Anal. Chem.* 76:1336–41.

326. Lazar, L.M., Karger, B.L. 2001. Microchip integrated analysis systems for electrospray analysis of complex peptide mixtures. In *Proceedings of the 5th Micro Total Analysis Systems Symposium*, Monterey, CA, October 21–25, pp. 219–21.

327. Yang, Z., Goto, H., Matsumoto, M., Maeda, R. 2000. Active micromixer for microfluidic systems using lead-zirconate-titanate (PZT)-generated ultrasonic vibration. *Electrophoresis* 21:116–19.

328. Laugere, F., Guijt, R.M., Bastemeijer, J., van der Steen, G., Berthold, A., Baltussen, E., Sarro, P., van Dedem, G.W.K., Vellekoop, M., Bossche, A. 2003. On-chip contactless four-electrode conductivity detection for capillary electrophoresis devices. *Anal. Chem.* 75:306–12.

329. Haab, B.B., Mathies, R.A. 1999. Single-molecule detection of DNA separations in microfabricated capillary electrophoresis chips employing focused molecular streams. *Anal. Chem.* 71:5137–45.

330. Chiem, N., Harrison, D.J. 1997. Microchip-based capillary electrophoresis for immunoassays: Analysis of monoclonal antibodies and theophylline. *Anal. Chem.* 69:373–78.

331. Breadmore, M.C., Wolfe, K.A., Arcibal, I.G., Leung, W.K., Dickson, D., Giordano, B.C., Power, M.E., Ferrance, J.P., Feldman, S.H., Norris, P.M., Landers, J.P. 2003. Microchip-based purification of DNA from biological samples. *Anal. Chem.* 75:1880–86.

332. Osbourn, D.M., Lunte, C.E. 2003. On-column electrochemical detection for microchip capillary electrophoresis. *Anal. Chem.* 75:2710–14.

333. Gottschlich, N., Jacobson, S.C., Culbertson, C.T., Ramsey, J.M. 2001. Two-dimensional electrochromatography/capillary electrophoresis on a microchip. *Anal. Chem.* 73:2669–74.

334. Shoji, S., Esashi, M. 1992. Micro flow cell for blood gas analysis realizing very small sample volume. *Sensors Actuators* 8B:205–8.

335. Timmer, B.H., Bomer, J.G., van Delft, K.M., Otjes, R.P., Olthuis, W., Bergveld, P., van den Berg, A. 2001. Fluorocarbon coated micromachined gas sampling device. In *Proceedings of the 5th Micro Total Analysis Systems Symposium*, Monterey, CA, October 21–25, pp. 381–82.

336. Chmela, E., Tijssen, R., Blom, M.T., Gardeniers, H., J.G.E., van den Berg, A. 2002. A chip system for size separation of macromolecules and particles by hydrodynamic chromatography. *Anal. Chem.* 74:3470–75.

337. Gulliksen, A., Solli, L., Karlsen, F., Rogne, H., Hovig, E., Nordstrom, T., Sirevag, R. 2004. Real-time nucleic acid sequence-based amplification in nanoliter volumes. *Anal. Chem.* 76:9–14.

338. Constantin, S., Freitag, R., Solignac, D., Sayah, A., Gijs, M.A.M. 2001. Utilization of the sol-gel technique for the development of novel stationary phases for capillary electrochromatography on a chip. *Sensors Actuators* 78B:267–72.

339. Culbertson, C.T., Jacobson, S.C., Ramsey, J.M. 2000. Microchip devices for high-efficiency separations. *Anal. Chem.* 72:5814–19.

340. Effenhauser, C.S., Manz, A., Widmer, H.M. 1993. Glass chips for high-speed capillary electrophoresis separation with submicrometer plate heights. *Anal. Chem.* 65:2637–42.

341. Yu, C., Mutlu, S., Selvaganapathy, P., Mastrangelo, C.H., Svec, F., Frechet, J.M.J. 2003. Flow control valves for analytical microfluidic chips without mechanical parts based on thermally responsive monolithic polymers. *Anal. Chem.* 75:1958–61.

342. Yu, C., Davey, M.H., Svec, F., Frechet, J.M.J. 2001. Monolithic porous polymer for on-chip solid-phase extraction and preconcentration prepared by photoinitiated in situ polymerization within a microfluidic device. *Anal. Chem.* 73:5088–96.

343. Uchiyama, K., Hibara, A., Sato, K., Hisamoto, H., Tokeshi, M., Kitamori, T. 2003. An interface chip connection between capillary electrophoresis and thermal lens microscope. *Electrophoresis* 24:179–84.

344. Peterman, M.C., Noolandi, J., Blumenkranz, M.S., Fishman, H.A. 2004. Fluid flow past an aperture in a microfluidic channel. *Anal. Chem.* 76:1850–56.

345. Manz, A., Harrison, D.J., Fettinger, J.C., Verpoorte, E., Lüdi, H., Widmer, H.M. 1991. Integrated electroosmotic pumps and flow manifolds for total chemical analysis systems. *Transducers* 91:939–41.

346. Jacobson, S.C., Ramsey, J.M. 1995. Microchip electrophoresis with sample stacking. *Electrophoresis* 16:481–86.

347. Song, S., Singh, A.K., Shepodd, T.J., Kirby, B.J. 2004. Microchip dialysis of proteins using in situ photopatterned nanoporous polymer membranes. *Anal. Chem.* 76:2367–73.

348. Toma, Y., Hatakeyama, M., Ichiki, K., Huang, H.L., Yamauchi, K., Watanabe, K., Kato, T. 1997. Fast atom beam etching of glass materials with contact and non-contact masks. *Jpn. J. Appl. Phys.* 36:7655–7659.

349. He, B., Tait, N., Regnier, F. 1998. Fabrication of nanocolumns for liquid chromatography. *Anal. Chem.* 70:3790–97.

350. Oki, A., Adachi, S., Takamura, Y., Onoda, H., Ito, Y., Horiike, Y. 2001. Electrophoresis velocity measurement of lymphocytes under suppression of pH change of PBS in microcapillary. In *Proceedings of the 5th Micro Total Analysis Systems Symposium*, Monterey, CA, October 21–25, pp. 505–6.

351. Kaji, N., Tezuka, Y., Takamura, Y., Ueda, M., Nishimoto, T., Nakanishi, H., Horiike, Y., Baba, Y. 2004. Separation of long DNA molecules by quartz nanopillar chips under a direct current electric field. *Anal. Chem.* 76:15–22.

352. Jacobson, S.C., Moore, A.W., Ramsey, J.M. 1995. Fused quartz substrates for microchip electrophoresis. *Anal. Chem.* 67:2059–63.

353. Hashimoto, M., Tsukagoshi, K., Nakajima, R., Kondo, K., Arai, A. 2000. Microchip capillary electrophoresis using on-line chemiluminescence detection. *J. Chromatogr.* 867A:271–79.

354. Hosokawa, K., Fujii, T., Endo, I. 1998. Hydrophobic microcapillary vent for pneumatic manipulation of liquid in μTAS. In *Micro Total Analysis Systems '98,* Banff, Canada, October 13–16, pp. 307–10.

355. Ichiki, T., Ujiie, T., Hara, T., Horiike, Y., Yasuda, K. 2001. On-chip cell sorter for single cell expression analysis. In *Proceedings of the 5th Micro Total Analysis Systems Symposium*, Monterey, CA, October 21–25, pp. 271–73.

356. Lu, H., Schmidt, M.A., Jensen, K.F. 2001. Photochemical reactions and on-line UV detection in micro-fabricated reactors. *Labchip* 1:22–28.

357. Liu, J., Enzelberger, M., Quake, S. 2002. A nanoliter rotary device for polymerase chain reaction. *Electrophoresis* 23:1531–36.

358. Zhang, Y., Timperman, A.T. 2003. Integration of nanocapillary arrays into microfluidic devices for use as analyte concentrators. *Analyst* 128:537–42.

359. Duffy, D.C., Schueller, O.J.A., Brittain, S.T., Whitesides, G.M. 1999. Rapid prototyping of microfluidic switches in poly(dimethyl siloxane) and their actuation by electro-osmotic flow. *J. Micromech. Microeng.* 9:211–17.

360. Slentz, B.E., Penner, N.A., Lugowska, E., Regnier, F. 2001. Nanoliter capillary electrochromatography columns based on collocated monolithic support structures molded in poly(dimethyl siloxane). *Electrophoresis* 22:3736–43.

361. Leclerc, E., Sakai, Y., Fujii, T. 2003. Cell culture in 3-dimensional microfluidic structure of PDMS (poly-dimethylsiloxane). *Biomed. Microdevices* 5:109–14.

362. Krulevitch, P., Benett, W., Hamilton, J., Maghribi, M., Rose, K. 2002. Polymer-based packaging platform for hybrid microfluidic systems. *Biomed. Microdevices* 4:301–8.

363. Wu, H., Odom, T.W., Whitesides, G.M. 2002. Reduction photolithography using microlens arrays: Applications in gray scale photolithography. *Anal. Chem.* 74:3267–73.

364. Hosokawa, K., Fujii, T., Endo, I. 1999. Handling of picoliter liquid samples in a poly(dimethylsiloxane)-based microfluidic device. *Anal. Chem.* 71:4781–85.

365. Liu, Y., Fanguy, J.C., Bledsoe, J.M., Henry, C.S. 2000. Dynamic coating using polyelectrolyte multilayers for chemical control of electroosmotic flow in capillary electrophoresis microchips. *Anal. Chem.* 72:5939–44.

366. Bodor, R., Madajova, V., Kaniansky, D., Masar, M., Johnck, M., Stanislawski, B. 2001. Isotachophoresis and isotachophoresis-zone electrophoresis separations of inorganic anions present in water samples on a planar chip with column-coupling separation channels and conductivity detection. *J. Chromatogr.* 916A:155–65.

367. Grzybowski, B.A., Haag, R., Bowden, N., Whitesides, G.M. 1998. Generation of micrometer-sized patterns for microanalytical applications using a laser direct-write method and microcontact printing. *Anal. Chem.* 70:4645–52.

368. Wheeler, A.R., Throndset, W.R., Whelan, R.J., Leach, A.M., Zare, R.N., Liao, Y.H., Farrell, K., Manger, I.D., Daridon, A. 2003. Microfluidic device for single-cell analysis. *Anal. Chem.* 75:3581–86.

369. Gallardo, B.S., Gupta, V.K., Eagerton, F.D., Jong, L.I., Craig, T., V.S., Shah, R.R., Abbott, N.L. 1999. Electrochemical principles for active control of liquids on submillimeter scales. *Science* 283:57–61.

370. Lin, Y.C., Li, M., Wu, C.C. 2004. Simulation and experimental demonstration of the electric field assisted electroporation microchip for in vitro gene delivery enhancement. *Labchip* 4:104–8.

371. Sniadecki, N.J., Lee, C.S., Sivanesan, P., DeVoe, D.L. 2004. Induced pressure pumping in polymer microchannels via field-effect flow control. *Anal. Chem.* 76:1942–47.

372. Mrksich, M., Chen, C.S., Xia, Y., Dike, L.E., Ingber, D.E., Whitesides, G.M., 1996. Controlling cell attachment on contoured surfaces with self-assembled monolayers of alkanethiolates on gold. *Proc. Natl. Acad. Sci. USA* 93:10775–78.

373. Lu, H., Jackman, R.J., Gaudet, S., Cardone, M., Schmidt, M.A., Jensen, K.F. 2001. Microfluidic devices for cell lysis and isolation of organelles. In *Proceedings of the 5th Micro Total Analysis Systems Symposium*, Monterey, CA, October 21–25, pp. 297–98.

374. Zhao, M., Crooks, R.M., Nguyen, U., Ricco, A.J., Zhu, Q. 2001. Electrode-integrated microfluidic plastic devices. *Electrochem. Soc. Proc.* 18:388–94.

375. Lacher, N.A., Lunte, S.M., Martin, R.S. 2004. Development of a microfabricated palladium decoupler/electrochemical detector for microchip capillary electrophoresis using a hybrid glass/poly(dimethylsiloxane) device. *Anal. Chem.* 76:2482–91.

376. Grass, B., Neyer, A., Johnck, M., Siepe, D., Eisenbeiss, F., Weber, G., Hergenroder, R. 2001. A new PMMA-microchip device for isotachophoresis with integrated conductivity detector. *Sensors Actuators* 72B:249–58.

377. Vogt, O., Grass, B., Weber, G., Hergenroder, R., Siepe, D., Neyer, A., Pohl, J.P. 2001. Characterization of sputtered thin film electrodes on PMMA microchips with electrochemical impedance spectroscopy and cyclic voltammetry. In *Proceedings of the 5th Micro Total Analysis Systems Symposium*, Monterey, CA, October 21–25, pp. 327–28.

378. Yoon, D.S., Lee, Y.S., Lee, Y., Cho, H.J., Sung, S.W., Oh, K.W., Cha, J.H., Lim, G. 2002. Precise temperature control and rapid thermal cycling in a micromachined DNA polymerase chain reaction chip. *J. Micromech. Microeng.* 12:813–23.

379. Lin, Y.C., Ho, H.C., Tseng, C.K., Hou, S.Q. 2001. A poly-methylmethacrylate electrophoresis microchip with sample preconcentrator. *J. Micromech. Microeng.* 11:189–94.

380. Kar, S., McWhorter, S., Ford, S.M., Soper, S.A. 1998. Piezoelectric mechanical pump with nanoliter per minute pulse-free flow delivery for pressure pumping in micro-channels. *Analyst* 123:1435–41.

381. Monahan, J., Gewirth, A.A., Nuzzo, R.G. 2001. A method for filling complex polymeric microfluidic devices and arrays. *Anal. Chem.* 73:3193–97.

382. Huh, D., Tung, Y.C., Wei, H.H., Grotberg, J.B., Skerlos, S.J., Kurabayashi, K., Takayama, S. 2002. Use of air-liquid two-phase flow in hydrophobic microfluidic channels for disposable flow cytometers. *Biomed. Microdevices* 4:141–49.

383. Zoval, J.V., Madou, M.J. 2004. Centrifuge-based fluidic platforms. *Proc. IEEE* 92:140–53.

384. Shelley, J.P., Lin, D.S.W., Kuo, J.S. Chiu, D.T. 2003. High radial acceleration in microvortices. *Nature* 425:38.

385. Alarie, J.P., Jacobson, S.C., Broyles, B.S., Mcknight, T.E., Culbertson, C.T., Ramsey, J.M. 2001. Electroosmotically induced hydraulic pumping on microchips. In *Proceedings of the 5th Micro Total Analysis Systems Symposium*, Monterey, CA, October 21–25, pp. 131–32.

386. Guijt, R.M., Lichtenberg, J., Baltussen, E., Verpoorte, E., de Rooij, N.F., van Dedem, G.W.K. 2001. Indirect electro-osmotic pumping for direct sampling from bioreactors. In *Proceedings of the 5th Micro Total Analysis Systems Symposium*, Monterey, CA, October 21–25, pp. 399–400.

387. Culbertson, C.T., Ramsey, R.S., Ramsey, J.M. 2000. Electroosmotically induced hydraulic pumping on microchips: Differential ion transport. *Anal. Chem.* 72:2285–91.

388. Kerby, M.B., Spaid, M., Wu, S., Parce, J.W., Chien, R.-L. 2002. Selective ion extraction: A separation method for microfluidic devices. *Anal. Chem.* 74:5175–83.

389. Chien, R.L., Bousse, L. 2002. Electroosmotic pumping in microchips with nonhomogeneous distribution of electrolytes. *Electrophoresis* 23:1862–69.

390. Takamura, Y., Onoda, H., Inokuchi, H., Adachi, S., Oki, A., Horiike, Y. 2001. Low-voltage electroosmosis pump and its application to on-chip linear stepping pneumatic pressure source. In *Proceedings of the 5th Micro Total Analysis Systems Symposium*, Monterey, CA, October 21–25, pp. 230–32.

391. Böhm, S., Olthuis, W., Bergveld, P. 1998. An integrated micromachined electrochemical pump and dosing system. *Biomed. Microdevices* 1:121–30.

392. Burns, M.A., Mastrangelo, C.H., Sammarco, T.S., Man, F.P., Webster, J.R., Johnson, B.N., Foerster, B., Jones, D., Fields, Y., Kaiser, A.R., Burke, D.T. 1996. Microfabricated structures for integrated DNA analysis. *Proc. Natl. Acad. Sci. USA* 93:5556–61.

393. Darhuber, A.A., Davis, J.M., Reisner, W.W., Troian, S.M. 2001. Thermocapillary migration of liquids on patterned surfaces: Design concept for microfluidic delivery. In *Proceedings of the 5th Micro Total Analysis Systems Symposium*, Monterey, CA, October 21–25, pp. 244–46.

394. Handique, K., Burke, D.T., Mastrangelo, C.H., Burns, M.A. 2001. On-chip thermopneumatic pressure for discrete drop pumping. *Anal. Chem.* 73:1831–38.

395. Tsai, J.H., Lin, L.W. 2002. A thermal-bubble-actuated micronozzle-diffuser pump. *J. Microelectromech. Syst.* 11:665–71.

396. Harmon, M.E., Tang, M., Frank, C.W. 2003. A microfluidic actuator based on thermoresponsive hydrogels. *Polymer* 44:4547–56.

397. Sohn, Y.-S., Goodey, A.P., Anslyn, E.V., McDevitt, J.T., Shear, J.B., Neikirk, D.P. 2001. Development of a micromachined fluidic structure for a biological and chemical sensor array. In *Proceedings of the 5th Micro Total Analysis Systems Symposium*, Monterey, CA, October 21–25, pp. 177–78.

398. Walker, G.M., Beebe, D.J. 2002. A passive pumping method for microfluidic devices. *Labchip* 2:131–34.

399. Jen, C.P., Lin, Y.C., Wu, W.D., Wu, C.Y., Wu, G.G., Chang, C.C. 2003. Improved design and experimental demonstration of a bi-directional microfluidic driving system. *Sensors Actuators* 96B:701–8.

400. Eijkel, J.C.T., Dalton, C., Hayden, C.J., Burt, J.P.H., Manz, A. 2003. Circular AC magnetohydrodynamic micropump for chromatographic applications. *Sensors Actuators* 92B:215–21.

401. Lemoff, A.V., Lee, A.P. 2003. An AC magnetohydrodynamic microfluidic switch for micro total analysis systems. *Biomed. Microdevices* 5:55–61.

402. Effenhauser, C.S., Harttig, H., Krämer, P. 2002. An evaporation-based disposable micropump concept for continuous monitoring applications. *Biomed. Microdevices* 4:27–33.

403. Goedecke, N., Eijkel, J., Manz, A. 2002. Evaporation driven pumping for chromatography application. *Labchip* 2:219–23.

404. Yang, Z., Matsumoto, S., Maeda, R. 2002. A prototype of ultrasonic micro-degassing device for portable dialysis system. *Sensors Actuators* 95A:274–80.

405. Schasfoort, R.B.M., Luttge, R., van den Berg, A. 2001. Magneto-hydrodynamically (MHD) directed flow in microfluidic networks. In *Proceedings of the 5th Micro Total Analysis Systems Symposium*, Monterey, CA, October 21–25, pp. 577–78.

406. Eijkel, J.C.T., Dalton, C., Hayden, C.J., Drysdale, J.A., Kwok, Y.C., Manz, A. 2001. Development of a micro system for circular chromatography using wavelet transform detection. In *Proceedings of the 5th Micro Total Analysis Systems Symposium*, Monterey, CA, October 21–25, pp. 541–42.

407. Lemoff, A.V., Lee, A.P. 2000. An AC magnetohydrodynamic microfluidic switch. In *Proceedings of the 4th Micro Total Analysis Systems Symposium,* Enschede, Netherlands, May 14–18, pp. 571–74.

408. Green, N.G. 2001. Integration of a solid state micropump and a sub-micrometre particle analyser/separator. In *Proceedings of the 5th Micro Total Analysis Systems Symposium*, Monterey, CA, October 21–25, pp. 545–46.

409. Paegel, B.M., Hutt, L.D., Simpson, P.C., Mathies, R.A. 2000. Turn geometry for minimizing band broadening in microfabricated capillary electrophoresis channels. *Anal. Chem.* 72:3030–37.

410. Ross, D., Locascio, L.E. 2003. Effect of caged fluorescent dye on the electroosmotic mobility in microchannels. *Anal. Chem.* 75:1218–20.

411. Shelby, J.P., Chiu, D.T. 2003. Mapping fast flows over micrometer-length scales using flow-tagging velocimetry and single-molecule detection. *Anal. Chem.* 75:1387–92.

412. Gosch, M., Blom, H., Holm, J., Heino, T., Rigler, R. 2000. Hydrodynamic flow profiling in microchannel structures by single molecule fluorescence correlation spectroscopy. *Anal. Chem.* 72:3260–65.

413. Mogensen, K.B., Kwok, Y.C., Eijkel, J.C.T., Petersen, N.J., Manz, A., Kutter, J.P. 2003. A microfluidic device with an integrated waveguide beam splitter for velocity measurements of flowing particles by Fourier transformation. *Anal. Chem.* 75:4931–36.

414. Chung, J.W., Grigoropoulos, C.P., Greif, R. 2003. Infrared thermal velocimetry in MEMS-based fluidic devices. *J. Microelectromech. Syst.* 12:365–72.

415. Kirby, B.J., Wheeler, A.R., Shepodd, T.J., Fruetel, J.A., Hasselbrink, E.F., Zare, R.N. 2001. A laser-polymerized thin film silica surface modification for suppression of cell adhesion and electroosmotic flow in microchannels. In *Proceedings of the 5th Micro Total Analysis Systems Symposium*, Monterey, CA, October 21–25, pp. 605–6.

416. Barker, S.L.R., Tarlov, M.J., Branham, M., Xu, J., Maccrehan, W., Gaitan, M., Locascio, L.E. 2000. Derivatization of plastic microfluidic device with polyelectrolyte multilayers. In *Proceedings of the 4th Micro Total Analysis Systems Symposium,* Enschede, Netherlands, May 14–18, pp. 67–70.

417. Barker, S.L.R., Tarlov, M.J., Canavan, H., Hickman, J.J., Locascio, L.E. 2000. Plastic microfluidic devices modified with polyelectrolyte multilayers. *Anal. Chem.* 72:4899–903.

418. Liu, Y., Ganser, D., Schneider, A., Liu, R., Grodzinski, P., Kroutchinina, N. 2001. Microfabricated polycarbonate CE devices for DNA analysis. In *Proceedings of the 5th Micro Total Analysis Systems Symposium*, Monterey, CA, October 21–25, pp. 119–20.

419. Khoury, C., Moorthy, J., Stremler, M.A., Moore, J.S., Beebe, D.J. 2000. TiO_2 surface modifications for light modulated control of flow velocity. In *Proceedings of the 4th Micro Total Analysis Systems Symposium,* Enschede, Netherlands, May 14–18, pp. 331–34.

420. Hu, S., Ren, X., Bachman, M., Sims, C.E., Li, G.P., Allbritton, N. 2002. Surface modification of poly(dimethylsiloxane) microfluidic devices by ultraviolet polymer grafting. *Anal. Chem.* 74:4117–23.

421. Henry, A.C., Waddell, E.A., Shreiner, R., Locascio, L.E. 2002. Control of electroosmotic flow in laser-ablated and chemically modified hot imprinted poly(ethylene terephthalate glycol) microchannels. *Electrophoresis* 23:791–98.

422. Buch, J.S., Wang, P.-C., DeVoe, D.L., Lee, C.S. 2001. Field-effect flow control in a polydimethylsiloxane-based microfluidic system. *Electrophoresis* 22:3902–7.

423. Schasfoort, R.B.M., Schlautmann, S., Hendrikse, J., van den Berg, A. 1999. Field-effect flow control for microfabricated fluidic networks. *Science* 286:942–45.

424. Polson, N.A., Hayes, M.A. 2000. Electroosmotic flow control of fluids on a capillary electrophoresis microdevice using an applied external voltage. *Anal. Chem.* 72:1088–92.

425. Hisamoto, H., Horiuchi, T., Tokeshi, M., Hibara, A., Kitamori, T. 2001. On-chip integration of neutral ionophore-based ion pair extraction reaction. *Anal. Chem.* 73:1382–86.

426. Tokeshi, M., Minagawa, T., Kitamori, T. 2000. Integration of a microextraction system on a glass chip: Ion-pair solvent extraction of Fe(II) with 4,7-diphenyl-1,10-phenanthrolinedisulfonic acid and tri-n-octylmethylammonium chloride. *Anal. Chem.* 72:1711–14.

427. Shaw, J., Nudd, R., Naik, B., Turner, C., Rudge, D., Benson, M., Garman, A. 2000. Liquid/liquid extraction systems using microcontactor arrays. In *Proceedings of the 4th Micro Total Analysis Systems Symposium,* Enschede, Netherlands, May 14–18, pp. 371–74.

428. Tokeshi, M., Minagawa, T., Uchiyama, K., Hibara, A., Sato, K., Hisamoto, H., Kitamori, T. 2002. Continuous-flow chemical processing on a microchip by combining microunit operations and a multiphase flow network. *Anal. Chem.* 74:1565–71.

429. Maruyama, T., Matsushita, H., Uchida, J.-I., Kubota, F., Kamiya, N., Goto, M. 2004. Liquid membrane operations in a microfluidic device for selective separation of metal ions. *Anal. Chem.* 76:4495–500.

430. Hibara, A., Nonaka, M., Hisamoto, H., Uchiyama, K., Kikutani, Y., Tokeshi, M., Kitamori, T. 2002. Stabilization of liquid interface and control of two-phase confluence and separation in glass microchips by utilizing octadecylsilane modification of microchannels. *Anal. Chem.* 74:1724–28.

431. Kuban, P., Dasgupta, P.K., Morris, K.A. 2002. Microscale continuous ion exchanger. *Anal. Chem.* 74:5667–75.

432. Kenis, P.J.A., Ismagilov, R.F., Whitesides, G.M. 1999. Microfabrication inside capillaries using multiphase laminar flow patterning. *Science* 285:83–85.

433. Zhao, B., Moore, J.S., Beebe, D.J. 2001. Surface-directed liquid flow inside microchannels. *Science* 291:1023–26.

434. Zhao, B., Moore, J.S., Beebe, D.J. 2003. Pressure-sensitive microfluidic gates fabricated by patterning surface free energies inside microchannels. *Langmuir* 19:1873–79.

435. Hisamoto, H., Shimizu, Y., Uchiyama, K., Tokeshi, M., Kikutani, Y., Hibara, A., Kitamori, T. 2003. Chemicofunctional membrane for integrated chemical processes on a microchip. *Anal. Chem.* 75:350–54.

436. Seiler, K., Fan, Z.H., Fluri, K., Harrison, D.J. 1994. Electroosmotic pumping and valveless control of fluid flow within a manifold of capillaries on a glass chip. *Anal. Chem.* 66:3485–91.

437. Murakami, Y., Kanekiyo, T., Kinpara, T., Tamiya, E. 2001. On-chip bypass structure for sample segment division and dilution. In *Proceedings of the 5th Micro Total Analysis Systems Symposium*, Monterey, CA, October 21–25, pp. 175–76.

438. Cheng, S.B., Skinner, C.D., Harrison, D.J. 1998. Integrated serial dilution on a microchip for immunoassay sample treatment and flow injection analysis. In *Micro Total Analysis Systems '98*, Banff, Canada, October 13–16, pp. 157–60.

439. Jacobson, S.C., McKnight, T.E., Ramsey, J.M. 1999. Microfluidic devices for electrokinetically driven parallel and serial mixing. *Anal. Chem.* 71:4455–59.

440. Dertinger, S.K.W., Chiu, D.T., Jeon, N.L., Whitesides, G.M. 2001. Generation of gradients having complex shapes using microfluidic networks. *Anal. Chem.* 73:1240–46.

441. Jeon, N.L., Dertinger, S.K.W., Chiu, D.T., Choi, I.S., Stroock, A.D., Whitesides, G.M. 2000. Generation of solution and surface gradients using microfluidic systems. *Langmuir* 16:8311–16.

442. Kamholz, A.E., Weigl, B.H., Finlayson, B.A., Yager, P. 1999. Quantitative analysis of molecular interaction in a microfluidic channel: The T-sensor. *Anal. Chem.* 71:5340–47.

443. Weigl, B.H., Yager, P. 1999. Microfluidic diffusion-based separation and detection. *Science* 283:346–47.

444. Weigl, B.H., Kriebel, J., Mayes, K., Yager, P., Wu, C.C., Holl, M., Kenny, M., Zebert, D. 1998. Simultaneous self-referencing analyte determination in complex sample solutions using microfabricated flow structure (T-sensors). In *Micro Total Analysis Systems '98*, Banff, Canada, October 13–16, pp. 81–84.

445. Yang, M., Yang, J., Li, C., Zhao, J. 2002. Generation of concentration gradient by controlled flow distribution and diffusive mixing in a microfluidic chip. *Labchip* 2:158–63.

446. Fosser, K.A., Nuzzo, R.G. 2003. Fabrication of patterned multicomponent protein gradients and gradient arrays using microfluidic depletion. *Anal. Chem.* 75:5775–82.

447. Mao, H., Holden, M.A., You, M., Cremer, P.S. 2002. Reusable platforms for high-throughput on-chip temperature gradient assays. *Anal. Chem.* 74:5071–75.

448. Mao, H., Yang, T., Cremer, P.S. 2002. A microfluidic device with a linear temperature gradient for parallel and combinatorial measurements. *J. Am. Chem. Soc.* 124:4432–35.

449. Wang, Y.-C., Choi, M.H., Han, J. 2004. Two-dimensional protein separation with advanced sample and buffer isolation using microfluidic valves. *Anal. Chem.* 76:4426–31.

450. Jeon, N.L., Chiu, D.T., Wargo, C.J., Wu, H.K., Choi, I.S., Anderson, J.R., Whitesides, G.M. 2002. Microfluidics section: Design and fabrication of integrated passive valves and pumps for flexible polymer 3-dimensional microfluidic systems. *Biomed. Microdevices* 4:117–21.

451. Zhao, B., Moore, J.S., Beebe, D.J. 2002. Principles of surface-directed liquid flow in microfluidic channels. *Anal. Chem.* 74:4259–68.

452. Liu, R.H., Qing, Y., Beebe, D.J. 2002. Fabrication and characterization of hydrogel-based microvalves. *J. Microelectromech. Syst.* 11:45–53.

453. Miller, R.A., Nazarov, E.G., Eiceman, G.A., King, A.T. 2001. A MEMS radio-frequency ion mobility spectrometer for chemical vapor detection. *Sensors Actuators* 91A:301–6.

454. Madou, M.J., Lee, L.J., Daunert, S., Lai, S.Y., Shih, C.H. 2001. Design and fabrication of CD-like microfluidic platforms for diagnostics: Microfluidic functions. *Biomed. Microdevices* 3:245–54.

455. Rehm, J.E., Shepodd, T.J., Hasselbrink, E.F., Jr. 2001. Mobile flow control elements for high-pressure micro-analytical systems fabricated using in-situ polymerization. In *Proceedings of the 5th Micro Total Analysis Systems Symposium*, Monterey, CA, October 21–25, pp. 227–29.

456. Hasselbrink, E.F., Jr., Shepodd, T.J., Rehm, J.E. 2002. High-pressure microfluidic control in lab-on-a-chip devices using mobile polymer monoliths. *Anal. Chem.* 74:4913–18.

457. Reichmuth, D.S., Shepodd, T.J., Kirby, B.J. 2004. On-chip high-pressure picoliter injector for pressure-driven flow through porous media. *Anal. Chem.* 76:5063–68.

458. Hua, S.Z., Sachs, F., Yang, D.X., Chopra, H.D. 2002. Microfluidic actuation using electrochemically generated bubbles. *Anal. Chem.* 74:6392–96.

459. Juncker, D., Schmid, H., Drechsler, U., Wolf, H., Wolf, M., Michel, B., de Rooij, N., Delamarche, E. 2002. Autonomous microfluidic capillary system. *Anal. Chem.* 74:6139–44.

460. Tashiro, K., Ikeda, S., Sekiguchi, T., Shoji, S., Makazu, H., Funatsu, T., Tsukita, S. 2001. A particle and biomolecules sorting micro flow system using thermal gelation of methyl cellulose solution. In *Proceedings of the 5th Micro Total Analysis Systems Symposium*, Monterey, CA, October 21–25, pp. 471–73.

461. Jannasch, H.W., Mcgill, P.R., Zdeblick, M., Erickson, J. 2001. Integrated micro-analyzers with frozen plug valves. In *Proceedings of the 5th Micro Total Analysis Systems Symposium*, Monterey, CA, October 21–25, pp. 529–30.

462. Liu, R.H., Yang, J.N., Lenigk, R., Bonanno, J., Grodzinski, P. 2004. Self-contained, fully integrated biochip for sample preparation, polymerase chain reaction amplification, and DNA microarray detection. *Anal. Chem.* 76:1824–31.

463. Choi, J.W., Oh, K.W., Han, A., Wijayawardhana, C.A., Lannes, C., Bhansali, S., Schlueter, K.T., Heineman, W.R., Halsall, H.B., Nevin, J.H., Helmicki, A.J., Henderson, H.T., Ahn, C.H. 2001. Development and characterization of microfluidic devices and systems for magnetic bead-based biochemical detection. *Biomed. Microdevices* 3:191–200.

464. Hartshorne, H., Ning, Y., Lee, W.E., Backhouse, C. 1998. Development of microfabricated valves for μTAS. In *Micro Total Analysis Systems '98,* Banff, Canada, October 13–16, pp. 379–81.

465. Chung, Y.C., Jen, C.P., Lin, Y.C., Wu, C.Y., Wu, T.C. 2003. Design of a recursively-structured valveless device for microfluidic manipulation. *Labchip* 3:168–72.

466. Bessoth, F.G., deMello, A.J., Manz, A. 1999. Microstructure for efficient continuous flow mixing. *Anal. Commun.* 36:213–15.

467. Kakuta, M., Hinsmann, P., Manz, A., Lendl, B. 2003. Time-resolved Fourier transform infrared spectrometry using a microfabricated continuous flow mixer: Application to protein conformation study using the example of ubiquitin. *Labchip* 3:82–85.

468. Xu, Y., Bessoth, F.G., Eijkel, J.C.T., Manz, A. 2000. On-line monitoring of chromium(III) using a fast micromachined mixer/reactor and chemiluminescence detection. *Analyst* 125:677–84.

469. Neils, C., Tyree, Z., Finlayson, B., Folch, A. 2004. Combinatorial mixing of microfluidic streams. *Labchip* 4:342–50.

470. Johnson, T.J., Locascio, L.E. 2002. Characterization and optimization of slanted well designs for microfluidic mixing under electroosmotic flow. *Labchip* 2:135–40.

471. Stroock, A.D., Dertinger, S.K., Whitesides, G.M., Ajdari, A. 2002. Patterning flows using grooved surfaces. *Anal. Chem.* 74:5306–12.

472. Lin, Y., Gerfen, G.J., Rousseau, D.L., Yeh, S.-R. 2003. Ultrafast microfluidic mixer and freeze-quenching device. *Anal. Chem.* 75:5381–86.

473. Biddiss, E., Erickson, D., Li, D. 2004. Heterogeneous surface charge enhanced micromixing for electrokinetic flows. *Anal. Chem.* 76:3208–13.

474. Song, H., Tice, J.D., Ismagilov, R.F. 2003. A Microfluidic system for controlling reaction networks in time. *Angew. Chem. Int. Ed. Engl.* 42:768–72.

475. Zheng, B., Tice, J.D., Ismagilov, R.F. 2004. Formation of droplets of alternating composition in microfluidic channels and applications to indexing of concentrations in droplet-based assays. *Anal. Chem.* 76:4977–82.

476. Tice, J.D., Song, H., Lyon, A.D., Ismagilov, R.F. 2003. Formation of droplets and mixing in multiphase microfluidics at low values of the Reynolds and the capillary numbers. *Langmuir* 19:9127–33.

477. Liu, R.H., Stremler, M.A., Sharp, K.V., Olsen, M.G., Santiago, J.G., Adrian, R.J., Aref, H., Beebe, D.J. 2000. Passive mixing in a three-dimensional serpentine microchannel. *J. Microelectromech. Syst.* 9:190–97.

478. Boer, G., Dodge, A., Fluri, K., van der Schoot, B.H., Verpoorte, E., de Rooij, N.F. 1998. Studies of hydrostatic pressure effects in electrokinetically driven μTAS. In *Micro Total Analysis Systems '98,* Banff, Canada, October 13–16, pp. 53–56.

479. Lettieri, G.L., Dodge, A., Boer, G., de Rooij, N.F., Verpoorte, E. 2003. A novel microfluidic concept for bioanalysis using freely moving beads trapped in recirculating flows. *Labchip* 3:34–39.

480. Oddy, M.H., Santiago, J.G., Mikkelsen, J.C. 2001. Electrokinetic instability micromixing. *Anal. Chem.* 73:5822–32.

481. Niu, X.Z., Lee, Y.K. 2003. Efficient spatial-temporal chaotic mixing in microchannels. *J. Micromech. Microeng.* 13:454–62.

482. Glasgow, I., Lieber, S., Aubry, N. 2004. Parameters influencing pulsed flow mixing in microchannels. *Anal. Chem.* 76:4825–32.

483. Glasgow, I., Aubry, N. 2003. Enhancement of microfluidic mixing using time pulsing. *Labchip* 3:114–20.

484. Chung, Y.C., Hsu, Y.L., Jen, C.P., Lu, M.C., Lin, Y.C. 2004. Design of passive mixers utilizing microfluidic self-circulation in the mixing chamber. *Labchip* 4:70–77.

485. Chou, H.P., Unger, M.A., Quake, S.R. 2001. A microfabricated rotary pump. *Biomed. Microdevices* 3:323–30.

486. Mariella, R.J. 2002. MEMS for bio-assays. *Biomed. Microdevices* 4:77–87.

487. Yang, Z., Matsumoto, S., Goto, H., Matsumoto, M., Maeda, R. 2001. Ultrasonic micromixer for microfluidic systems. *Sensors Actuators* 93A:266–72.

488. Kim, E.S., Zhu, X. 1998. Microfluidic motion generation with acoustic waves. *Sensors Actuators* 66A:355–60.

489. Yaralioglu, G.G., Wygant, I.O., Marentis, T.C., Khuri-Yakub, B.T. 2004. Ultrasonic mixing in microfluidic channels using integrated transducers. *Anal. Chem.* 76:3694–98.

490. Yuen, P.K., Li, G.S., Bao, Y.J., Müller, U.R. 2003. Microfluidic devices for fluidic circulation and mixing improve hybridization signal intensity on DNA arrays. *Labchip* 3:46–50.

491. Vreeland, W.N., Locascio, L.E. 2003. Using bioinspired thermally triggered liposomes for high-efficiency mixing and reagent delivery in microfluidic devices. *Anal. Chem.* 75:6906–11.

492. Bohm, S., Greiner, K., Schlautmann, S., de Vries, S., van den Berg, A. 2001. A rapid vortex micromixer for studying high-speed chemical reactions. In *Proceedings of the 5th Micro Total Analysis Systems Symposium*, Monterey, CA, October 21–25, pp. 25–27.

493. He, B., Burke, B.J., Zhang, X., Zhang, R., Regnier, F.E. 2001. A picoliter-volume mixer for microfluidic analytical systems. *Anal. Chem.* 73:1942–47.

494. Svasek, P., Jobst, G., Urban, G., Svasek, E. 1996. Dry film resist based fluid handling components for μTAS. In *Micro Total Analysis Systems '96,* Basel, November 19–22, pp. 27–29.

495. Bökenkamp, D., Desai, A., Yang, X., Tai, Y.-C., Marzluff, E.M., Mayo, Stephen L. 1998. Microfabricated silicon mixers for submillisecond quench-flow analysis. *Anal. Chem.* 70:232–36.

496. Rohr, T., Yu, C., Davey, M.H., Svec, F., Frechet, J.M.J. 2001. Porous polymer monoliths: Simple and efficient mixers prepared by direct polymerization in the channels of microfluidic chips. *Electrophoresis* 22:3959–67.

497. Larsen, U.D., Branebjerg, J., Blankenstein, G. 1996. Fast mixing by parallel multilayer lamination. In *Micro Total Analysis Systems '96,* Basel, November 19–22, pp. 228–30.

498. Mensinger, H., Richter, Th., Hessel, V., Döpper, J., Ehrfeld, W. 1994. Microreactor with integrated static mixer and analysis system. In *Micro Total Analysis Systems '94,* University of Twente, Netherlands, November 21–22, pp. 237–43.

499. Fujii, T., Hosokawa, K., Shoji, S., Yotsumoto, A., Nojima, T., Endo, I. 1998. Development of a microfabricated biochemical workbench—Improving the mixing efficiency. In *Micro Total Analysis Systems '98,* Banff, Canada, October 13–16, pp. 173–76.

500. Manz, A., Bessoth, F., Kopp, M.U. 1998. Continuous flow versus batch processing–A few examples. In *Micro Total Analysis Systems '98,* Banff, Canada, October 13–16, pp. 235–40.

501. Losey, M.W., Jackman, R.J., Firebaugh, S.L., Schmidt, M.A., Jensen, K.F. 2002. Design and fabrication of microfluidic devices for multiphase mixing and reaction. *J. Microelectromech. Syst.* 11:709–17.

502. Lu, L., Ryu, K.S., Liu, C. 2001. A novel microstirrer and arrays for microfluidic mixing. In *Proceedings of the 5th Micro Total Analysis Systems Symposium*, Monterey, CA, October 21–25, pp. 28–30.

503. Hong, C.-C., Choi, J.-W., Ahn, C.H. 2001. A novel in-plane passive micromixer using Coanda effect. In *Proceedings of the 5th Micro Total Analysis Systems Symposium*, Monterey, CA, October 21–25, pp. 31–33.

504. Oddy, M.H., Santiago, J.G., Mikkelsen, J.C. 2001. Electrokinetic instability micromixers. In *Proceedings of the 5th Micro Total Analysis Systems Symposium*, Monterey, CA, October 21–25, pp. 34–36.

505. Choi, J.-W., Hong, C.-C., Ahn, C.H. 2001. An electrokinetic active micromixer. In *Proceedings of the 5th Micro Total Analysis Systems Symposium*, Monterey, CA, October 21–25, pp. 621–22.

506. Sundaram, N., Tafti, D.K. 2004. Evaluation of microchamber geometries and surface conditions for electrokinetic driven mixing. *Anal. Chem.* 76:3785–93.

507. Liu, R.H., Ward, M., Bonanno, J., Ganser, D., Athavale, M., Grodzinski, P. 2001. Plastic in-line chaotic micromixer for biological applications. In *Proceedings of the 5th Micro Total Analysis Systems Symposium*, Monterey, CA, October 21–25, pp. 163–64.

508. Wong, S.H., Bryant, P., Ward, M., Wharton, C. 2003. Investigation of mixing in a cross-shaped micromixer with static mixing elements for reaction kinetics studies. *Sensors Actuators* 95B:414–24.

509. Jen, C.P., Wu, C.Y., Lin, Y.C., Wu, C.Y. 2003. Design and simulation of the micromixer with chaotic advection in twisted microchannels. *Labchip* 3:77–81.

510. Schönfeld, F., Hessel, V., Hofmann, C. 2004. An optimised split-and-recombine micro-mixer with uniform chaotic mixing. *Labchip* 4:65–69.

511. Vijayendran, R.A., Motsegood, K.M., Beebe, D.J., Leckband, D.E. 2003. Evaluation of a three-dimensional micromixer in a surface-based biosensor. *Langmuir* 19:1824–28.

512. Morgan, H., Green, N.G., Hughes, M.P., Monaghan, W., Tan, T.C. 1997. Large-area travelling-wave dielectrophoresis particle separator. *J. Micromech. Microeng.* 7:65–70.

513. Hong, C.C., Choi, J.W., Ahn, C.H. 2004. A novel in-plane passive microfluidic mixer with modified Tesla structures. *Labchip* 4:109–13.

514. Lu, L.H., Ryu, K.S., Liu, C. 2002. A magnetic microstirrer and array for microfluidic mixing. *J. Microelectromech. Syst.* 11:462–69.

515. Liu, R.H., Yang, J.N., Pindera, M.Z., Athavale, M., Grodzinski, P. 2002. Bubble-induced acoustic micromixing. *Labchip* 2:151–57.

516. Woias, P., Hauser, K., Yacoub-George, E. 2000. An active silicon micromixer for µTAS applications. In *Proceedings of the 4th Micro Total Analysis Systems Symposium*, Enschede, Netherlands, May 14–18, pp. 277–82.

517. Yang, Z., Goto, H., Matsumoto, M., Yada, T. 1998. Micro mixer incorporated with piezoelectrically driven valveless micropump. In *Micro Total Analysis Systems '98*, Banff, Canada, October 13–16, pp. 177–80.

518. Wheeler, A.R., Moon, H., Kim, C.-J., Loo, J.A., Garrell, R.L. 2004. Electrowetting-based microfluidics for analysis of peptides and proteins by matrix-assisted laser desorption/ionization mass spectrometry. *Anal. Chem.* 76:4833–38.

519. Dittrich, P.S., Schwille, P. 2002. Spatial two-photon fluorescence cross-correlation spectroscopy for controlling molecular transport in microfluidic structures. *Anal. Chem.* 74:4472–79.

520. Santiago, J.G. 2001. Electroosmotic flows in microchannels with finite inertial and pressure forces. *Anal. Chem.* 73:2353–65.

521. Van Theemsche, A., Deconinck, J., Van den Bossche, B., Bortels, L. 2002. Numerical solution of a multi-ion one-potential model for electroosmotic flow in two-dimensional rectangular microchannels. *Anal. Chem.* 74:4919–26.

522. Hu, L., Harrison, J.D., Masliyah, J.H. 1999. Numerical model of electrokinetic flow for capillary electrophoresis. *J. Colloid Interface Sci.* 215:300–12.

523. Poppe, H. 2002. Mass transfer in rectangular chromatographic channels. *J. Chromatogr.* 948A:3–17.

524. Chen, C.H., Santiago, J.G. 2002. A planar electroosmotic micropump. *J. Microelectromech. Syst.* 11:672–83.

525. Cummings, E.B., Griffiths, S.K., Nilson, R.H., Paul, P.H. 2000. Conditions for similitude between the fluid velocity and electric field in electroosmotic flow. *Anal. Chem.* 72:2526–32.

526. Hu, Y., Werner, C., Li, D. 2003. Electrokinetic transport through rough microchannels. *Anal. Chem.* 75:5747–58.

527. Devasenathipathy, S., Santiago, J.G., Takehara, K. 2002. Particle tracking techniques for electrokinetic microchannel flows. *Anal. Chem.* 74:3704–13.

528. Qian, S., Bau, H.H. 2002. A chaotic electroosmotic stirrer. *Anal. Chem.* 74:3616–25.

529. Fu, L.-M., Yang, R.-J., Lee, G.-B., Liu, H.-H. 2002. Electrokinetic injection techniques in microfluidic chips. *Anal. Chem.* 74:5084–91.

530. Patankar, N.A., Hu, H.H. 1998. Numerical simulation of electroosmotic flow. *Anal. Chem.* 70:1870–81.

531. Ren, L.Q., Li, D.Q. 2002. Theoretical studies of microfluidic dispensing processes. *J. Colloid Interface Sci.* 254:384–95.

532. Ermakov, S.V., Jacobson, S.C., Ramsey, J.M. 1998. Computer simulations of electrokinetic transport in microfabricated channel structures. *Anal. Chem.* 70:4494–504.

533. Zholkovskij, E.K., Masliyah, J.H., Czarnecki, J. 2003. Electroosmotic dispersion in microchannels with a thin double layer. *Anal. Chem.* 75:901–9.

534. Footz, T., Somerville, M.J., Tomaszewski, R., Elyas, B., Backhouse, C.J. 2004. Integration of combined heteroduplex/restriction fragment length polymorphism analysis on an electrophoresis microchip for the detection of hereditary haemochromatosis. *Analyst* 129:25–31.

535. Molho, J.I., Herr, A.E., Mosier, B.P., Santiago, J.G., Kenny, T.W., Brennen, R.A., Gordon, G.B., Mohammadi, B. 2001. Optimization of turn geometries for microchip electrophoresis. *Anal. Chem.* 73:1350–60.

536. Griffiths, S.K., Nilson, R.H. 2002. Design and analysis of folded channels for chip-based separations. *Anal. Chem.* 74:2960–67.

537. Griffiths, S.K., Nilson, R.H. 2000. Band spreading in two-dimensional microchannel turns for electrokinetic species transport. *Anal. Chem.* 72:5473–82.

538. Culbertson, C.T., Jacobson, S.C., Ramsey, J.M. 1998. Dispersion sources for compact geometries on microchips. *Anal. Chem.* 70:3781–89.

539. Dutta, D., Leighton, D.T., Jr. 2002. A low dispersion geometry for microchip separation devices. *Anal. Chem.* 74:1007–16.

540. Griffiths, S.K., Nilson, R.H. 2001. Modeling electrokinetic transport for the design and optimization of microchannel systems. In *Proceedings of the 5th Micro Total Analysis Systems Symposium*, Monterey, CA, October 21–25, pp. 456–58.

541. Fiechtner, G.J., Cummings, E.B. 2003. Faceted design of channels for low-dispersion electrokinetic flows in microfluidic systems. *Anal. Chem.* 75:4747–55.

542. Griffiths, S.K., Nilson, R.H. 2001. Low-dispersion turns and junctions for microchannel systems. *Anal. Chem.* 73:272–78.

543. Mitnik, L., Carey, L., Burger, R., Desmarais, S., Koutny, L., Wernet, O., Matsudaira, P., Ehrlich, D. 2002. High-speed analysis of multiplexed short tandem repeats with an electrophoretic microdevice. *Electrophoresis* 23:719–26.

544. Weigl, B.H., Holl, M.R., Schutte, D., Brody, J.P., Yager, P. 1996. Diffusion-based optical chemical detection in silicon flow structure. In *Micro Total Analysis Systems '96,* Basel, November 19–22, pp. 174–84.

545. Cabrera, C.R., Finlayson, B., Yager, P. 2001. Formation of natural pH gradients in a microfluidic device under flow conditions: Model and experimental validation. *Anal. Chem.* 73:658–66.

546. Bai, X., Josserand, J., Jensen, H., Rossier, J.S., Girault, H.H. 2002. Finite element simulation of pinched pressure-driven flow injection in microchannels. *Anal. Chem.* 74:6205–15.

547. Effenhauser, C.S., Paulus, A., Manz, A., Widmer, H.M. 1994. High-speed separation of antisense oligonucleotides on a micromachined capillary electrophoresis device. *Anal. Chem.* 66:2949–53.

548. Liu, S., Shi, Y., Ja, W.W., Mathies, R.A. 1999. Optimization of high-speed DNA sequencing on microfabricated capillary electrophoresis channels. *Anal. Chem.* 71:566–73.

549. Alarie, J.P., Jacobson, S.C., Ramsey, J.M. 2001. Electrophoretic injection bias in a microchip valving scheme. *Electrophoresis* 22:312–17.

550. Seiler, K., Harrison, D.J., Manz, A. 1993. Planar glass chips for capillary electrophoresis: Repetitive sample injection, quantitation, and separation efficiency. *Anal. Chem.* 65:1481–88.

551. von Heeren, F., Verpoorte, E., Manz, A., Thormann, W. 1996. Micellar electrokinetic chromatography separations and analyses of biological samples on a cyclic planar microstructure. *Anal. Chem.* 68:2044–53.

552. Shultz-Lockyear, L.L., Colyer, C.L., Fan, Z.H., Roy, K.I., Harrison, D.J. 1999. Effects of injector geometry and sample matrix on injection and sample loading in integrated capillary electrophoresis devices. *Electrophoresis* 20:529–38.

553. Crabtree, H.J., Cheong, E.C.S., Tilroe, D.A., Backhouse, C.J. 2001. Microchip injection and separation anomalies due to pressure effects. *Anal. Chem.* 73:4079–86.

554. Woolley, A.T., Mathies, R.A. 1995. Ultra-high-speed DNA sequencing using capillary electrophoresis chips. *Anal. Chem.* 67:3676–80.

555. Herr, A.E., Singh, A.K. 2004. Photopolymerized cross-linked polyacrylamide gels for on-chip protein sizing. *Anal. Chem.* 76:4727–33.

556. Zhang, C., Manz, A. 2001. Narrow sample channel injectors for capillary electrophoresis on microchips. *Anal. Chem.* 73:2656–62.

557. Simpson, P.C., Roach, D., Woolley, A.T., Thorsen, T., Johnston, R., Sensabaugh, G.F., Mathies, R.A. 1998. High-throughput genetic analysis using microfabricated 96-sample capillary array electrophoresis microplates. *Proc. Natl. Acad. Sci. USA* 95:2256–61.

558. Bousse, L., Minalla, A., West, J. 2000. Novel injection schemes for ultra-high speed DNA separations. In *Proceedings of the 4th Micro Total Analysis Systems Symposium,* Enschede, Netherlands, May 14–18, pp. 415–18.

559. Deshpande, M., Greiner, K.B., West, J., Gilbert, J.R., Bousse, L., Minalla, A. 2000. Novel designs for electrokinetic injection in μTAS. In *Proceedings of the 4th Micro Total Analysis Systems Symposium,* Enschede, Netherlands, May 14–18, pp. 339–42.

560. Dang, F., Zhang, L., Jabasini, M., Kaji, N., Baba, Y. 2003. Characterization of electrophoretic behavior of sugar isomers by microchip electrophoresis coupled with videomicroscopy. *Anal. Chem.* 75:2433–39.

561. Sinton, D., Ren, L.Q., Xuan, X.C., Li, D.Q. 2003. Effects of liquid conductivity differences on multicomponent sample injection, pumping and stacking in microfluidic chips. *Labchip* 3:173–79.

562. Lin, Y.C., Wu, W.M., Fan, C.S. 2004. Design and simulation of sample pinching utilizing microelectrodes in capillary electrophoresis microchips. *Labchip* 4:60–64.

563. Bessoth, F.G., Naji, O.P., Eijkel, J.C.T., Manz, A. 2002. Towards an on-chip gas chromatograph: The development of a gas injector and a dc plasma emission detector. *J. Anal. Atom. Spectrom.* 17:794.

564. Jacobson, S.C., Ermakov, S.V., Ramsey, J.M. 1999. Minimizing the number of voltage sources and fluid reservoirs for electrokinetic valving in microfluidic devices. *Anal. Chem.* 71:3273–76.

565. Ramsey, J.D., Jacobson, S.C., Culbertson, C.T., Ramsey, J.M. 2003. High-efficiency, two-dimensional separations of protein digests on microfluidic devices. *Anal. Chem.* 75:3758–64.

566. Slentz, B.E., Penner, N.A., Regnier, F. 2002. Sampling bias at channel junctions in gated flow injection on chips. *Anal. Chem.* 74:4835–40.

567. Chen, S., Lin, Y., Wang, L., Lin, C., Lee, G. 2002. Flow-through sampling for electrophoresis-based microchips and their applications for protein analysis. *Anal. Chem.* 74:5146–53.

568. Lin, Y.H., Lee, G.B., Li, C.W., Huang, G.R., Chen, S.H. 2001. Flow-through sampling for electrophoresis-based microfluidic chips using hydrodynamic pumping. *J. Chromatogr.* 937A:115–25.

569. Vahey, P.G., Park, S.H., Marquardt, B.J., Xia, Y.N., Burgess, L.W., Synovec, R.E. 2000. Development of a positive pressure driven micro-fabricated liquid chromatographic analyzer through rapid-prototyping with poly(dimethylsiloxane): Optimizing chromatographic efficiency with sub-nanoliter injections. *Talanta* 51:1205–12.

570. Bai, X.X., Lee, H.J., Rossier, J.S., Reymond, F., Schafer, H., Wossner, M., Girault, H.H. 2002. Pressure pinched injection of nanolitre volumes in planar micro-analytical devices. *Labchip* 2:45–49.

571. O'Neill, A.P., O'Brien, P., Alderman, J., Hoffman, D., McEnery, M., Murrihy, J., Glennon, J.D. 2001. On-chip definition of picolitre sample injection plugs for miniaturised liquid chromatography. *J. Chromatogr.* 924A:259–63.

572. Reichmuth, D.S., Shepodd, T.J., Kirby, B.J. 2003. RP-HPLC microchip separations with subnanoliter on-chip pressure injections. In *Proceedings of the 7th Micro Total Analysis Systems Symposium,* Squaw Valley, CA, October 5–9, pp. 1021–24.

573. Blom, M.T., Chmela, E., Gardeniers, J.G.E., Tijssen, R., Elwenspoek, M., van den Berg, A. 2002. Design and fabrication of a hydrodynamic chromatography. *Sensors Actuators* 82B:111–16.

574. Chmela, E., Blom, M.T., Gardeniers, J.G.E., van den Berg, A., Tijssen, R. 2002. A pressure driven injection system for an ultra-flat chromatographic microchannel. *Labchip* 2:235–41.

575. O'Neill, A.P., O'Brien, P., Alderman, J., Hoffman, D., McEnery, M., Murrihy, J., Glennon, J.D. 2001. On-chip definition of picolitre sample injection plugs for miniaturised liquid chromatography. *J. Chromatogr.* 924A:259–63.

576. Lapos, J.A., Ewing, A.G. 2000. Injection of fluorescently labeled analytes into microfabricated chips using optically gated electrophoresis. *Anal. Chem.* 72:4598–602.

577. Xu, H., Roddy, T.P., Lapos, J.A., Ewing, A.G. 2002. Parallel analysis with optically gated sample introduction on a multichannel microchip. *Anal. Chem.* 74:5517–22.

578. Lichtenberg, J., Verpoorte, E., de Rooij, N.F. 2001. Sample preconcentration by field amplification stacking for microchip-based capillary electrophoresis. *Electrophoresis* 22:258–71.

579. Bharadwaj, R., Santiago, J.G. 2001. Optimization of field amplified sample stacking on a microchip. In *Proceedings of the 5th Micro Total Analysis Systems Symposium,* Monterey, CA, October 21–25, pp. 613–14.

580. Liu, Y., Foote, R.S., Jacobson, S.C., Ramsey, R.S., Ramsey, J.M. 2000. Transport number mismatch induced stacking of swept sample zones for microchip-based sample concentration. In *Proceedings of the 4th Micro Total Analysis Systems Symposium,* Enschede, Netherlands, May 14–18, pp. 295–98.

581. Lichtenberg, J., Daridon, A., Verpoorte, E., de Rooij, N.F. 2000. Combination of sample pre-concentration and capillary electrophoresis on-chip. In *Proceedings of the 4th Micro Total Analysis Systems Symposium,* Enschede, Netherlands, May 14–18, pp. 307–10.

582. Harrison, D.J., Manz, A., Fan, Z., Lüdi, H., Wildmer, H.M. 1992. Capillary electrophoresis and sample injection systems integrated on a planar glass chip. *Anal. Chem.* 64:1926–32.

583. Palmer, J., Burgi, D.S., Munro, N.J., Landers, J.P. 2001. Electrokinetic injection for stacking neutral analytes in capillary and microchip electrophoresis. *Anal. Chem.* 73:725–31.

584. Jung, B., Bharadwaj, R., Santiago, J.G. 2003. Thousandfold signal increase using field-amplified sample stacking for on-chip electrophoresis. *Electrophoresis* 24:3476–83.

585. Yang, H., Chien, R.L. 2001. Sample stacking in laboratory-on-a-chip devices. *J. Chromatogr.* 924A:155–63.

586. Kutter, J.P., Jacobson, S.C., Ramsey, J.M. 2000. Solid phase extraction on microfluidic device. *J. Microcolumn Separations* 12: 93–97.

587. Jemere, A.B., Oleschuk, R.D., Ouchen, F., Fajuyigbe, F., Harrison, D.J. 2002. An integrated solid-phase extraction system for sub-picomolar detection. *Electrophoresis* 23:3537–44.

588. Stachowiak, T.B., Rohr, T., Hilder, E.F., Peterson, D.S., Yi, M.Q., Svec, F., Fréchet, J.M.J. 2003. Fabrication of porous polymer monoliths covalently attached to the walls of channels in plastic microdevices. *Electrophoresis* 24:3689–93.

589. Ueno, Y., Horiuchi, T., Niwa, O. 2002. Air-cooled cold trap channel integrated in a microfluidic device for monitoring airborne BTEX with an improved detection limit. *Anal. Chem.* 74:1712–17.

590. Khandurina, J., Jacobson, S.C., Waters, L.C., Foote, R.S., Ramsey, J.M. 1999. Microfabricated porous membrane structure for sample concentration and electrophoretic analysis. *Anal. Chem.* 71:1815–19.

591. Kuo, T.-C., Cannon, D.M., Jr., Bohn, P.W., Sweedler, J.V. 2003. Nanocapillary array interconnects for gated analyte injections and electrophoretic separations in multilayer microfluidic architectures. *Anal. Chem.* 75:2224–30.

592. Kuo, T.-C., Cannon, D.M., Jr., Chen, Y., Tulock, J.J., Shannon, M.A., Sweedler, J.V., Bohn, P.W. 2003. Gateable nanofluidic interconnects for multilayered microfluidic separation systems. *Anal. Chem.* 75:1861–67.

593. Kuo, T.C., Cannon, D.M., Jr., Shannon, M.A., Bohn, P.W., Sweedler, J.V. 2003. Hybrid three-dimensional nanofluidic/microfluidic devices using molecular gates. *Sensors Actuators* 102A:223–33.

594. Song, S., Singh, A.K., Kirby, B.J. 2004. Electrophoretic concentration of proteins at laser-patterned nanoporous membranes in microchips. *Anal. Chem.* 76:4589–92.

595. Hofmann, O., Viorin, G., Niedermann, P., Manz, A. 2001. 3D-flow confinement for efficient sample delivery: Application to immunoassays on planar optical waveguides. In *Proceedings of the 5th Micro Total Analysis Systems Symposium*, Monterey, CA, October 21–25, pp. 133–34.

596. Jandik, P., Weigl, B.H., Kessler, N., Cheng, J., Morris, C.J., Schulte, T., Avdalovic, N. 2002. Initial study of using a laminar fluid diffusion interface for sample preparation in high-performance liquid chromatography. *J. Chromatogr.* 954A:33–40.

597. Ross, D., Locascio, L.E. 2002. Microfluidic temperature gradient focusing. *Anal. Chem.* 74:2556–64.

598. Lambertus, G., Elstro, A., Sensenig, K., Potkay, J., Agah, M., Scheuering, S., Wise, K., Dorman, F., Sacks, R. 2004. Design, fabrication, and evaluation of microfabricated columns for gas chromatography. *Anal. Chem.* 76:2629–37.

599. Naji, O.P., Bessoth, F.G., Manz, A. 2001. Novel injection methods for miniaturised gas chromatography. In *Proceedings of the 5th Micro Total Analysis Systems Symposium*, Monterey, CA, October 21–25, pp. 655–57.

600. Lehmann, U., Krusemark, O., Muller, J. 2000. Micro machined gas chromatograph based on a plasma polymerised stationary phase. In *Proceedings of the 4th Micro Total Analysis Systems Symposium,* Enschede, Netherlands, May 14–18, pp. 167–70.

601. Hannoe, S., Sugimoto, I., Katoh, T. 1998. Silicon-micromachined separation columns coated with amino acid films for an integrated on-chip gas chromatograph. In *Micro Total Analysis Systems '98,* Banff, Canada, October 13–16, pp. 145–48.

602. Frye-Mason, G., Kottenstette, R.J., Heller, E.J., Matzke, C.M., Casalnuovo, S.A., Lewis, P.R., Manginell, R.P., Schubert, W.K., Hietala, V.M., Shul, R.J. 1998. Integrated chemical analysis systems for gas phase CW agent detection. In *Micro Total Analysis Systems '98,* Banff, Canada, October 13–16, pp. 477–81.

603. Frye-Mason, G., Kottenstette, R., Mowry, C., Morgan, C., Manginell, R., Lewis, P., Matzke, C., Dulleck, G., Anderson, L., Adkins, D. 2001. Expanding the capabilities and applications of gas phase miniature chemical analysis systems (μChemLab). In *Proceedings of the 5th Micro Total Analysis Systems Symposium*, Monterey, CA, October 21–25, pp. 658–60.

604. Jacobson, S.C., Culbertson, C.T., Daler, J.E., Ramsey, J.M. 1998. Microchip structures for submillisecond electrophoresis. *Anal. Chem.* 70:3476–80.

605. Ludwig, M., Kohler, F., Belder, D. 2003. High-speed chiral separations on a microchip with UV detection. *Electrophoresis* 24:3233–38.

606. Tang, T., Badal, M.Y., Ocvirk, G., Lee, W.E., Bader, D.E., Bekkaoui, F., Harrison, D.J. 2002. Integrated microfluidic electrophoresis system for analysis of genetic materials using signal amplification methods. *Anal. Chem.* 74:725–33.

607. Lin, Y.-C., Wu, C.-Y. 2001. Design of moving electric field driven capillary electrophoresis chips. In *Proceedings of the 5th Micro Total Analysis Systems Symposium*, Monterey, CA, October 21–25, pp. 553–54.

608. Badal, M.Y., Wong, M., Chiem, N., Salimi-Moosavi, H., Harrison, D.J. 2001. Developing a routine coating method for multichannel flow networks on a chip using pyrolyzed poly(dimethylsiloxane). In *Proceedings of the 5th Micro Total Analysis Systems Symposium*, Monterey, CA, October 21–25, pp. 535–36.

609. Fruetel, J., Renzi, R., Crocker, R., VanderNoot, V., Stamps, J., Shokair, I., Yee, D. 2001. Application of microseparation arrays to the detection of biotoxins in aerosol backgrounds. In *Proceedings of the 5th Micro Total Analysis Systems Symposium*, Monterey, CA, October 21–25, pp. 523–24.

610. Hutt, L.D., Glavin, D.P., Bada, J.L., Mathies, R.A. 1999. Microfabricated capillary electrophoresis amino acid chirality analyzer for extraterrestrial exploration. *Anal. Chem.* 71:4000–6.

611. Tian, H.J., Landers, J.P. 2002. Hydroxyethylcellulose as an effective polymer network for DNA analysis in uncoated glass microchips: Optimization and application to mutation detection via heteroduplex analysis. *Anal. Biochem.* 309:212–23.

612. Sanders, J.C., Breadmore, M.C., Kwok, Y.C., Horsman, K.M., Landers, J.P. 2003. Hydroxypropyl cellulose as an adsorptive coating sieving matrix for DNA separations: Artificial neural network optimization for microchip analysis. *Anal. Chem.* 75:986–94.

613. Xu, F., Jabasini, M., Baba, Y. 2002. DNA separation by microchip electrophoresis using low-viscosity hydroxypropylmethylcellulose-50 solutions enhanced by polyhydroxy compounds. *Electrophoresis* 23:3608–14.

614. Song, L.G., Fang, D.F., Kobos, R.K., Pace, S.J., Chu, B. 1999. Separation of double-stranded DNA fragments in plastic capillary electrophoresis chips by using $E_{99}P_{69}E_{99}$ as separation medium. *Electrophoresis* 20:2847–55.

615. Brahmasandra, S.N., Ugaz, V.M., Burke, D.T., Mastrangelo, C.H., Burns, M.A. 2001. Electrophoresis in microfabricated devices using photopolymerized polyacrylamide gels and electrode-defined sample injection. *Electrophoresis* 22:300–11.

616. Munro, N.J., Snow, K., Kant, J.A., Landers, J.P. 1999. Molecular diagnostics on microfabricated electrophoretic devices: From slab gel- to capillary- to microchip-based assays for T- and B-cell lymphoproliferative disorders. *Clin. Chem.* 45:1906–17.

617. Chen, Y.-H., Wang, W.-C., Young, K.-C., Chang, T.-Y., Chen, S.-H. 1999. Plastic microchip electrophoresis for analysis of PCR products of hepatitis C virus. *Clin. Chem.* 45:1938–43.

618. Shediac, R., Pizarro, S.A., Singh, A.K. 2003. Ultra fast on chip separations of cytokines by SDS-PAGE using UV-initiated polyacrylamide. In *Proceedings of the 7th Micro Total Analysis Systems Symposium*, Squaw Valley, CA, October 5–9, pp. 971–74.

619. Tabuchi, M., Kuramitsu, Y., Nakamura, K., Baba, Y. 2003. A 15-s protein separation employing hydrodynamic force on a microchip. *Anal. Chem.* 75:3799–805.

620. Wallenborg, S.R., Bailey, C.G. 2000. Separation and detection of explosives on a microchip using micellar electrokinetic chromatography and indirect laser-induced fluorescence. *Anal. Chem.* 72:1872–78.

621. Wang, J., Pumera, M., Collins, G., Opekar, F., Jelínek, I. 2002. A chip-based capillary electrophoresis-contactless conductivity microsystem for fast measurements of low-explosive ionic components. *Analyst* 127:719–23.

622. Lu, Q., Collins, G.E., Smith, M., Wang, J. 2002. Sensitive capillary electrophoresis microchip determination of trinitroaromatic explosives in nonaqueous electrolyte following solid phase extraction. *Anal. Chim. Acta* 469:253–60.

623. Kirby, B.J., Wheeler, A.R., Zare, R.N., Fruetel, J.A., Shepodd, T.J. 2003. Programmable modification of cell adhesion and zeta potential in silica microchips. *Labchip* 3:5–10.

624. Rodriguez, I., Lee, H.K., Li, S.F.Y. 1999. Microchannel electrophoretic separation of biogenic amines by micellar electrokinetic chromatography. *Electrophoresis* 20:118–26.

625. Wang, J., Chatrathi, M.P., Mulchandani, A., Chen, W. 2001. Capillary electrophoresis microchips for separation and detection of organophosphate nerve agents. *Anal. Chem.* 73:1804–8.

626. Hofmann, O., Che, D., Cruickshank, K.A., Mueller, U.R. 1999. Adaptation of capillary isoelectric focusing to microchannels on a glass chip. *Anal. Chem.* 71:678–86.

627. Rodríguez, I., Jin, L.J., Li, S.F.Y. 2000. High-speed chiral separations on microchip electrophoresis devices. *Electrophoresis* 21:211–19.

628. Skelley, A.M., Mathies, R.A. 2004. Chiral separation of fluorescamine-labeled amino acids using microfabricated capillary electrophoresis devices for extraterrestrial exploration. *J. Chromatogr.* 1021A:191–99.

629. Wheeler, A., Morishima, K., Kirby, B., Leach, A., Zare, R.N. 2001. CATH.a neuron cell analysis on a chip with micellar electrokinetic chromatography. In *Proceedings of the 5th Micro Total Analysis Systems Symposium*, Monterey, CA, October 21–25, pp. 311–12.

630. Masar, M., Kaniansky, D., Bodor, R., Johnck, M., Stanislawski, B. 2001. Determination of organic acids and inorganic anions in wine by isotachophoresis on a planar chip. *J. Chromatogr.* 916A:167–74.

631. Kaniansky, D., Masar, M., Bielcikova, J., Ivanyi, F., Eisenbeiss, J., Stanislawski, B., Grass, B., Neyer, A., Johnck, M. 2000. Capillary electrophoresis separations on a planar chip with the column-coupling configuration of the separation channels. *Anal. Chem.* 72:3596–604.

632. Masar, M., Zuborova, M., Bielcikova, J., Kaniansky, D., Johnck, M., Stanislawski, B. 2001. Conductivity detection and quantitation of isotachophoretic analytes on a planar chip with on-line coupled separation channels. *J. Chromatogr.* 916A:101–11.

633. Prest, J.E., Baldock, S.J., Fielden, P.R., Goddard, N.J., Brown, B.J.T. 2003. Miniaturised isotachophoretic analysis of inorganic arsenic speciation using a planar polymer chip with integrated conductivity detection. *J. Chromatogr.* 990A:325–34.

634. Vreeland, W.N., Williams, S.J., Barron, A.E., Sassi, A.P. 2003. Tandem isotachophoresis-zone electrophoresis via base-mediated destacking for increased detection sensitivity in microfluidic systems. *Anal. Chem.* 75:3059–65.

635. Kurnik, R.T., Boone, T.D., Nguyen, U., Ricco, A.J., Williams, S.J. 2003. Use of floating electrodes in transient isotachophoresis to increase the sensitivity of detection. *Labchip* 3:86–92.

636. Wainright, A., Nguyen, U.T., Bjornson, T., Boone, T.D. 2003. Preconcentration and separation of double-stranded DNA fragments by electrophoresis in plastic microfluidic devices. *Electrophoresis* 24:3784–92.

637. Bodor, R., Zúborová, M., Ölvecká, E., Madajová, V., Masár, M., Kaniansky, D., Stanislawski, B. 2001. Isotachophoresis and isotachophoresis-zone electrophoresis of food additives on a chip with column-coupling separation channels. *J. Separ. Sci.* 24:802–9.

638. Prest, J.E., Baldock, S.J., Fielden, P.R., Goddard, N.J., Treves-Brown, B.J. 2002. Bidirectional isotachophoresis on a planar chip with integrated conductivity detection. *Analyst* 127:1413–19.

639. Oleschuk, R.D., Shultz-Lockyear, L.L., Ning, Y., Harrison, D.J. 2000. Trapping of bead-based reagents within microfluidic systems. On-chip solid-phase extraction and electrochromatography. *Anal. Chem.* 72:585–90.

640. Finot, M., Jemere, A.B., Oleschuk, R.D., Takahashi, L., Harrison, D.J. 2001. High throughput pharmaceutical formulation evaluation and analysis using capillary electrochromatography on a microfluidic chip. In *Proceedings of the 5th Micro Total Analysis Systems Symposium*, Monterey, CA, October 21–25, pp. 480–82.

641. Penner, N.A., Slentz, B.E., Regnier, F. 2001. Capillary electrochromatography on microchips with in situ fabricated particles. In *Proceedings of the 5th Micro Total Analysis Systems Symposium*, Monterey, CA, October 21–25, pp. 559–60.

642. Broyles, B.S., Jacobson, S.C., Ramsey, J.M. 2001. Sample concentration and separation on microchips. In *Proceedings of the 5th Micro Total Analysis Systems Symposium*, Monterey, CA, October 21–25, pp. 537–38.

643. Kutter, J.P., Jacobson, S.C., Matsubara, N., Ramsey, J.M. 1998. Solvent-programmed microchip open-channel electrochromatography. *Anal. Chem.* 70:3291–97.

644. He, B., Regnier, F. 1998. Microfabricated liquid chromatography columns based on collocated monolith support structures. *J. Pharm. Biomed. Anal.* 17:925–32.

645. Vervoort, N., Billen, J., Gzil, P., Baron, G.V., Desmet, G. 2004. Importance and reduction of the sidewall-induced band-broadening effect in pressure-driven microfabricated columns. *Anal. Chem.* 76:4501–7.

646. Slentz, B.E., Penner, N.A., Regnier, F.E. 2002. Capillary electrochromatography of peptides on microfabricated poly(dimethylsiloxane) chips modified by cerium(IV)-catalyzed polymerization. *J. Chromatogr.* 948A:225–33.

647. Shediac, R., Ngola, S.M., Throckmorton, D.J., Anex, D.S., Shepodd, T.J., Singh, A.K. 2001. Reversed-phase electrochromatography of amino acids and peptides using porous polymer monoliths. *J. Chromatogr.* 925A:251–63.

648. Burggraf, N., Manz, A., Verpoorte, E., Effenhauser, C.S., Widmer, H.M., de Rooij, N.F. 1994. A novel approach to ion separation in solution: Synchronized cyclic capillary electrophoresis (SCCE). *Sensors Actuators* 20B:103–10.

649. Manz, A., Bousse, L., Chow, A., Metha, T., Kopf-Sill, A., Parce, W.J. 2001. Synchronized cyclic capillary electrophoresis using channels arranged in a triangle and low voltages. *Fres. J. Anal. Chem.* 371:195–201.

650. von Heeren, F., Verpoorte, E., Manz, A., Thormann, W. 1996. Characterization of electrophoretic sample injection and separation in a gel-filled cyclic planar microstructure. *J. Microcolumn Separations* 8:373–81.

651. Backhouse, C.J., Gajdal, A., Pilarski, L.M., Crabtree, H.J. 2003. Improved resolution with microchip-based enhanced field inversion electrophoresis. *Electrophoresis* 24:1777–86.

652. Raymond, D.E., Manz, A., Widmer, H.M. 1996. Continuous separation of high molecular weight compounds using a microliter volume free-flow electrophoresis microstructure. *Anal. Chem.* 68:2515–22.

653. Zhang, C.-X., Manz, A. 2003. High-speed free-flow electrophoresis on chip. *Anal. Chem.* 75:5759–66.

654. Ro, K.W., Lim, K., Hahn, J.H. 2001. PDMS microchip for precolumn reaction and micellar electrokinetic chromatography of biogenic amines. In *Proceedings of the 5th Micro Total Analysis Systems Symposium*, Monterey, CA, October 21–25, pp. 561–62.

655. Wang, J., Chatrathi, M.P., Tian, B. 2000. Micromachined separation chips with a precolumn reactor and end-column electrochemical detector. *Anal. Chem.* 72:5774–78.

656. Giordano, B.C., Jin, L., Couch, A.J., Ferrance, J.P., Landers, J.P. 2004. Microchip laser-induced fluorescence detection of proteins at submicrogram per milliliter levels mediated by dynamic labeling under pseudonative conditions. *Anal. Chem.* 76:4705–14.

657. Colyer, C.L., Mangru, S.D., Harrison, D.J. 2000. Microchip-based capillary electrophoresis of human serum proteins. *J. Chromatogr.* 781A:271–76.

658. Fluri, K., Fitzpatrick, G., Chiem, N., Harrison, D.J. 1996. Integrated capillary electrophoresis devices with an efficient postcolumn reactor in planar quartz and glass chips. *Anal. Chem.* 68:4285–90.

659. Liu, Y., Foote, R.S., Jacobson, S.C., Ramsey, R.S., Ramsey, J.M. 2000. Electrophoretic separation of proteins on a microchip with noncovalent, postcolumn labeling. *Anal. Chem.* 72:4608–13.

660. Cowen, S., Craston, D.H. 1994. An on-chip miniature liquid chromatography system: Design, construction and characterization. In *Micro Total Analysis Systems '94*, University of Twente, Netherlands, November 21–22, pp. 295–98.

661. McEnery, M., Tan, A., Alderman, J., Patterson, J., O'Mathuna, S.C., Glennon, J.D. 2000. Liquid chromatography on-chip: Progression towards μ-total analysis system. *Analyst* 125:25–27.

662. Seki, M., Yamada, M., Ezaki, R., Aoyama, R., Hong, J.W. 2001. Chromatographic separation of proteins on a PDMS-polymer chip by pressure flow. In *Proceedings of the 5th Micro Total Analysis Systems Symposium*, Monterey, CA, October 21–25, pp. 48–50.

663. Effenhauser, C.S., Manz, A., Widmer, H.M. 1995. Manipulation of sample fractions on a capillary electrophoresis chip. *Anal. Chem.* 67:2284–87.

664. Hong, J.W., Hagiwara, H., Fujii, T., Machida, H., Inoue, M., Seki, M., Endo, I. 2001. Separation and collection of a specified DNA fragment by chip-based CE system. In *Proceedings of the 5th Micro Total Analysis Systems Symposium*, Monterey, CA, October 21–25, pp. 113–14.

665. Khandurina, J., Chován, T., Guttman, A. 2002. Micropreparative fraction collection in microfluidic devices. *Anal. Chem.* 74:1737–40.

666. Rocklin, R.D., Ramsey, R.S., Ramsey, J.M. 2000. A microfabricated fluidic device for performing two-dimensional liquid-phase separations. *Anal. Chem.* 72:5244–49.

667. Li, Y., Buch, J.S., Rosenberger, F., DeVoe, D.L., Lee, C.S. 2004. Integration of isoelectric focusing with parallel sodium dodecyl sulfate gel electrophoresis for multidimensional protein separations in a plastic microfluidic network. *Anal. Chem.* 76:742–48.

668. Herr, A.E., Molho, J.I., Drouvalakis, K.A., Mikkelsen, J.C., Utz, P.J., Santiago, J.G., Kenny, T.W. 2003. On-chip coupling of isoelectric focusing and free solution electrophoresis for multidimensional separations. *Anal. Chem.* 75:1180–87.

669. Buch, J.S., Li, Y., Rosenberger, F., DeVoe, D.L., Lee, C.S. 2003. Two-dimensional genomic and proteomic separations in a plastic microfluidic network. In *Proceedings of the 7th Micro Total Analysis Systems Symposium*, Squaw Valley, CA, October 5–9, pp. 477–80.

670. Lapos, J.A., Manica, D.P., Ewing, A.G. 2002. Dual fluorescence and electrochemical detection on an electrophoresis microchip. *Anal. Chem.* 74:3348–53.

671. Liang, Z., Chiem, N., Ocvirk, G., Tang, T., Fluri, K., Harrison, D.J. 1996. Microfabrication of a planar absorbance and fluorescence cell for integrated capillary electrophoresis devices. *Anal. Chem.* 68:1040–46.

672. Roos, P., Skinner, C.D. 2003. A two bead immunoassay in a micro fluidic device using a flat laser intensity profile for illumination. *Analyst* 128:527–31.

673. Foquet, M., Korlach, J., Zipfel, W.R., Webb, W.W., Craighead, H.G. 2004. Focal volume confinement by submicrometer-sized fluidic channels. *Anal. Chem.* 76:1618–26.

674. Qi, S.Z., Liu, X.Z., Ford, S., Barrows, J., Thomas, G., Kelly, K., McCandless, A., Lian, K., Goettert, J., Soper, S.A. 2002. Microfluidic devices fabricated in poly(methyl methacrylate) using hot-embossing with integrated sampling capillary and fiber optics for fluorescence detection. *Labchip* 2:88–95.

675. Zugel, S.A., Burke, B.J., Regnier, F.E., Lytle, F.E. 2000. Electrophoretically mediated microanalysis of leucine aminopeptidase using two-photon excited fluorescence detection on a microchip. *Anal. Chem.* 72:5731–35.

676. Hibara, A., Saito, T., Kim, H.-B., Tokeshi, M., Ooi, T., Nakao, M., Kitamori, T. 2002. Nanochannels on a fused-silica microchip and liquid properties investigation by time-resolved fluorescence measurements. *Anal. Chem.* 74:6170–76.

677. Fister, J.C., III, Jacobson, S.C., Davis, L.M., Ramsey, J.M. 1998. Counting single chromophore molecules for ultrasensitive analysis and separations on microchip devices. *Anal. Chem.* 70:431–37.

678. Jiang, G., Attiya, S., Ocvirk, G., Lee, W.E., Harrison, D.J. 2000. Red diode laser induced fluorescence detection with a confocal microscope on a microchip for capillary electrophoresis. *Biosens. Bioelectronics* 14:861–69.

679. Soper, S.A., Ford, S.M., Xu, Y.C., Qi, S.Z., McWhorter, S., Lassiter, S., Patterson, D., Bruch, R.C. 1999. Nanoliter-scale sample preparation methods directly coupled to polymethylmethacrylate-based microchips and gel-filled capillaries for the analysis of oligonucleotides. *J. Chromatogr.* 853A:107–20.

680. Peng, X.Y., Li, P.C.H. 2004. A three-dimensional flow control concept for single-cell experiments on a microchip. 1. Cell selection, cell retention, cell culture, cell balancing, and cell scanning. *Anal. Chem.* 76:5273–81.

681. Zhang, T., Fang, Q., Wang, S.L., Fangl, Z.L. 2003. Enhancement of signal-to-noise levels by synchronized dual wavelength modulation for light emitting diode fluorimetry in microfluidic systems. In *Proceedings of the 7th Micro Total Analysis Systems Symposium*, Squaw Valley, CA, October 5–9, p. 387.

682. Huang, Z., Munro, N., Huehmer, A.F.R., Landers, J.P. 1999. Acousto-optical deflection-based laser beam scanning for fluorescence detection on multichannel electrophoretic microchips. *Anal. Chem.* 71:5309–14.

683. Ferrance, J., Landers, J.P. 2001. Exploiting sensitive laser-induced fluorescence detection on electrophoretic microchips for executing rapid clinical diagnostics. *Luminescence* 16:79–88.

684. Sanders, J.C., Huang, Z.L., Landers, J.P. 2001. Acousto-optical deflection-based whole channel scanning for microchip isoelectric focusing with laser-induced fluorescence detection. *Labchip* 1:167–72.

685. Wang, S., Morris, M.D. 2000. Plastic microchip electrophoresis with analyte velocity modulation. Application to fluorescence background rejection. *Anal. Chem.* 72:1448–52.

686. Huang, L.R., Cox, E.C., Austin, R.H., Sturm, J.C. 2003. Tilted Brownian ratchet for DNA analysis. *Anal. Chem.* 75:6963–67.

687. Prinz, C., Tegenfeldt, J.O., Austin, R.H., Cox, E.C., Sturm, J.C. 2002. Bacterial chromosome extraction and isolation. *Labchip* 2:207–12.

688. Roulet, J., Völkel, R., Herzig, H.P., Verpoorte, E. 2002. Performance of an integrated microoptical system for fluorescence detection in microfluidic systems. *Anal. Chem.* 74:3400–7.

689. Camou, S., Fujita, H., Fujii, T. 2003. PDMS 2D optical lens integrated with microfluidic channels: Principle and characterization. *Labchip* 3:40–45.

690. Kamei, T., Paegel, B.M., Scherer, J.R., Skelley, A.M., Street, R.A., Mathies, R.A. 2003. Integrated hydrogenated amorphous SI photodiode detector for microfluidic bioanalytical devices. *Anal. Chem.* 75:5300–5.

691. Webster, J.R., Burns, M.A., Burke, D.T., Mastrangelo, C.H. 2001. Monolithic capillary electrophoresis device with integrated fluorescence detector. *Anal. Chem.* 73:1622–26.

692. Chabinyc, M.L., Chiu, D.T., McDonald, J.C., Stroock, A.D., Christian, J.F., Karger, A.M., Whitesides, G.M. 2001. An integrated fluorescence detection system in poly(dimethylsiloxane) for microfluidic applications. *Anal. Chem.* 73:4491–98.

693. Namasivayam, V., Lin, R.S., Johnson, B., Brahmasandra, S., Razzacki, Z., Burke, D.T., Burns, M.A. 2004. Advances in on-chip photodetection for applications in miniaturized genetic analysis systems. *J. Micromech. Microeng.* 14:81–90.

694. Hofmann, O., Viorin, G., Niedermann, P., Manz, A. 2004. Three-dimensional microfluidic confinement for efficient sample delivery to biosensor surfaces. Application to immunoassays on planar optical waveguides. *Anal. Chem.* 74:5243–50.

695. Chronis, N., Lee, L.P. 2004. Total internal reflection-based biochip utilizing a polymer-filled cavity with a micromirror sidewall. *Labchip* 4:125–30.

696. Munro, N.J., Huang, Z., Finegold, D.N., Landers, J.P. 2002. Indirect fluorescence detection of amino acids on electrophoretic microchips. *Anal. Chem.* 72:2765–73.

697. Sirichai, S., de Mello, A.J. 2000. A capillary electrophoresis microchip for the analysis of photographic developer solutions using indirect fluorescence detection. *Analyst* 125:133–37.

698. Crabtree, H.J., Kopp, M.U., Manz, A. 1999. Shah convolution Fourier transform detection. *Anal. Chem.* 71:2130–38.

699. Kwok, Y.C., Manz, A. 2001. Characterization of Shah convolution Fourier transform detection. *Analyst* 126:1640–44.

700. Kwok, Y.C., Manz, A. 2001. Shah convolution differentiation Fourier transform for rear analysis in microchip capillary electrophoresis. *J. Chromatogr.* 924A:177–86.

701. Kwok, Y.C., Manz, A. 2001. Shah convolution Fourier transform detection: Multiple-sample injection technique. *Electrophoresis* 22:222–29.

702. Eijkel, J.C.T., Kwok, Y.C., Manz, A. 2001. Wavelet transform for Shah convolution velocity measurements of single particles and solutes in a microfluidic chip. *Labchip* 1:122–26.

703. McReynolds, J.A., Edirisinghe, P., Shippy, S.A. 2002. Shah and sine convolution Fourier transform detection for microchannel electrophoresis with a charge coupled device. *Anal. Chem.* 74:5063–70.

704. McReynolds, J.A., Shippy, S.A. 2004. Comparison of Hadamard transform and signal-averaged detection for microchannel electrophoresis. *Anal. Chem.* 76:3214–21.

705. Fister, J.C., III, Jacobson, S.C., Ramsey, J.M. 1999. Ultrasensitive cross-correlation electrophoresis on microchip devices. *Anal. Chem.* 71:4460–64.

706. Mogensen, K.B., Petersen, N.J., Hubner, J., Kutter, J.P. 2001. Monolithic integration of optical waveguides for absorbance detection in microfabricated electrophoresis devices. *Electrophoresis* 22:3930–38.

707. Ro, K.W., Shim, B.C., Lim, K., Hahn, J.H. 2001. Integrated light collimating system for extended optical-path-length absorbance detection in microchip-based capillary electrophoresis. In *Proceedings of the 5th Micro Total Analysis Systems Symposium*, Monterey, CA, October 21–25, pp. 274–76.

708. Snakenborg, D., Mogensen, K.B., Kutter, J.P. 2003. Optimization of signal-to-noise ratio in absorbance detection by integration of microoptical components. In *Proceedings of the 7th Micro Total Analysis Systems Symposium*, Squaw Valley, CA, October 5–9, pp. 841–44.

709. Wolk, J., Spaid, M., Jensen, M., MacReynolds, R., Steveson, K., Chien, R. 2001. Ultraviolet absorbance spectroscopy in a 3-dimensional microfluidic chip. In *Proceedings of the 5th Micro Total Analysis Systems Symposium*, Monterey, CA, October 21–25, pp. 367–68.

710. Verpoorte, E., Manz, A., Luedi, H., Bruno, A.E., Maystre, F., Krattiger, B., Widmer, H.M., Van der Schoot, B.H., De Rooij, N.F. 1992. A silicon flow cell for optical detection in miniaturized total chemical analysis systems. *Sensors Actuators* 6B:66–70.

711. Salimi-Moosavi, H., Jiang, Y., Lester, L., McKinnon, G., Harrison, D.J. 2000. A multireflection cell for enhanced absorbance detection in microchip-based capillary electrophoresis devices. *Electrophoresis* 21:1291–99.

712. Duggan, M.P., McCreedy, T., Aylott, J.W. 2003. A non-invasive analysis method for on-chip spectrophotometric detection using liquid-core waveguiding within a 3D architecture. *Analyst* 128:1336–40.

713. Hulme, J.P., Fielden, P.R., Goddard, N.J. 2004. Fabrication of a spectrophotometric absorbance flow cell using injection-molded plastic. *Anal. Chem.* 76:238–43.

714. Llobera, A., Wilke, R., Büttgenbach, S. 2004. Poly(dimethylsiloxane) hollow Abbe prism with microlenses for detection based on absorption and refractive index shift. *Labchip* 4:24–27.

715. Ueno, Y., Horiuchi, T., Tomita, M., Niwa, O., Zhou, H., Yamada, T., Honma, I. 2002. Separate detection of BTX mixture gas by a microfluidic device using a function of nanosized pores of mesoporous silica adsorbent. *Anal. Chem.* 74:5257–62.

716. Eijkel, J.C.T., Stoeri, H., Manz, A. 2000. A dc microplasma on a chip employed as an optical emission detector for gas chromatography. *Anal. Chem.* 72:2547–52.

717. Eijkel, J.C.T., Stoeri, H., Manz, A. 1999. A molecular emission detector on a chip employing a direct current microplasma. *Anal. Chem.* 71:2600–6.

718. Eijkel, J.C.T., Stoeri, H., Manz, A.J. 2000. An atmospheric pressure dc glow discharge on a microchip and its application as a molecular emission detector. *J. Anal. Atom. Spectrom.* 15:297–300.

719. Jenkins, G., Manz, A. 2002. A miniaturized glow discharge applied for optical emission detection in aqueous analytes. *J. Micromech. Microeng.* 12:N19–22.

720. Mangru, S.D., Harrison, D.J. 1998. Chemiluminescence detection in integrated post-separation reactors for microchip-based capillary electrophoresis and affinity electrophoresis. *Electrophoresis* 19:2301–7.

721. Richter, T., Shultz-Lockyear, L.L., Oleschuk, R.D., Bilitewski, U., Harrison, D.J. 2002. Bi-enzymatic and capillary electrophoretic analysis of non-fluorescent compounds in microfluidic devices: Determination of xanthine. *Sensors Actuators* 81B:369–76.

722. Greenway, G.M., Nelstrop, L.J., Port, S.N. 2000. Tris(2,2-bipyridyl)ruthenium (II) chemiluminescence in a microflow injection system for codeine determination. *Anal. Chim. Acta* 405:43–50.

723. Tyrrell, E., Gibson, C., MacCraith, B.D., Gray, D., Byrne, P., Kent, N., Burke, C., Paull, B. 2004. Development of a micro-fluidic manifold for copper monitoring utilising chemiluminescence detection. *Labchip* 4:384–90.

724. Hashimoto, M., Tsukagoshi, K., Nakajima, R., Kondo, K. 1999. Chemiluminescence detection in microchip capillary electrophoresis. *Chem. Lett.* 99:781–82.

725. Arora, A., Eijkel, J.C.T., Morf, W.E., Manz, A. 2001. A wireless electrochemiluminescence detector applied to direct and indirect detection for electrophoresis on a microfabricated glass device. *Anal. Chem.* 73:3282–88.

726. Arora, A., de Mello, A.J., Manz, A. 1997. Sub-microliter electrochemiluminescence detector—A model for small volume analysis systems. *Anal. Commun.* 34:393–95.

727. Hsueh, Y.-T., Smith, R.L., Northrup, M.A. 1996. A microfabricated, electrochemiluminescence cell for the detection of amplified DNA. *Sensors Actuators* 33B:110–14.

728. Zhan, W., Alvarez, J., Sun, L., Crooks, R.M. 2003. A multichannel microfluidic sensor that detects anodic redox reactions indirectly using anodic electrogenerated chemiluminescence. *Anal. Chem.* 75:1233–38.

729. Burggraf, N., Krattiger, B., de Rooij, N.F., Manz, A., de Mello, A.J. 1998. Holographic refractive index detector for application in microchip-based separation systems. *Analyst* 123:1443–47.

730. Costin, C.D., Synovec, R.E. 2002. A microscale-molecular weight sensor: Probing molecular diffusion between adjacent laminar flows by refractive index gradient detection. *Anal. Chem.* 74:4558–65.

731. Sato, K., Kawanishi, H., Tokeshi, M. 1999. Sub-zeptomole detection in a microfabricated glass channel by thermal-lens microscopy. *Anal. Sci.* 15:525.

732. Tokeshi, M., Uchida, M., Hibara, A., Sawada, T., Kitamori, T. 2001. Determination of subyoctomole amounts of nonfluorescent molecules using a thermal lens microscope: Subsingle-molecule determination. *Anal. Chem.* 73:2112–16.

733. Tokeshi, M., Uchida, M., Uchiyama, K., Sawada, T., Kitamori, T. 1999. Single- and countable-molecule detection of non-fluorescent molecules in liquid phase. *J. Luminesc.* 83–84:261–64.

734. Kikutani, Y., Hisamoto, H., Tokeshi, M., Kitamori, T. 2004. Micro wet analysis system using multi-phase laminar flows in three-dimensional microchannel network. *Labchip* 4:328–32.

735. Sato, K., Tokeshi, M., Sawada, T., Kitamori, T. 2000. Molecular transport between two phases in a microchannel. *Anal. Sci.* 16:455.

736. Kapur, R., Giuliano, K.A., Campana, M., Adams, T., Olson, K., Jung, D., Mrksich, M., Vasudevan, C., Taylor, D.L. 1999. Streamlining the drug discovery process by integrating miniaturization, high throughput screening, high content screening, and automation on the CellChip™ System. *Biomed. Microdevices* 2:99–109.

737. Kim, H.B., Hagino, T., Sasaki, N., Kitamori, T. 2003. Ultrasensitive detection of electrochemical reactions by thermal lens microscopy for microchip chemistry. In *Proceedings of the 7th Micro Total Analysis Systems Symposium*, Squaw Valley, CA, October 5–9, pp. 817–20.

738. Walker, P.A., III, Morris, M.D., Burns, M.A., Johnson, B.N. 1998. Isotachophoretic separations on a microchip. Normal Raman spectroscopy detection. *Anal. Chem.* 70:3766–769.

739. Keir, R., Igata, E., Arundell, M., Smith, W.E., Graham, D., McHugh, C., Cooper, J.M. 2002. SERRS. In situ substrate formation and improved detection using microfluidics. *Anal. Chem.* 74:1503–8.

740. Kawazumi, H., Yasunaga, S., Ogino, K., Maeda, H., Gobi, K.V., Miura, N. 2003. Compact and multiple surface-plasmon-resonance immunosensor for sub-ppb-level small molecules. In *Proceedings of the 7th Micro Total Analysis Systems Symposium*, Squaw Valley, CA, October 5–9, pp. 813–15.

741. Iwasaki, Y., Niwa, O. 2003. Integration of heterogeneous biosensors on microfluidic chip using surface plasmon resonance. In *Proceedings of the 7th Micro Total Analysis Systems Symposium*, Squaw Valley, CA, October 5–9, pp. 833–36.

742. Wegner, G.J., Lee, H.J., Corn, R.M. 2003. Characterization and optimization of peptide arrays for the study of epitope-antibody interactions using surface plasmon resonance imaging. *Anal. Chem.* 74:5161–68.

743. Hinsmann, P., Frank, J., Svasek, P., Harasek, M., Lendl, B. 2001. Design, simulation and application of a new micromixing device for time resolved infrared spectroscopy of chemical reactions in solution. *Labchip* 1:16–21.

744. Pan, T., Kelly, R.T., Asplund, M.C., Woolley, A.T. 2004. Fabrication of calcium fluoride capillary electrophoresis microdevices for on-chip infrared detection. *J. Chromatogr.* 1027A:231–35.

745. Woolley, A.T., Lao, K., Glazer, A.N., Mathies, R.A. 1998. Capillary electrophoresis chips with integrated electrochemical detection. *Anal. Chem.* 70:684–88.

746. Backofen, U.B., Matysik, F., Lunte, C.E. 2002. A chip-based electrophoresis system with electrochemical detection and hydrodynamic injection. *Anal. Chem.* 74:4054–59.

747. Baldwin, R.P., Roussel, T.J., Jr., Crain, M.M., Bathlagunda, V., Jackson, D.J., Gullapalli, J., Conklin, J.A., Pai, R., Naber, J.F., Walsh, K.M., Keynton, R.S. 2002. Fully integrated on-chip electrochemical detection for capillary electrophoresis in a microfabricated device. *Anal. Chem.* 74:3690–97.

748. Martin, R.S., Gawron, A.J., Lunte, S.M. 2000. Dual-electrode electrochemical detection for poly(dimethylsiloxane)-fabricated capillary electrophoresis microchips. *Anal. Chem.* 72:3196–202.

749. Wang, J., Tian, B., Sahlin, E. 1999. Integrated electrophoresis chips/amperometric detection with sputtered gold working electrodes. *Anal. Chem.* 71:3901–4.

750. Martin, R.S., Ratzlaff, K.L., Huynh, B.H., Lunte, S.M. 2002. In-channel electrochemical detection for microchip capillary electrophoresis using an electrically isolated potentiostat. *Anal. Chem.* 74:1136–43.

751. Jackson, D.J., Naber, J.F., Roussel, T.J., Jr., Crain, M.M., Walsh, K.M., Keynton, R.S., Baldwin, R.P. 2003. Portable high-voltage power supply and electrochemical detection circuits for microchip capillary electrophoresis. *Anal. Chem.* 75:3643–49.

752. Garcia, C.D., Henry, C.S. 2003. Direct determination of carbohydrates, amino acids, and antibiotics by microchip electrophoresis with pulsed amperometric detection. *Anal. Chem.* 75:4778–83.

753. Wang, J., Tian, B., Sahlin, E. 1999. Micromachined electrophoresis chips with thick-film electrochemical detectors. *Anal. Chem.* 71:5436–40.

754. Zeng, Y., Chen, H., Pang, D., Wang, Z., Cheng, J. 2002. Microchip capillary electrophoresis with electrochemical detection. *Anal. Chem.* 74:2441–45.

755. Hinkers, H., Conrath, N., Czupor, N., Frebel, H., Hüwel, S., Köckemann, K., Trau, D., Wittkampf, M., Chemnitius, G., Haalck, L., Meusel, M., Cammann, K., Knoll, M., Spener, F., Rospert, M., Kakerow, R., Köster, O., Lerch, T., Mokwa, W., Woias, P., Richter, M., Abel, T., Mexner, L. 1996. Results of the development of sensors and µTAS-modules. In *Micro Total Analysis Systems '96*, Basel, November 19–22, pp. 110–12.

756. Olofsson, J., Pihl, J., Sinclair, J., Sahlin, E., Karlsson, M., Orwar, O. 2004. A microfluidics approach to the problem of creating separate solution environments accessible from macroscopic volumes. *Anal. Chem.* 76:4968–76.

757. Martin, R.S., Kikura-Hanajiri, R., Lacher, N.A., Lunte, S.M. 2001. Studies to improve the performance of electrochemical detection for microchip electrophoresis. In *Proceedings of the 5th Micro Total Analysis Systems Symposium*, Monterey, CA, October 21–25, pp. 325–26.

758. Rossier, J.S., Roberts, M.A., Ferrigno, R., Girault, H.H. 1999. Electrochemical detection in polymer microchannels. *Anal. Chem.* 71:4294–99.

759. Kurita, R., Hayashi, K., Fan, X., Yamamoto, K., Kato, T., Niwa, O. 2002. Microfluidic device integrated with pre-reactor and dual enzyme-modified microelectrodes for monitoring in vivo glucose and lactate. *Sensors Actuators* 87B:296–303.

760. Wang, J., Chen, G., Chatrathi, M.P., Fujishima, A., Tryk, D.A., Shin, D. 2003. Microchip capillary electrophoresis coupled with a boron-doped diamond electrode-based electrochemical detector. *Anal. Chem.* 75:935–39.

761. Sun, X., Yan, J., Yang, X., Wang, E. 2004. Electrochemical detector based on sol-gel-derived carbon composite material for capillary electrophoresis microchips. *Electrophoresis* 25:3455–60.

762. Gawron, A.J., Martin, R.S., Lunte, S.M. 2001. Fabrication and evaluation of a carbon-based dual-electrode detector for poly(dimethylsiloxane) electrophoresis chips. *Electrophoresis* 22:242–48.

763. Martin, R.S., Gawron, A.J., Fogarty, B.A., Regan, F.B., Dempsey, E., Lunte, S.M. 2001. Carbon paste-based electrochemical detectors for microchip capillary electrophoresis/electrochemistry. *Analyst* 126:277–80.

764. Hebert, N.E., Kuhr, W.G., Brazill, S.A. 2003. A microchip electrophoresis device with integrated electrochemical detection: A direct comparison of constant potential amperometry and sinusoidal voltammetry. *Anal. Chem.* 75:3301–7.

765. Hebert, N.E., Snyder, B., McCreery, R.L., Kuhr, W.G., Brazill, S.A. 2003. Performance of pyrolyzed photoresist carbon films in a microchip capillary electrophoresis device with sinusoidal voltammetric detection. *Anal. Chem.* 75:4265–71.

766. Tantra, R., Manz, A. 2000. Integrated potentiometric detector for use in chip-based flow cells. *Anal. Chem.* 72:2875–78.

767. Badr, I.H.A., Johnson, R.D., Madou, M.J., Bachas, L.G. 2002. Fluorescent ion-selective optode membranes incorporated onto a centrifugal microfluidics platform. *Anal. Chem.* 74:5569–75.

768. Hüller, J., Pham, M.T., Howitz, S. 2003. Thin layer copper ISE for fluidic microsystem. *Sensors Actuators* 91B:17–20.

769. Ferrigno, R., Lee, J.N., Jiang, X., Whitesides, G.M. 2004. Potentiometric titrations in a poly(dimethylsiloxane)-based microfluidic device. *Anal. Chem.* 76:2273–80.

770. Darling, R.B., Yager, P., Weigl, B., Kriebel, J., Mayes, K. 1998. Integration of microelectrodes with etched microchannels for in-stream electrochemical analysis. In *Micro Total Analysis Systems '98*, Banff, Canada, October 13–16, pp. 105–8.

771. Kaltenpoth, G., Schnabel, P., Menke, E., Walter, E.C., Grunze, M., Penner, R.M. 2003. Multimode detection of hydrogen gas using palladium-covered silicon μ-channels. *Anal. Chem.* 75:4756–65.

772. Pumera, M., Wang, J., Opekar, F., Jelinek, I., Feldman, J., Lowe, H., Hardt, S. 2002. Contactless conductivity detector for microchip capillary electrophoresis. *Anal. Chem.* 74:1968–71.

773. Tanyanyiwa, J., Hauser, P.C. 2002. High-voltage capacitively coupled contactless conductivity detection for microchip capillary electrophoresis. *Anal. Chem.* 74:6378–82.

774. Baldock, S.J., Fielden, P.R., Goddard, N.J., Prest, J.E., Brown, B.J.T. 2003. Integrated moulded polymer electrodes for performing conductivity detection on isotachophoresis microdevices. *J. Chromatogr.* 990A:11–22.

775. Wang, J., Chen, G., Muck, A., Jr. 2003. Movable contactless-conductivity detector for microchip capillary electrophoresis. *Anal. Chem.* 75:4475–79.

776. Figeys, D., Gypi, S.P., McKinnon, G., Aebersold, R. 1998. An integrated microfluidics-tandem mass spectrometry system for automated protein analysis. *Anal. Chem.* 70:3728–34.

777. Wang, Y.X., Cooper, J.W., Lee, C.S., DeVoe, D.L. 2004. Efficient electrospray ionization from polymer microchannels using integrated hydrophobic membranes. *Labchip* 4:363–67.

778. Lion, N., Gellon, J.O., Jensen, H., Girault, H.H. 2003. On-chip protein sample desalting and preparation for direct coupling with electrospray ionization mass spectrometry. *J. Chromatogr.* 1003A:11–19.

779. Keetch, C.A., Hernandez, H., Sterling, A., Baumert, M., Allen, M.H., Robinson, C.V. 2003. Use of a microchip device coupled with mass spectrometry for ligand screening of a multi-protein target. *Anal. Chem.* 75:4937–41.

780. Liu, H., Felten, C., Xue, Q., Zhang, B., Jedrzejewski, P., Karger, B.L., Foret, F. 2000. Development of multichannel devices with an array of electrospray tips for high-throughput mass spectrometry. *Anal. Chem.* 72:3303–10.

781. Yuan, C.-H., Shiea, J. 2001. Sequential electrospray analysis using sharp-tip channels fabricated on a plastic chip. *Anal. Chem.* 73:1080–83.

782. Guber, A.E., Dittrich, H., Heckele, M., Herrmann, D., Musliga, A., Pfleging, W., Schaller, T. 2001. Polymer micro needles with through-going capillaries. In *Proceedings of the 5th Micro Total Analysis Systems Symposium*, Monterey, CA, October 21–25, pp. 155–56.

783. Licklider, L., Wang, X.-Qi., Desai, A., Tai, Y.-C., Lee, T.D. 2000. A micromachined chip-based electrospray source for mass spectrometry. *Anal. Chem.* 72:367–75.

784. Kim, J.-S., Knapp, D.R. 2001. Miniaturized multichannel electrospray ionization emitters on poly(dimethylsiloxane) microfluidic devices. *Electrophoresis* 22:3993–99.

785. Kim, J.S., Knapp, D.R. 2001. Microfabrication of polydimethylsiloxane electrospray ionization emitters. *J. Chromatogr.* 924A:137–45.

786. Le Gac, S., Arscott, S., Rolando, C. 2003. A planar microfabricated nanoelectrospray emitter tip based on a capillary slot. *Electrophoresis* 24:3640–47.

787. Svedberg, M., Veszelei, M., Axelsson, J., Vangbo, M., Nikolajeff, F. 2003. Fabrication of open PDMS electrospray tips integrated with microchannels using replication from a nickel master. In *Proceedings of the 7th Micro Total Analysis Systems Symposium*, Squaw Valley, CA, October 5–9, pp. 375–78.

788. Kameoka, J., Orth, R., Ilic, B., Czaplewski, D., Wachs, T., Craighead, H.G. 2002. An electrospray ionization source for integration with microfluidics. *Anal. Chem.* 74:5897–901.

789. Ssenyange, S., Taylor, J., Harrison, D.J., McDermott, M.T. 2004. A glassy carbon microfluidic device for electrospray mass spectrometry. *Anal. Chem.* 76:2393–97.

790. Schultz, G.A., Corso, T.N., Prosser, S.J., Zhang, S. 2000. A fully integrated monolithic microchip electrospray device for mass spectrometry. *Anal. Chem.* 72:4058–63.

791. Griss, P., Stemme, G. 2003. Side-opened out-of-plane microneedles for microfluidic transdermal liquid transfer. *J. Microelectromech. Syst.* 12:296–301.

792. Liu, J., Tseng, K., Garcia, B., Lebrilla, C.B., Mukerjee, E., Collins, S., Smith, R. 2001. Electrophoresis separation in open microchannels. A method for coupling electrophoresis with MALDI-MS. *Anal. Chem.* 73:2147–51.

793. Gustafsson, M., Hirschberg, D., Palmberg, C., Jornvall, H., Bergman, T. 2004. Integrated sample preparation and MALDI mass spectrometry on a microfluidic compact disk. *Anal. Chem.* 76:345–50.

794. Hirschberg, D., Jagerbrink, T., Samskog, J., Gustafsson, M., Stahlberg, M., Alvelius, G., Husman, B., Carlquist, M., Jornvall, H., Bergman, T. 2004. Detection of phosphorylated peptides in proteomic analyses using microfluidic compact disk technology. *Anal. Chem.* 76:5864–71.

795. Slyadnev, M.N., Tanaka, Y., Tokeshi, M., Kitamori, T. 2001. Non-contact temperature measurement inside microchannel. In *Proceedings of the 5th Micro Total Analysis Systems Symposium*, Monterey, CA, October 21–25, pp. 361–62.

796. Firebaugh, S.L., Jensen, K.F., Schmidt, M.A. 2002. Miniaturization and integration of photoacoustic detection. *J. Appl. Phys.* 92:1555–63.

797. Massin, C., Vincent, F., Homsy, A., Ehrmann, K., Boero, G., Besse, P.A., Daridon, A., Verpoorte, E., de Rooij, N.F., Popovic, R.S. 2003. Planar microcoil-based microfluidic NMR probes. *J. Mag. Reson.* 164:242–55.

798. Walton, J.H., de Ropp, J.S., Shutov, M.V., Goloshevsky, A.G., McCarthy, M.J., Smith, R.L., Collins, S.D. 2003. A micromachined double-tuned NMR microprobe. *Anal. Chem.* 75:5030–36.

799. Figeys, D., Chris, L., Lorne, T., Aebersold, R. 1998. Microfabricated device coupled with an electrospray ionization quadrupole time-of-flight mass spectrometer: Protein identifications based on enhanced-resolution mass spectrometry and tandem mass spectrometry data. *Rapid Commun. Mass Spectrom.* 12:1435–44.

800. Chan, J.H., Timperman, A.T., Qin, D., Aebersold, R. 1999. Microfabricated polymer devices for automated sample delivery of peptides for analysis by electrospray ionization tandem mass spectrometry. *Anal. Chem.* 71:4437–44.

801. Figeys, D., Ning, Y., Aebersold, R. 1997. A microfabricated device for rapid protein identification by microelectrospray ion trap mass spectrometry. *Anal. Chem.* 69:3153–60.

802. Zhang, B., Liu, H., Karger, B.L., Foret, F. 1999. Microfabricated device for capillary electrophoresis-electrospray mass spectrometry. *Anal. Chem.* 71:3258–64.

803. Figeys, D., Aebersold, R. 1998. Nanoflow solvent gradient delivery from a microfabricated device for protein identifications by electrospray ionization mass spectrometry. *Anal. Chem.* 70:3721–27.

804. Deng, Y., Henion, J. 2001. Chip-based capillary electrophoresis/mass spectrometry determination of carnitines in human urine. *Anal. Chem.* 73:639–46.

805. Wang, C., Oleschuk, R., Ouchen, F., Li, J., Thibault, P., Harrison, D.J. 2000. Integration of immobilized trypsin bead beds for protein digestion within a microfluidic chip incorporating capillary electrophoresis separations and an electrospray mass spectrometry interface. *Rapid Commun. Mass Spectrom.* 15:1377–83.

806. Deng, Y., Zhang, H., Henion, J. 2001. Chip-based quantitative capillary electrophoresis/mass spectrometry determination of drugs in human plasma. *Anal. Chem.* 73:1432–39.

807. Lazar, L.M., Ramsey, R.S., Ramsey, J.M. 2001. On-chip proteolytic digestion and analysis using "wrong-way-round" electrospray time-of-flight mass spectrometry. *Anal. Chem.* 73:1733–39.

808. Kameoka, J., Craighead, H.G., Zhang, H., Henion, J. 2001. A polymeric microfluidic chip for CE/MS determination of small molecules. *Anal. Chem.* 73:1935–41.

809. Li, J., Kelly, J.F., Chernuschevich, I., Harrison, D.J., Thibault, P. 2000. Separation and identification of peptides from gel-isolated membrane proteins using a microfabricated device for combined capillary electrophoresis/nanoelectrospray mass spectrometry. *Anal. Chem.* 72:599–609.

810. Lazar, L.M., Ramsey, R.S., Sundberg, S., Ramsey, J.M. 1999. Subattomole-sensitivity microchip nanoelectrospray source with time-of-flight mass spectrometry detection. *Anal. Chem.* 71:3627–31.

811. Xu, N., Lin, Y., Hofstadler, S.A., Matson, D., Call, C.J., Smith, R.D. 1998. A microfabricated dialysis device for sample cleanup in electrospray ionization mass spectrometry. *Anal. Chem.* 70:3553–56.

812. Li, J., Thibault, P., Bings, N.H., Skinner, C.D., Wang, C., Colyer, C., Harrison, D.J. 1999. Integration of microfabricated devices to capillary electrophoresis-electrospray mass spectrometry using a low dead volume connection: Application to rapid analyses of proteolytic digests. *Anal. Chem.* 71:3036–45.

813. Zhang, B., Foret, F., Karger, B.L. 2000. A microdevice with integrated liquid junction for facile peptide and protein analysis by capillary electrophoresis/electrospray mass spectrometry. *Anal. Chem.* 72:1015–22.

814. Chen, S.-H., Sung, W.-C., Lee, G.-B., Lin, Z.-Y., Chen, P.-W., Liao, P.-C. 2001. A disposable poly(methylmethacrylate)-based microfluidic module for protein identification by nanoelectrospray ionization-tandem mass spectrometry. *Electrophoresis* 22:3972–77.

815. Xiang, F., Lin, Y., Wen, J., Matson, D.W., Smith, R.D. 1999. An integrated microfabricated device for dual microdialysis and online ESI-ion trap mass spectrometry for analysis of complex biological samples. *Anal. Chem.* 71:1485–90.

816. Ramsey, R.S., Ramsey, J.M. 1997. Generation electrospray from microchip devices using electroosmotic pumping. *Anal. Chem.* 69:1174–78.

817. Gao, J., Xu, J., Locascio, L.E., Lee, C.S. 2001. Integrated microfluidic system enabling protein digestion, peptide separation, and protein identification. *Anal. Chem.* 73:2648–55.

818. Xue, Q., Foret, F., Dunayevskiy, Y.M., Zavracky, P.M., McGruer, N.E., Karger, B.L. 1997. Multichannel microchip electrospray mass spectrometry. *Anal. Chem.* 69:426–30.

819. Ekström, S., Önnerfjord, P., Nilsson, J., Bengtsson, M., Laurell, T., Marko-Varga, G. 2000. Integrated microanalytical technology enabling rapid and automated protein identification. *Anal. Chem.* 72:286–93.

820. Brivio, M., Fokkens, R.H., Verboom, W., Reinhoudt, D.N., Tas, N.R., Goedbloed, M., van den Berg, A. 2002. Integrated microfluidic system enabling (bio)chemical reactions with on-line MALDI-TOF mass spectrometry. *Anal. Chem.* 74:3972–76.

821. Jiang, Y., Wang, P.-C., Locascio, L.E., Lee, C.S. 2001. Integrated plastic microfluidic devices with ESI-MS for drug screening and residue analysis. *Anal. Chem.* 73:2048–53.

822. Killeen, K., Yin, H.F., Sobek, D., Brennen, R., van de Goor, T. 2003. Chip-LC/MS: HPLC-MS using polymer microfluidics. In *Proceedings of the 7th Micro Total Analysis Systems Symposium*, Squaw Valley, CA, October 5–9, pp. 481–84.

823. Yang, Y., Kameoka, J., Wachs, T., Henion, J.D., Craighead, H.G. 2004. Quantitative mass spectrometric determination of methylphenidate concentration in urine using an electrospray ionization source integrated with a polymer microchip. *Anal. Chem.* 76:2568–74.

824. Zamfir, A., Vakhrushev, S., Sterling, A., Niebel, H.J., Allen, M., Peter-Katalinic, J. 2004. Fully automated chip-based mass spectrometry for complex carbohydrate system analysis. *Anal. Chem.* 76:2046–54.

825. Wilding, P., Pfahler, J., Bau, H.H., Zemel, J.N., Kricka, L.J. 1994. Manipulation and flow of biological fluids in straight channels micromachined in silicon. *Clin. Chem.* 40:43–47.

826. Sutton, N., Tracey, M.C., Johnston, I.D. 1997. A novel instrument for studying the flow behaviour of erythrocytes through microchannels simulating human blood capillaries. *Microvasc. Res.* 53:272–81.

827. Stemme, G., Kittilsland, G. 1998. New fluid filter structure in silicon fabricated using a self-aligning technique. *Appl. Phys. Lett.* 53:1566–68.

828. He, B., Tan, L., Regnier, F. 1999. Microfabricated filters for microfluidic analytical systems. *Anal. Chem.* 71:1464–68.

829. Russom, A., Ahmadian, A., Andersson, H., Van Der Wijngaart, W., Lundeberg, J., Uhlen, M., Stemme, G., Nilsson, P. 2001. SNP analysis by allele-specific pyrosequencing extension in a micromachined filter-chamber device. In *Proceedings of the 5th Micro Total Analysis Systems Symposium*, Monterey, CA, October 21–25, pp. 22–24.

830. Andersson, H., Ahmadian, A., van der Wijngaart, W., Nilsson, P., Enoksson, P., Uhlen, M., Stemme, G. 2000. Micromachined flow-through filter-chamber for solid phase DNA analysis. In *Proceedings of the 4th Micro Total Analysis Systems Symposium*, Enschede, Netherlands, May 14–18, pp. 473–76.

831. Andersson, H., van der Wijngaart, W., Enoksson, P., Stemme, G. 2000. Micromachined flow-through filter-chamber for chemical reactions on beads. *Sensors Actuators* 67B:203–8.

832. Moorthy, J., Beebe, D.J. 2003. In situ fabricated porous filters for microsystems. *Labchip* 3:62–66.

833. Denoual, M., Aoki, K., Mita-Tixier, A., Fujita, H. 2003. A microfluidic device for long-term study of individual cells. In *Proceedings of the 7th Micro Total Analysis Systems Symposium*, Squaw Valley, CA, October 5–9, pp. 531–34.

834. Klauke, N., Smith, G., Cooper, J. 2002. Microsystems technology for cell screening in new medicines discovery. In *Proceedings of the 6th Micro Total Analysis Systems Symposium*, Nara, Japan, November 3–7, pp. 853–56.

835. Walker, G.M., Piston, D.W., McGuinness, P.O., Rocheleau, J.V. 2003. A microfluidic device for partical surface treatment of islets of Langerhans. In *Proceedings of the 7th Micro Total Analysis Systems Symposium,* Squaw Valley, CA, October 5–9, pp. 543–46.

836. Brody, J.P., Osborn, T.D., Forster, F.K., Yager, P. 1996. A planar microfabricated fluid filter. *Sensors Actuators* 54A:704–8.

837. Zhu, L., Zhang, Q., Feng, H.H., Ang, S., Chau, F.S., Liu, W.T. 2004. Filter-based microfluidic device as a platform for immunofluorescent assay of microbial cells. *Labchip* 4:337–41.

838. Wolff, A., Perch-Nielsen, I.R., Larsen, U.D., Friis, P., Goranovic, G., Poulsen, C.R., Kutter, J.P., Telleman, P. 2003. Integrating advanced functionality in a microfabricated high-throughput fluorescent-activated cell sorter. *Labchip* 3:22–27.

839. Kikuchi, Y., Sato, K., Ohki, H., Kaneko, T. 1992. Optically accessible microchannels formed in a single-crystal silicon substrate for studies of blood rheology. *Microvasc. Res.* 44:226–40.

840. Yang, M., Li, C., Yang, J. 2002. Cell docking and on-chip monitoring of cellular reactions with a controlled concentration gradient on a microfluidic device. *Anal. Chem.* 74:3991–4001.

841. Li, C.W., de Camprieu, L., Li, P.C.H. 2000. Electric voltage study of mammalian cells trapped at the weir structure in a microfabricated biochip. In *83rd Conference of Canadian Society for Chemistry*, Calgary, Canada, May, abstract 0076.

842. Irimia, D., Tompkins, R.G., Toner, M. 2004. Single-cell chemical lysis in picoliter-scale closed volumes using a microfabricated device. *Anal. Chem.* 76:6137–43.

843. Yun, K.S., Yoon, E. 2003. A micro/nano-fluidic chip-based micro-well array for high-throughput cell analysis and drug screening. In *Proceedings of the 7th Micro Total Analysis Systems Symposium*, Squaw Valley, CA, October 5–9, pp. 861–64.

844. Chiem, N., Lockyear-Shultz, L., Andersson, P., Skinner, C., Harrision, D.J. 2000. Room temperature bonding of micromachined glass devices for capillary electrophoresis. *Sensors Actuators* 63B:147–52.

845. Seong, G.H., Heo, J., Crooks, R.M. 2003. Measurement of enzyme kinetics using a continuous-flow microfluidic system. *Anal. Chem.* 75:3161–67.

846. Sato, K., Tokeshi, M., Odake, T., Kimura, H., Ooi, T., Nakao, M., Kitamori, T. 2000. Integration of an immunosorbent assay system: Analysis of secretory human immunoglobulin A on polystyrene beads in a microchip. *Anal. Chem.* 72:1144–47.

847. Parce, J.W., Owicki, J.C., Kercso, K.M., Sigal, G.B., Wada, H.G., Muir, V.C., Bousse, L.J., Ross, K.L., Sikic, B.I., McConnell, H.M. 1989. Detection of cell-affecting agents with a silicon biosensor. *Science* 246:243–47.

848. Bousse, L., McReynolds, R.J., Kirk, G., Dawes, T., Lam, P., Bemiss, W.R., Parce, J.W. 1993. Micromachined multichannel systems for the measurement of cellular metabolism. *Transducers* 93:916–20.

849. Bousse, L., McReynolds, R. 1994. Micromachined flow-through measurement chambers using LAPS chemical sensors. In *Micro Total Analysis Systems '94*, University of Twente, Netherlands, November 21–22, pp. 127–38.

850. Matsubara, Y., Murakami, Y., Kinpara, T., Morita, Y., Yokoyama, K., Tamiya, E. 2001. Allergy sensor using animal cells with microfluidics. In *Proceedings of the 5th Micro Total Analysis Systems Symposium*, Monterey, CA, October 21–25, pp. 299–300.

851. Tamaki, E., Sato, K., Tokeshi, M., Sato, K., Aihara, M., Kitamori, T. 2002. Single-cell analysis by a scanning thermal lens microscope with a microchip: Direct monitoring of cytochrome c distribution during apoptosis process. *Anal. Chem.* 74:1560–64.

852. Kantak, A.S., Gale, B.K., Lvov, Y., Jones, S.A. 2003. Platelet function analyzer: Shear activation of platelets in microchannels. *Biomed. Microdevices* 5:207–15.

853. Gao, J., Yin, X.F., Fang, Z.L. 2004. Integration of single cell injection, cell lysis, separation and detection of intracellular constituents on a microfluidic chip. *Labchip* 4:47–52.

854. Chen, C.S., Mrksich, M., Huang, S., Whitesides, G.M., Ingber, D.E. 1998. Micropatterned surfaces for control of cell shape, position, and function. *Biotechnol. Prog.* 14:356–63.

855. Khademhosseini, A., Suh, K.Y., Jon, S., Eng, G., Yeh, J., Chen, G.-J., Langer, R. 2004. A soft lithographic approach to fabricate patterned microfluidic channels. *Anal. Chem.* 7:3675–81.

856. Martinoia, S., Bove, M., Tedeso, M., Margesin, B., Grattarola, M. 1999. A simple microfluidic system for patterning populations of neurons on silicon micromachined substrates. *J. Neurol. Methods* 87:35–44.

857. Tokano, H., Sul, J., Mazzanti, M.L., Doyle, R.T., Haydon, P.G., Porter, M.D. 2002. Micropatterned substrates: Approach to probing intercellular communication pathways. *Anal. Chem.* 74:4640–46.

858. Takayama, S., McDonald, J.C., Ostuni, E., Liang, M.N., Kenis, P.J.A., Ismagilov, R.F., Whitesides, G.M. 1999. Patterning cells and their environments using multiple laminar fluid flows in capillary networks. *Proc. Natl. Acad. Sci. USA* 96:5545–48.

859. Takayama, S., Ostuni, E., LeDuc, P., Naruse, K., Ingber, D.E., Whitesides, G.M. 2001. Laminar flows: Subcellular positioning of small molecules. *Nature* 411:1016.

860. Hediger, S., Fontannaz, J., Sayah, A., Hunziker, W., Gijs, M.A.M. 2000. Biosystem for the culture and characterisation of epithelial cell tissues. *Sensors Actuators* 63B:63–73.

861. Koh, W.-G., Itle, L.J., Pishko, M.V. 2003. Molding of hydrogel microstructures to create multiphenotype cell microarrays. *Anal. Chem.* 75:5783–89.

862. Tan, W., Desai, T.A. 2004. Layer-by-layer microfluidics for biomimetic three-dimensional structures. *Biomaterials* 25:1355–64.

863. Moriguchi, H., Wakamoto, Y., Sugio, Y., Takahashi, K., Inoue, I., Yasuda, K. 2002. An agar-microchamber cell-cultivation system: Flexible change of microchamber shapes during cultivation by photo-thermal etching. *Labchip* 2:125–32.

864. Peng, X.Y., Li, P.C.H. 2004. A three-dimensional flow control concept for single-cell experiments on a microchip. 2. Fluorescein diacetate metabolism and calcium mobilization in a single yeast cell as stimulated by glucose and pH changes. *Anal. Chem.* 76:5282–92.

865. Lettieri, G.-L., Verpoorte, E., de Rooij, N.F. 2001. Affinity-based bioanalysis using freely moving beads as matrices for heterogeneous assays. In *Proceedings of the 5th Micro Total Analysis Systems Symposium*, Monterey, CA, October 21–25, pp. 503–4.

866. Arai, F., Ichikawa, A., Ogawa, M., Fukuda, T., Horio, K., Itoigawa, K. 2001. High-speed separation system of randomly suspended single living cells by laser trap and dielectrophoresis. *Electrophoresis* 22:283–88.

867. Zhang, H.C., Tu, E., Hagen, N.D., Schnabel, C.A., Paliotti, M.J., Hoo, W.S., Nguyen, P.M., Kohrumel, J.R., Butler, W.F., Chachisvillis, M., Marchand, P.J. 2004. Time-of-flight optophoresis analysis of live whole cells in microfluidic channels. *Biomed. Microdevices* 6:11–21.

868. Chen, H., Acharya, D., Gajraj, A., Meiners, J.-C. 2003. Robust interconnects and packaging for microfluidic elastomeric chips. *Anal. Chem.* 75:5287–91.

869. Fiedlerr, S., Shirley, S.G., Schnelle, T., Fuhr, G. 1998. Dielectrophoretic sorting of particles and cells in a microsystem. *Anal. Chem.* 70:1909–15.

870. Müller, T., Gradl, G., Howitz, S., Shirley, S., Schnelle, Th., Fuhr, G. 1999. A 3-D microelectrode system for handling and caging single cells and particles. *Biosens. Bioelectronics* 14:247–56.

871. Voldman, J., Gray, M.L., Toner, M., Schmidt, M.A. 2002. A microfabrication-based dynamic array cytometer. *Anal. Chem.* 74:3984–90.

872. Grad, G., Müller, T., Pfennig, A., Shirley, S., Schnelle, T., Fuhr, G. 2000. New micro devices for single cell analysis, cell sorting and cloning-on-a-chip: the Cytocon™ instrument. In *Proceedings of the 4th Micro Total Analysis Systems Symposium,* Enschede, Netherlands, May 14–18, pp. 443–46.

873. Fuhr, G., Wagner, B. 1994. Electric field mediated cell manipulation, characterization and cultivation in highly conductive media. In *Micro Total Analysis Systems '94*, University of Twente, Netherlands, November 21–22, pp. 209–14.

874. Huang, Y., Yang, J.M., Hopkins, P.J., Kassegne, S., Tirado, M., Forster, A.H., Reese, H. 2003. Separation of simulants of biological warfare agents from blood by a miniaturized dielectrophoresis device. *Biomed. Microdevices* 5:217–25.

875. Dürr, M., Kentsch, J., Müller, T., Schnelle, T., Stelzle, M. 2003. Microdevices for manipulation and accumulation of micro- and nanoparticles by dielectrophoresis. *Electrophoresis* 24:722–31.

876. Kicka, L.J., Faro, I., Heyner, S., Garside, W.T., Fitzpatrick, G., Wilding, P. 1995. Micromachined glass-glass microchips for in vitro fertilization. *Clin. Chem.* 41:1358–59.

877. Kricka, L.J., Fortina, P., Panaro, N.J., Wilding, P., Alonso-Amigo, G., Becker, H. 2002. Fabrication of plastic microchips by hot embossing. *Labchip* 2:1–4.

878. Cho, B.S., Schuster, T.G., Zhu, X., Chang, D., Smith, G.D., Takayama, S. 2003. Passively driven integrated microfluidic system for separation of motile sperm. *Anal. Chem.* 75:1671.

879. Li, P.C.H., Harrision, D.J. 1997. Transport, manipulation, and reaction of biological cells on-chip using electrokinetic effects. *Anal. Chem.* 69:1564–68.

880. McClain, M.A., Culbertson, C.T., Jacobson, S.C., Ramsey, J.M. 2001. Single cell lysis on microfluidic devices. In *Proceedings of the 5th Micro Total Analysis Systems Symposium*, Monterey, CA, October 21–25, pp. 301–2.

881. Salimi-Moosavi, H., Szarka, R., Andersson, P., Smith, R., Harrision, D.J. 1998. Biology lab-on-a chip for drug screening. In *Micro Total Analysis Systems '98,* Banff, Canada, October 13–16, pp. 69–72.

882. McClain, M.A., Culbertson, C.T., Jacobson, S.C., Ramsey, J.M. 2001. Flow cytometry of *Escherichia coli* on microfluidic devices. *Anal. Chem.* 73:5334–38.

883. Culbertson, C.T., Alarie, J.P., McClain, M.A., Jacobson, S.C., Ramsery, J.M. 2001. Rapid cellular assays on microfabricated fluidic device. In *Proceedings of the 5th Micro Total Analysis Systems Symposium*, Monterey, CA, October 21–25, pp. 285–86.

884. Sohn, L.L., Saleh, O.A., Facer, G.R., Beavis, A.J., Allan, R.S., Notterman, D.A. 2000. Capacitance cytometry: Measuring biological cells one by one. *Proc. Natl. Acad. Sci. USA* 97:10687–90.

885. Gómez, R., Bashir, R., Bhunia, A.K. 2002. Microscale electronic detection of bacterial metabolism. *Sensors Actuators* 86B:198–208.

886. Gawad, S., Schild, L., Renaud, Ph. 2001. Micromachined impedance spectroscopy flow cytometer for cell analysis and particle sizing. *Labchip* 1:76–82.

887. Wang, Z., Ali, J.E., Engelund, M., Gotsæd, T., Perch-Nielsen, I.R., Mogensen, K.B., Snakenborg, D., Kutter, J.P., Wolff, A. 2004. Measurements of scattered light on a microchip flow cytometer with integrated polymer based optical elements. *Labchip* 4:372–78.

888. Krüger, J., Singh, K., O'Neill, A., Jackson, C., Morrison, A., O'Brien, P. 2002. Development of a microfluidic device for fluorescence activated cell sorting. *J. Micromech. Microeng.* 12:486–94.

889. Schrum, D.P., Culbertson, C.T., Jacobson, S.C., Ramsey, J.M. 1999. Microchip flow cytometry using electrokinetic focusing. *Anal. Chem.* 71:4173–77.

890. Fu, A.Y., Chou, H., Spence, C., Arnold, F.H., Quake, S.R. 2002. An integrated microfabricated cell sorter. *Anal. Chem.* 74:2451–57.

891. Nieuwenhuis, J.H., Bastemeijer, J., Sarro, P.M., Vellekoop, M.J. 2003. Integrated flow-cells for novel adjustable sheath flows. *Labchip* 3:56–61.

892. Lao, A.I.K., Trau, D., Hsing, I. 2002. Miniaturized flow fractionation device assisted by a pulsed electric field for nanoparticle separation. *Anal. Chem.* 74:5364–69.

893. Lu, H., Gaudet, S., Schmidt, M.A., Jensen, K.F. 2004. A microfabricated device for subcellular organelle sorting. *Anal. Chem.* 76:5705–12.

894. Edwards, T., Gale, B.K., Frazier, A.B. 2001. Microscale purification systems for biological sample preparation. *Biomed. Microdevices* 3:211–18.

895. Deng, T., Prentiss, M., Whitesides, G.M. 2002. Fabrication of magnetic microfiltration systems using soft lithography. *Appl. Phys. Lett.* 80:461–63.

896. Fuhr, G. 1996. Examples of three-dimensional micro-structures for handling and investigation of adherently growing cells and sub-micron particles. In *Micro Total Analysis Systems '96*, Basel, November 10–22, p. 39.

897. Verhaegen, K., Simaels, J., Van Driessche, W., Baert, K., Sansen, W., Puers, B., Hermans, L., Mertens, R. 1999. A biomedical microphysiometer. *Biomed. Microdevices* 2:93–98.

898. Borenstein, J.T., Terai, H., King, K.R., Weinberg, E.J., Kaazempur-Mofrad, M.R., Vacanti, J.P. 2002. Microfabrication technology for vascularized tissue engineering. *Biomed. Microdevices* 4:167–75.

899. Welle, A., Gottwald, E. 2002. UV-based patterning of polymeric substrates for cell culture applications. *Biomed. Microdevices* 4:33–41.

900. Desai, T.A., Deutsch, J., Motlagh, D., Tan, W., Russell, B. 1999. Microtextured cell culture platforms: Biomimetic substrates for the growth of cardiac myocytes and fibroblasts. *Biomed. Microdevices* 2:123–29.

901. Thiébaud, P., Lauer, L., Knoll, W., Offenhäusser, A. 2002. PDMS device for patterned application of microfluids to neuronal cells arranged by microcontact printing. *Biosens. Bioelectronics* 17:87–93.

902. Walker, G.M., Zeringue, H.C., Beebe, D.J. 2004. Microenvironment design considerations for cellular scale studies. *Labchip* 4:91–97.

903. Strömberg, A., Karlsson, A., Ryttsén, F., Davidson, M., Chiu, D.T., Orwar, O. 2001. Microfluidic device for combinatorial fusion of liposomes and cells. *Anal. Chem.* 73:126–30.

904. Strike, D.J., Fiaccabrino, G.C., Koudelka-Hep, M., de Rooij, N.F. 2000. Enzymatic microreactor using Si, glass and EPON SU-8. *Biomed. Microdevices* 2:175–78.

905. Huang, Y., Rubinsky, B. 2003. Flow-through micro-electroporation chip for high efficiency single-cell genetic manipulation. *Sensors Actuators* 104A:205–12.

906. Olofsson, J., Nolkrantz, K., Ryttsén, F., Lambie, B.A., Weber, S.G., Orwar, O. 2003. Single-cell electroporation. *Curr. Opin. Biotechnol.* 14:29–34.

907. Obeid, P.J., Christopoulos, T.K., Crabtree, H.J., Backhouse, C.J. 2003. Microfabricated device for DNA and RNA amplification by continuous-flow polymerase chain reaction and reverse transcription-polymerase chain reaction with cycle number selection. *Anal. Chem.* 75:288–95.

908. Folch, A., Toner, M. 1998. Cellular micropatterns on biocompatible materials. *Biotechnol. Prog.* 14:388–92.

909. Munaka, T., Kanai, M., Abe, H., Fujiyama, Y., Sakamoto, T., Mahara, A., Yamayoshi, A., Nakanishi, H., Shoji, S., Murakami, A. 2003. In situ cell monitoring on a microchip using time-resolved fluorescence anisotropy analysis. In *Proceedings of the 7th Micro Total Analysis Systems Symposium*, Squaw Valley, CA, October 5–9, pp. 283–86.

910. Schultze, P., Ludwig, M., Kohler, F., Belder, D. 2005. Deep UV laser induced fluorescent detection of unlabeled drugs and proteins in microchip electrophoresis. *Anal. Chem.* 77:1325–29.

911. Dimalanta, E.T., Lim, A., Runnheim, R., Lamers, C., Churas, C., Forrest, D.K., de Pablo, J.J., Graham, M.D., Coppersmith, S.N., Goldstein, S., Schwartz, D.C. 2004. A microfluidic system for large DNA molecule arrays. *Anal. Chem.* 76:5293–301.

912. Sharma, S., Johnson, R.W., Desai, T.A. 2004. Evaluation of the stability of nonfouling ultrathin poly(ethylene glycol) films for silicon-based microdevices. *Langmuir* 20:348–58.

913. Wolfe, K.A., Breadmore, M.C., Ferrance, J.P., Power, M.E., Conroy, J.F., Norris, P.M., Landers, J.P. 2001. Toward a microchip-based solid-phase extraction method for isolation of nucleic acids. *Electrophoresis* 23:727–33.

914. Chung, Y.C., Jan, M.S., Lin, Y.C., Lin, J.H., Cheng, W.C., Fan, C.Y. 2004. Microfluidic chip for high efficiency DNA extraction. *Labchip* 4:141–47.

915. Hong, J.W., Studer, V., Hang, G., Anderson, W.F., Quake, S.R. 2004. A nanoliter-scale nucleic acid processor with parallel architecture. *Nat. Biotechnol.* 22:435–39.

916. Wilding, P., Shoffner, M.A., Kricka, L.J. 1994. PCR in a silicon microstructure. *Clin. Chem.* 40:1815–18.

917. Cheng, J., Shoffner, M.A., Hvichia, G.E., Kricka, L.J., Wilding, P. 1996. Chip PCR. II. Investigation of different PCR amplification systems in microfabricated silicon-glass chips. *Nucleic Acids Res.* 24:380–85.

918. Shoffner, M.A., Cheng, J., Hvichia, G.E., Kricka, L.J., Wilding, P. 1996. Chip PCR. I. Surface passivation of microfabricated silicon-glass chips for PCR. *Nucleic Acids Res.* 24:375–79.

919. Wilding, P., Kricka, L.J., Cheng, J., Hvichia, G. 1998. Integrated cell isolation and polymerase chain reaction analysis using silicon microfilter chambers. *Anal. Biochem.* 257:95–100.

920. Daniel, J.H., Moore, D.F., Iqbal, S., Millington, R.B., Lowe, C.R., Leslie, D.L., Lee, M.A., Pearce, M.J. 1998. Silicon microchambers for DNA amplification. *Sensors Actuators* 71A:81–88.

921. Zou, Q.B., Miao, Y.B., Chen, Y., Sridhar, U., Chong, C.S., Chai, T.C., Tie, Y., Teh, C.H.L., Lim, T.M., Heng, C.K. 2002. Micro-assembled multi-chamber thermal cycler for low-cost reaction chip thermal multiplexing. *Sensors Actuators* 102A:114–21.

922. Zhao, Z., Cui, Z., Cui, D.F., Xia, S.H. 2003. Monolithically integrated PCR biochip for DNA amplification. *Sensors Actuators* 108A:162–67.

923. Lee, D.S., Park, S.H., Yang, H., Chung, K.H., Yoon, T.H., Kim, S.J., Kim, K., Kim, Y.T. 2004. Bulk-micromachined submicroliter-volume PCR chip with very rapid thermal response and low power consumption. *Labchip* 4:401–7.

924. Woolley, A.T., Hadley, D., Landre, P., de Mello, A.J., Mathies, R.A., Northrup, M.A. 1996. Functional integration of PCR amplification and capillary electrophoresis in a microfabricated DNA analysis device. *Anal. Chem.* 68:4081–86.

925. Khandurina, J., McKnight, T.E., Jacobson, S.C., Waters, L.C., Foote, R.S., Ramsey, J.M. 2000. Integrated system for rapid PCR-based DNA analysis in microfluidic devices. *Anal. Chem.* 72:2995–3000.

926. Lagally, E.T., Medintz, I., Mathies, R.A. 2001. Single-molecule DNA amplification and analysis in an integrated microfluidic device. *Anal. Chem.* 73:565–70.

927. Rodriguez, I., Lesaicherre, M., Tie, Y., Zou, Q.B., Yu, C., Singh, J., Meng, L.T., Uppili, S., Li, S.F.Y., Gopalakrishnakone, P., Selvanayagam, Z.E. 2003. Practical integration of polymerase chain reaction amplification and electrophoretic analysis in microfluidic devices for genetic analysis. *Electrophoresis* 24:172–78.

928. Sniadecki, N.J., Chang, R., Beamesderfer, M., Lee, C.S., DeVoe, D.L. 2003. Field effect flow control in a polymer T-Intersection microfluidic network. In *Proceedings of the 7th Micro Total Analysis Systems Symposium*, Squaw Valley, CA, October 5–9, pp. 899–902.

929. Water, L.C., Jacobson, S.C., Kroutchinina, N., Khandurina, J., Foote, R.S., Ramsey, J.M. 1998. Multiple sample PCR amplification and electrophoretic analysis on a microchip. *Anal. Chem.* 70:5172–76.

930. Waters, L.C., Jacobson, S.C., Kroutchinina, N., Khandurina, J., Foote, R.S., Ramsey, J.M. 1998. Microchip device for cell lysis, multiplex PCR amplification, and electrophoretic sizing. *Anal. Chem.* 70:158–62.

931. Cheng, J., Waters, L.C., Fortina, P., Hvichia, G., Jacobson, S.C., Ramsey, J.M., Kricka, L.J., Wilding, P. 1998. Degenerate oligonucleotide primed-polymerase chain reaction and capillary electrophoretic analysis of human DNA on microchip-based devices. *Anal. Biochem.* 257:101–6.

932. Taylor, T.B., Harvey, S.E., Albin, M., Lebak, L., Ning, Y.B., Mowat, I., Schuerlein, T., Principe, E. 1998. Process control for optimal PCR performance in glass microstructures. *Biomed. Microdevices* 1:65–70.

933. Perch-Nielsen, I.R., Bang, D.D., Poulsen, C.R., Ali, J.E., Wolff, A. 2003. Removal of PCR inhibitors using dielectrophoresis as a selective filter in a microsystem. *Labchip* 3:212–16.

934. Munro, N.J., Hühmer, A.F.R., Landers, J.P. 2001. Robust polymeric microchannel coating for microchip-based analysis of neat PCR products. *Anal. Chem.* 73:1784–94.

935. Giordano, B.C., Copeland, E.R., Landers, J.P. 2001. Towards dynamic coating of glass microchip chambers for amplifying DNA via the polymerase chain reaction. *Electrophoresis* 22:334–40.

936. Lou, X.J., Panaro, N., Wilding, P., Fortina, P., Kricka, L.J. 2004. Increased amplification efficiency of microchip-based PCR by dynamic surface passivation. *Biotechniques* 36:248–50.

937. Liu, Y., Rauch, C.B., Stevens, R.L., Lenigk, R., Yang, J., Rhine, D.B., Grodzinski, P. 2002. DNA amplification and hybridization assays in integrated plastic monolithic devices. *Anal. Chem.* 74:3063–70.

938. Liu, Y., Ganser, D., Schneider, A., Liu, R., Grodzinski, P., Kroutchinina, N. 2001. Microfabricated polycarbonate CE devices for DNA analysis. *Anal. Chem.* 73:4196–201.

939. Lenigk, R., Liu, R.H., Athavale, M., Chen, Z.J., Ganser, D., Yang, J.N., Rauch, C., Liu, Y.J., Chan, B., Yu, H.N., Ray, M., Marrero, R., Grodzinski, P. 2002. Plastic biochannel hybridizaton device: A new concept for microfluidic DNA array. *Anal. Biochem.* 311:40–49.

940. Yu, X.M., Zhang, D.C., Li, T., Hao, L., Li, X.H. 2003. 3-D microarrays biochip for DNA amplification in polydimethylsiloxane (PDMS) elastomer. *Sensors Actuators* 108A:103–7.

941. Liu, J., Hansen, C., Quake, S.R. 2003. Solving the world-to-chip interface problem with a microfluidic matrix. *Anal. Chem.* 75:4718–23.

942. Burns, M.A., Johnson, B.N., Brahmasandra, S.N., Handique, K., Webster, J., Krishnan, M., Sammarco, T.S., Man, P.M., Jones, D., Heldsinger, D., Mastrangelo, C.H., Burke, D.T. 1998. An integrated nanoliter DNA analysis device. *Science* 282:484–87.

943. Albin, M., Kowallis, R., Picozza, E., Raysberg, Y., Sloan, C., Winn-Deen, E., Woudenberg, T., Zupfer, J. 1996. Micromachining and microgenetics: What are they and where do they work together? *Transducers* 96:253–57.

944. Ibrahim, M.S., Lofts, R.S., Jahrling, P.B., Henchal, E.A., Weedn, V.W., Northrup, M.A. Belgrader, P., 1998. Real-time microchip PCR for detecting single-base differences in viral and human DNA. *Anal. Chem.* 70:2013–17.

945. Taylor, M.T., Belgrader, P., Joshi, R., Kintz, G.A., Northrup, M.A. 2001. Fully automated sample preparation for pathogen detection performed in a microfluidic cassette. In *Proceedings of the 5th Micro Total Analysis Systems Symposium*, Monterey, CA, October 21–25, pp. 670–72.

946. Kopp, M.U., de Mello, A.J., Manz, A. 1998. Chemical amplification: Continuous-flow PCR on a chip. *Science* 280:1046–48.

947. Kopp, M.U., Luechinger, M.B., Manz, A. 1998. Continuous flow PCR on a chip. In *Micro Total Analysis Systems '98,* Banff, Canada, October 13–16, pp. 7–10.

948. Chou, C.F., Changrani, R., Roberts, P., Sadler, D., Lin, S., Mulholland, A., Swami, N., Terbrueggen, R., Zenhausern, F. 2001. A miniaturized cyclic PCR device. In *Proceedings of the 5th Micro Total Analysis Systems Symposium*, Monterey, CA, October 21–25, pp. 151–52.

949. Zhang, Q.T., Wang, W.H., Zhang, H.S., Wang, Y.L. 2002. Temperature analysis of continuous-flow micro-PCR based on FEA. *Sensors Actuators* 82B:75–81.

950. Bu, M.Q., Melvin, T., Ensell, G., Wilkinson, J.S., Evans, A.G.R. 2003. Design and theoretical evaluation of a novel microfluidic device to be used for PCR. *J. Micromech. Microeng.* 13:S125–30.

951. Cheng, J., Shoffner, M.A., Mitchelson, K.R., Kricka, L.J., Wilding, P. 1996. Analysis of ligase chain reaction products amplified in a silicon-glass chip using capillary electrophoresis. *J. Chromatogr.* 732A:151–58.

952. Tang, T., Ocvirk, G., Harrison, D.J. 1997. Iso-thermal DNA reactions and assays in microfabricated capillary electrophoresis systems. *Transducers* 97:523–26.

953. Hataoka, Y., Zhang, L., Mori, Y., Tomita, N., Notomi, T., Baba, Y. 2004. Analysis of specific gene by integration of isothermal amplification and electrophoresis on poly(methyl methacrylate) microchips. *Anal. Chem.* 76:3689–93.

954. Htaoka, Y., Notomi, T., Baba, Y. 2002. Integrated microsystem of isothermal amplification of DNA and electrophoresis on a microfabricated plastic chip for detection of specific gene and analysis of genetic materials. In *Proceedings of the 6th Micro Total Analysis Systems Symposium*, 6th Nara, Japan, November 3–7, pp. 215–16.

955. Esch, M.B., Locascio, L.E., Tarlov, M.J., Durst, R.A. 2001. Detection of visible *Cryptosporidium parvum* using DNA-modified liposomes in a microfluidic chip. *Anal. Chem.* 73:2952–58.

956. Harrison, D.J., Majid, E., Attiya, S., Jiang, G. 2001. Enhancing the microfluidic toolbox for functional genomics and recombinant DNA methods. In *Proceedings of the 5th Micro Total Analysis Systems Symposium*, Monterey, CA, October 21–25, pp. 10–12.

957. Jiang, G., Harrison, D.J. 2000. mRNA isolation for cDNA library construction on a chip. In *Proceedings of the 4th Micro Total Analysis Systems Symposium*, Enschede, Netherlands, May 14–18, pp. 537–40.

958. Wang, Y., Vaidya, B., Farquar, H.D., Stryjewski, W., Hammer, R.P., McCarley, R.L., Soper, S.A., Cheng, Y.-W., Barany, F. 2003. Microarrays assembled in microfluidic chips fabricated from poly(methyl methacrylate) for the detection of low-abundant DNA mutations. *Anal. Chem.* 75:1130–40.

959. Fan, Z.H., Mangru, S., Granzow, R., Heaney, P., Ho, W., Dong, Q., Kumar, R. 1999. Dynamic DNA hybridization on a chip using paramagnetic beads. *Anal. Chem.* 71:4851–59.

960. Seong, G.H., Zhan, W., Crooks, R.M. 2002. Fabrication of microchambers defined by photopolymerized hydrogels and weirs within microfluidic systems: Application to DNA hybridization. *Anal. Chem.* 74:3372–77.

961. Olsen, K.G., Ross, D.J., Tarlov, M.J. 2002. Immobilization of DNA hydrogel plugs in microfluidic channels. *Anal. Chem.* 74:1436–41.

962. Zangmeister, R.A., Tarlov, M.J. 2004. DNA displacement assay integrated into microfluidic channels. *Anal. Chem.* 76:3655–59.

963. Keramas, G., Perozziello, G., Geschke, O., Christensen, C.B.V. 2004. Development of a multiplex microarray microsystem. *Labchip* 4:152–58.

964. Grodzinski, P., Liu, R.H., Chen, B., Blackwell, J., Liu, Y., Rhine, D., Smekal, T., Ganser, D., Romero, C., Yu, H., Chan, T., Kroutchinina, N. 2001. Development of plastic microfluidic devices for sample preparation. *Biomed. Microdevices* 3:275283.

965. Noerholm, M., Bruus, H., Jakobsen, M.H., Telleman, P., Ramsing, N.B. 2004. Polymer microfluidic chip for online monitoring of microarray hybridizations. *Labchip* 4:28–37.

966. Salas-Solano, O., Schmalzing, D., Koutny, L., Buonocore, S., Adourian, A., Matsudaira, P., Ehrlich, D. 2000. Optimization of high-performance DNA sequencing on short microfabricated electrophoretic devices. *Anal. Chem.* 72:3129–37.

967. Schmalzing, D., Adourian, A., Koutny, L., Ziaugra, L., Matsudaira, P., Ehrlich, D. 1998. DNA sequencing on microfabricated electrophoresis devices. *Anal. Chem.* 70:2303–10.

968. Koutny, L., Schmalzing, D., Salas-Solano, O., El-Difrawy, S., Adourian, A., Buonocore, S., Abbey, K., McEwan, P., Matsudaira, P., Ehrlich, D. 2000. Eight hundred-base sequencing in a microfabricated electrophoretic device. *Anal. Chem.* 72:3388–91.

969. Boone, T.D., Ricco, A.J., Gooding, P., Bjornson, T.O., Singh, S., Xiao, V., Gibbons, I., Williams, S.J., Tan, H. 2000. Sub-microliter assays and DNA analysis on plastic microfluidics. In *Proceedings of the 4th Micro Total Analysis Systems Symposium*, Enschede, Netherlands, May 14–18, pp. 541–44.

970. Schmalzing, D., Belenky, A., Novotny, M.A., Koutny, L., Salas-Solano, O., El-Difrawy, S., Adourian, A., Matsudaira, P., Ehrlich, D. 2000. Microchip electrophoresis: A method for high-speed SNP detection. *Nucleic Acids Res.* 28:e43.

971. Paegel, B.M., Yeung, S.H., Mathies, R.A. 2002. Microchip bioprocessor for integrated nanovolume sample purification and DNA sequencing. *Anal. Chem.* 74:5092–98.

972. Paegel, B.M., Emrich, C.A., Wedemayer, G.J., Scherer, J.R., Mathies, R.A. 2002. High throughput DNA sequencing with a microfabricated 96-lane capillary array electrophoresis bioprocessor. *Proc. Natl. Acad. Sci. USA* 99:574–79.

973. Ehrlich, D., Adourian, A., Barr, C., Breslau, D., Buonocore, S., Burger, R., Carey, L., Carson, S., Chiou, J., Dee, R., Desmarais, S., El-Difrawy, S., King, R., Koutny, L., Lam, R., Matsudaira, P., Mitnik-Gankin, L., O'Neil, T., Novotny, M., Saber, G., Salas-Solano, O., Schmalzing, D., Srivastava, A., Vazquez, M. 2001. BioMEMS-768 DNA sequencer. In *Proceedings of the 5th Micro Total Analysis Systems Symposium*, Monterey, CA, October 21–25, pp. 16–18.

974. Woolley, A.T., Sensabaugh, G.F., Mathies, R.A. 1997. High-speed DNA genotyping using microfabricated capillary array electrophoresis chips. *Anal. Chem.* 69:2181–86.

975. Fan, Z.H., Tan, W., Tan, H., Qiu, X.C., Boone, T.D., Kao, P., Ricco, A.J., Desmond, M., Bay, S., Hennessy, K. 2001. Plastic microfluidic devices for DNA sequencing and protein separations. In *Proceedings of the 5th Micro Total Analysis Systems Symposium*, Monterey, CA, October 21–25, pp. 19–21.

976. Liu, S., Ren, H., Gao, Q., Roach, D.J., Loder, R.T., Jr., Armstrong, T.M., Mao, Q., Blaga, L., Barker, D.L., Jovanovich, S.B. 2000. Parallel DNA sequencing on microfabricated electrophoresis chips. In *Proceedings of the 4th Micro Total Analysis Systems Symposium,* Enschede, Netherlands, May 14–18, pp. 477–80.

977. Shi, Y., Simpson, P.C., Scherer, J.R., Wexler, D., Skibola, C., Smith, M.T., Mathies, R.A. 1999. Radial capillary array electrophoresis microplate and scanner for high-performance nucleic acid analysis. *Anal. Chem.* 71:5354–61.

978. Medintz, I., Wong, W.W., Sensabaugh, G., Mathies, R.A. 2000. High speed single nucleotide polymorphism typing of a hereditary haemochromatosis mutation with capillary array electrophoresis microplates. *Electrophoresis* 21:2352–58.

979. Scherer, J.R., Paegel, B.M., Wedemayer, G.J., Emrich, C.A., Lo, J.E., Medintez, I.L., Mathies, R.A. 2001. High-pressure gel loader for capillary array electrophoresis microchannel plates. *Biotechniques* 31:1150–53.

980. Emrich, C.A., Tian, H., Medintz, I.L., Mathies, R.A. 2002. Microfabricated 384-lane capillary array electrophoresis bioanalyzer for ultrahigh-throughput genetic analysis. *Anal. Chem.* 74:5076–83.

981. Sung, W.C., Lee, G.B., Tzeng, C.C., Chen, S.H. 2001. Plastic microchip electrophoresis for genetic screening: The analysis of polymerase chain reactions products of fragile X (CGG)n alleles. *Electrophoresis* 22:1188–93.

982. Mueller, O., Hahnenberger, K., Dittmann, M., Yee, H., Dubrow, R., Nagle, R., Ilsley, D. 2000. A microfluidic system for high-speed reproducible DNA sizing and quantitation. *Electrophoresis* 21:128–34.

983. Schmalzing, D., Koutny, L., Chisholm, D., Adourian, A., Matsudaira, P., Ehrlich, D. 1999. Two-color multiplexed analysis of eight short tandem repeat loci with an electrophoretic microdevice. *Anal. Biochem.* 270:148–52.

984. Schmalzing, D., Koutny, L., Adourian, A., Belgrader, P., Matsudaira, P., Ehrlich, D. 1997. DNA typing in thirty seconds with a microfabricated device. *Proc. Natl. Acad. Sci. USA* 94:10273–78.

985. Chou, H., Spence, C., Scherer, A., Quake, S. 1999. A microfabricated device for sizing and sorting DNA molecules. *Proc. Natl. Acad. Sci. USA* 96:11–13.

986. Wabuyele, M.B., Ford, S.M., Stryjewski, W., Barrow, J., Soper, S.A. 2001. Single molecule detection of double-stranded DNA in poly(methylmethacrylate) and polycarbonate microfluidic devices. *Electrophoresis* 22:3939–48.

987. Volkmuth, W.D., Austin, R.H. 1992. DNA electrophoresis in microlithographic arrays. *Nature* 358:600–2.

988. Volkmuth, W.D., Duke, T., Austin, R.H., Cox, E.C. 1995. Trapping of branched DNA in microfabricated structures. *Proc. Natl. Acad. Sci. USA* 92:6887–91.

989. Cabodi, M., Chen, Y.F., Turner, S., Craighead, H. 2001. Laterally asymmetric diffusion array with out-of-plane sample injection for continuous sorting of DNA molecules. In *Proceedings of the 5th Micro Total Analysis Systems Symposium*, Monterey, CA, October 21–25, pp. 103–4.

990. Han, J., Craighead, H.G. 2000. Separation of long DNA molecules in a microfabricated entropic trap array. *Science* 288:1026–29.

991. Dorfman, K.D., Brenner, H. 2002. Modeling DNA electrophoresis in microfluidic entropic trapping devices. *Biomed. Microdevices* 4:237–44.

992. Baba, M., Sano, T., Iguchi, N., Iida, K., Sakamoto, T., Kawaura, H. 2003. DNA size separation using artificially nanostructured matrix. *Appl. Phys. Lett.* 83:1468–70.

993. Doyle, P.S., Futterer, C., Minc, N., Goubault, E., Bibette, J., Viovy, J.L. 2002. Self-assembled magnetic colloids for DNA separation in microfluidic devices. In *Proceedings of the 6th Micro Total Analysis Systems Symposium*, Nara, Japan, November 3–7, pp. 48–50.

994. Doyle, P.S., Bibette, J., Bancaud, A., Viovy, J.L. 2002. Self-assembled magnetic matrices for DNA separation chips. *Science* 295:2237.

995. Minc, N., Futterer, C., Dorfman, K.D., Bancaud, A., Gosse, C., Goubault, C., Viovy, J.-L. 2004. Quantitative microfluidic separation of DNA in self-assembled magnetic matrixes. *Anal. Chem.* 76:3770–76.

996. Tabuchi, M., Ueda, M., Kaji, N., Yamasaki, Y., Nagasaki, Y., Yoshikawa, K., Kataoka, K., Baba, Y. 2004. Nanospheres for DNA separation chips. *Nat. Biotechnol.* 22:337–40.

997. Northrup, M.A., Ching, M.T., White, R.M., Watson, R.T. 1993. DNA amplification with a microfabricated reaction chamber. *Transducers* 93:924–26.

998. Yu, H., Sethu, P., Chan, T., Kroutchinina, N., Blackwell, J., Mastrangelo, C.H., Grodzinski, P. 2000. A miniaturized and integrated plastic thermal chemical reactor for genetic analysis. In *Proceedings of the 4th Micro Total Analysis Systems Symposium,* Enschede, Netherlands, May 14–18, pp. 545–48.

999. Lee, D.S., Park, S.H., Yang, H., Yoon, T.H., Kim, S.J., Kim, H., Shin, Y.B., Kim, K., Kim, Y.T. 2003. Submicroliter-volume PCR chip with fast thermal response and very low power consumption. In *Proceedings of the 7th Micro Total Analysis Systems Symposium,* Squaw Valley, CA, October 5–9, pp. 187–90.

1000. Shin, Y.S., Cho, K.C., Lim, S.H., Chung, S., Park, S.J., Chung, C., Han, D.C., Chang, J.K. 2003. PDMS-based micro PCR chip with Parylene coating. *J. Micromech. Microeng.* 13:768–74.

1001. Yang, J.N., Liu, Y.J., Rauch, C.B., Stevens, R.L., Liu, R.H., Lenigk, R., Grodzinski, P. 2002. High sensitivity PCR assay in plastic micro reactors. *Labchip* 2:179–87.

1002. Obeid, P.J., Christopoulos, T.K. 2003. Continuous-flow DNA and RNA amplification chip combined with laser-induced fluorescence detection. *Anal. Chim. Acta* 494:1–9.

1003. Schneegass, I., Bräutigam, R., Köhler, J.M. 2001. Miniaturized flow-through PCR with different template types in a silicon chip thermocycler. *Labchip* 1:42–49.

1004. Chiem, N.H., Harrison, D.J. 1998. Monoclonal antibody binding affinity determined by microchip-based capillary electrophoresis. *Electrophoresis* 19:3040–44.

1005. Harrison, D.J., Fluri, K., Chiem, N., Tang, T., Fan, Z. 1996. Micromachining chemical and biochemical analysis and reaction systems on glass substrates. *Sensors Actuators* 33B:105–9.

1006. Koutny, L.B., Schmalzing, D., Taylor, T.A., Fuchs, M. 1996. Microchip electrophoretic immunoassay for serum cortisol. *Anal. Chem.* 68:18–22.

1007. Cheng, S.B., Skinner, C.D., Taylor, J., Attiya, S., Lee, W.E., Picelli, G., Harrison, D.J. 2001. Development of a multichannel microfluidic analysis system employing affinity capillary electrophoresis for immuno-assay. *Anal. Chem.* 73:1472–79.

1008. Taylor, J., Picelli, G., Harrison, D.J. 2001. An evaluation of the detection limits possible for competitive capillary electrophoretic immunoassays. *Electrophoresis* 22:3699–708.

1009. Wang, J., Ibanez, A., Chatrathi, M.P., Escarpa, A. 2001. Electrochemical enzyme immunoassays on microchip platforms. *Anal. Chem.* 73:5323–27.

1010. Abad-Villar, E.M., Tanyanyiwa, J., Fernandez-Abedul, M.T., Costa-Garcia, A., Hauser, P.C. 2004. Detection of human immunoglobulin in microchip and conventional capillary electrophoresis with contactless conductivity measurements. *Anal. Chem.* 76:1282–88.

1011. Martynova, L., Locascio, L.E., Gaitain, M., Kramer, G.W., Christensen, R.G., MacCrehan, W.A. 1997. Fabrication of plastic microfluid channels by imprinting methods. *Anal. Chem.* 69:4783–89.

1012. Lim, T.-K., Ohta, H., Matsunaga, T. 2003. Microfabricated on-chip-type electrochemical flow immuno-assay system for the detection of histamine released in whole blood samples. *Anal. Chem.* 75:3316–21.

1013. Bromberg, A., Mathies, R.A. 2003. Homogeneous immunoassay for detection of TNT and its analogues on a microfabricated capillary arrary electrophoresis chip. In *Proceedings of the 7th Micro Total Analysis Systems Symposium,* Squaw Valley, CA, October 5–9, pp. 979–82.

1014. Yager, P., Bell, D., Brody, J.P., Qin, D., Cabrera, C., Kamholz, A., Weigl, B. 1998. Applying microfluidic chemical analytical systems to imperfect samples. In *Micro Total Analysis Systems '98,* Banff, Canada, October 13–16, pp. 207–12.

1015. Hatch, A., Kamholz, A.E., Hawkins, K.R., Munson, M.S., Schilling, E.A., Weigl, B.H., Yager, P. 2001. A rapid diffusion immunoassay in a T-sensor. *Nat. Biotechnol.* 19:461–65.

1016. Choi, J.W., Oh, K.W., Thomas, J.H., Heineman, W.R., Halsall, H.B., Nevin, J.H., Helmicki, A.J., Henderson, H.T., Ahn, C.H. 2002. An integrated microfluidic biochemical detection system for protein analysis with magnetic bead-based sampling capabilities. *Labchip* 2:27–30.

1017. Dodge, A., Fluri, K., Verpoorte, E., de Rooij, N.F. 2001. Electrokinetically driven microfluidic chips with surface-modified chambers for heterogeneous immunoassays. *Anal. Chem.* 73:3400–9.

1018. Fluri, K., Lettieri, G.L., Schoot, B.H.V.D., Verpoorte, E., de Rooij, N.F. 1998. Chip-based heterogeneous immunoassay for clinical diagnostic applications. In *Micro Total Analysis Systems '98,* Banff, Canada, October 13–16, pp. 347–50.

1019. Dodge, A., Fluri, K., Linder, V., Lettieri, G., Linchtenberg, J., Verpoorte, E., de Rooij, N.F. 2000. Valveless, sealed microfluidic device for automated heterogeneous immunoassay: Design and operational considerations. In *Proceedings of the 4th Micro Total Analysis Systems Symposium,* Enschede, Netherlands, May 14–18, pp. 407–10.

1020. Rossier, J.S., Girault, H.H. 2001. Enzyme linked immunosorbent assay on a microchip with electrochemical detection. *Labchip* 1:153–57.

1021. Sato, K., Tokeshi, M., Kimura, H., Kitamori, T. 2001. Determination of carcinoembryonic antigen in human sera by integrated bead-bed immunoassay in a microchip for cancer diagnosis. *Anal. Chem.* 73:1213–18.

1022. Sato, K., Yamanaka, M., Takahashi, H., Uchiyama, K., Tokeshi, M., Katou, H., Kimura, H., Kitamori, T. 2001. Integrated immunoassay system using multichannel microchip for simultaneous determination. In *Proceedings of the 5th Micro Total Analysis Systems Symposium*, Monterey, CA, October 21–25, pp. 511–12.

1023. Sato, K., Yamanaka, M., Takahashi, H., Tokeshi, M., Kimura, H., Kitamori, T. 2002. Microchip-based immunoassay system with branching multichannels for simultaneous determination of interferon-γ. *Electrophoresis* 23:734–39.

1024. Marquette, C.A., Blum, L.J. 2004. Direct immobilization in poly(dimethylsiloxane) for DNA, protein and enzyme fluidic biochips. *Anal. Chim. Acta* 506:127–32.

1025. Delamarche, E., Bernard, A., Schmid, H., Michel, B., Biebuyck, H. 1997. Patterned delivery of immunoglobulins to surfaces using microfluidic networks. *Science* 276:779–81.

1026. Sapsford, K.E., Charles, P.T., Patterson, C.H., Jr., Ligler, F.S. 2002. Demonstration of four immunoassay formats using the array biosensor. *Anal. Chem.* 74:1061–68.

1027. Bernard, A., Michel, B., Delamarche, E. 2001. Micromosaic immunoassays. *Anal. Chem.* 73: 8–12.

1028. Rowe, C.A., Scruggs, S.B., Feldstein, M.J., Golden, J.P., Ligler, F.S. 1999. An array immunosensor for simultaneous detection of clinical analytes. *Anal. Chem.* 71:433–39.

1029. Yang, T., Jung, S., Mao, H., Cremer, P.S. 2001. Fabrication of phospholipid bilayer-coated microchannels for on-chip immunoassays. *Anal. Chem.* 73:165–69.

1030. Yakovleva, J., Davidsson, R., Lobanova, A., Bengtsson, M., Eremin, S., Laurell, T., Emnéus, J. 2002. Microfluidic enzyme immunoassay using silicon microchip with immobilized antibodies and chemiluminescence detection. *Anal. Chem.* 74:2994–3004.

1031. Saleh, O.A., Sohn, L.L. 2003. Direct detection of antibody-antigen binding using an on-chip artificial pore. *Proc. Natl. Acad. Sci. USA* 100:820–24.

1032. Oki, A., Adachi, S., Takamura, Y., Ishihara, K., Kataoka, K., Ichiki, T., Honike, Y. 2000. Glucose measurement in blood serum injected by electroosmosis into phospholipid polymer coated microcapillary. In *Proceedings of the 4th Micro Total Analysis Systems Symposium,* Enschede, Netherlands, May 14–18, pp. 403–6.

1033. Duffy, D.C., McDonald, J.C., Schueller, O.J.A., Whitesides, G.M. 1998. Rapid prototyping of microfluidic systems in poly(dimethylsiloxane). *Anal. Chem.* 70:4974–84.

1034. Bousse, L., Mouradian, S., Minalla, A., Yee, H., Williams, K., Dubrow, R. 2001. Protein sizing on a chip. *Anal. Chem.* 73:1207–12.

1035. Giordano, B.C., Couch, A.J., Ahmadzadeh, H., Jin, L.J., Landers, J.P. 2001. Dynamic labeling of protein-sodium dodecyl sulfate (SDS) complexes for laser induced fluorescence (LIF) detection on microchips. In *Proceedings of the 5th Micro Total Analysis Systems Symposium*, Monterey, CA, October 21–25, pp. 109–10.

1036. Macounova, K., Cabrera, C.R., Holl, M.R., Yager, P. 2000. Generation of natural pH gradients in microfluidic channels for use in isoelectric focusing. *Anal. Chem.* 72:3745–51.

1037. VanderNoot, V.A., Hux, G., Schoeniger, J., Shepodd, T. 2001. Isoelectric focusing using electrokinetically-generated pressure mobilization. In *Proceedings of the 5th Micro Total Analysis Systems Symposium*, Monterey, CA, October 21–25, pp. 127–28.

1038. Yager, P., Cabrera, C., Hatch, A., Hawkins, K., Holl, M., Kamholz, A., Macounova, K., Weigl, B.H. 2000. Analytical devices based on transverse transport in microchannels. In *Proceedings of the 4th Micro Total Analysis Systems Symposium,* Enschede, Netherlands, May 14–18, pp. 15–18.

1039. Li, Y., DeVoe, D.L., Lee, C.S. 2003. Dynamic analyte introduction and focusing in plastic microfluidic devices for proteomic analysis. *Electrophoresis* 24:193–99.

1040. Macounová, K., Cabrera, C.R., Yager, P. 2001. Concentration and separation of proteins in microfluidic channels on the basis of transverse IEF. *Anal. Chem.* 73:1627–33.

1041. Mao, Q., Pawliszyn, J. 1999. Demonstration of isoelectric focusing on an etched quartz chip with UV absorption imaging detection. *Analyst* 124:637–41.

1042. Duffy, D.C., Gillis, H.L., Lin, J., Sheppard, N.F., Jr., Kellogg, G.J. 1999. Microfabricated centrifugal microfluidic systems: Characterization and multiple enzymatic assays. *Anal. Chem.* 71:4669–78.

1043. Mao, H., Yang, T., Cremer, P.S. 2002. Design and characterization of immobilized enzymes in microfluidic systems. *Anal. Chem.* 74:379–85.

1044. Holden, M.A., Jung, S.-Y., Cremer, P.S. 2004. Patterning enzymes inside microfluidic channels via photoattachment chemistry. *Anal. Chem.* 76:1838–43.

1045. Hadd, A.G., Jacobson, S.C., Ramsey, J.M. 1999. Microfluidic assays of acetylcholinesterase inhibitors. *Anal. Chem.* 71:5206–12.

1046. Boone, T.D., Hooper, H.H. 1998. Multiplexed, disposable, plastic microfluidic systems for high-throughput applications. In *Micro Total Analysis Systems '98,* Banff, Canada, October 13–16, pp. 257–60.

1047. Moser, I., Jobst, G., Svasek, P., Varahram, M., Urban, G. 1997. Rapid liver enzyme assay with miniaturized liquid handling system comprising thin film biosensor array. *Sensors Actuators* 44B:377–80.

1048. Burke, B.J., Regnier, F.E. 2003. Stopped-flow enzyme assays on a chip using a microfabricated mixer. *Anal. Chem.* 75:1786–91.

1049. Hadd, A.G., Raymond, D.E., Halliwell, J.W., Jacobson, S.C., Ramsey, J.M. 1997. Microchip device for performing enzyme assays. *Anal. Chem.* 69:3407–12.

1050. Cohen, C.B., Chin-Dixon, E., Jeong, S., Nikiforov, T.T. 1999. A microchip-based enzyme assay for protein kinase A. *Anal. Biochem.* 273:89–97.

1051. Schilling, E.A., Kamholz, A.E., Yager, P. 2002. Cell lysis and protein extraction in a microfluidic device with detection by a fluorogenic enzyme assay. *Anal. Chem.* 74:1798–804.

1052. Schilling, E.A., Kamholz, A.E., Yager, P. 2001. Cell lysis and protein extraction in a microfluidic device with detection by a fluorogenic enzyme assay. In *Proceedings of the 5th Micro Total Analysis Systems Symposium*, Monterey, CA, October 21–25, pp. 265–67.

1053. Heo, J., Thomas, K.J., Seong, G.H., Crooks, R.M. 2003. A microfluidic bioreactor based on hydrogel-entrapped *E. coli*: Cell viability, lysis, and intracellular enzyme reactions. *Anal. Chem.* 75:22–26.

1054. Wang, J., Chatrathi, M.P., Tian, B. 2001. Microseparation chips for performing multienzymatic dehydrogenase/oxidase assays: Simultaneous electrochemical measurement of ethanol and glucose. *Anal. Chem.* 73:1296–300.

1055. Wang, J., Chatrathi, M.P., Ibañez, A. 2001. Glucose biochip: Dual analyte response in connection to two pre-column enzymatic reactions. *Analyst* 126:1203–6.

1056. Zhan, W., Seong, G.H., Crooks, R.M. 2002. Hydrogel-based microreactors as a functional component of microfluidic systems. *Anal. Chem.* 74:4647–52.

1057. Hosokawa, K., Sato, K., Ichikawa, N., Maeda, M. 2003. Power-free microfluidic pumping by air-evacuated PDMS. In *Proceedings of the 7th Micro Total Analysis Systems Symposium*, Squaw Valley, CA, October 5–9, pp. 499–502.

1058. Gottschlich, N., Culbertson, C.T., McKnight, T.E., Jacobson, S.C., Ramsey, J.M. 2000. Integrated microchip-device for the digestion, separation and postcolumn labeling of proteins and peptides. *J. Chromotogr.* 745B:243–49.

1059. Kato, K.S., Kato, M., Ishihara, K., Toyo'oka, T. 2004. An enzyme-immobilization method for integration of biofunctions on a microchip using a water-soluble amphiphilic phospholipid polymer having a reacting group. *Labchip* 4:4–6.

1060. Peterson, D.S., Rohr, T., Svec, F., Fréchet, J.M.J. 2002. Enzymatic microreactor-on-a-chip: Protein mapping using trypsin immobilized on porous polymer monoliths molded in channels of microfluidic devices. *Anal. Chem.* 74:4081–88.

1061. Ekstroem, S., Oennerfjord, P., Nilsson, J., Bengtsson, M., Laurell, T., Marko-Varga, G. 2000. Integrated microanalytical technology enabling rapid and automated protein identification. *Anal. Chem.* 72:286–93.

1062. Sakai-Kato, K., Kato, M., Toyo'oka, T. 2003. Creation of an on-chip enzyme reactor by encapsulating trypsin in sol-gel on a plastic microchip. *Anal. Chem.* 75:388–93.

1063. Henry, A.C., Tutt, T.J., Galloway, M., Davidson, Y.Y., McWhorter, C.S., Soper, S.A., McCarley, R.L. 2000. Surface modification of poly(methyl methacrylate) used in the fabrication of microanalytical devices. *Anal. Chem.* 72:5331–37.

1064. Jemere, A.B., Oleschuk, R.D., Harrison, D.J. 2002. Integrated size exclusion and reversed phase. In *Proceedings of the 6th Micro Total Analysis Systems Symposium*, Nara, Japan, November 3–7, pp. 16–18.

1065. Schwarz, M.A., Galliker, B., Fluri, K., Kappes, T., Hauser, P.C. 2001. A two-electrode configuration for simplified amperometric detection in a microfabricated electrophoretic separation device. *Analyst* 126:147–51.

1066. Valussi, S., Manz, A. 2002. Biochemical applications: Electronic field assisted extraction and focusing of fingerprint residues by means of a microfluidic device. In *Proceedings of the 6th Proceedings Micro Total Analysis Systems Symposium*, Nara, Japan, November 3–7, pp. 865–67.

1067. Wang, J., Chatrathi, M.P. 2003. Microfabricated electrophoresis chip for bioassay of renal markers. *Anal. Chem.* 75:525–29.

1068. Manica, D.P., Lapos, J.A., Jones, A.D., Ewing, A.G. 2001. Dual electrochemical and optical detection on a microfabricated electrophoresis chip. In *Proceedings of the 5th Micro Total Analysis Systems Symposium*, Monterey, CA, October 21–25, pp. 262–64.

1069. Hu, L.G., Harrison, J.D., Masliyah, J.H. 1999. Numerical model of electrokinetic flow for capillary electrophoresis. *Anal. Commun.* 36:305.

1070. Fanguy, J.C., Henry, C.S. 2002. The analysis of uric acid in urine using microchip capillary electrophoresis with electrochemical detection. *Electrophoresis* 23:767–73.

1071. Liu, Y., Vickers, J.A., Henry, C.S. 2004. Simple and sensitive electrode design for microchip electrophoresis/electrochemistry. *Anal. Chem.* 76:1513–17.

1072. Hayashi, K., Iwasaki, Y., Krita, R., Horiuchi, T., Sunagawa, K., Niwa, O. 2003. On-line electrochemical device for highly sensitive monitoring of biomolecules in blood. In *Proceedings of the 7th Micro Total Analysis Systems Symposium*, Squaw Valley, CA, October 5–9, pp. 865–67.

1073. Kurita, R., Hayashi, K., Horiuchi, T., Niwa, O., Maeyama, K., Tanizawa, K. 2002. Differential measurement with a microfluidic device for the highly selective continuous measurement of histamine released from rat basophilic leukemia cells (RBL-2H3). *Labchip* 2:34–38.

1074. Tokuyama, T., Fujii, S.I., Abo, M., Okubo, A. 2002. Cellular analysis systems of histamine release on a microchip. In *Proceedings of the 6th Micro Total Analysis Systems Symposium*, Nara, Japan, November 3–7, pp. 832–34.

1075. Wakida, S.I., Wu, X.L., Akama, K., Motoshige, T., Yoshino, K., Matsuoka, K., Niki, E. 2002. High throughput stress marker assay using polymer microchip electrophoresis with laser induced fluorescence detection. In *Proceedings of the 6th Micro Total Analysis Systems Symposium*, Nara, Japan, November 3–7, pp. 210–11.

1076. Schwarz, M.A., Hauser, P.C. 2001. Fast chiral on-chip separations with amperometric detection. In *Proceedings of the 5th Micro Total Analysis Systems Symposium*, Monterey, CA, October 21–25, pp. 547–48.

1077. Goto, M., Sato, K., Tokeshi, M., Kitamori, T. 2003. Development of microchip-based bioassay system using cultured cells. In *Proceedings of the 7th Micro Total Analysis Systems Symposium*, Squaw Valley, CA, October 5–9, pp. 785–88.

1078. Huang, Z., Sanders, J.C., Dunsmor, C., Ahmadzadeh, H., Landers, J.P. 2001. A method for UV-bonding in the fabrication of glass electrophoretic microchips. *Electrophoresis* 22:3924–29.

1079. Cantafora, A., Blotta, I., Bruzzese, N., Calandra, S., Bertolini, S. 2001. Rapid sizing of microsatellite alleles by gel electrophoresis on microfabricated channels: Application to the D19S394 tetranucleotide repeat for cosegregation study of familial hypercholesterolemia. *Electrophoresis* 22:4012–15.

1080. Gottwald, E., Muller, O., Polten, A. 2001. Semiquantitative reverse transcription-polymerase chain reaction with the Agilent 2100 Bioanalyzer. *Electrophoresis* 22:4016–22.

1081. Nemoda, Z., Ronai, Z., Szekely, A., Kovacs, E., Shandrick, S., Guttman, A., Sasvari-Szekely, M. 2001. High-throughput genotyping of repeat polymorphism in the regulatory region of serotonin transporter gene by gel microchip electrophoresis. *Electrophoresis* 22:4008–11.

1082. Sanders, J.C., Breadmore, M.C., Mitchell, P.S., Landers, J.P. 2002. A simple PDMS-based electro-fluidic interface for microchip electrophoretic separations. *Analyst* 127:1558–63.

1083. Nishimoto, T., Fujiyama, Y., Abe, H., Kanai, M., Nakanishi, H., Arai, A. 2000. Microfabricated CE chips with optical slit for UV absorption detection. In *Proceedings of the 4th Micro Total Analysis Systems Symposium*, Enschede, Netherlands, May 14–18, pp. 395–98.

1084. Gustafsson, M., Hirschberg, D., Palmberg, C., Jornvall, H., Bergman, T. 2004. Integrated sample preparation and MALDI mass spectrometry on a microfluidic compact disk. *Anal. Chem.* 76:345–50.

1085. Sato, K., Yamanaka, M., Tokesih, M., Morishima, K., Kitamori, T. 2003. Multichannel micro elisa system. In *Proceedings of the 7th Micro Total Analysis Systems Symposium*, Squaw Valley, CA, October 5–9, pp. 781–82.

1086. Qiu, C.X., Harrison, D.J. 2001. Integrated self-calibration via electrokinetic solvent proportioning for microfluidic immunoassays. *Electrophoresis* 22:3949–58.

1087. Quake, S. 2003. Biological large scale integration. In *Proceedings of the 7th Micro Total Analysis Systems Symposium*, Squaw Valley, CA, October 5–9, pp. 1–4.

1088. Kerby, M., Chieng, R.-L. 2001. A fluorogenic assay using pressure-driven flow on a microchip. *Electrophoresis* 22:3916–23.

1089. Xue, Q., Wainright, A., Gangakhedkar, S., Gibbons, I. 2001. Multiplexed enzyme assays in capillary electrophoretic single-use microfluidic devices. *Electrophoresis* 22:4000–7.

1090. Jobst, G., Moser, I., Svasek, P., Svasek, E., Varahram, M., Urban, G. 1996. Rapid liver enzyme assay with low cost µTAS. In *Proceedings of the 2nd Micro Total Analysis Systems Symposium*, Basel, November 19–22, p. 221.

1091. Takao, H., Noda, T., Ashiki, M., Miyamura, K., Sawada, K., Ishida, M. 2001. A silicon microchip for blood hemoglobin measurement using multireflection structure. In *Proceedings of the 5th Micro Total Analysis Systems Symposium*, Monterey, CA, October 21–25, pp. 363–64.

1092. Schluter, M., Mammitzsch, S., Martens, M., Gasso, S., Lilienhof, H.J. 2003. Micro fluidic immunoassy chip with integrated liquid handling. In *Proceedings of the 7th Micro Total Analysis Systems Symposium*, Squaw Valley, CA, October 5–9, pp. 275–78.

1093. Jesson, G., Kylberg, G., Andersson, P. 2003. A versatile macro-to-micro dispensing system. In *Proceedings of the 7th Micro Total Analysis Systems Symposium*, Squaw Valley, CA, October 5–9, pp. 155–58.

1094. Schmitt, H., Brecht, A., Gauglitz, G. 1996. An integrated system for microscale affinity measurements. In *Proceedings of the 2nd Micro Total Analysis Systems Symposium*, Basel, November 19–22, pp. 104–9.

1095. Hattori, A., Yamaguchi, H., Yamaguchi, J., Matsuoka, Y., Kanki, S., Fukuzawa, T., Miwa, T., Totama, M., Tokeshi, M., Kitamori, T. 2003. Practical studies on compact photo-thermal lens spectroscopy detection system with micro chemical chip. In *Proceedings of the 7th Micro Total Analysis Systems Symposium*, Squaw Valley, CA, October 5–9, pp. 359–62.

1096. Li, Z.-P., Tsunoda, H., Okano, K., Nagai, K., Kambara, H. 2003. Microchip electrophoresis of tagged probes incorporated with one-colored ddNTP for analyzing single-nucleotide polymorphisms. *Anal. Chem.* 75:3345–51.

1097. Rong, W., Kutter, J.P. 2001. On-line chemiluminescence detection of bioprocesses using polymer-based microchips with immobilized enzymes. In *Proceedings of the 5th Micro Total Analysis Systems Symposium*, Monterey, CA, October 21–25, pp. 181–82.

1098. Forssen, L., Elderstig, H., Eng, L., Nordling, M. 1994. Integration of an amperometric glucose sensor in a µ-TAS. In *Micro Total Analysis Systems '94*, University of Twente, Netherlands, November 21–22, pp. 203–7.

1099. Blankenstein, G., Scampavia, L., Branebjerg, J., Larsen, U.D., Ruzicka, J. 1996. Flow switch for analyte injection and cell/particle sorting. In *Proceedings of the 2nd Micro Total Analysis Systems Symposium*, Basel, November 19–22, pp. 82–84.

1100. Xu, Z.R., Fang, Z.L. 2004. Composite poly(dimethylsiloxane)/glass microfluidic system with an immobilized enzymatic particle-bed reactor and sequential sample injection for chemiluminescence determinations. *Anal. Chim. Acta* 507:129–35.

1101. Yamada, M., Sugiyama, N., Seki, M. 2002. Microfluidic reactor array for multistep droplet reactions. In *Proceedings of the 6th Micro Total Analysis Systems Symposium*, Nara, Japan, November 3–7, pp. 43–44.

1102. Zimmermann, S., Fienbork, D., Wasilik, M., Liepmann, D. 2002. A novel in-device enzyme immobilizatoin method for BIOMENS, demonstrated for a continuous glucose monitor. In *Proceedings of the 6th Micro Total Analysis Systems Symposium*, Nara, Japan, November 3–7, pp. 449–51.

1103. Jorgensen, A.M., Petersen, D., Geschke, O. 2002. An integrated chemiluminescence detector for measuring enzymatically generated hydrogen peroxide. In *Proceedings of the 6th Micro Total Analysis Systems Symposium*, Nara, Japan, November 3–7, pp. 891–93.

1104. Dempsey, E., Diamond, D., Smyth, M.R., Urban, G., Jobst, G., Moser, I., Verpoorte, E.M.J., Manz, A., Widmer, H.M., Rabenstein, K., Freaney, R. 1997. Design and development of a miniaturised total chemical analysis system for on-line lactate and glucose monitoring in biological samples. *Anal. Chim. Acta* 346:341–49.

1105. Suzuki, S., Shimotsu, N., Honda, S., Arai, A., Nakanishi, H. 2001. Rapid analysis of amino sugars by microchip electrophoresis with laser-induced fluorescence detection. *Electrophoresis* 22:4023–31.

1106. Jakeway, S.C., De Mello, A.J. 2001. A single point evanescent wave probe for on-chip refractive index detection. In *Proceedings of the 5th Micro Total Analysis Systems Symposium*, Monterey, CA, October 21–25, pp. 347–48.

1107. Lendl, B., Schindler, R., Frank, J., Kellner, R., Drott, J., Laurell, T. 1997. Fourier transform infrared detection in miniaturized total analysis systems for sucrose analysis. *Anal. Chem.* 69:2877–81.

1108. Shiomoide, K., Mawatari, K., Mukaiyama, S., Fukui, H. 2002. Bio-chemical analysis on microfabricated polymer. In *Proceedings of the 6th Proceedings Symposium*, Nara, Japan, November 3–7, pp. 918–21.

1109. Wang, J., Chatrathi, M.P., Tian, B., Polsky, R. 2000. Microfabricated electrophoresis chips for simultaneous bioassays of glucose, uric acid, ascorbic acid, and acetaminophen. *Anal. Chem.* 72:2514–18.

1110. Greenwood, P.A., Greenway, G.M. 2001. Development of a µTAS screening device for drug analysis by chemiluminescence. In *Proceedings of the 5th Micro Total Analysis Systems Symposium,* Monterey, CA, October 21–25, pp. 343–44.

1111. Nelstrop, L.J., Greenway, G.M. 1998. Investigation of chemiluminescent microanalytical systems. In *Micro Total Analysis Systems '98,* Banff, Canada, October 13–16, pp. 355–58.

1112. Harrison, D.J., Chiem, N. 1996. Microchip lab for biochemical analysis. In *Proceedings of the 2nd Micro Total Analysis Systems Symposium,* Basel, November 19–22, pp. 31–33.

1113. Wicks, D.A., Li, P.C.H. 2004. Separation of fluorescent derivatives of hydroxyl-containing small molecules on a microfluidic chip. *Anal. Chim. Acta* 507:107–14.

1114. Daridon, A., Thronset, W., Liau, I., Farrell, K., Tseng, F., Javadi, S., Manger, I. 2002. A programmable cell assay platform for kinetic studies of a single cell. In *Proceedings of the 6th Micro Total Analysis Systems Proceedings Symposium,* Nara, Japan, November 3–7, pp. 21–33.

1115. Morris, C.J., Daiber, T., Weigl, B.H., Kesler, N., Bardell, R.L. 2000. New assays based on laminar fluid diffusion interfaces. Results from prototype and product testing. In *Proceedings of the 4th Micro Total Analysis Systems Symposium,* Enschede, Netherlands, May 14–18, pp. 688–90.

1116. Chudy, M., Wróblewaski, W., Dybko, A., Brzózka, Z. 2001. PMMA/PDMS based microfluidic system with optical detection for total heavy metals concentration assessment. In *Proceedings of the 5th Micro Total Analysis Systems Symposium,* Monterey, CA, October 21–25, pp. 521–22.

1117. Greenway, G.M., Nelstrop, L.J., McCreedy, T., Greenwood, P. 2000. Luminol chemiluminescence systems for metal analysis by µTAS. In *Proceedings of the 4th Micro Total Analysis Systems Symposium,* Enschede, Netherlands, May 14–18, pp. 363–66.

1118. Fielden, P.R., Baldock, S.J., Goddard, N.J., Pickering, L.W., Prest, J.E., Snook, R.D., Brown, B.J.T., Vaireanu, D.I. 1998. A miniaturized planar isotachophoresis separation device for transition metals with integrated conductivity detection. In *Micro Total Analysis Systems '98,* Banff, Canada, October 13–16, pp. 323–26.

1119. Tokeshi, M., Minagawa, T., Kitamori, T. 2000. Integration of a microextraction system. Solvent extraction of a Co-2-nitroso-5-dimethylaminophenol complex on a microchip. *J. Chromatogr.* 894A:19–23.

1120. Liu, B.F., Ozaki, M., Terabe, S., Utsumi, Y., Hattori, T. 2002. Comparison of different strategies of chemiluminescence detection for microchip system fabricated in poly(dimethylsiloxane). In *Proceedings of the 6th Micro Total Analysis Systems Symposium,* Nara, Japan, November 3–7, pp. 293–95.

1121. Song, Q.J., Greenway, G.M., McCreedy, T. 2000. Interfacing microchip CE with ICPMS for elements. In *Proceedings of the 4th Micro Total Analysis Systems Symposium,* Enschede, Netherlands, May 14–18, pp. 22–24.

1122. Jenkins, G., Manz, A. 2001. Optical emission detection of liquid analytes using a micro-machined d.c. glow-discharge device at atmospheric pressure. In *Proceedings of the 5th Micro Total Analysis Systems Symposium,* Monterey, CA, October 21–25, pp. 349–50.

1123. Fiehn, H., Howitz, S., Pham, M.T., Vopel, T., Bürger, M., Wegner, T. 1994. Components and technology for a fluidic-isfet-microsystem. In *Micro Total Analysis Systems '94,* University of Twente, Netherlands, November 21–22, pp. 289–93.

1124. Prest, J.E., Baldock, S.J., Bektas, N., Fielden, P.R., Treves Brown, B.J. 1999. Single electrode conductivity detection for electrophoretic separation systems. *J. Chromatogr.* 836A:59–65.

1125. Luttge, R., Gardeniers, J.G.E., Vrouwe, E.X., van den Berg, A. 2003. Microneedle array interface to CE on chip. In *Proceedings of the 7th Micro Total Analysis Systems Symposium,* Squaw Valley, CA, October 5–9, pp. 511–14.

1126. Chen, H., Fang, Q., Yin, X.F., Fang, Z.L. 2002, A multiphase laminar flow diffusion chip with ion selective electrode detection. In *Proceedings of the 6th Micro Total Analysis Symposium,* Nara, Japan, November 3–7, 371–73.

1127. Guijt, R.M., Baltussen, E., Van der Steen, G., Schasfoort, R.B.M., Schlautmann, S., Billiet, H.A.H., Frank, J., Van Dedem, G.W.K., Van den Berg, A. 2001. New approaches for fabrication of microfluidic capillary electrophoresis devices with on-chip conductivity detection. *Electrophoresis* 22:235–41.

1128. Vrouwe, E., Luttge, R., van den Berg, A. 2002. Measuring lithium in whole blood using capillary. In *Proceedings of the 6th Micro Total Analysis Systems Symposium,* Nara, Japan, November 3–7, pp. 178–80.

1129. Vroume, E.X., van den Berg, A. 2003. Destacking loading conditions on a CE chip for measuring samples with a high matrix concentration. In *Proceedings of the 7th Micro Total Analysis Systems Symposium,* October 5–9, pp. 89–92.

1130. Usui, S., Fujii, T. 2003. Development of in situ flow-through analyzer of Mn^{2+} in seawater with a PDMS microfluidic device. In *Proceedings of the 7th Micro Total Analysis Systems Symposium*, Squaw Valley, CA, October 5–9, pp. 291–94.

1131. Wang, J., Pumera, M. 2003. Nonaqueous electrophoresis microchip separations: Conductivity detection in UV-absorbing solvents. *Anal. Chem.* 75:341–45.

1132. Timmer, B.H., van Delft, K.M., Otjes, R.P., Olthuis, W., van den Berg, A. 2004. Miniaturized measurement system for ammonia in air. *Anal. Chim. Acta* 507:137–43.

1133. Timmer, B.H., van Delft, K.M., Otjes, R.P., Olthuis, W., Bergveld, P., van den Berg, A. 2003. Towards a miniaturized ambient ammonia detection system. In *Proceedings of the 7th Micro Total Analysis Systems Symposium*, Squaw Valley, CA, October 5–9, pp. 81–84.

1134. Laugere, F., van der Steen, G., Bastemeijer, J., Guijt, R.M., Sarro, P.M., Vellekoop, M.J., Bossche, A. 2002. Separation and detection of organic acids in a CE microdevice with contractless four-electrode conductivity detection. In *Proceedings of the 6th Micro Total Analysis Systems Symposium*, Nara, Japan, November 3–7, pp. 491–93.

1135. Sorouraddin, M.H., Hibara, A., Proskurnin, M.A., Kitamori, T. 2000. Integrated FIA for the determination of ascorbic acid and dehydroscorbic acid in microfabricated glass-channel by thermal-lens microscopy. *Anal. Sci.* 16:1033.

1136. Prest, J.E., Baldock, S.J., Fielden, P.R., Goddard, N.J., Brown, B.J.T. 2003. Determination of inorganic selenium species by miniaturised isotachophoresis on a planar polymer chip. *Anal. Bioanal. Chem.* 376:78–84.

1137. Stanislawski, B., Kaniansky, D., Masar, M., Johnck, M. 2002. Design principles, performance and perspectives of a complete miniaturized electrophoretic instrument. In *Proceedings of the 6th Micro Total Analysis Systems Symposium*, Nara, Japan, November 3–7, pp. 350–52.

1138. Bodor, R., Kaniansky, D., Masár, M., Silleová, K., Stanislawski, B. 2002. Determination of bromate in drinking water by zone electrophoresis-isotachophoresis on a column-coupling chip with conductivity detection. *Electrophoresis* 23:3630–37.

1139. Weber, G., Johnck, M., Siepe, D., Neyer, A., Hergenroder, R. 2000. Capillary electrophoresis with direct and contactless conductivity detection on a polymer microchip. In *Proceedings of the 4th Micro Total Analysis Systems Symposium,* Enschede, Netherlands, May 14–18, pp. 383–86.

1140. Kim, D.J., Cho, W.H., Ro, K.W., Hahn, J.H. 2001. Microchip-based simultaneous on-line monitoring for Cr(III) and Cr(VI) using highly efficient chemiluminescence detection. In *Proceedings of the 5th Micro Total Analysis Systems Symposium*, Monterey, CA, October 21–25, pp. 525–26.

1141. Baldock, S.J., Bektas, N., Fielden, P.R., Goddard, N.J., Pickering, L.W., Prest, J.E., Snook, R.D., Brown, B.J.T., Vaireanu, D.I. 1998. Isotachophoresis on planar polymeric substrates. In *Micro Total Analysis Systems '98,* Banff, Canada, October 13–16, pp. 359–62.

1142. Kikura-Hanajiri, R., Martin, R.S., Lunte, S.M. 2002. Indirect measurement of nitric oxide production by monitoring nitrate and nitrite using microchip electrophoresis with electrochemical detection. *Anal. Chem.* 74:6370–77.

1143. van der Schoot, B.H., Verpoorte, E.M.J., Jeanneret, S., Manz, A., de Rooij, N.F. 1994. Microsystems for analysis in flowing solutions. In *Micro Total Analysis Systems '94*, University of Twente, Netherlands, November 21–22, pp. 181–90.

1144. Chudy, M., Prokaryn, P., Dybko, A., Wroblewski, W., Brzozka, Z. 2002. PDMS microstructures integrated with detection. In *Proceedings of the 6th Micro Total Analysis Systems Symposium*, Nara, Japan, November 3–7, pp. 392–94.

1145. Verpoorte, E., Manz, A., Widmer, H.M., van der Schoot, B., de Rooji, N.F. 1993. A three-dimensional micro flow system for a multi-step chemical analysis. *Transducers* 93:939–42.

1146. Chudy, M., Prokaryn, P., Dybko, A., Brzozka, Z. 2003. Temperature controlled microfluidic PDMS reactor. In *Proceedings of the 7th Micro Total Analysis Systems Symposium*, Squaw Valley, CA, October 5–9, pp. 171–74.

1147. Fujii, S.I., Tokuyama, T., Abo, M., Okubo, A. 2003. Development of a reflected light fluorescence unit for the microfluidic detection system. In *Proceedings of the 7th Micro Total Analysis Systems Symposium*, Squaw Valley, CA, October 5–9, pp. 391–94.

1148. Masaki, H., Susaki, H., Uchiyama, K., Ito, M., Korenaga, T. 2002. Development of micro detector for benzo[a] pyrene monitoring. In *Proceedings of the 6th Micro Total Analysis Systems Symposium*, Nara, Japan, November 3–7, pp. 527–29.

1149. Lu, C.J., Tian, W.C., Steinecker, W.H., Guyon, A., Agah, M., Oborny, M.C., Sacks, R.D., Wise, K.D., Pang, S.W., Zeller, E.T. 2003. Functionally integrated MEMS micro gas chromatograph subsystem. In *Proceedings of the 7th Micro Total Analysis Systems Symposium*, Squaw Valley, CA, pp. 411–16.

1150. Firebaugh, S.L., Jensen, K.F., Schmidt, M.A. 2000. Miniaturization and integration of photoacoustic detection with a microfabricated chemical reactor system. In *Proceedings of the 4th Micro Total Analysis Systems Symposium,* Enschede, Netherlands, May 14–18, pp. 49–52.

1151. Horiuchi, T., Ueno, Y., Niwa, O. 2001. Micro-fluidic device for detection and identification of aromatic VOCs by optical method. In *Proceedings of the 5th Micro Total Analysis Systems Symposium*, Monterey, CA, October 21–25, pp. 527–28.

1152. Fintschenko, Y., Choi, W.Y., Ngola, S.M. 2001. Chip electrochromatography of polycyclic aromatic hydrocarbons on an acrylate-based UV-initiated porous polymer monolith. *Fres. J. Anal. Chem.* 371:174–81.

1153. Wang, J., Pumera, M. 2002. Dual conductivity/amperometric detection system for microchip capillary electrophoresis. *Anal. Chem.* 74:5919–23.

1154. Chen, C.H., Lin, H., Lele, S.K., Santiago, J.G. 2000. Electrokinetic microflow instability with conductivity gradients. In *Proceedings of the 4th Micro Total Analysis Systems Symposium,* Enschede, Netherlands, May 14–18, pp. 983–87.

1155. Hilmi, A., Luong, J.H.T. 2000. Micromachined electrophoresis chips with electrochemical detectors for analysis of explosive compounds in soil and groundwater. *Environ. Sci. Technol.* 34:3046–50.

1156. Wang, J., Chatrathi, M.P., Tian, B.M., Polsky, R. 2000. Capillary electrophoresis chips with thick-film amperometric detectors: Separation and detection of hydrazine compounds. *Electroanalysis* 12:691–94.

1157. Ocvirk, G., Tang, T., Harrison, J.D. 1998. Optimization of confocal epifluorescence microscopy for microchip-based miniaturized total analysis systems. *Analyst* 123:1429–34.

1158. Ruano, J.M., Benoit, V., Aitchison, J.S., Cooper, J.M. 2000. Flame hydrolysis deposition of glass on silicon for the integration of optical and microfluidic devices. *Anal. Chem.* 72:1093–97.

1159. Wang, J., Pumera, M., Collins, G.E., Mulchandani, A. 2002. Measurements of chemical warfare agent degradation products using an electrophoresis microchip with contactless conductivity detector. *Anal. Chem.* 74:6121–25.

1160. Reshni, K.A., Morris, M.D., Johnson, B.N., Burns, M.A. 1998. On-line detection of electrophoretic separations on a microchip by Raman spectroscopy. In *Micro Total Analysis Systems '98,* Banff, Canada, October 13–16, pp. 109–12.

1161. Endo, T., Okuyama, A., Matsubara, Y., Kobayashi, M., Morita, Y., Mizukami, H., Tamiya, E. 2003. Monitoring of coplanar polychlorinated biphenyls (Co-PCB) by the multi flow antibody chip. In *Proceedings of the 7th Micro Total Analysis Systems Symposium*, Squaw Valley, CA, October 5–9, pp. 567–70.

1162. Floyd, T.M., Schmidt, M.A., Jensen, K.F. 2001. A silicon microchip for infrared transmission kinetics studies of rapid homogeneous liquid reactions. In *Proceedings of the 5th Micro Total Analysis Systems Symposium*, Monterey, CA, October 21–25, pp. 277–79.

1163. Rossier, J.S., Vollet, C., Carnal, A., Lagger, G., Gobry, V., Girault, H.H., Michel, P. 2002. Plasma etched polymer microelectrochemical systems. *Labchip* 2:145–50.

1164. Schumacher, J., Ranft, M., Wilhelm, T., Dahint, R., Grunze, M. 1998. Chemical analysis based on environmentally sensitive hydrogels and optical diffraction. In *Micro Total Analysis Systems '98,* Banff, Canada, October 13–16, pp. 61–64.

1165. Jackman, R.J., Queeney, K.T., Herzig-Marx, R., Schmidt, M.A., Jensen, K.F. 2001. Integration of multiple internal reflection (MIR) infrared spectroscopy with silicon-based chemical microreactors. In *Proceedings of the 5th Micro Total Analysis Systems Symposium*, Monterey, CA, October 21–25, pp. 345–46.

1166. Sirichai, S., de Mello, A.J. 2001. A capillary electrophoresis chip for the analysis of print and film photographic developing agents in commercial processing solutions using indirect fluorescence detection. *Electrophoresis* 22:348–54.

1167. Berger, R., Gerber, C., Gimzewski, J.K. 1996. Nanometers, picowatts, femtojoules: Thermal analysis and optical spectroscopy using micromechanics. In *Proceedings of the 2nd Micro Total Analysis Systems Symposium*, Basel, November 19–22, pp. 74–77.

1168. Liu, J., Yan, J., Yang, X., Wang, E. 2003. Miniaturized tris(2,2'-bipyridyl)ruthenium(II) electrochemiluminescence detection cell for capillary electrophoresis and flow injection analysis. *Anal. Chem.* 75:3637–42.

1169. Ito, T., Sobue, K., Ohya, S. 2002. Water glass bonding for micro-total analysis system. *Sensors Actuators* B 811:87–195.

1170. Harrison, D.J., Andersson, P.E., Li, P.C.H., Chiem, N., Tang, T., Smith, R., Szarka, R., Tran, T. 1997. Integration of biochemical and cellular reactions with separation on-chip. Paper presented at 19th International Symposium on Capillary Chromatography and Electrophoresis, May 18–22.

1171. Zeringue, H.C., Beebe, D.J., Wheeler, M.B. 2001. Removal of cumulus from mammalian zygotes using microfluidic techniques. *Biomed. Microdevices* 3:219–24.

1172. Yuen, P.K., Despa, M., Li, C.C., Dejneka, M.J. 2003. Microbarcode sorting device. Labchip 3:198–201.

1173. Shelby, J.P., Mutch, S.A., Chiu, D.T. 2004. Direct manipulation and observation of the rotational motion of single optically trapped microparticles and biological cells in microvortices. *Anal. Chem.* 76:2492–97.

1174. Ozkan, M., Wang, M., Ozkan, C., Flynn, R., Esener, S. 2003. Optical manipulation of objects and biological cells in microfluidic devices. *Biomed. Microdevices* 5:61–67.

1175. Carlson, R.H., Gabel, C., Chan, S., Austin, R.H. 1998. Activation and sorting of human white blood cells. *Biomed. Microdevices* 1:39–47.

1176. McClain, M.A., Culbertson, C.T., Jacobson, S.C., Allbritton, N.L., Sims, C.E., Ramsey, J.M. 2003. Microfluidic devices for the high-throughput chemical analysis of cells. *Anal. Chem.* 75:5646–55.

1177. Yamashita, K., Yamaguchi, Y., Miyazaki, M., Nakamura, H., Shimizu, H., Maeda, H. 2004. Sequence-selective DNA detection using multiple laminar streams: A novel microfluidic analysis method. *Labchip* 4:1–3.

1178. Roubi, M. 1999. Sizing, sorting DNA one piece at a time. *Chem. Eng. News* 78:5–6.

Glossary

4-AAP: 4-Aminoantipyrine
Ab: Antibody
AC: Alternating current
Acetaminophen: 4-Acetamidophenol or paracetamol
AChE: Acetylcholinesterase
ACN: Acetonitrile
ACTH: Adrenocorticotropin
ADH: Alcohol dehydrogenase
Adrenaline: See *epinephrine*
Ag: Antigen
AIN: 2,2'-Azobioisobutylonitrile
ALP: Alkaline phosphatase
ALT: Alanine aminotransferase (same as GPT)
AML: Acute myeloid leukemia
AMP: Adenosine monophosphate
ANS: 8-Anilino-1-naphthanesulfonic acid
APP: 4-Aminophenyl phosphate
APTES: 3-Aminopropyltriethoxysilane
APTS: 8-Amino-1,3,6-pyrenetrisulfonic acid
AST: Aspartate aminotransferase (same as GOT)
ASV: Anodic stripping voltammetry
ATP: Adenosine triphosphate
ATR: Absorbance total internal reflection
Au: Gold
B4F: Biotin 4-fluorescein
BAEC: Bovine aortic endothelial cell
BaP: Benzo[a]pyrene
BCE: Bovine adrenal capillary endothelial cell
BCECF: 2',7'-Bis(2-carboxyethyl)carboxyfluorescein
BCQ: (Acryloylaminopropyl)trimethylammonium chloride
BEE: 2-(2-Butoxyethoxy)ethanol
β-Gal: β-Galactosidase
BK7: A borosilicate crown glass
BME: Benzoin methyl ether or α-methoxy-α-phenylacetophenone
BOD: Boron-doped diamond
BODIPY: 4,4-Difluoro-5,7-dimethyl-4-bora-3a,4a-diaza-s-indacene
BOE: Buffered oxide etch (NH_4F/HF)
bp: Base pair
BRCA: Breast cancer
BSA: Bovine serum albumin
BSM: Bovine submaxillary mucin
BTX: Mixture of benzene, toluene, and xylene
4-CN: 4-Chloro-1-naphthol
C18: See *ODS*
CA: Carbonic anhydrase
CAE: Capillary array electrophoresis

CAT: Catechol

Catechol: 1,2-Benzenediol or CAT

CBQCA: 3-(4-Carboxybenzoyl)quinoline-2-carboxaldehyde

CCD: Charge-coupled device

CD-3: 4-Amino-3-methyl-N-ethyl-N(beta-methane sulfonamidoethyl)aniline in print developing solution (Kodak RA-4)

CD-4: 4-(N-ethyl-N-hydroxyethyl)-2-methylphenylenediamine in film developing solution (Kodak C-41)

cDOPA: Hydrazinomethyldihydroxyphenylalanine

CE: Capillary electrophoresis

CEA: Carcinoembryonic antigen

CFTR: Cystic fibrosis transmembrane conductance regulator

CGE: Capillary gel electrophoresis

CHCA: α-Cyano-4-hydroxycinnamic acid

CHO: Chinese hamster ovary cells

CL: Chemiluminescence

CMOS: Complementary metal oxide semiconductor

Co: Cobalt

COC: Cyclic olefin copolymer: amorphous copolymer of 2-norbornene and ethylene (Zeonor)

COMOSS: Collated monolithic support structure

cP: Centipoise (a viscosity unit)

CPM: Courmarinylphenylmaleimide (to react with thiocholine to produce a fluorescent thioether)

CPT: Cycling probe technology

Cr: Chromium

CRP: Human C-reactive protein (a cardiac marker)

CTAB: Cetyltrimethylammonium bromide

CTAC: Cetyltrimethylammonium chloride

CV: Cyclic voltammetry

CVD: Chemical vapor deposition

CW: Chemical warfare

CZE: Capillary zone electrophoresis

2,4-DNT: 2,4-Dinitrotoluene

2,6-DNT: 2,6-Dinitrotoluene

D: Diffusion coefficient

DA: Dopamine (or 3-hydroxytyramine)

DABSY: 4-Dimethylaminoazobenzene-4'-sulfonyl group

DC: Direct current

DCF: Dichlorofluorescein

DDi: D-dimer (a marker of sepsis and thrombotic disorder)

DEP: Dielectrophoresis

DHAB: 2,2'-Dihydroxyazobenzene

DiFMU: 6,8-Difluoro-4-methylumbelliferone or 6,8-difluoro-7-hydroxycoumarin

DIG: Digoxigenin

DMD: Duchenne muscular dystrophy

DMDA: N,N-Dimethyldodecylamine

DMPA: 2,2-Dimethoxy-2-phenylacetophenone (or Irg 651)

DNA: Deoxyribonucleic acid

DNB: 1,3-Dinitrobenzene

DNP: Dinitrophenyl

DNT: Dinitrotoluene

Dns: Dansylated

DOPA: Dihydroxyphenylalanine
Dopamine: 3,4-Dihydroxyphenethylamine
DOP-PCR: Degenerate oligonucleotide-primed PCR
DOPS: 3,4-Dihydroxyphenylserine
DRIE: Deep reactive ion etching
DS: Dextran sulfate
ds: Double stranded
DTSSP: 3,3-Dithiopropionic acid bis-*N*-hydroxysulfosuccinimide ester
DTT: Dithiothreitol
E: Electric field strength
ECDM: Electrochemical discharge machining
ECL: Electrochemiluminescence
E. coli: Escherichia coli
EDP: Ethylenediamine/pyrocatechol
EDTA: Ethylenediaminetetraacetic acid
EFGF: Electric field gradient focusing
EGFP: Enhanced green fluorescent protein
EGP: Epidermal growth factor
EK: Electrokinetic
ELISA: Enzyme-linked immunosorbent assay
EOF: Electroosmotic flow
EPDMA: Epoxy poly(dimethylacrylamide)
Epinephrine: Adrenaline or 4-[1-hydroxy-2-(methylamino)ethyl]-1,2-benzenediol hydrochloride
EPW: EPD and water
ESI: Electrospray ionization
EtBr: Ethidium bromide
EWOD: Electrowetting-on-dielectric
FACD: Fluorescently activated cell sorting
$F_c\gamma R$: F_c gamma receptor
FDA: Fluorescein diacetate
FDTS: Fluorododecyltrichlorosilane
FEA: Finite element analysis
FEP: Fluoroethylenepropylene copolymer
FID: Flame ionization detector
FITC: Fluorescein isothiocyanate
FMLP: Formyl-methionyl-leucyl-phenylalanine
FQ: 3-(2-Furoyl)quinoline-2-carboxaldehyde
FSCE: Free solution capillary electrophoresis
FT: Fourier transform
G-6-PDH: Glucose-6-phosphate dehydrogenase
GC: Gas chromatography
GOT: Glutamic oxaloacetic transaminase (same as AST)
GOx: Glucose oxidase
GPT: Glutamic pyruvic transaminase (same as ALT)
GSH: Glutathione
5-HT: 5-Hydroxytryptamine or serotonin
H: Plate height
HBP: 2-Hydroxybiphenyl
HCV: Hepatitis C virus
H_{det}: Plate height due to detector size
HDF: Hydrodynamic flow

H_{diff}: Plate height due to axial diffusion

HDL: High-density lipoprotein

He: Helium

HEC: Hydroxyethylcellulose

HEK: Human embryonic kidney cell

HEMA: Hydroxyethylmethacrylate

HEPES: N-(2-Hydroxyethyl)piperazine-N'-(2-ethanesulfonic acid)

HepG2: Human hepatocellular carcinoma cell

HFE: A gene for hemochromatosis

HFTS: Heptadecafluoro-1,1,2,2-tetrahydrodecyltrichlorosilane

Hg: Mercury

H_{geo}: Plate height due to channel turn geometry

HHC: Hereditary hemochromatosis

H_{inj}: Plate height due to injector size

H_{Joule}: Plate height due to Joule heating

H_m: Plate height due to mass transfer

H_{mc}: Plate height due to sorption/desorption micellar kinetics

HMCV: Hydrophobic microcapillary vent

HMPP: 2-Hydroxy-2-methylpropiophenone

HPC: Hydroxylpropylcellulose

HPMC: Hydroxypropylmethylcellulose

HPPA: 3-(p-Hydroxyphenyl)-propionic acid

HPV: Human papillomavirus

HQS: 8-Hydroxyguinline-5-sulfonic acid

HRP: Horseradish peroxidase

HSA: Human serum albumin

Hsp: Heat shock protein

HT: Hadamard transform

HTTLPR: 5-HT transporter linked polymorphic region

HV: High voltage

HVA: Homovanillic acid (a dopamine metabolite)

IBE: Ion beam etching

IEF: Isoelectric focusing

IFN-γ: Interferon gamma

IGF: Insulin growth factor

IgG: Immunoglobulin G

IMPA: Isopropylmethylphosphonic acid

IPA: Isopropylalcohol

IR: Infrared

Irg 651: 2,2-Dimethoxy-2-phenylacetophenone. See also *DMPA*

ISFET: Ion-selective field-effect transistor

Isoproterenol: 1-[3',4'-Dihydroxyphenyl]-2-isopropylaminoethanol

ITO: Indium-tin oxide

ITP: Isotachophoresis

IVF: *In vitro* fertilization

Kapton: Polyimide (Dupont)

KB: Human KB cancer cell

L: Capillary length

LAMP: Loop-mediated isothermal amplification

LAP: Leucine aminopeptidase

LAPS: Light-addressable potentiometric sensor

LC: Liquid chromatography

LCR: Ligase chain reaction

L_{det}: Length of detector

LDL: Low-density lipoprotein

l-Dopa: (3,4-Dihydroxy-phenylalanine)

LDR: Ligase detection reaction

LE: Leading elecrolyte

LED: Light-emitting diode

LHRH: Luteinizing hormone releasing hormone

LIF: Laser-induced fluorescence

LIGA: Lithographie galvanoformung und abformung

L_{inj}: Length of injector

LOD: Limit of detection

LOX: Lactate oxidase

LPA: Linear polyacrylamide

LPS: Lipopolysaccharide

LTCC: Low-temperature cofired ceramic (an aluminum borosilicate)

Luminol.: 5-Amino-2,3-dihydro-1,4-phthalazinedione

µ: Electrical mobilità

MALDI: Matrix-assisted laser desorption ionization

MECC: Micellar electrokinetic capillary chromatography

MES: 2-(*N*-Morpholino)-ethanesulfonic acid (used as a buffer)

MHEC: Methylhydroxyethylcellulose

MIP: Microwave-induced plasma

MMA: Methylmethacrylate

MPA: Methylphosphonic acid

MPC: Methacrylocyloxyethylphosphorylcholine

MS: Mass spectrometry

MSH: Melanocyte-stimulating hormone (a melanocyte is a pigment cell in the skin)

MT: Methoxytyramine (a dopamine metabolite)

MTHFR: Methylenetetrahydrofolate reductase

MW: Molecular weight

Mylar: PET (Dupont)

4-NT: 4-Nitrotoluene

N: Number of theoretical plates

NA: Noradrenaline

NAD$^+$: Nicotine adenosine dinucleotide (oxidized form)

NADH: Nicotine adenosine dinucleotide (reduced form)

NAM: *N*-(9-Acridiyl)maleimide

NASBA: Nucleic acid sequence-based amplification

NBD-F: 4-Fluoro-7-nitro-2,1,3-benzoxadiazole

NDA: Naphthalene-2,3-dicarboxaldehyde

NN: 2-Nitroso-1-naphthol

NO: Nitric oxide

NSB: Nonspecific binding

ODS: Octadecylsilane or C18

ODTMS: Octyldecyltrimethoxysilane

OEP: Octae-hylporphyrin (forms complex with Pb^{2+})

OES: Optical emission spectroscopy

OPA: *o*-Phthaldialdehyde

OTS: Octadecyltricholorosilane

PA: Protein A

PAAH: Poly(allylamine hydrochloride)

PAD: Pulsed amperometric detection

PAGE: Polyacrylamide gel electrophoresis

PAH: Polyaromatic hydrocarbon

PAP: *p*-Aminophenol

PAPP: *p*-Aminophenol phosphate

Paracetamol: Acetaminophen or 4-acetamidophenol

PB: Polybrene

PBS: Phosphate buffered saline

PC: Polycarbonate (or Lexan)

PCB: Polychlorinated biphenyl

PCR: Polymerase chain reaction

Pd: Palladium

PDDA: Polydimethyldiallylammonium

PDI: Protein disulfide isomerase

PDLA: Poly(d-lactic acid)

PDMA: Poly(*N,N'*-dimethylacrylamide)

PDMS: Poly(dimethylsiloxane)

p(dT): Poly(deoxythymidine)

PE: Polyethylene

PECVD: Plasma-enhanced CVD

PEG: Polyethylene glycol or polyethylene oxide

PEG-DA: Poly(ethylene glycol) diacrylate

PEI: Polyethylenimine

PEM: Polyelectrolyte multilayer

PEO: Polyethylene oxide or polyethylene glycol

PET: Poly(ethylene terephthalate) or Mylar

PETG: Poly(ethylene terephthalate glycol) or Vivak or copolyester: PI

Propidium iodide: PIN: Positive-intrinsic-negative (a type of photodiode)

PKA: Protein kinase A

pK_a: Negative logarithm of the acid dissociation constant

PLA: Poly(lactic acid)

Pluronics: Triblock copolymer polyethylene oxide–polypropylene oxide–polyethylene oxide (PEO-PPO-PEO)

PMMA: Poly(methylmethacrylate) or Plexiglas or Lucite or acrylic

PMPA: Pinacolylmethylphosphonic acid

PMT: photomultiplier tube

PNB: *p*-Nitrobenzyl

PP: Polypropylene

PPO: Poly(propylene oxide)

PPX-PPF: Copolymer of *p*-xylylene and *p*-xylylenecarboxylic acid pentafluorophenol ester

PS: Polystyrene

PSA: Prostate-specific antigen

PSS: Poly(styrene sulfate)

Pt: Platinum

PTFE: Polytetrafluoroethylene

PTG: Poly(ethylene terephthalate) or Mylar

PU: polyurethane

PVA: Polyvinyl alcohol

PVDF: Poly(vinylidene fluoride)

PVP: Poly(vinylpyrrolidone)

PVPD: poly(vinylpyridine)

PZT: Lead zirconate titanate

QTOF: Quadrupole time of flight

R6G: Rhodamine 6G

RBC: Red blood cell

RBG: Resorufin-β-d-galactopyranoside (a substrate for β-Gal)

RBL-2H3: Rat basophilic leukemia 2H3 cell or mast cell

r_c**:** Radius of curvature

RDX: Hexahydro-1,3,5-trinitro-1,3,5-triazine

rf: Radio frequency

RH: Relative humidity

RIE: Reactive ion etching

RNA: Ribonucleic acid

RP-HPLC: Reverse-phase HPLC

RSD: Relative standard deviation

RT: Reverse transcription

SAM: Self-assembled monolayer

SCOFT: Shah convolution Fourier transform

SDA: Strand displacement amplification

SDS: Sodium dodecyl sulfate

SEC: Size-exclusion chromatography

SERRS: Surface-enhanced resonance Raman scattering

s-IgA: Human secretory immunoglobulin A

SIN-1: 3-Morpholinosydnonimine (a NO-releasing compound that is a metabolite of the vasodilator molsidomine)

SNP: Single nucleotide polymorphism

SPE: Solid-phase extraction

ss: Single stranded

SV: Sinusoidal voltammetry

SVR: Surface-to-volume ratio

TAE buffer: Tris/acetate/EDTA buffer

TAMRA: 5-Carboxytetramethylrhodamine

TAPS: N-Tris(hydroxymethyl)methyl-3-aminopropanesulfonic acid

TBA: Tetrabutylammonium

TBR: Dichlorotris(2,2'-bypyridyl)ruthenium (II) hydrate or $Ru(bpy)^{2+}$

TCD: Thermal conductivity detector

TCM: Traditional Chinese medicine

TCPTP: Human T cell protein tyrosine phosphatase

TDPO: Bis[2-(3,6,9,-trioxadecanyloxycarbonyl)-4-nitrophenyl]oxalate

TE: Terminating electrolyte

TEAA: Triethylammonium acetate

Teflon AF: Amorphous Teflon or copolymer of PTFE and 2,2-bis(trifluoro-methyl)-4,5-difluoro-1,3-dioxide

Teryl: 2,4,6-Trinitrophenyl-N-methylnitramine

TFA: Trifluoroacetic acid

TGF: Temperature gradient focusing

TGGE: Temperature gradient gel electrophoresis

Ti: Titanium

TIRFM: Total internal reflection fluorescence microscopy

TLM: Thermal lens microscopy

TMA: Tetramethylammonium

TMAH: Tetramethylammonium hydroxide

TMPM: 3-(Trimethyoxysilyl)propyl methacrylate

TNB: 1,3,5-Trinitrobenzene

TNT: 2,4,6-Trinitrotoluene

TO: Thiazole orange

TO6: (*N,N'*-Tetramethylpropanediamino)propylthiazole orange

TOAOH: Tetraoctylammonium hydroxide

TOF: Time of flight

TPA: Tripropylamine

TPF: Two-photon fluorescence

TPM: 3-(Trichlorosilyl)propyl methacrylate

TPR: Dichlorotris(1,10-phenanthroline)ruthenium (II) hydrate or $Ru(phen)^{2+}$

TREKS: Transradial electrokinetic selection

TRITC: Tetramethylrhodamine isothiocyanate

TSM: Thickness shear mode

Tween-20: Polyoxyethylene-sorbitan monolaurate

Uracil: Same as urasil

WBC: White blood cell

XPS: X-ray photoelectron spectroscopy

Zeonex: Amorphous homopolymer of norbornene

Zeonor: Amorphous copolymer of 2-norbornene and ethylene (cyclic olefin copolymer)

Appendix: Analytical Applications of Microfluidic Technology

Class of Compounds	Compound	Sample Matrix	Concentration	Microchip	Separation Mode	Detection Mode	Ref.
Amino acid	Cysteine		200 μM	PDMS-PDMS	CZE	Dual amp (C)	763
	Glycine		150 μM	PDMS-PDMS	CZE	Dual amp (C)	763
	19 amino acids		10 mM	Glass	CZE	LIF (TRITC)	339
	19 amino acids	In urine	32.9 μM LOD (average)	Glass	CZE	Indirect LIF (fluo 0.5 mM)	683, 696
	Amino acids		140–220 μM	Glass	CZE	LIF (NBD)	313
	Amino acids		90–100 μm	Quartz	CEC	Indirect LIF	1064
	d/l amino acids	Meteorite	1 mM	Glass	CD-MECC	LIF (FITC)	610, 628
	d/l Dns-Phe		0.39 μM LOD, 1–100 μM range	PDMS	CE	CL	168
	Arginine		0.24 mM	PDMS	CE	Amp (Au)	752
	Histidine		0.70 mM	PDMS	CE	Amp (Au)	752
	Cysteine		0.03 mM	PDMS	CE	Amp (Au)	752
	Glycine			PDMS-PDMS	CZE	Amp (C fiber)	750
	Glycine, tryptophan, and histidine		1 mM	Glass	CZE	Amp (Cu/Pt)	1065
	Phenylalanine		2 μM	Fused silica	CZE	LIF (OPA)	658
	Proline		1.2 μM LOD, 5–600 μM range	PDMS-glass	CE	ECL	273
	Valine, alanine, aspartic acid		300 μM	Si-Pyrex	MECC	ECL (TBR)	727
	Aspartic acid	Fingerprint	75 fmol	Glass	CE	Fluo	1066
	Valine, leucine		2.5 μM LOD	Glass	MECC (SDS)	Amp (Au on C)	655
	Dansyl lysine		10 μM_LOD	Quartz	CE	CL	353
	Alanine		8.0 nM LOD	PMMA	CE	Conductivity	208
	Creatine	Urine	250 μM	Glass	CE	Amp (Au)	1067
	Creatinine	Urine	250 μM	Glass	CE	Amp (Au)	1067
	Selenoamino acids (Cys, Met)		1–30 μM	PMMA	ITP-CE	Conductivity	637
Biogenic amine	Nortriptyline, amitriptyline	Antidepressant	0.10–0.15 mg/mL	Quartz	CEC	UV (239 nm)	154
	Catechol					Amp/LIF	1068
	Catechol			PDMS		Amp (C)	757
	Catechol		0.47 μM LOD	Plastic (Plexiglas)	CZE	Amp (Pd)	205
	Catechol		100 μM	PDMS-glass		Dual amp (Au)	748
	Catechol		4 μM	PDMS-glass		1st-stage amp (Au)	748

Analyte	Sample	LOD/Concentration	Substrate	Separation	Detection	Ref.
Catechol		4–5 µM	Glass		Amp (Pt)	747
Catechol	Spiked cerebral spinal fluid	110 µM	Glass	CZE	Amp	670
Catechol		4 µM LOD	PDMS-PDMS	CZE	Amp (C fiber)	750
Catechol		0.78 µM LOD	Glass	CZE	Amp (C)	753
Catechol		0.5 µM	PDMS-PDMS	CZE	Amp (C fiber)	762
Catechol		0.24 µM	Glass	CE	Amp (Au)	272
Catechol		100 µM	PMMA	CE	Amp (C)	223
Catechol		12 µM LOD	Glass	CZE	Amp (Pt)	745
Catechol		5–200 µM (1 µM LOD)	PDMS-PDMS	CZE	Dual amp (C)	763
Catechol		4 µM LOD	Glass	CE	Amp (Pt)	751
Catechol		25 µM	LTCC	CE	Amp (Pt)	1069
Catechol		1.8 µM LOD, 10–500 µM range	PDMS-glass	CE	Amp (Au)	318
Dopamine		3.7 µM LOD	Glass	CZE	Amp (Pt)	745
Dopamine		0.29 µM LOD	PMMA	CZE	Amp (Pd)	205
Dopamine		0.24 µM	Glass		Amp (C)	754
Dopamine		4–5 µM	Glass		Amp (Pt)	747
Dopamine	Spiked cerebral spinal fluid	50 µM	Glass	CZE	Amp	670
Dopamine		0.38 µM LOD	Glass	CZE	Amp (C)	753
Dopamine		20 µM	PDMS	CZE	Amp	365
Dopamine		20–200 µM (1 µM LOD)	Glass	CZE	Amp (Au)	749
Dopamine		10 µM	Glass	CZE	Amp (Au/Pt)	1065
Dopamine		3.4 µM LOD	PDMS-glass	CE	Amp (Au)	764
Dopamine		0.47 µM LOD	PDMS-glass	CE	SV (Au)	764
Dopamine		0.1 µM LOD	PDMS-quartz	CE	SV (C)	765
Dopamine		1–165 µM	PDMS-glass	CZE	Amp (Pt)	1070
Dopamine		100 µM	PMMA	CE	Amp (C)	223
Dopamine		0.1 µM	PDMS	CE	Amp (Pt)	1071
Dopamine		100 µM	Glass	Nil	ECL	728
Dopamine		0.125 µM LOD, 0.25–50 µM	PDMS-glass	CE	Amp	242
Dopamine		0.025 µM	PDMS-glass	CE	Amp (C)	332
Dopamine		0.5 µM	PDMS-glass	CE	Amp	375

—continued

Class of Compounds	Compound	Sample Matrix	Concentration	Microchip	Separation Mode	Detection Mode	Ref.
	Dopamine		4 μM LOD	Glass	CE	Amp (Pt)	751
	Dopamine		1.2 μM LOD, 10–500 μM range	PDMS-glass	CE	Amp (Au)	318
	Dopamine		25 μM	LTCC	CE	Amp (Pt)	1069
	Dopamine	With ascorbic and uric acids interferents	1 nM	Glass	Nil	Amp (Au)	1072
	Histamine	Human blood	0.2–2 μg/mL	PMMA	Nil	Amp (C)	1012
	Histamine	Rat mast cells	25 nM LOD, 0.5–500 μM range	Glass	Nil	Amp	1073
	Histamine	Rat mast cells	0.06–1 mM	PDMS	Nil	Fluo	1074
	Histamine	Soy sauce	2.95 μM LOD	Glass	MECC	Fluo	624
	Putrescine	Soy sauce	2.95 μM LOD	Glass	MECC	Fluo	624
	Spermine	Soy sauce	2.95 μM LOD	Glass	MECC	Fluo	624
	l-Dopa		23 μM LOD	PDMS-glass	CE	Amp (Au)	764
	l-Dopa		8.7 μM LOD	PDMS-glass	CE	SV (Au)	764
	Isoproterenol		3.7 μM LOD	PDMS-glass	CE	Amp (Au)	764
	Isoproterenol		0.59 μM LOD	PDMS-glass	CE	SV (Au)	764
	Isoproterenol		0.5 μM LOD, 0.3–50 μM range	PDMS-quartz	CE	SV (C)	765
	Isoproterenol		1.3 μM	Glass	CZE	Amp (Au)	749
	Epinephrine or adrenaline		6.5 μM LOD	Glass	CZE	Amp (Pt)	745
	Epinephrine or adrenaline		0.1 mM	Glass	CE	Amp (C)	754
	d/l-Adrenaline		1 μM LOD	Glass	MECC	Amp (Au)	120
	Epinephrine		10 μM	Glass	CZE	Amp (Au/Pt)	1065
	Epinephrine			PDMS-PDMS	CZE	Amp (C)	762
	Epinephrine		0.515 μM LOD	PDMS-quartz	CZE	SV(C)	765
	Epinephrine		2.1 μM	PDMS-glass	CE	Amp	375
	d/l-Noradrenaline		1 μM LOD	Glass	MECC	Amp (Au)	120
	Noradrenaline		100 mM	PDMS-glass	CZE	Amp (C)	746
	Noradrenaline		10 μM	Glass	CZE	Amp (Au/Pt)	1065
	Stress marker	Human saliva		PMMA	MECC	LIF	1075
	Cortisol	In serum	10–600 μg/L	Fused silica	CZE	LIF (FITC)	1006
	d/l-Metanephrine				CE	Amp	1076
	d/l-Metanephrine		1 μM LOD	Glass	MECC	Amp (Au)	120

Class	Analyte	Sample	LOD/Note	Material	Separation mode	Detection	Ref.
	d/l-Normetanephrine		1 µM LOD	Glass	MECC	Amp (Au)	120
	d/l DOPS		10 µM LOD	Glass	MECC	Amp (Au)	120
	d/l Dopa		10 µM LOD	Glass	MECC	Amp (Au)	120
	l-Dopa		100 mM	PDMS-glass	CZE	Amp (C)	746
	l-Dopa		100 µM	Glass	CZE	Amp (Au)	749
	l-Dopa		3.6 µM LOD	PDMS-quartz	CE	SV (C)	765
	d/l cDOPA		10 µM LOD	Glass	MECC	Amp (Au)	120
	Serotinin			PDMS-PDMS	CZE	Amp (C)	762
	Serotinin		0.1 mM	Glass	CZE	Amp (C)	754
	5-Hydroxyindole-3-acetic acid	Serotinin metabolite		PDMS-PDMS	CZE	Amp	762
Nucleic acids	Nitric oxide	Macrophage	0.1 µM LOD	Quartz	Nil	TLM	1077
	DNA		40 fmol LOD	Si-Pyrex		ECL (TBR)	100
	pBR322	HaeIII digests		Glass	CGE (HEC)	LIF (YO-PRO-1)	1078
	pBR322	HaeIII digests		Glass	CGE (HEC)	LIF (YO-PRO-1)	682
	ΦX174	HaeIII digests	75 fg of 72 bp fragment	Parylene C	CGE (HEC)	LIF (SYBR Green I)	691
	ΦX174	HaeIII digests		Glass	CGE (cross-linked PA)	LIF (YOYO-1)	615
	HLA-DQa			Glass	CGE	LIF (TO6)	315
	LDL-receptor gene			Glass	CGE	LIF	1079
	Oligo p(dT)10–25			Glass	CGE (LPA)	LIF	663
	PBluescript	100–1,000 bp		Glass	CGE	Fluoro burst counting	329
	Hsp72 mRNA	RT-PCR product		Glass	CGE	LIF	1080
	Salmonella DNA			Glass	CGE (HEC)	Indirect amp (TPR)	745
	Serotonin transporter gene (5-HTTLPR)			Glass	CGE (agarose-LPA)	LIF (EtBr)	1081
	BRCA gene	Human blood		Glass	CGE	LIF	791
	BRCA1 and 2 genes	Human serum		Glass	CGE	LIF	683
	Viral DNA			Plastic	CGE	LIF	683
	DMD gene			PDMS-glass	CE	LIF	1082
	Tetanus toxin C fragment		20 nM LOD	Glass	SDS-PAGE	Fluo	555

—continued

Class of Compounds	Compound	Sample Matrix	Concentration	Microchip	Separation Mode	Detection Mode	Ref.
Nucleotides	Uracil			Quartz	CEC	UV (254 nm)	154
	Uracil					UV	1083
	ATP	Jurkat cells	0.16 µM	PDMS-quartz	Nil	LIF (fluo 3)	840
	ATP	Rabbit RBC	2 µM	PDMS-glass	Nil	CL (luciferin)	169
	Guanine		11.6 µM LOD	Glass	CE	Amp (C)	1084
Peptides	Des-Tyr-Leu-enkephalin		450 µM	PDMS-PDMS	CZE	Dual amp (C)	763
	Des-Tyr-Leu-enkephalin			PDMS-PDMS	CZE	Amp (C)	762
	Tri-, tetra-, penta-, octa-peptides		300 nM	Glass	IEF	LIF (Cy5)	626
	TyrGlyGly		340 µM	PDMS-PDMS	CZE	Dual amp (C)	763
	F-src, F-calc		50 µM	PDMS	CE	Fluo	254
	Heart failure marker peptide		0.01 pg/ml	Quartz		TLM	1085
Proteins (antibody)	Alpha-lactalbumin		5 µM	PDMS-PDMS	CZE	LIF	692
	Anti-estradiol		4.3 nM LOD	Glass (soda lime)	CZE	LIF	1007
	Anti-rabbit IgG			PDMS-Si		LIF (Cy5)	694
	Monoclonal anti-BSA IgG	Mouse ascites fluid	34–134 µg/mL	Glass	CZE	LIF	330
	Mouse IgG		10–60 µg/mL	Glass	CZE	CL (HRP/luminol)	720
	Mouse anti-BSA IgG	Mouse ascites fluid	46.8 µg/mL	Glass	CZE	LIF	1086
	IgM		34 ng/ml LOD	PMMA	CE	Contactless conductivity	1010
	IgG	Rat hybridoma	5 mg/L	PMMA	Nil	Fluo	226
	Sheep IgM		17 nM	PMMA	Nil	Fluo	173
	Mouse IgG		0.25 fg/mL	Glass	CE	Amp (C)	781
	Mouse IgG		50–100 ng/mL	Glass	Nil	Amp	1016
	s-IgA	Human saliva	200 µg/mL	Glass	Nil	TLM	846
Proteins (enzyme)	Insulin	Islet of Langerhans	3 nM	Glass	CE	Fluo	139
	Insulin			Quartz	SEC	LIF	1064
	Carbonic anhydrase		30 nM	Plastic	CZE	LIF	1034
	Carbonic anhydrase		5 µM	PDMS-PDMS	CZE	LIF	692
	Cytochrome *c* peroxidase	*E. coli*		PDMS	Nil	Fluo	1087

	Analyte	Sample	Concentration / LOD	Glass (soda lime)	Pressure flow	LIF (DiFMU)	
	Human T-cell protein tyrosine phosphatase	Human T-cell					1088
	Phosphatase			Glass	Nil	Fluo	1088
	Kinase			Plastic	CZE	LIF (FITC)	1089
	Alkaline phosphatase		0.1 mg/ml	Plastic	Nil	UV (430 nm for pNP)	1042
	GPT, GOT	Serum	10–300 U/L	Glass	Nil	Amp	1090
	GPT, GOT	Serum	6–192 U/L	Glass	Nil	Amp	1047
	Leucine aminopeptidase		30 nM LOD	Quartz	CE	Fluo	675
Other proteins	Ribonuclease B (bovine pancreas)		200 μg	PMMA	CE	Fluo	560
	Bovine alpha-lactalbumin		85 nM LOD	Glass	CZE	LIF (NanoOrange)	659
	Bovine beta-lactoglobulin A and B		70 nM LOD	Glass	CZE	LIF (NanoOrange)	659
	BSA		0.1 mg/ml	Fused silica	CZE	LIF (OPA)	658
	CEA	In human serum		Quartz	Nil	TLM	1021
	Cyano-methemoglobin					UV	1091
	D-dimer		0.1 nM LOD, 0.1–100 nM range	PET	Nil	Amp (C)	1020
	D-dimer			PDMS	Nil	LIF	1028
	CRP			Glass	CGE	LIF	459
	CRP		10 ng/ml LOD	SU-8	Nil	CL	1092
	HSA			Glass	Nil	LIF	443
	HSA			Glass	CGE	LIF	442
	Myoglobin		30 μg/ml or 2.4 ng LOD	Quartz	IEF	UV absorption imaging	1041
	Myoglobin		4–85 nM	PC	Nil	Fluo	1093
	pI marker 6.6		0.3 μg/ml or 24 pg	Quartz	IEF	UV absorption imaging	1041
	Thrombin		0.5 5 μg/ml	Au-Cu channel	Nil	Optical	1094
	IFN-γ		0.01 ng/ml LOD	Quartz	Nil	TLM	1023
	IFN-γ		0.01–0.1 ng/ml	Glass	Nil	TLM	1095
	p53	Human blood	100 nM	PMMA	CGE	LIF	1096
Sugars	Fructose and galactose		1 mM	Glass	CZE	Amp (Cu/Pt)	1065
	Glucose					CL	1097
	Glucose		1–80 mM	Quartz	Nil	Amp (Au)	1098

—continued

Class of Compounds	Compound	Sample Matrix	Concentration	Microchip	Separation Mode	Detection Mode	Ref.
	Glucose		0.01–10 mM	Si-Pyrex	Nil	CL	1099
	Glucose		1–25 mM	Si-Pyrex	Nil	Amp (Au)	281
	Glucose	Red wine	0.7 mM	Glass	CZE	Amp (Au/C)	1054
	Glucose		6 µM LOD	Glass	CZE	Amp (Au/C)	1109
	Glucose		10 µM LOD	PDMS-glass	Nil	CL	1100
	Glucose			PDMS			1101
	Glucose			Si-Pyrex	Nil	Amp	1102
	Glucose		0.1–1.0 g/L	Si-glass	Nil	CL	1103
	Glucose		2 mM	PDMS	Nil	Fluo	1057
	Glucose		100 µM	PDMS	CE	Amp (Cu-C)	761
	Glucose	Plasma	0.1–35 mM	Si-glass	—	Amp (Pt)	1104
	Glucose		2–10 mM	PDMS	Nil	Colorimeter	260
	Glucose		90 µM LOD, 90–1,800 µM range	PDMS-glass	CE	Amp (C)	752
	Mannose		150 µM LOD	PDMS-glass	CE	Amp (C)	752
	Glucose		0.75–1.25 mM	Glass	CE	Amp (Au)	18
	Glucose		2.7 µM LOD, 20–1,000 µM range	PDMS-glass	CE	Amp (C)	318
	Glucose	Rat brain	2.3 µM LOD, 5–5,000 µM range	Glass	Nil	Amp (C)	759
	Amino sugars	From BSM	12.5 µM	Quartz	CZE	LIF (NBD-F)	1105
	Raffinose		33 mM	Glass (soda lime)	CZE	RI	729
	Sucrose					RI	1106
	Sucrose		33 mM	Glass (soda lime)	CZE	RI	729
	Sucrose		10–100 mM	Si	Nil	FTIR	1107
	Sucrose	Serum	1 mM	Glass	CZE	Amp (Cu/Pt)	1065
Lipids	Cholesterol		50–400 mg/L	PMMA	Nil	TLM	1108
	LDL and HDL			Pyrex	CZE	LIF	313
Drugs	Acetaminophen		100 µM	Glass	CZE	Conductivity (contactless)	773
	Acetaminophen			Glass	CZE	Amp (Au/Pt)	1109
	Acetaminophen		16.5 mM	Si		UV (254 nm)	88
	Atropine	Alkaloid		Glass		CL	1110
	Atropine		3.8 nM LOD, 5–1,000 µM range	Glass	Nil	CL (TBR)	283
	Pethidine		77 nM LOD, 77 nM–1 mM range	Glass	Nil	CL (TBR)	283

Category	Analyte	Sample	Concentration	Material	Mode	Detection	Ref
	Caffeine		12.9 mM	Si monolithic		UV (254 nm)	88
	Codeine					CL	1111
	Codeine		0.5–2 µM, 0.83 µM LOD	Glass		CL (TBR, 620 nm)	722
	Hydroquinone		40 µM	PDMS	CZE	Amp	365
	Ibuprofen		100 µM	Glass	CZE	Conductivity (contactless)	773
	Ketoprofen	Anti-inflammatory	9.8 mM	Si		UV (254 nm)	88
	Methylphenidate	Human urine	0.4–800 ng/ml	COC	Nil	MS	823
	Pethidine	From morphine			CZE	CL	1110
	Theophylline		10–20 µg/ml	Glass	CZE	LIF	1112
	Theophylline	In serum	7–20 µg/ml	Glass	CZE/MECC	LIF	551
	Theophylline	In serum	2.5–40 µg/ml	Glass	CZE	LIF	330
	Thyroxine	Human serum	0.03–0.24 µg/ml	Quartz	CE	LIF	152
	Xanthine		14.3 µM LOD	Glass	CE	Amp (C)	1084
	Xanthine		1 mM	Glass	CE	Indirect LIF	721
	Xanthine		1 mM	Glass	Nil	CL (luminol)	721
	Penicillin		5 µM LOD	PDMS-glass	CE	Amp (C)	752
	Ampicillin		5 µM LOD	PDMS-glass	CE	Amp (C)	752
	Glycyrrhin, isoliquiritigenin		1.88 mM	Glass	CE	Fluo	1113
Cations	Aluminum		30 ppb LOD	Quartz	CZE	LIF (HQS)	352
	Aluminum		0.6 mM	Styrol	Nil	Fluo	103
	Barium		10 µM–0.1 M	Glass	Nil	Potent-ISE	766
	Calcium	Muscle cell		Glass	Nil	Acoustic wave	133
	Calcium	Jurkat cell		PDMS	Nil	Fluo	1114
	Calcium		20 µM	Plastic	Nil	Fluo	1115
	Calcium, manganese, nickel, zinc, lanthanum, Nd^{3+}, Gd^{3+}		0.8 mM	PS	ITP	Contactless conductivity	774
	Cadmium			Quartz	CZE	UV	1116
	Cadmium		57 ppb LOD			LIF (HQS)	352
	Cobalt					CL	1117
	Cobalt					Conductivity	1118

—continued

Class of Compounds	Compound	Sample Matrix	Concentration	Microchip	Separation Mode	Detection Mode	Ref.
	Cobalt	As-Co complex in water	0.1–1 μM	Quartz	Toluene extraction	TLM	1119
	Cobalt		18 nM LOD	Quartz	Nil	TLM	428
	Cobalt		0.2 mM	PDMS	CE	CL	1120
	Cobalt		0.493 μM LOD	PDMS	CE	CL (luminol)	168
	Cobalt		0.04 nM LOD, 0.1–1,000 nM range	Glass	Nil	CL (luminol)	284
	Cobalt		0.2 μM LOD	Pyrex	Nil	TLM	734
	Chromium		6.8 mM LOD	Glass	CZE	Conductivity (contactless)	773
	Chromium		1–100 μM	Si-glass	Nil	CL (luminol)	468
	Chromium		100 μg/L	PDMS-glass	CE	ICPMS	1121
	Copper					Optical emission	1122
	Copper					UV	1116
	Copper	Water sample	pCu 0.3–5	Si-glass	Nil	Potentiometric (ISE)	768
	Copper		20 μg/L LOD, 0–150 μg/L range	PMMA	Nil	CL	723
	Iron		1–21 μM	Quartz	Nil	TLM	426
	Iron		0.8 μM LOD	Pyrex	Nil	TLM	734
	H^+ (pH)			Si	Nil	Potent (ISFET)	1123
	H^+ (pH)		1–9	Si-Pyrex	Nil	Amp	334
	Mercury					UV	1116
	Potassium					Conductivity	128
	Potassium		100 μM–1 M	Quartz	Nil	TLM	425
	Potassium		2.52 mM	PDMS-PDMS	ITP	Conductivity	1124
	Potassium			Plastic disk	Nil	Potent	767
	Potassium		0.49 mM LOD	Glass	CZE	Conductivity (contactless)	773
	Potassium	Blood	500 μM LOD	Glass	CZE	Conductivity	1127
	Potassium		10 mM	PDMS	CE	Conductivity	1125
	Potassium		0.1–100 mM	PDMS-glass	Nil	ISE	1126
	Potassium		18 μM LOD	Glass	CE	Contactless conductivity	141

Analyte	Sample/Notes	LOD	Material	Method	Detection	Ref.
Potassium		1.28 μM LOD	PMMA	CE	Contactless conductivity	772
Potassium		6.2 μM LOD	PMMA	CE	Contactless conductivity	621
Potassium		5 μM LOD	Glass	—	Contactless conductivity	328
Potassium		5 μM LOD	Glass	—	Contactless conductivity	328
Potassium		0.7 μM LOD, 20–2,000 μM range	Glass	CE	Contactless conductivity	145
Sodium		2.0 μM LOD	PMMA	CE	Contactless conductivity	772
Lithium	Blood	500 μM LOD	Glass	CZE	Conductivity	128
Lithium			Pyrex	CE	Conductivity	1127
Lithium			Glass	CE	Conductivity	1128
Lithium	With sodium interferent	0.5–5 mM		CE	Conductivity	1129
Lithium, sodium		1 mM LOD	Glass	—	Contactless conductivity	328
Magnesium		0.35 mM LOD	Glass	CZE	Conductivity (contactless)	773
Magnesium		1 mM	PS	ITP	Contactless conductivity	774
Manganese	Seawater	1–20 nM	PDMS-glass	—	Conductivity	1118
Manganese					CL	1130
Manganese		2.1 mM LOD	Glass	CZE	Conductivity (contactless)	773
Sodium					Conductivity	1139
Sodium			Si-Pyrex		Conductivity (contactless)	770
Sodium		100 μM–1 M	Quartz	Nil	TLM	425
Sodium		2.17 mM	PDMS-PDMS	ITP	Conductivity	1124
Sodium		0.41 mM LOD	Glass	CZE	Conductivity (contactless)	773
Sodium		1–50 mM, range 500 μM LOD	Glass	CZE	Conductivity	1127

—*continued*

Class of Compounds	Compound	Sample Matrix	Concentration	Microchip	Separation Mode	Detection Mode	Ref.
	Sodium, potassium		0.7 mM	PMMA	CE	Contactless conductivity	775
	Sodium	Blood	10 mM	PDMS	CE	Conductivity	1125
	Sodium, potassium, lithium		100 μM	PET	CE	Conductivity	238
	Sodium, ammonium, $CH_3NH_3^+$		1 mM	PMMA	CE	Conductivity	223
	Sodium, ammonium, lithium		0.01 M	PMMA	ITP (bidirectional)	Conductivity (contactless)	638
	Conductivity (contactless)						
	Sulfate, nitrate, fluoride		0.01 M	PMMA	ITP (bidirectional)	Conductivity (contactless)	638
	Ammonium		3.2 μM LOD	PMMA	CE	Contactless conductivity	621
	Ammonium		0.7 mM	PMMA	CE	Contactless conductivity	775
	Organoammonium (TMA, DMDA, TBA)		500 μM	Glass	Nonaqueous CE	Contactless conductivity	1131
	$CH_3NH_3^+$		5.8 μM LOD	PMMA	CE	Contactless conductivity	621
	Ammonium	Air	10 μM–0.1 M	Si	Nil	Potent (ISFET)	1123
	Ammonium	Air	1 ppm LOD	Glass	Nil	Conductivity	1132
	Ammonium	Air	500 ppb	Glass	Nil	Conductivity	1133
	Nickel				Nil	Conductivity	1118
	Nickel		2.5–50 μM	Quartz	Nil	TLM	735
	Lead					UV	1116
	Lead	Saliva, blood	100 ppm	Si-Pyrex		ASV	770
	Lead		0.2 ppb LOD	Polyimide		ASV	229
	Lead	As OEP complex in benzene	97–780 pM or 0.4–3.4 molecules	Quartz	Nil	TLM	732
	Zinc					UV	1116
	Zinc		46 ppb LOD	Quartz	CZE	LIF (HQS)	352
	Zinc		2.8 mM LOD	Glass	CZE	Conductivity (contactless)	773

Class	Analyte	Sample/Note	LOD/Range	Material	Method	Detection	Ref.
Anions	Acetic acid		1 mM	Glass	CE	Conductivity	1134
	Acetate	Red wine		PMMA	ITP	Conductivity	376
	Ascorbic acid		14.2 mM	Si monolithic		UV (254 nm)	88
	Ascorbic acid	Only 1st-stage oxidation		PDMS-glass		Dual amp (Au)	748
	Ascorbic acid		5 µM LOD	PDMS-glass	CZE	Amp (C)	746
	Ascorbic acid		5 µM LOD	Glass	CZE	Amp (Au/C)	1065
	Ascorbic acid		100 µM	Glass	CZE	Amp (Au/Pt)	1109
	Ascorbic acid		19.0 µM LOD	Glass	CE	Amp (C)	1084
	Ascorbic acid	Vitamin C tablet	0.1 µM	Quartz	Nil	TLM	1135
	Dehydroascorbic	Urine		Quartz	Nil	TLM	1135
	Arsenate (III)		4.8 mg/L LOD	PMMA	ITP	Contactless conductivity	633
	Arsenate (V)		1.8 mg/L LOD	PMMA	ITP	Contactless conductivity	633
	Selenate (III)		0.52 mg/L LOD, 1–25 mg/L range	PMMA	ITP	Contactless conductivity	1136
	Selenate (VI)		0.65 mg/L LOD, 0.5–25 mg/L range	PMMA	ITP	Contactless conductivity	1136
	Benzoate	Ketchup	10 µM		ITP-ZE	Conductivity	1137
	Bromate	Drinking water	0.2 µM		ITP-ZE	Conductivity	1137
	Bromate	Drinking water	10 ppb LOD	PMMA	ITP-CE	Conductivity	1138
	Chloride		1 mM–1 M	PC		Conductivity	1139
	Chloride		20–30 µM	PMMA	ITP	Absorbance	453
	Chloride		8.7 µM LOD	PMMA	CE	Conductivity	366
	Chloride		1 mM	PMMA	CE	Contactless conductivity	621
	Chloride		6.4 µM LOD	PMMA	CE	Contactless conductivity	775
	Chloride	Drinking water		PMMA	CE	Contactless conductivity	772
	Chlorite	Drinking water	3 µM		ITP-ZE	Conductivity	1137
	Citric acid		10 µM LOD	Glass	CZE	Conductivity	1127
	Citric acid	Red wine		PMMA	ITP	Conductivity	376

—continued

Class of Compounds	Compound	Sample Matrix	Concentration	Microchip	Separation Mode	Detection Mode	Ref.
	Citric acid	White wine	12–42 mg/L	PMMA	ITP	Conductivity	630
	Citric acid		100 μM	Glass	CZE	Conductivity (contactless)	773
	Citric acid		1 mM	Glass	CE	Conductivity	1134
	Chloroacetate	Drinking water	2 μM		ITP-ZE	Conductivity	1137
	Dichloroacetate	Drinking water	1 μM		ITP-ZE	Conductivity	1137
	Chromate		100 μg/L	PDMS-glass	CE	CL	1140
	Chromate			Si-Pyrex	CE	ICPMS	1121
	CO_2		30–60 mmHg		Nil	Amp	334
	Fluoride	Drinking water	2 μM		ITP-ZE	Conductivity	1137
	Fluoride		8.1 μM LOD	PMMA	CE	Contactless conductivity	772
	Fluoride					Conductivity	1139
	Fluoride					Conductivity	1141
	Fluoride	Tap, mineral, river water	10 μM	PMMA	ITP-CZE	Conductivity	631
	Fluoride		0.5–0.7 μM	PMMA	ITP-CZE	Conductivity	366
	Fluoride		10 pM	PS	ITP	Contactless conductivity	774
	Fumaric acid		5 μM LOD	Glass	CZE	Conductivity	1127
	Fumaric acid		1 mM	Glass	CE	Conductivity	1134
	Fumaric, citric, succinic, pyruvic, acetic, lactic acid		1 mM LOD	PMMA	CE	Contactless conductivity	621
	Glutamic acid	Red wine		PMMA	ITP	Conductivity	376
	Lactic acid	Red wine		PMMA	ITP	Conductivity	376
	Lactic acid	White wine	12–42 mg/L	PMMA	ITP	Conductivity	630
	Lactic acid		100 μM	Glass	CZE	Contactless conductivity	773
	Lactic acid		1 mM	Glass	CE	Conductivity	1134
	Lactate	Plasma	0.05–15 mM	Si-glass	—	Amp (Pt)	1104
	Lactate	Rat brain	2.3 μM LOD, 5–5,000 μM range	Glass	Nil	Amp (C)	759
	Lactate		0.1–1 mM	BK7-glass	—	SPR	741

Analyte	Sample	Concentration/LOD	Substrate	Separation	Detection	Ref.
Malic acid	White wine	12–42 mg/L	PMMA	ITP	Conductivity	630
Malic acid	Red wine	10 µM LOD	Glass	CZE	Conductivity	1127
Malic acid			PMMA	ITP	Conductivity	376
Nitrate		20–30 µM	PMMA	ITP	Conductivity	128
Nitrate			PMMA	Nil	Conductivity	366
Nitrate		10 µM–0.1 M	Si		Potent (ISFET)	1123
Nitrate		7.2 µM LOD	PMMA	CE	Contactless conductivity	621
Nitrate		1 mM	PMMA	CE	Contactless conductivity	775
Nitrate		1 mM	Glass	CE	Conductivity	279
Nitrite		50 pM	PS	ITP	Contactless conductivity	774
Nitrite			PMMA		Amp (C)	757
Nitrite		10 µM	PMMA	ITP-CZE	Conductivity	631
Nitrite	Tap, mineral, river water	0.5–0.7 µM	PMMA	ITP-CZE	Conductivity	366
Nitrite		10 µM	PDMS-PDMS	CZE	Amp (C fiber)	750
Nitrite			Plastic disk	Nil	Potent	767
Nitrite	For NO-releasing compound SIN-1	10–250 µM range, 1 µM LOD	PDMS-PDMS	CZE	Amp (C)	1142
Oxalic acid			PMMA	ITP	Conductivity	376
Perchlorate	Red wine	6.2 µM LOD	PMMA	CE	Contactless conductivity	621
Perchlorate		0.5 mM	PMMA	CE	Contactless conductivity	775
Phosphate		1 mM	Glass	CE	Conductivity	279
Phosphate		9.67–96.7 µM	Si	Nil	Absorbance	1143
Phosphate		0.01–1.0 mM	Si	Nil	Amp	755
Phosphate			PMMA		Conductivity	128
Phosphate		10 µM	PMMA	ITP-CZE	Conductivity	631
Phosphate	White wine	12–42 mg/L	PMMA	ITP	Conductivity	630
Phosphate	Tap, mineral, river water	0.5–0.7 µM	PMMA	ITP-CZE	Conductivity	366

—continued

Class of Compounds	Compound	Sample Matrix	Concentration	Microchip	Separation Mode	Detection Mode	Ref.
	Phosphate	Drinking water	2 µM		ITP-ZE	Conductivity	1137
	Phosphate			PDMS	Nil	Colorimetric	1144
	Phosphate		38.7–96.7 µM	Si	Nil	Absorbance (660 nm)	1145
	Phosphate		0.25–3.0 µg/ml	Si-glass		UV	1146
	Pyruvic acid		1 mM	Glass	CE	Conductivity	1134
	Salicylic acid		100 µM	Glass	CZE	Conductivity (contactless)	773
	Succinic acid	Red wine	1 mM	PMMA	ITP	Conductivity	376
	Succinic acid			Glass	CE	Conductivity	1134
	Sulfate		20–30 µM	PMMA	ITP	Conductivity	366
	Sulfate		5.1 µM LOD	PMMA	CE	Contactless conductivity	772
	Sulfite		500 µM	PDMS		Fluo	1147
	Tartaric acid	White wine	12–42 mg/L	PMMA	ITP	Conductivity	630
	Tartaric acid		5 µM LOD	Glass	CZE	Conductivity	1127
	Tartaric acid	Red wine		PMMA	ITP	Conductivity	376
	Uric acid		5 µM LOD	Glass	CZE	Amp (Au/C)	1109
	Uric acid	Urine	15–110 µM	PDMS-glass	CZE	Amp (Pt)	1070
	Uric acid	Urine	100 µM	Glass	CE	Amp (Au)	1067
Hydrocarbons	Benzene		10 ppm	Pyrex	Nil	UV	715
	BaP		100 µg/L	Quartz	Nil	Fluo	1148
	BaP		17 nM	Quartz	CEC	LIF (325 nm)	148
	BaP		0.01–100 ng/ml	PMMA-PDMS	Nil	SPR	740
	Hexane		10^{-12} g/s or 800 ppb	Glass	GC	Optical emission (519 nm)	716
	Hydrocarbons						
	Methane		10^{-13} g/s or 400 ppb	Si	GC	TCD	1
	Methane		10^{-12} g/s or 600 ppm LOD	Glass	Nil	Optical emission	718
	Methane			Glass	Nil	Optical emission (CH)	717
	o-Xylene		10 ppm	Pyrex	Nil	UV	715
	m-Xylene		6 ppb	Si-Pyrex	GC	Chemiresistor	1149

	Analyte	Concentration	Material	Method	Detection	Ref
	Styrene	6 ppb	Si-Pyrex	GC	Chemiresistor	1149
	Nonane	6 ppb	Si-Pyrex	GC	Chemiresistor	1149
	Mesitylene	6 ppb	Si-Pyrex	GC	Chemiresistor	1149
	3-Octanone	6 ppb	Si-Pyrex	GC	Chemiresistor	1149
	Propane				Photoacoustic	1150
	Decane	6 ppb	Si-Pyrex	GC	Chemiresistor	1149
	Toluene	10 ppm	Pyrex	Nil	UV	715
	Toluene	6 ppb	Si-Pyrex	GC	Chemiresistor	1149
	VOC				UV	1151
	Anthracene	3.1 nM	Quartz	CEC	LIF (325 nm)	148
	Pyrene	1.0 nM	Quartz	CEC	LIF (325 nm)	148
	1,2-Benzofluorene	8.1 nM	Quartz	CEC	LIF (325 nm)	148
	PAH	10 mM	Pyrex	CEC	LIF	338
	PAH	10 mM	Fused silica	CEC	LIF	647
Phenols	Aminophenol	5 µM LOD, 5–1,200 µM range	PET	CE	Amp (C)	191
	Chlorophenols	100 µM	Glass	CZE	Amp (Au/Pt)	1065
	Phenol	6.38 µM LOD	Glass	CE	Amp (C)	1084
	Phenol	0.2 ppb LOD	Polyimide	Nil	Amp	229
	2-Chlorophenol	9.37 µM LOD	Glass	CE	Amp (C)	1084
	2,4-Dichlorophenol	19.0 µM LOD	Glass	CE	Amp (C)	1084
	Hydroquinone	0.5 µM	PDMS-glass	CE	Amp (C)	332
	Phenol	100 µM LOD	Glass	MECC	Amp (BOD)	760
	2-Chlorophenol	100 µM LOD	Glass	MECC	Amp (BOD)	760
	2,4-Dichlorophenol	200 µM LOD	Glass	MECC	Amp (BOD)	760
	2,3-Dichlorophenol	200 µM LOD	Glass	MECC	Amp (BOD)	760
	Paraoxon	10 ppm LOD	Glass	MECC	Amp (BOD)	760
	Methyl parathion	10 ppm LOD	Glass	MECC	Amp (BOD)	760
Explosives	Explosives	Spiked soil 1 ppm	Glass (borofloat)	MECC	Indirect LIF (5 µM Cy7)	620
	Explosives			MECC	Indirect LIF	289
	Explosives (2,4-DNT and 2,6-DNT)	2–15 mM	Glass	MECC	Conductivity-amp	1153
	Explosives (4-NT)	20 mg/L	Glass	MECC	Amp (C)	753
	Explosives (DNB, TNT)	2 mg/L	Glass	MECC	Amp (C)	753
		0.6 mg/L	Glass	MECC	Amp (C)	753

—continued

Class of Compounds	Compound	Sample Matrix	Concentration	Microchip	Separation Mode	Detection Mode	Ref.
Explosives (DNB, DNT, TNT)	TNB		15 ppm LOD	Glass	MECC	Amp (BOD)	760
	TNT	Seawater	0.25 µg/L	Glass	Nonaqueous CE	Absorbance (vis)	622
	Tetryl	Seawater	0.34 µg/L	Glass	Nonaqueous CE	Absorbance (vis)	622
	TNT	Seawater	0.19 µg/L	Glass	Nonaqueous CE	Absorbance (vis)	622
	2,4-DNT		24 µg/L	Glass	MECC	Amp (Au)	272
	2,3DNT		33 µg/L	Glass	MECC	Amp (Au)	272
	2,6-DNT		35 µg/L	Glass	MECC	Amp (Au)	272
			36 µg/L	Glass	MECC	Amp (Au)	272
	TNB-Fl		1 ng/ml LOD	Glass	CE	Fluo	1154
	TNT-Fl		1–300 ng/ml	Glass	CE	Fluo	1154
	2,4-DNT		150 µg/L LOD	Glass	MECC	Amp (Au)	1155
	2,6-DNT		160 µg/L LOD	Glass	MECC	Amp (Au)	1155
	2,3-DNT		150 µg/L LOD	Glass	MECC	Amp (Au)	1155
	TNT		200 µg/L LOD	Glass	MECC	Amp (Au)	1155
	RDX		110 µg/L LOD	Glass	MECC	Amp (Au)	1155
	TNT		1 ng/ml	Glass	CE	LIF	285
	Hydrazine		5.5 µM LOD	Glass	CE	Amp (C)	1084
	Hydrazine, methylhydrazine, dimethylhydrazine phenylhydrazine		1.5 µM LOD, 20–200 µM range	Glass	CE	Amp (Pd)	1156
Fluorescent	FITC		6 µM	Glass		UV/LIF	671
	Fluorescein		300 fM	Glass	CZE	LIF	1157
	Fluorescein		25 nM	PDMS-PDMS	CZE	LIF	692
	Fluorescein		10 µM	Si-Pyrex	CZE	UV absorbance (488 nm)	706
	Fluorescein		10 µM	Glass	CZE	LIF	990
	Fluorescein		63 pM	Glass	CZE	LIF	1166
	Fluorescein		0.5 mM	Glass	CZE	Indirect LIF (fluorescein 0.5 mM)	696
	Fluorescein		20 nM	Glass	CE	LIF	704
	Fluorescein		17 nM	Glass	CE	Fluo	690
	Cy5		3.3 nM	Glass		LIF	688

Category	Analyte	Sample	LOD	Material	Separation	Detection	Ref.
	Cy5		20 pM LOD	Glass	Nil	LIF	1158
	Organophosphate nerve agents		48–86 µg/L LOD	Glass	CZE	Conductivity	1159
Organophosphorus compounds	Paraoxon, methylparaoxon, parathions, fernitrothion	Spiked river water	30–60 µM	Glass	MECC	Amp (C)	625
	Paraquat, diquat					Raman	1160
	Paraquat, diquat		0.23 µM	Glass	ITP	Raman	738
	Dimethyl methyl phosphate					SAW	602
	Methyl phosphonic acid (MPA)		200 ppm	PMMA	CE	Contactless conductivity	775
	IMPA		200 ppm	PMMA	CE	Contactless conductivity	775
	PMPA		400 ppm	PMMA	CE	Contactless conductivity	775
	Methylammonium		0.7 ppm	PMMA	CE	Contactless conductivity	775
Organohalogen compounds	Trichloroethylene		6 ppb	Si-Pyrex	GC	Chemiresistor	1149
	Perfluoroethylene		6 ppb	Si-Pyrex	GC	Chemiresistor	1149
	4-Methoxy-3,3,4-trichlorobiphenyl	Coplanar PCB	0.1 pg/ml	PDMS	Nil	Fluo	1161
Others	Butylacetate		6 ppb	Si-Pyrex	GC	Chemiresistor	1149
	Glycerol		742 µM	Silica	Nil	RI	151
	Methyl formate					IR	1162
	Ferrocene		500 µM	Polyimide-PET	Nil	Voltammetry	1163
	Ferrocene carboxylic acid		3 µM	Plastic		CV (C) and chronoamp (C)	758
	Citrulline		150 µM	PDMS-PDMS	CZE	Dual amp (C)	763
	Ethanol					Diffraction	1164
	Ethanol	Red wine	200 mM	Glass	CZE	Amp (Au/C)	1054
	Ethyl acetate					Infrared	1165
	Beta-naphthylamine		30 nM LOD	Quartz	CE	Fluo	675
	Phenytoin		8.1 nM LOD	Mylar-glass	Nil	Fluo	1015
	N-Acetylglucosamine		33 mM	Glass (soda lime)	CZE	RI	729

—continued

Class of Compounds	Compound	Sample Matrix	Concentration	Microchip	Separation Mode	Detection Mode	Ref.
	Penicillamine		200 µM	PDMS-PDMS	CZE	Dual amp (C)	763
	CD-3	Fresh or seasoned developer solutions	0.17 mM LOD	Glass	CZE	Indirect LIF (5 mM fluorescein)	1166
	CD-3	Photographic developer solution	5–20 mg/L	Glass	CZE	Indirect LIF (2 mM fluorescein)	697
	CD-4	Fresh or seasoned developer solutions	0.39 mM LOD	Glass	CZE	Indirect LIF (5 mM fluorescein)	1166
	Aliphatic amines		500 µM	Glass	Nonaqueous CE	Contactless conductivity	1131
	Alkylphenones		0.3–0.6 mg/ml	Quartz	CEC	UV (240 nm)	154
	Atrazine		3.7–209 pM	Si	Nil	CL (HRP/luminol)	1030
	Azo dyes		1 nM or 10 fmol LOD		Nil	SERRS	739
	O_2		20–100%	Si-Pyrex	Nil	Amp	334
	Water		25–70% RH	Si	Nil	Microcantilever, optical deflection	1167

Index

T - #0387 - 101024 - C4 - 254/178/23 - PB - 9781439818558 - Gloss Lamination